DRINKING WATER: CONTAMINATION, TOXICITY AND TREATMENT

DRINKING WATER: CONTAMINATION, TOXICITY AND TREATMENT

JAVIER D. ROMERO AND
PABLO S. MOLINA
EDITORS

Nova Science Publishers, Inc.
New York

For permission to use material from this book please contact us:
Telephone 631-231-7269; Fax 631-231-8175
Web Site: http://www.novapublishers.com

LIBRARY OF CONGRESS CATALOGING-IN-PUBLICATION DATA

Drinking water : contamination, toxicity, and treatment / Javier D. Romero and Pablo S. Molina, editors.
p. cm.
Includes bibliographical references and index.
ISBN 978-1-60456-747-2 (hardcover)
1. Drinking water--Purification. 2. Drinking water--Contamination. 3. Water--Pollution--Toxicology. I. Romero, Javier D. II. Molina, Pablo S.
TD430.D768 2008
628.1'62--dc22 2008023357

Published by Nova Science Publishers, Inc. ✦ New York

CONTENTS

Preface		**vii**
Chapter 1	Toxin Contamination of Surface and Subsurface Water Bodies Connected with Lake Vico's Watershed (Central Italy) *Roberto Mazza, Giuseppe Capelli, Pamela Teoli, Milena Bruno, Valentina Messineo, Serena Melchiorre and Antonio Di Corcia*	**1**
Chapter 2	Drinking Water Contamination with Metals *M.S. Gimenez, S.M. Alvarez, E. Larregle and A.M. Calderoni*	**101**
Chapter 3	Current and Emerging Microbiology Issues of Potable Water in Developed Countries *William J. Snelling, Roy D. Sleator, Catherine D. Carrillo, Colm J. Lowery, John E. Moore, John P. Pezacki and James S.G. Dooley*	**121**
Chapter 4	The Reliability of the Grab Sample for the Determination of Fluoride in Potable Water Supplies *J.A. Armstrong and S.A. Katz*	**153**
Chapter 5	Retention of Microbial Pathogens by Filtration *Vitaly Gitis and Elizabeth Arkhangelsky*	**161**
Chapter 6	Effective Removal of Low Concentrations of Arsenic and Lead and the Monitoring of Molecular Removal Mechanism at Surface *Yasuo Izumi*	**199**
Chapter 7	Detoxification of Steroidal Hormones in the Aquatic Environment *Tomoaki Nishida and Hideo Okamura*	**213**

Chapter 8 Influence of Mineralized/Mined Areas in the Quality of Waters Destined for the Production of Drinking Water **223**
A. Ordóñez, R. Alvarez, E. De Miguel, J. Loredo and F. Pendás

Chapter 9 Electrochemical Stripping Analysis in Water Quality Control **243**
Jaroslava Švarc-Gajić

Chapter 10 Identification and Quantification of ^{224}Ra, ^{226}Ra and ^{228}Ra in the Hydrogeochemical Environment of Southern New Jersey: A Review **297**
J.A. Armstrong, S.A. Katz and T.L. Paulson

Chapter 11 Pumping Test Data Analysis and Characterization of Deep Aquifer in the Lower Delta Floodplain at Madaripur, Southern Bangladesh **313**
Anwar Zahid, Jeffrey L. Imes, M. Qumrul Hassan, David W. Clark and Satish C. Das

Chapter 12 The Influence of pH on Sorption/Desorption Equilibrium of Cd^{2+} and Pb^{2+} in a Bottom Sediment-Water System **329**
Vladislav Chrastný, Michael Komárek and Aleš Vaněk

Index **339**

PREFACE

Water of sufficient quality to serve as drinking water is termed potable water whether it is used as such or not. Although many sources are utilized by humans, some contain disease vectors or pathogens and cause long-term health problems if they do not meet certain water quality guidelines. Water that is not harmful for human beings is sometimes called safe water, water which is not contaminated to the extent of being unhealthy. The available supply of drinking water is an important criterion of carrying capacity, the population level that can be supported by planet Earth.

Typically water supply networks deliver single or multiple qualities of water, whether it is to be used for drinking, washing or landscape irrigation; one counterexample is urban China, where drinking water can be optionally delivered by a separate tap.

This new book focuses on contamination, toxicity and treatment of drinking water.

Chapter 1 - Cyanotoxins are a new class of freshwater risk factors that are widespread all over the world and whose health implications are not yet fully understood. The environmental fate of these toxins, which are produced by many cyanobacterial species, holds consequences not only for water bodies where blooms occur, but also for public water sources. Indeed, drinking water wells may be contaminated by toxins capable of migrating through groundwater and causing toxic blooms in lakes and ponds.

In Italy, toxic cyanobacterial blooms are largely diffused in lakes, fish ponds, channels and even along sinuous rivers. However, their impact on health has been so far poorly investigated.

In order to assess the level of contamination of groundwater around lakes with toxic algal populations, a multidisciplinary study was conducted in the 2005-2007 period on the entire recharge area of Lake Vico, a volcanic crater lake in central Italy with a *Planktothrix rubescens* population producing hepatotoxic microcystins. The Lake, which is part of a natural reservoir, has significant environmental and scenic value and supplies drinking water to two towns. It has no tributaries, a long renewal time and tendency to accumulate polluting substances.

During a 18-month monitoring survey, waters from 5 stations in the Lake and 11 wells in its watershed were monthly sampled in order to investigate the dynamics of the *P. rubescens* population and groundwater flowpaths. Nutrient storage and eutrophication of the Lake's watershed and subsurface hydrogeological system were characterized quantitatively and qualitatively.

Possible recommendations for remediation of Lake Vico's waters should be focused on reduction of nutrient levels deriving from farming and residential sewage. The reduction could be achieved through actions based on agronomical technique improvement and sewage discharge regulations.

Chapter 2 - It is estimated that contamination of drinking water with heavy metals has considerable impact on the health of the world population. The supply of high quality drinking water of is, therefore, necessary to develop and apply suitable processes which allow the reduction of hazardous metals - arsenic, cadmium, lead - below the international standards set for drinking water. Due to the discharge of untreated or insufficiently treated industrial waste waters or waste disposal, many water resources exhibit increased concentrations of heavy metals. The contamination of drinking water with metals represents a serious problem in human health that results in direct toxic effects causing nephropathies, hepatic necrosis, pulmonary emphysema, osteoporosis, hormone alterations, exacerbation of autoimmune disease, and its well-known carcinogenic effect. The adverse consequences of exposure to water contamination on reproductive organs have been widely considered; some of them can be the long-term impairment of neurobehavioral status, and alteration in complex behaviors, such as learning. Particular attention has to be paid to the fact that the treatment of drinking water - due the large volumes needed - must be selective when eliminating heavy metals, and must not eliminate other components which should remain in the water.

Chapter 3 - Water is vital for life, for commercial and industrial purposes and for leisure activities in the daily lives of the world's population. Diarrhoeal disease associated with consumption of poor quality water is one of the leading causes of morbidity and mortality in developing countries (especially in children <5 years old). In developed countries, whilst potable water is not a leading cause of death, it still can still pose a significant health risk. Water quality is assessed using a number of criteria, e.g., microbial load and nutrient content which affects microbial survival, as well as aesthetic factors such as odor. In water systems, the presence of disinfectant, low temperatures, flow regimes and low organic carbon sources do not appear to be conducive to microbial persistence. However, frequently this is not the case. A variety of human pathogens can be transmitted orally by water and in the developed world water quality regulations require that potable water contains no microbial pathogens.

Chlorine dioxide is a safe, relatively effective biocide that has been widely used for drinking water disinfection for 40 years. Providing the water is of low turbidity, standard chlorination procedures are sufficient to prevent the spread of planktonic bacteria along water mains. However, despite this, bacterial contamination of water distribution systems is well documented, with growth typically occuring on surfaces, including pipe walls and sediments. Rivers, streams and lakes are all important sources of drinking water and are used routinely for recreational purposes. However, due to fouling by farm and wild animals, these sources can be contaminated with microbes, e.g., chlorine-resistant *Cryptosporidium* oocysts, no matter how pristine the source or well maintained the water delivery system. The high incidence of *Cryptosporidium* in surface water sources underlines the need for frequent monitoring of the parasite in drinking water. The use of coliforms as indicator organisms, although considered relevant to most cases, is not without limitations, and is thus not a completely reliable parameter of water safety, e.g., *Campylobacter* contamination cannot be accurately predicted by coliform enumeration. Furthermore, the presence of biofilms and bacterial interactions with protozoa in water facilitate increased resistance to antimicrobial agents and procedures such as disinfectants and heating, e.g., Legionnaires' disease caused by *Legionella pneumophila*.

The high cost of waterborne disease outbreaks should be considered in decisions regarding water utility improvement and treatment plant construction. The control of human illnesses associated with water would be aided by a greater understanding of the interactions between water-borne protozoa and bacterial pathogens, which until relatively recently have been overlooked.

Chapter 4 - Water samples were collected three times each week for five weeks from the potable water supplies at domestic sites in five Southern New Jersey communities. The sampling regimen was designed to reflect periods of high and low water usage.

None of the water supplies was fluoridated, but fluoride is known to be a naturally-occurring water contaminant in some southern New Jersey domestic water supplies.

The fluoride ion concentrations of the samples were determined by ion selective electrode potentiometry without prior distillation. The analytical error associated with the method was evaluated by repetitive measurements on reference samples, and the sampling errors were identified by statistical analysis of the results obtained from the samples.

The statistical analysis of the data showed the fluoride concentrations varied from site to site and from time to time. In some cases, the time-to-time variations exceeded the "two sigma" limits at a given site. This observation raises a serious question about the reliability of a grab sample for the determination of fluoride, and possibly other contaminants, in potable water supplies.

Chapter 5 - Microbial contamination is the biggest concern for drinking water suppliers. A wide variety of microbial pathogens including enteric viruses, *E.coli* 0157:H7 and *Cryptosporidium parvum* are spread out through water sources even in developed countries such as United States. Hence, the primary goal of water treatment process is to ensure that the drinking water is free of pathogenic viruses, bacteria and protozoa. Because no single treatment process can be expected to remove all of the different types of pathogens, the international health authorities has promulgated a multiple barrier approach. The idea is to use at least one physical and one chemical process to ensure pathogen attenuation to a safe level.

One of the most employed physical water treatment processes is filtration. Filtration is a common name for a process of physical retention of pathogenic microorganisms by size rejection or by accumulation of the pathogens within its media. Water treatment facilities employ either granular media filters or membrane filters. Four of most known filtration processes are slow sand, rapid granular, microfiltration and ultrafiltration. Degree of microbial retention by a filter depends on pathogen's transport, accumulation and inactivation. Each process has many variables/characteristics and for the same application a retention level may differ by orders of magnitude. Some of the variables are macroscopic and others are microscopic, their interplay and relative influence on the degree of retention are discussed, and useful conclusions are drawn.

Chapter 6 - New sorbents were investigated for the effective removal of low concentrations of arsenic and lead to adjust to modern worldwide environmental regulation of drinking water (10 ppb). Mesoporous Fe oxyhydroxide synthesized using dodecylsulfate was most effective for initial 200 ppb of As removal, especially for more hazardous arsenite for human's health. Hydrotalcite-like layered double hydroxide consisted of Fe and Mg was most effective for initial 55 ppb of Pb removal.

The molecular removal mechanism is critical for environmental problem and protection because valence state change upon removal of e.g., As on sorbent surface from environmental water may detoxify arsenite to less harmful arsenate. It is also because the evaluation of desorption rates is important to judge the efficiency of reuse of sorbents. To monitor the low concentrations of arsenic and lead on sorbent surface, selective X-ray absorption fine structure (XAFS) spectroscopy was applied for arsenic and lead species adsorbed, free from the interference of high concentrations of Fe sites contained in the sorbents and to selectively detect toxic As^{III} among the mixture of As^{III} and As^{V} species in sample.

Oxidative adsorption mechanism was demonstrated on Fe-montmorillonite and mesoporous Fe oxyhydroxide starting from As^{III} species in aqueous solution to As^{V} by making complex with unsaturated $FeO_x(OH)_y$ sites at sorbent surface. Coagulation mechanism was demonstrated on double hydroxide consisted of Fe and Mg from the initial 1 ppm of Pb^{2+} aqueous solution whereas the mechanism was simple ion exchange reaction when the initial Pb^{2+} concentrations were as low as 100 ppb.

Chapter 7 - Lignin is one of the most abundant organic polymers, probably only second to cellulose among the renewable organic resources on Earth. In contrast to cellulose and hemicellulose in wood, lignin is resistant to degradation by microorganisms. But ligninolytic activity has been recognized by certain microorganisms. At present, the white rot fungi are the best known and the most lignin-degrading microorganisms.

There is a great interest in the white rot fungi and their ligninolytic enzymes, because their industrial potential for degrading and detoxifying recalcitrant environmental pollutions. Considerable concern has recently been expressed that steroidal hormones (estrogens) excreted into the environment by humans, domestic or farm animals, and other wildlife, in part via sewage treatment plants, may be disruptive to the endocrine systems. This chapter will review the authors' recent research on detoxification of natural steroidal estrogens (17β-estradiol and estrone) and synthetic one (17α-ethynylestradiol), focusing on the contamination and environmental toxicity of these estrogens and the detoxification of these estrogens by the treatment with ligninolytic enzymes from white rot fungi.

Chapter 8 - Surface and groundwater quality is often altered in naturally mineralized areas. Mining activities can enhance these effects since they involve a number of processes that impact water quality. The type of water contamination is highly dependent on the mineralization, the reactants used in the ore processing activities and the type of mining: i) *Industrial minerals and rocks:* the absence of reactive mineral phases and chemical treatment implies that the typical affection to water bodies is caused by suspended solids; ii) *Metallic mining:* water contacting open-cut and underground mining of base metal deposits can show low pH values (*Acid Mine Drainage*) and high concentrations of many chemicals, leached from the ore and host rock (sulphate, Fe, Pb, Hg, As and many others), as well as substances used to extract metals from the ore (flotation reagents and leaching compounds); iii) *Fossil fuel mining:* In coal mining, low pH and high sulphate and metal content in water are very common, mainly due the oxidation of pyrite present in the coal bed. The most outstanding feature of waters related to oil and gas mining is the presence of brines (high salinity), CH_4 and other hydrocarbons, as well as H_2S. Abandoned pits and mine shafts are sometimes used for water supply after mine closure. Additionally, leachates from mining areas and/or spoil heaps, and affected groundwater are frequently received by water bodies which are used for drinking water production. In most cases, the input of water affected by mining and metallurgical works is negligible when compared with the total volume of input from the catchment and, therefore, an admissible risk is obtained as a result of a risk assessment in these sites. This is illustrated in several study cases presented in this paper, describing how different types of mining sites in Spain affect water bodies used for drinking purposes. That is the case of the "El Aramo" mine on a Cu-Co-Ni deposit in Asturias: groundwater sampled at a nearby spring contained more than 4.9 mg l^{-1} Cu, well above the national standard for drinking water. Considering that there is a significant catchment point for drinking water downstream of the mining works, the presence of high As and Cu concentrations in some periods of the year can pose a potential health risk. The old "Santa Agueda" mine in León extracted an As ore; mining and smelting wastes remain in spoil heaps, exposed to weathering, and the leachates, containing up to 0.9 mg l^{-1} As, flow directly to a water reservoir constructed for production of hydraulic energy and drinking water. Mercury mining districts in NW Spain, including several mine sites, contribute mine drainage and leachates (containing up to 18 mg·l^{-1} As) to a catchment sometimes used for drinking water supply. Additionally, human and animal consume of mine-affected water in rural areas has been detected. These and other cases are analysed from a holistic point of view in this paper.

Chapter 9 - Water quality is defined by the presence and quantity of the contaminants, nutrients and by the physical and chemical properties, such as pH and conductivity. Man significantly influences all these factors. Toxicants enter the natural waters as a result of anthropogenic and natural processes, retain in the ecosystem and enter the marine food chain. Among them, heavy metals, recognised as high toxic water contaminants which are not biodegradable, attract the attention of the analysts, because they further proliferate their negative effect to other environmental compartments, finally reaching men. Accurate determination of metal traces in water is very important because even smallest concentration can have profound influence on human health. Heavy metal traces are nowadays most often determined with the application of atomic absorption spectrometry (AAS), inductively coupled plasma spectrometry (ICP) and electrochemical stripping techniques (ESA). The application of certain technique depends not only on general analytical requirements (sensitivity, reproducibility, accuracy, selectivity), but also on specific ones, such as the price of the instrumentation and analysis cost, simplicity and safety, required sample size, sample preparation procedure, etc. Taking into consideration these requirements, flameless atomic absorption spectrometry and electrochemical stripping techniques fulfil most of the criteria for the technique selection.

In past few decades electrochemical stripping analysis becomes one of the most frequently exploited techniques for trace metal determination in various samples. It is most often used for the determination of toxic metals in environmental samples (water, soil and air), biological samples, food and pharmaceutical product. Its application for the determination of different organic and pharmacologically active compounds, mostly with the application of chemically-modified electrodes has risen as well. The reason for ESA so strongly competing with the most commonly used atomic absorption spectrometry, regarding metal determination, are multiple. High sensitivity, the possibility of simultaneous determination of more then one analyte and features of on-line analysis for continual monitoring are just some of the attributes. Analysis itself is non-destructive and can be performed in some cases directly (simple liquid sample) with the modest price. Selectivity, regarding the physico-chemical and thermodynamic properties, important for element speciation and for understanding their reactivity, toxicity and transport, is one of the distinguishing parameters among the other instrumental techniques.

The intention of this chapter is to present the basic principles of different electrochemical stripping techniques and its significance in water analysis. Interferences, difficulties, advantages and disadvantages will be discussed in order to reveal a place of this attractive technique in analytical practice, with emphasis on water quality control. In water analysis, more than in any other type of the sample, the advantages of the technique are expressed in its full light and will be given through the review of the developed and applied methods for the determination of different elements in different types of the water.

Part of the research work confirming the attributes of chronopotentiometric stripping technique will be presented in order to promote its wider application and to break misapprehensions. Namely, voltammetric stripping techniques have a leading position and a stabile place among other stripping techniques, due to their highest sensitivity. Chronopotentiometric stripping analysis have been ignored for a long time, but with the development of the digital techniques, when time measurement became more accurate, they reached the sensitivity of the competitive voltammetry. Nevertheless, its application remained dissimulated. The work describing the development of various systems for the chronopotentiometric stripping determination of problematic analytes such as mercury and arsenic is elaborated in this chapter. Different flow-through electrochemical cells for mercury determination are described as a result of the research work focused on the development of the methods which could be used for continual monitoring.

Chapter 10 - Investigations employing revised procedures that were able to measure short-lived radionuclides detected elevated levels of radioactivity in some potable water supplies from Dover Township, Ocean County, New Jersey. Subsequently, the New Jersey Department of Environmental Protection and the United States Geological Survey initiated collaborative investigations to better understand the source(s) and level(s) of radioactivity in the hydrogeochemical environment of Southern New Jersey. The results of these surveys confirmed some parts of the Kirkwood - Cohansey Aquifer in Atlantic, Burlington, Camden, Cape May, Cumberland, Gloucester, Mannmouth, Ocean and Salem Counties were found to contain ^{228}Ra, ^{226}Ra and ^{224}Ra concentrations in excess of 3 pCi/L.

The radiological origins of ^{228}Ra and ^{224}Ra were attributed to the decay of ^{232}Th while the decay of ^{238}U was the proposed source of ^{226}Ra. The hydrogeological origin of both thorium and uranium was ground water leaching of the quartz sand found in the Kirkwood-Cohansey Aquifer. The hydrogeo-chemistry of radium paralleled that of calcium in the relatively soft (from 30 - 60 ppm $CaCO_3$), slightly acidic (pH from 4 to 5) ground water.

Samples were collected from nearly a hundred wells in the Kirkwood-Cohansey Aquifer for measuring of the concentrations of ^{228}Ra, ^{226}Ra and ^{224}Ra as well as gross alpha activity. Results for ^{228}Ra ranged from < 0.5 to 12.8 pCi/L. Those for ^{226}Ra ranged from < 0.5 to 17.4 pCi/L, and those for ^{224}Ra ranged from < 0.5 to 16.8 pCi/L. Gross α activity measurements made within 48 hours of collection exceeded 15 pCi/L in nearly half of the samples examined.

The short lived ^{224}Ra ($t_{½}$ = 3.66 d) made significant contributions to the observed values for gross alpha activity. These observations are important in improving the understanding of the potential the health effects associated with the consumption of water containing ^{228}Ra, ^{226}Ra and ^{224}Ra.

Chapter 11 - Because of arsenic contamination in shallow (5-50m depth) groundwater in the deltaic and floodplain areas of Bangladesh, characterization of deep aquifers for sustainable drinking water use is becoming an important issue. At Madaripur town, the municipal water supply is facing quality degradation problems because of high arsenic, iron and chloride content in the upper aquifer, and salinity tendency even at greater depth. In this study a 30-m clay layer, identified as a confining unit, is encountered from 130m to 160m below land surface and separates the deep aquifer (generally containing small or no appreciable concentrations of arsenic) from the shallow problematic aquifer. The Hantush-Jacob solution (1955) for a leaky confined aquifer was determined to be the best-fit solution to aquifer-test data collected from a deep borehole. The best fit curve match to two observation wells yielded an average aquifer storage of 0.0003 to 0.006, aquifer transmissivity of 4500 m^2/day, vertical to lateral aquifer hydraulic conductivity ratio of 1.0, and a confining unit vertical hydraulic conductivity of 0.044 to 0.93 m/day. The analysis verifies that the aquifer can yield significant quantities of water to wells with small drawdown and little potential to draw possibly contaminated water from the shallow aquifer.

Chapter 12 - The behavior of metal/metalloid pollutants can play an important role in water quality, especially in the case of drinking water reservoirs. Bottom sediments are a good sorption medium for many organic and inorganic pollutants. Metals and metalloids of natural or anthropogenic origin present in aquatic environments can be bound to sediments and preconcentrated there. Changes of geochemical conditions of the water environment can trigger metal/metalloid release from the sediments. Among many, pH is a very important factor influencing the sorption/desorption equilibrium. The aim of this work was to quantify the influence of pH on the sorption/desorption equilibrium of Cd and Pb. For this purpose, laboratory batch experiments were carried out. The scheme consisted of pH- and concentration gradients of the studied metals added in soluble forms to the water-sediment suspension. The sorption/desorption behavior of the studied metals under a pH gradient of 2-12 was shown in appropriate concentration gradients of 0.001 – 0.1 and 0.01 – 1.0 ppm for Cd and Pb, respectively.

In: Drinking Water: Contamination, Toxicity and Treatment ISBN: 978-1-60456-747-2
Editors: J. D. Romero and P. S. Molina

Chapter 1

TOXIN CONTAMINATION OF SURFACE AND SUBSURFACE WATER BODIES CONNECTED WITH LAKE VICO'S WATERSHED (CENTRAL ITALY)

Roberto Mazza[1], Giuseppe Capelli[1], Pamela Teoli[1], Milena Bruno[2], Valentina Messineo[2], Serena Melchiorre[2] and Antonio Di Corcia[3]
[1]Dipartimento Ambiente e Prevenzione Primaria, Istituto Superiore di Sanità, Viale Regina Elena, 299 – 00161, Roma, Italia
[2]Dipartimento di Scienze Geologiche, Università degli Studi "Roma Tre", Largo San Leonardo Murialdo , 1 - 00100, Roma, Italia
[3]Dipartimento di Chimica, Università di Roma ''La Sapienza'', Piazzale Aldo Moro, 5 - 00185, Roma, Italia

INTRODUCTION

Cyanotoxins are a new class of freshwater risk factors that are widespread all over the world and whose health implications are not yet fully understood. The environmental fate of these toxins, which are produced by many cyanobacterial species, holds consequences not only for water bodies where blooms occur, but also for public water sources. Indeed, drinking water wells may be contaminated by toxins capable of migrating through groundwater and causing toxic blooms in lakes and ponds.

In Italy, toxic cyanobacterial blooms are largely diffused in lakes, fish ponds, channels and even along sinuous rivers. However, their impact on health has been so far poorly investigated.

In order to assess the level of contamination of groundwater around lakes with toxic algal populations, a multidisciplinary study was conducted in the 2005-2007 period on the entire recharge area of Lake Vico, a volcanic crater lake in central Italy with a *Planktothrix rubescens* population producing hepatotoxic microcystins. The Lake, which is part of a natural reservoir, has significant environmental and scenic value and supplies drinking water to two towns. It has no tributaries, a long renewal time and tendency to accumulate polluting substances.

During a 18-month monitoring survey, waters from 5 stations in the Lake and 11 wells in its watershed were monthly sampled in order to investigate the dynamics of the *P. rubescens* population and groundwater flowpaths. Nutrient storage and eutrophication of the Lake's watershed and subsurface hydrogeological system were characterized quantitatively and qualitatively.

Moreover, a statistical methodology was developed to determine nutrient loads in the study area, identifying the polluting sources in soil and surface water and integrating the collected data with information about local human and natural systems (climate, soil gradation, surface and subsurface water circulation).

Groundwater and lacustrine water data and nutrient contents confirmed the high pollution of the Lake. Microcystins (MCs) in the Lake proved to reach values of up to 6.5 μg/L, with high concentrations in October at a depth of 15 m and fish contamination in spring. Five different MCs were detected. During the fall-winter months, MCs were found to migrate into 10 out of 11 water wells, hitting values of up to 123 ng/L and confirming the capability of these pervasive toxins to migrate through groundwater, as reported for other lakes of central Italy.

Possible recommendations for remediation of Lake Vico's waters should be focused on reduction of nutrient levels deriving from farming and residential sewage. The reduction could be achieved through actions based on agronomical technique improvement and sewage discharge regulations.

1. Geology

The study area, which extends for about 300 km^2, is located north of Latium (near Viterbo) and makes part of the Cimino-Vico Volcanic District.

The products from the Cimino volcano are older than the ones of the Vico volcano.

The Cimino Volcanic District lies in a pre-Apennine sector with NW-SE- and NE-trending faults (Figure 1.1). From the Messinian to the Late Pliocene, the structural depressions of this District were filled with some thousands of meters of sediments (La Torre et al., 1981).

The overlying volcanic deposits consist of domes and trachydacitic ignimbrites, covered by olivine-latitic lava deposits.

Subsequently, acidic viscous magmas rose to the surface, forming domes and laccoliths. The development of domes was associated with violent explosive episodes leading to the emplacement of the wide ignimbritic plateau occurring in the Cimino area. At present, over 50 hills may be identified in this area; each of them is due to the accumulation of rhyolitic-trachydacitic lava domes. Many of these domes are likely to be buried under the ignimbritic cover or to have been destroyed by their own explosive activity (Figures 1.2-1.3).

In this Complex, the domes are arranged in a semi-radial and semi-annular pattern with respect to the dome of Mt. Cimino, which represents the apical portion of the laccolith (Cimarelli and De Rita, 2006). The activity of the Cimino District ended with the emission of latitic or olivine-latitic lava flows, extending as far as 10 km from their emission center.

The K/Ar age of the volcanic products belonging to this complex of domes ranges between 1.35 ± 0.075 and 0.95 ± 0.2 Ma (Nicoletti, 1969).

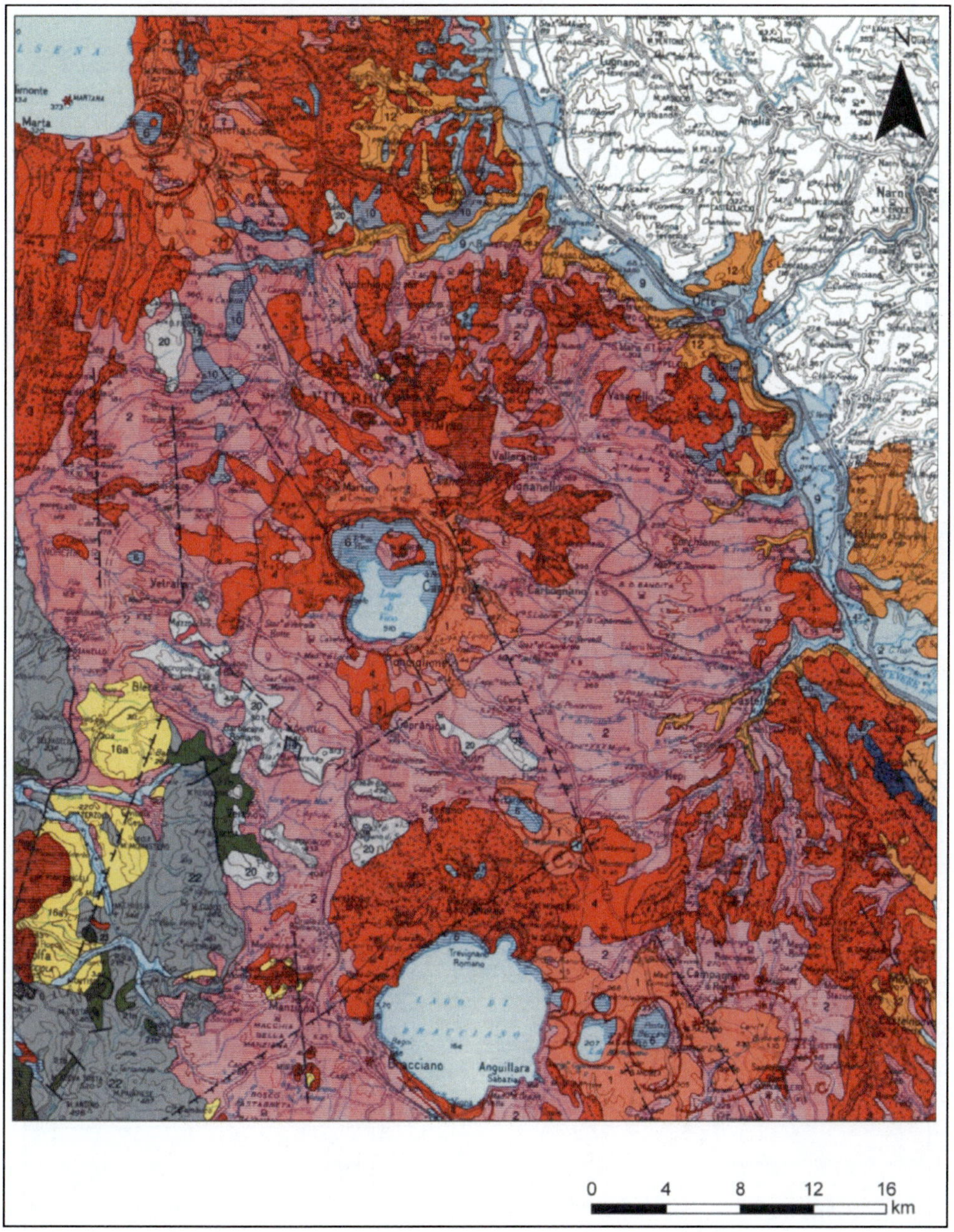

Figure 1.1. Structural geological map (Bigi et al., 1988). Legend: 1-Hydromagmatic pyroclastites;2-Pyroclastic flows;3-Fall-out deposits; 4-Undersaturated lavas;5-Acidic volcanites;6-Recent covers (Holocene); 7-Fluvio-lacustrine deposits (Holocene); 8- Debris (Holocene, Pleistocene); 9-Alluvia (Holocene, Pleistocene); 10-Travertines (Holocene, Pleistocene); 12-Heterogeneous clastic deposits (Pleistocene); 16a-Dominant clays with interbedded sands and gravels (Pliocene); 18- "Varicolori" argillites, marly limestones (Eocene, Upper Cretaceous); 19-Sandstones "Tolfa" Unit (Oligocene); 20-Calcarenitic turbidites (Tolfa Unit, Upper and Middle Eocene); 22-Argillitic succession with interbedded calcareous lithoids (Upper Cretaceous); 30-Micritic limestones with chert, marls, cherty limestones (Lower Cretaceous pp, Middle Lias – Transition facies) 32-Massive limestone (Lower Lias, Shelf-to-basin transitional calcareous units).

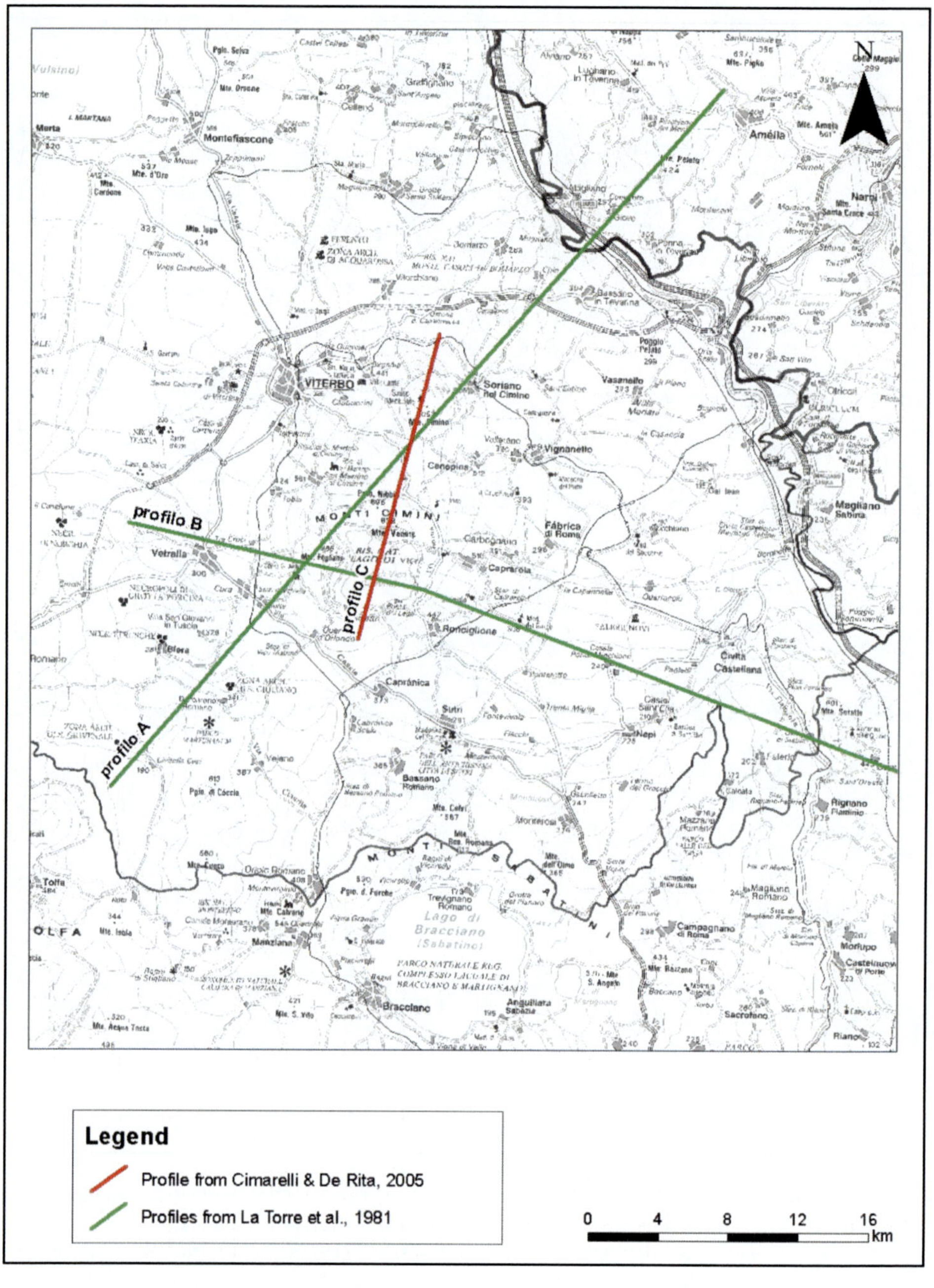

Figure 1.2. Map with location of geological profiles.

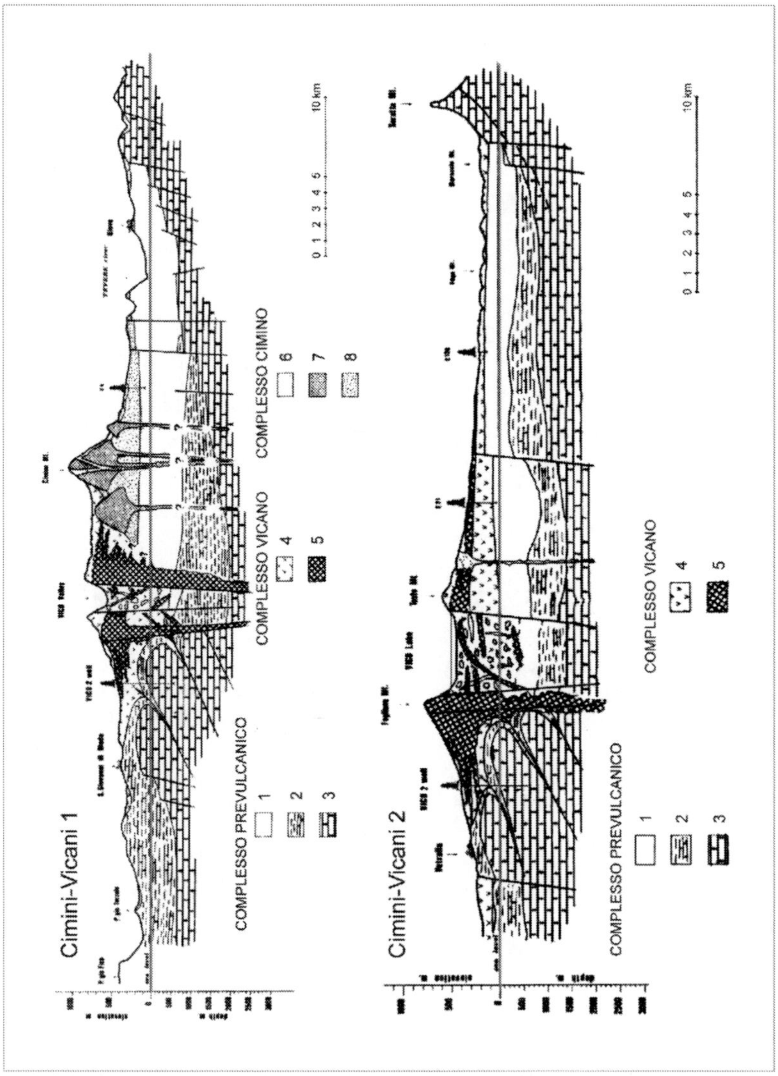

"Cimini-Vicani 1" and "Cimini-Vicani 2" geological profiles. ***Legend:*** **1** Formations of the neo-autochthonous cycle (Miocene-Pliocene-Quaternary); **2**: Allochthonous flysch facies formations (Upper Cretaceous-Oligocene); **3**: Mesozoic carbonate basement; **4**: Tuffs and lavas; **5**: Lavas and tuffs; **6**: Lava flows; **7**: Lava domes; **8**: Ignimbrites (from La Torre et al., 1981)

Figure 1.3. Profiles from La Torre et al., 1981

The activity of the Vico Volcanic District (started about 800,000 years ago and ended less than 90,000 years ago) mainly developed from a central edifice, the Vico Volcano. This is a typical strato-volcano whose terminal portion is truncated by a south-eccentric caldera. A secondary edifice (Mt. Venere volcano) originates from the bottom of the caldera. The Vico Volcano had dominantly explosive activity and alkali-potassic chemistry. The Authors distinguished 4 main stages of activity, with maximum paroxysm about 150,000 years ago. The landscape of the Vico District has a wide flat area, composed of flyschoid sediments and located NW of the central edifice. Sediments of a similar nature also make up the bedrock of the Volcano (De Rita et al., 1993a; De Rita et al. 1993b; De Rita et al., 1993c), (Figure 1.4).

The Vico Volcanic District was formed through the emplacement of large volumes of material from a single emission center.

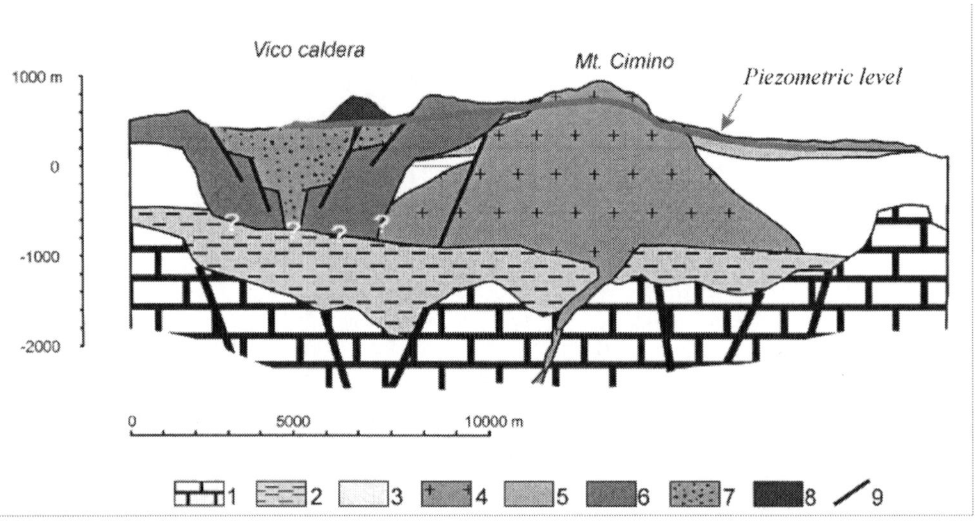

Schematic geological section of the Cimini Mts. *Legend:* 1: Mesozoic limestones; 2) Cenozoic siliciclastic sediments; 3: Plio-Pleistocene marine sediments; 4: Cimini Dome Complex, trachydacitic lavas; 5: Cimini Dome Complex, Ignimbrites; 6: Alkaline-potassic deposits of the Vico caldera; 7: Volcanic fills of the Vico Caldera; 8: Scoria cone of Mt. Venere; 9: Faults and fractures (from Cimarelli & De Rita, 2006)

Figure 1.4. Profile from Cimarelli & De Rita, 2006.

In this sector of northern Latium, the volcanic cover is a typical feature. Sporadic outcrops of sedimentary formations from the pre-volcanic landscape are only visible in its deep valleys or tectonic culminations.

Bertagnini and Sbrana (1986) summarized the Vico Volcano activity in 4 main stages, which started 0.9 to 0.095 Ma ago and which emplaced numerous formations (Mattias and Ventriglia, 1970) in the following sequence:

Stage I (0.9-0.4 Ma) – the activity began with emission of latitic-trachybasaltic lavas, followed by Plinian eruptions ("Tufi stratificati varicolori" of Vico) and emission of trachytic lavas from different eruptive centers. .

Stage II (0.33-0.2 Ma) – dominantly phonolitic-tephritic, tephritic-phonolitic and trachyphonolitic lava activity constructing the central edifice. This is strato-volcano, larger than the present one and located in the Vico Valley, between Mt. Fogliano and Poggio Nibbio, in the north-western sector of the caldera (Locardi, 1965).

Stage III (0.2-0.15 Ma) – this stage involved Plinian eruptions combined with intense explosive activity and repeated caldera collapses. Huge amounts of material became emplaced in this stage: ignimbrites A and B (Locardi, 1965) or the "Tufo Grigio a scorie nere" grey tuff with black scoriae (Mattias and Ventriglia, 1970) and ignimbrite C (Locardi, 1965) or the "Tufo Rosso a scorie nere" red tuff with black scoriae (Mattias and Ventriglia, 1970).

Stage IV (0.14-0.095 Ma) – a first stage with deposition of vesicular tuffs and massive tuffs was followed by emission of typically hydromagmatic deposits and pyroclastic surges.

2. HYDROGEOLOGY

The Cimino-Vico Hydrogeological Unit is composed of multiple depositional series and of a complex pre-volcanic bedrock with low permeability. Geophysical studies (ENEL, 1994) and wells for geothermal exploration (ENEL VDAG-URM, 1994) evidenced a structural high of the bedrock with two culminations: one lying under Mt. Cimino, north of Lake Vico, and one at Mt. Calvello, south of the same Lake (Figure 2.2). Between these two bedrock elevations lies the volcano-tectonic depression which accommodates the lacustrine basin. So far, no studies have provided information on the depth of the roof of the low-permeability bedrock lying beneath the central caldera. What is certain is that, a few kilometers away from the caldera, the roof of the bedrock lies at about 100 m above sea level. The reliefs hosting the watershed area of the Lake and the aquifers of the hydrogeological basin feeding it mainly consist of (Figure 2.1):

- *Complex of talus cones and debris*: slope debris or alluvial fans and landslide deposits. This complex is from very permeable to averagely permeable, depending on the nature of the slope feeding it. The permeability of the landslide bodies is related to the lithology of the slope and thus extremely variable.
- *Alluvial complex*: present and recent, occasionally terraced, sandy, clayey and sandy-gravelly alluvia and colluvial and eluvial covers; reworked tuffs and minutely stratified tuffites of the Vico depression; shore sands. The permeability of this complex ranges from average-low to average.
- *Complex of stratified tuffs and phreato-magmatic facies:* stratified tuffs, tuffites and earthy tuffs; occasional diatomiferous and lacustrine deposits, pumice levels and lapilli of variable size, ultra-fine ash and products of alteration of the peripheral belts of the various pyroclastic formations; terminal eruptive stages consisting of pyroclastic breccias and stratified tuffs (mostly cineritic) with lapilli and lava blocks that are densely stratified and embedding small tuffitic and limnic levels from palustrine sedimentation. This complex accommodates lithoformations of very different origin. In hydrogeological terms, the permeability of this complex is very low to low.
- *Complex of lavas, laccoliths and debris cones.* The permeability of this complex, which arises from an extensive net of fractures or porosity, ranges from average-high to high. Vertical permeability diminishes in places with cineritic or lahar levels.
- *Complex of pozzolanas*: deposits from pyroclastic flow, generally massive and chaotic, prevalently lithoid. The complex includes the ignimbrites and tuffs identified by the Authors. Its permeability goes from average to average-high and is principally due to porosity and subordinately to fracturing. Vertical permeability is affected by the occurrence of very extensive and often meter-scale palaeosols. Permeability is low where the volcanites have undergone a zeolitization process.
- *Complex of terrigenous deposits*: marly-arenaceous allochthonous flysh of Tolfa, consisting of rhythmically alternating clays, marls and sandstones; grey-blue silty clays of Pliocene age and clays with gypsum of Miocene age; mostly argillaceous deposits of marine and shelf-to-basin transitional facies; dominantly silty-clayey deposits of palustrine, lacustrine and brackish facies. The complex has low or very

low permeability, which affects groundwater circulation in the entire volcanic domain. Locally, minor shallow aquifers (chiefly arenaceous levels and alteration covers) are encountered.

- *Complex of debris limestones*: fissured and karstified limestones that are intercalated to a variable extent with marly and argillaceous cherty deposits with low permeability. The complex has good permeability values.

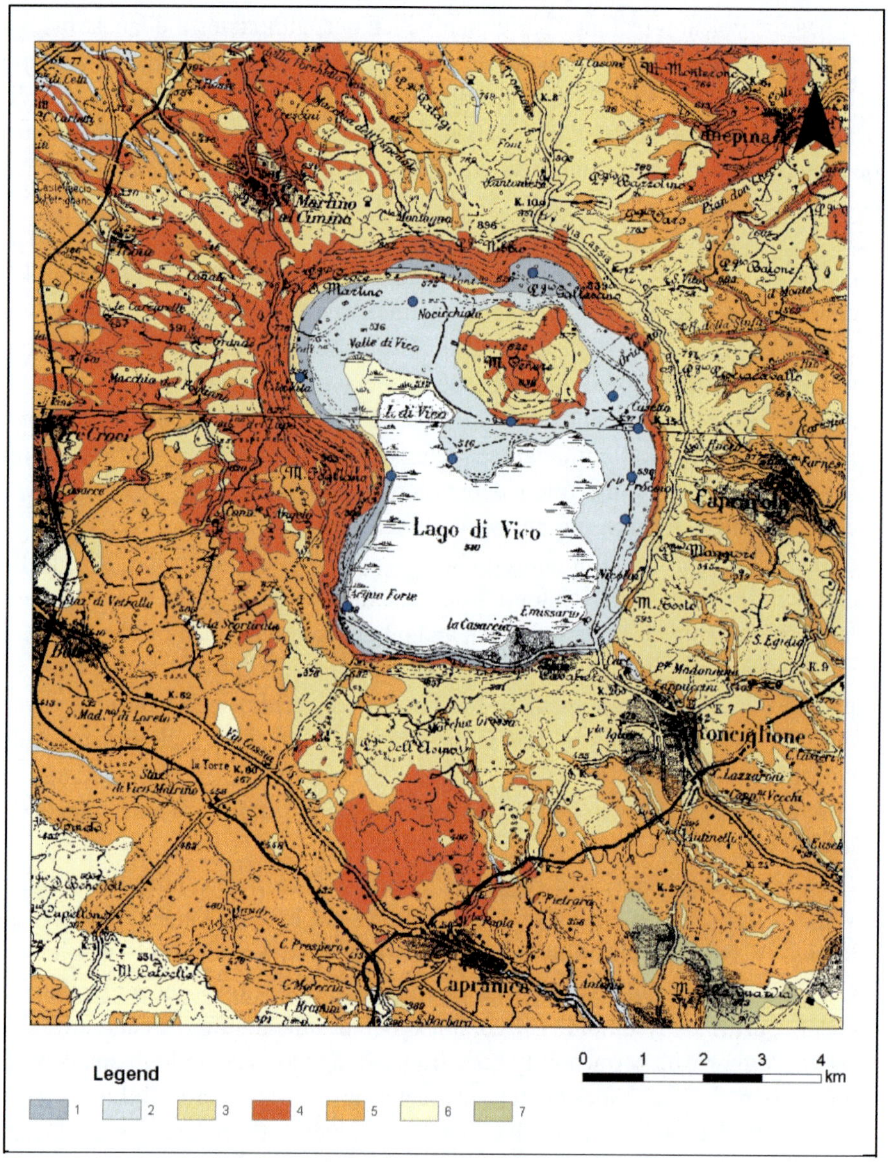

Figure 2.1. Map of hydrogeological complexes. Legend: 1- Complex of talus cones and debris;2- Alluvial complex; 3- Complex of stratified tuffs and phreato-magmatic facies;4- Complex of lavas, laccoliths and debris cones;5- Complex of pozzolanas; 6- Complex of terrigenous deposits; 7-Complex of debris limestones.

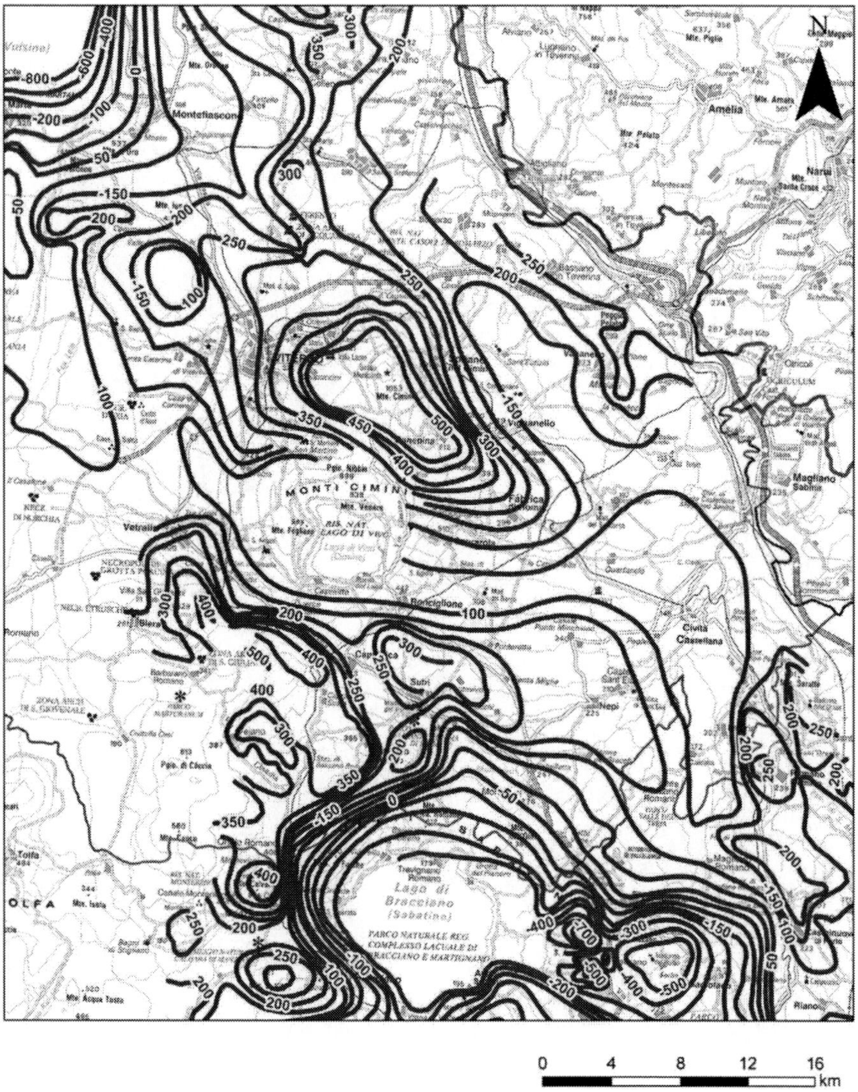

Figure 2.2. Map of isobaths of the pre-volcanic bedrock, ENEL 1994).

Lake Vico's watershed is endorheic. However, an outlet stream, which has been in existence for some millennia, discharges the Lake's water into the Rio Vicano stream through a tunnel (renovated in 2000). Some dozens of meters of lacustrine sediment are likely to have accumulated in the Lake's volcano-tectonic depression approximately 90,000 years after the end of volcanic activity. Debris deposits and slope debris overlap or interdigitate into these sedimentary deposits. The depth of the wells surveyed along the Lake's perimeter varies from 30 to 50 m. Therefore, the wells are assumed to draw more water from the debris covers than from the regional aquifer. Under this hydro-structural assumption and disregarding rainfall, the Lake's watershed is mainly recharged by overflow of the regional groundwater above the lacustrine deposits and the groundwater flowpath originates near Mt. Cimino and the north-eastern water divide (Figures 2.3-2.5).

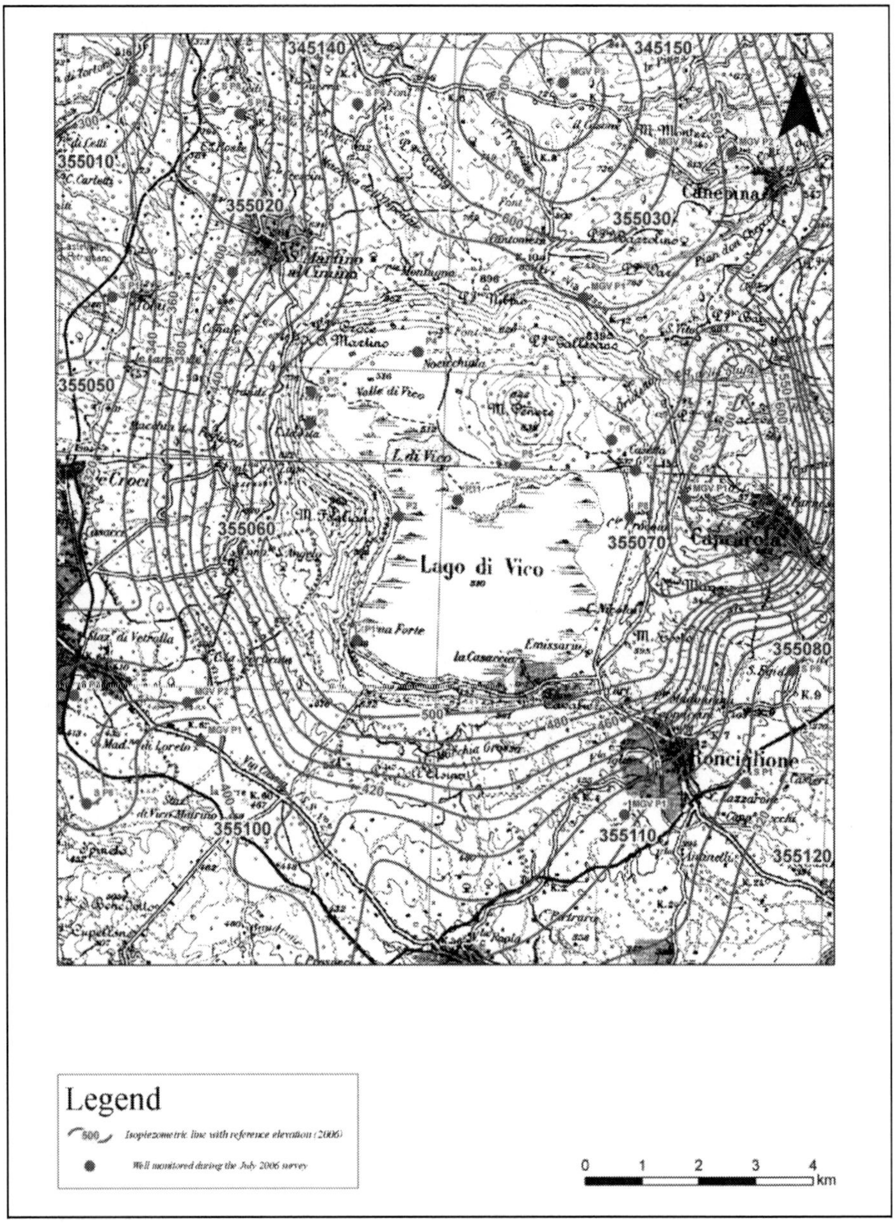

Figure 2.3. Piezometric map 2006.

Unlike other volcanic lakes of Latium, Lake Vico is not likely to be recharged from the bottom, i.e. through drainage from the confined aquifers below it. Conversely, water losses from the lacustrine basin (regardless of those from its outlet stream or evaporation) take place by overflow towards east (into the Treia river basin) and west (into the Marta river basin) above the lacustrine sediments and, to a lesser extent, by percolation through the same sediments. Under this schematic interpretation, the modes of circulation and renewal of the lacustrine waters have an impact on the flows of eutrophic substances or, more generally, of pollutants from and to the Lake.

In the past decades, several authors (including researchers from "Laboratorio di Idrogeologia" – Università "Roma Tre", Capelli et al., 2005) produced piezometric maps of the Cimino-Vico area.

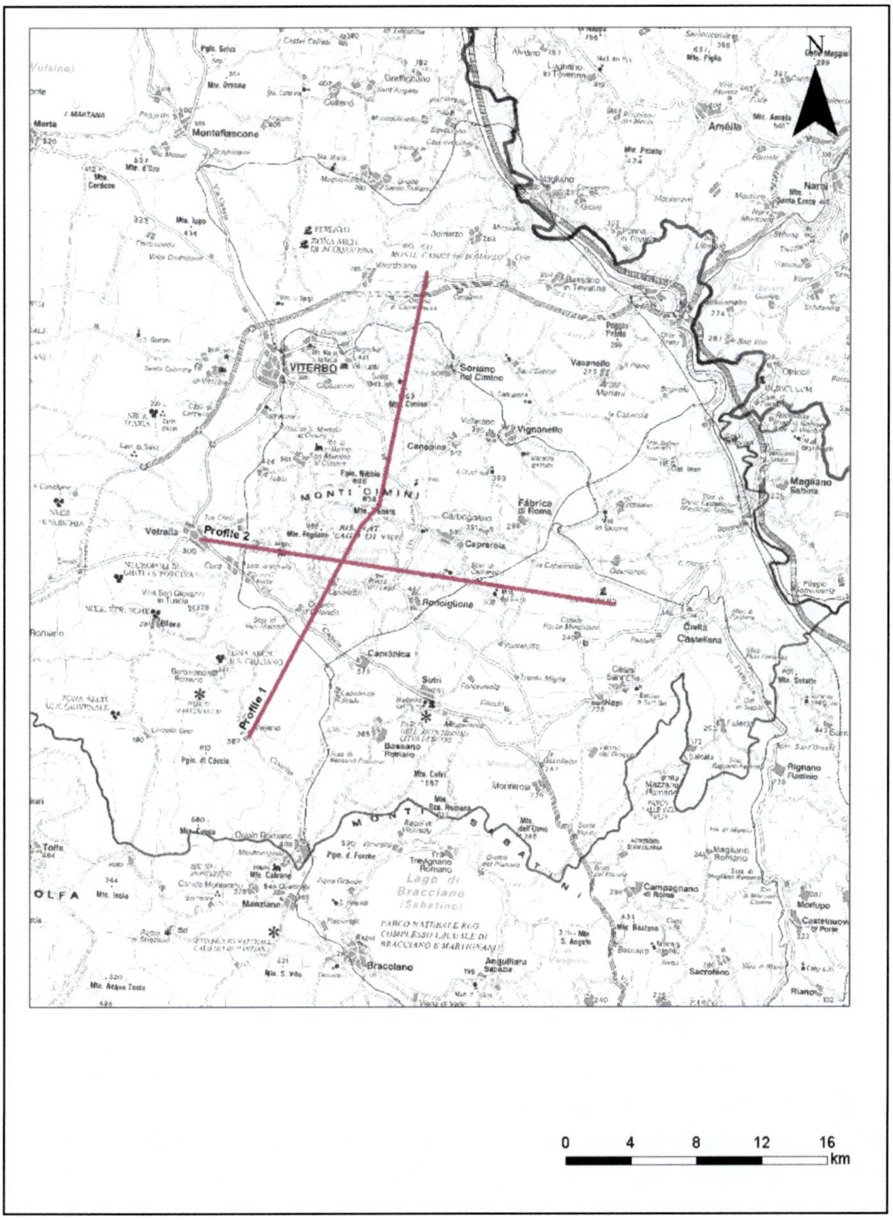

Figure 2.4. Map with location of hydrogeological profiles.

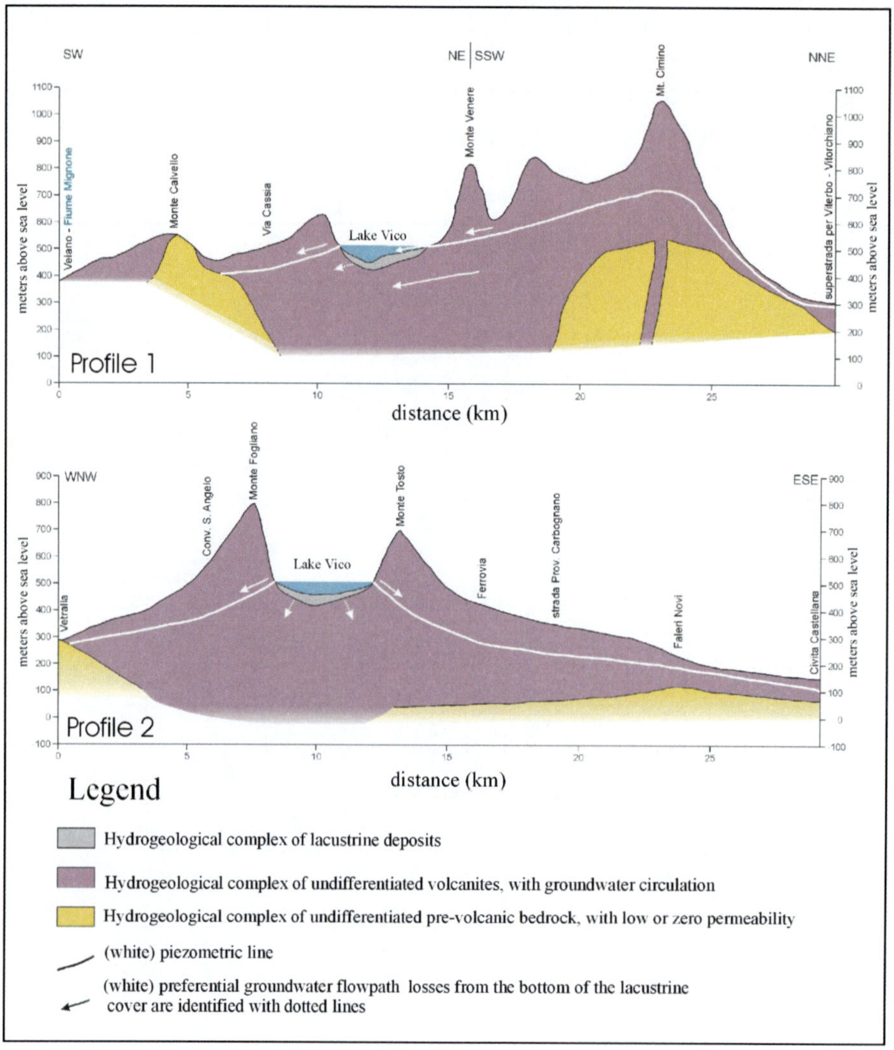

Figure 2.5. Hydrogeological profiles.

The 2005-2006 monitoring survey was designed to meet the requirements of this study. The morphology of the roof of the saturated zone remained practically unaltered over the years, while saturation elevations only changed at local level. The main piezometric high (Figs. 2.3 and 2.5) is located at Mt. Cimino and corresponds to the high of the above mentioned pre-volcanic bedrock. A secondary piezometric high lies west of Caprarola, against the water divide of the watershed area. The 2006 map (supported by about 32 monitoring points) incorporates the data from the 2002 surveys in the periphery of the study area. The hydrogeological basin recharging the Lake was identified on the basis of the morphology of the saturated zone. Integrated reading of piezometric map, geological-structural map and profiles can help understand the previously described groundwater flow assumptions.

3. Hydrology

To substantiate environmental and hydrogeological considerations about Lake Vico, it is worth recalling a few morphometric and hydrological data. Reference will be made to one of the most reliable bathymetric maps of Lake Vico reported in the literature (Figure 3.1).

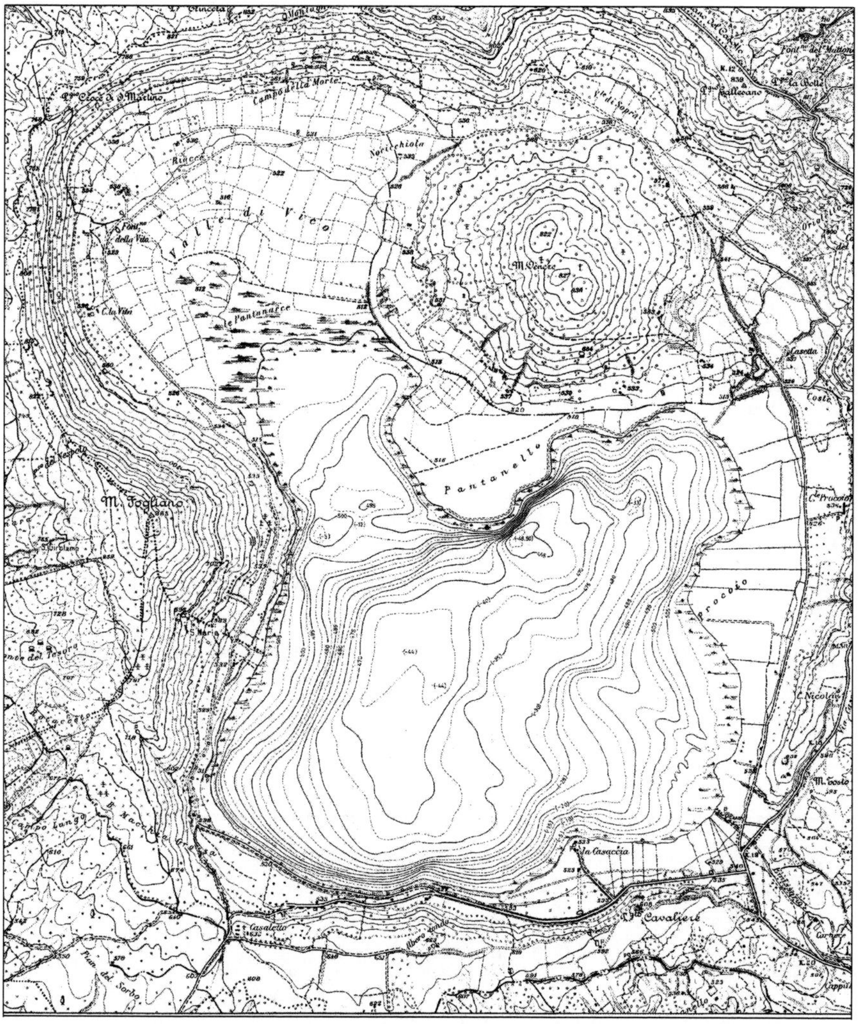

Figure 3.1. Bathymetric map of Lake Vico(Barbanti et al., 1968).

The bathymetric map of Lake Vico shows two significant depressions. One of them is located northwards, against the Pantanello plain under Mt. Venere. At the foot of a steep submerged scarp, this depression reaches a depth of 48.5 m. The second depression has a more central location at the piedmont of Mt. Fogliano and reaches a depth of 44 m. The surveys identified with code "L5" were made in this second depression. This is one of the most reliable maps because: i) it was built by a specialized institute ("Istituto Italiano di Idrobiologia", Pallanza); ii) it includes the traces of the survey profiles; and iii) it describes the methodology of the survey.

Table 3.1. Morphometric data (Carollo et al., 1974, modified)

PARAMETER	VALUE
Average elevation of lake surface	510 m above sea level
Perimeter	16.9 km
SL average lake surface area	12.1 km2
Maximum depth	48.5 m above sea level
Average depth	21.58 m above sea level
Lake volume	260,767 x 106 m3
SB Surface area of watershed, excluding lake	28.84 km2
SG Surface area of hydrogeological basin, excluding lake	33.29 km2
Ratio of watershed surface area to hydrological basin surface area (SL/ SL+ SB)	0.29

Table 3.1 displays the morphometric data of the Lake which were regarded as the most significant for the purposes of this study. The data were freely drawn from "Indagini Limnologiche sui laghi di Bolsena, Bracciano e Trasimeno", a publication issued by "Istituto di Ricerca sulle Acque – CNR" (IRSA, 1974).

The data obtained from SIMN (Servizio Idrografico e Mareografico Nazionale – National Hydrographic and Mareographic Service) suggest the hydrometric heights of the Lake at its outlet (before its water flows into the outlet stream tunnel) vs. precipitation into its watershed (Figure 3.2).

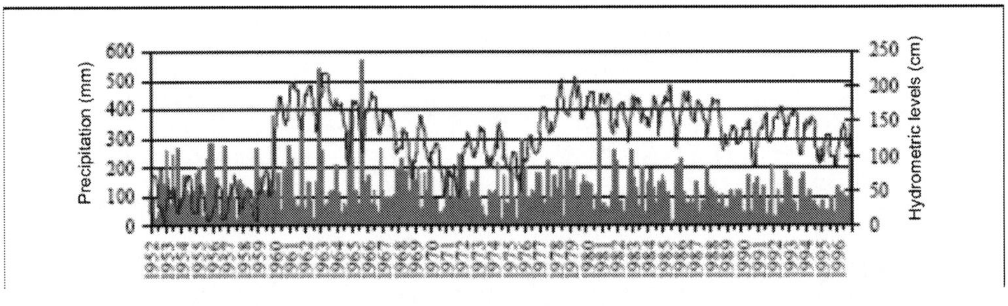

Figure 3.2. Hydrogram at the Lake's outlet (Dragoni et al., 2002).

The graph clearly shows a sudden shift in the average water level of the Lake in 1958 that is not correlated with change in precipitation. The shift is likely to be due to the correction of a systematic error in the measurements or to a different management of flows in the outlet stream, which are controlled through a gate. Therefore, the water levels prior to 1959 should be neglected. The same graph also displays an immediate response of the hydrometric level of the Lake to precipitation. Hydrometric levels cannot be compared with discharges, as no H/Q calibration curve has ever been drawn.

Nonetheless, the Authors of this study have measured the discharges of the outlet stream (at the outlet of the Lake) or of the Rio Vicano stream at different significant elevations for several decades. To the best of their knowledge, these are the only available discharge data. Table 3.2 exhibits discharge values vs. survey elevation..

Table 3.2. Discharges of the outlet stream

Discharges of Lake Vico and of Rio Vicano stream			
Elevation (m a.s.l.)	Discharge (l/s)	Date	Name
510	116	13/09/1988	outlet
490	34	17/01/1989	Rio Vicano
410	100	01/11/1982	Rio Vicano
300	33	18/10/2002	Rio Vicano
300	120	15/03/2002	Rio Vicano
300	86	01/07/2002	Rio Vicano
300	106	31/01/2003	Rio Vicano

Table 3.3. Withdrawals by aqueducts

Withdrawals from the Lake. PRGA Aqueduct Master Plan, Viterbo, 1998			
PRGA code	Elevation m a. s. l.	Discharge (l/s)	Municipality
806	510	9,8	Caprarola
806/A	510	30	Ronciglione

Withdrawals from wells (PRGA, Viterbo, 1998)			
PRGA code	Elevation m a. s. l.	Discharge (l/s)	Municipality
960	550	6	Caprarola
961	700	5	Caprarola
1037	674	10	San Martino C.

The values in the Table infer a discharge of the outlet stream of 30-150 l/s. Direct withdrawal by the municipal aqueducts of Ronciglione, Caprarola and San Martino al Cimino is to be added to these values, measured at the outlet of the watershed.

Table 3.3 indicates that total withdrawal from the Lake is about 40 l/s and that withdrawal from wells is 21 l/s. Therefore, 61 l/s are abstracted from the watershed supplying the Lake. Supposing an average discharge of the outlet stream of about 50 l/s, total discharge at the outlet of the Lake is equal to roughly 110 l/s.

4. METHODOLOGIES

Monthly sampling surveys were carried out to define the hydrogeological and hydrochemical parameters of the Lake and of its aquifers. Reliance was made on: i) 5 stations (4 riparian and 1 central), selected in agreement with the Department of Chemistry, University of Rome "La Sapienza" (the central station was a fixed buoy of 1 m^3); and ii) 11 points of the piezometric network (Figure 4.1). The survey was conducted from September 2005 to February 2007, excluding the month of August. Another survey (only in July 2006) was carried out to measure piezometric and physico-chemical parameters (pH, conductivity, temperature) in another 21 wells around Lake Vico. This survey updated piezometric data up to 2006 (Figure 2.3).

The following parameters were determined in the wells:

- pH, temperature (°C), static level (m above sea level), conductivity (μS/cm) (*in situ*);
- nitrates (μg/l), phosphates (μg/l), NTot (μg/l), PTot (μg/l) (in the laboratory).

The following parameters were determined in the Lake:

- pH, temperature (°C), static level (m above sea level), conductivity (μS/cm) (*in situ*);
- nitrates (μg/l), phosphates (μg/l), NTot (μg/l), PTot(μg/l) (in the laboratory);
- dissolved oxygen (%, mg/l), dissolved salts (g/l) (*in situ*).

The laboratories involved in the survey were different, in that it was deemed useful to compare multiple methodologies (using different equipment and protocols) and multiple operators with a view to minimizing systematic or random errors. The following Table lists the laboratories participating in the study in the various months.

Samples were collected with a manual sampler both in the Lake and in the wells. In the central station of the Lake, sampling was made every 5 m down to the –30m isobath; in riparian stations, sampling was made at shallow depth (-30 cm from the Lake's surface). For determining chemical elements, water was collected with 1000 cc glass and PVC bottles, sterilized after each sampling. The bottles were stored in a portable cool box (temperature below 4°C) and timely delivered to the laboratories.

In some cases (e.g. samples of October 2005), when the water samples were expected to be stored for a considerable time, they were frozen to stabilize their chemical and bacterial activity.

The glassware used for analyses was prepared according to a standard procedure:

- soaking in soapy water for at least one night;
- rinsing under running water, distilled water and, finally, deionized water;
- drying, keeping away from air and contamination sources.

The detergent was chosen among non-ionic products, free of the elements to be analyzed.

Table 4.1. Laboratories participating in the study on chemical elements

Laboratory	Institution	Position of Operator	Period of Work	Equipment Used
Geochemical Laboratory	Department of Geological Sciences, University of Rome "Roma TRE"	Post-Graduate student	September 05-December 05	Spectrophotometer: Perkin Elmer Perkin Elmer UV/VIS Lambda 25, supported by UV software - Win Lab, with usable optical path of 1 and 5 cm
Chemical Laboratory	Department of Chemistry, Univerisity of Rome "La Sapienza"	Laboratory Specialist	January 06 - February 07	- DX-100 Dionex Ion Chromatograph; - Cary 50 Win UV Varian - ICP – OES VARIAN VISTA MPX CCD simultaneous with Cetac ultrasound nebulizer - PURELAB-PLUS USF ELGA water purification system
Chemical Laboratory	Department of mechanical and industrial engineering, University of Rome "Roma TRE"	Laboratory Specialist	January 07-February 07	Spectrophotometer: Perkin Elmer Perkin Elmer UV/VIS Lambda 25, supported by UV software - Win Lab, with usable optical path of 1 and 5 cm

The physico-chemical parameters of the Lake, such as temperature, pH, conductivity, dissolved salts and dissolved oxygen, were measured with a multiparametric depth probe (Hydrolab, surveyor 4 model). Vertical logs were taken on a monthly basis; in riparian stations and wells, measures were taken at shallow depth via standard probes.

Main Physico-Chemical Characteristics of the Lake's Waters Through Experimental Monitoring Surveys

The monitoring and sampling surveys and laboratory tests conducted in the study period yielded a substantial volume of parametric data, taking into account the number of stations used.

To make results more readily understandable, values have been reported not only in tables but also in graphs of different type.

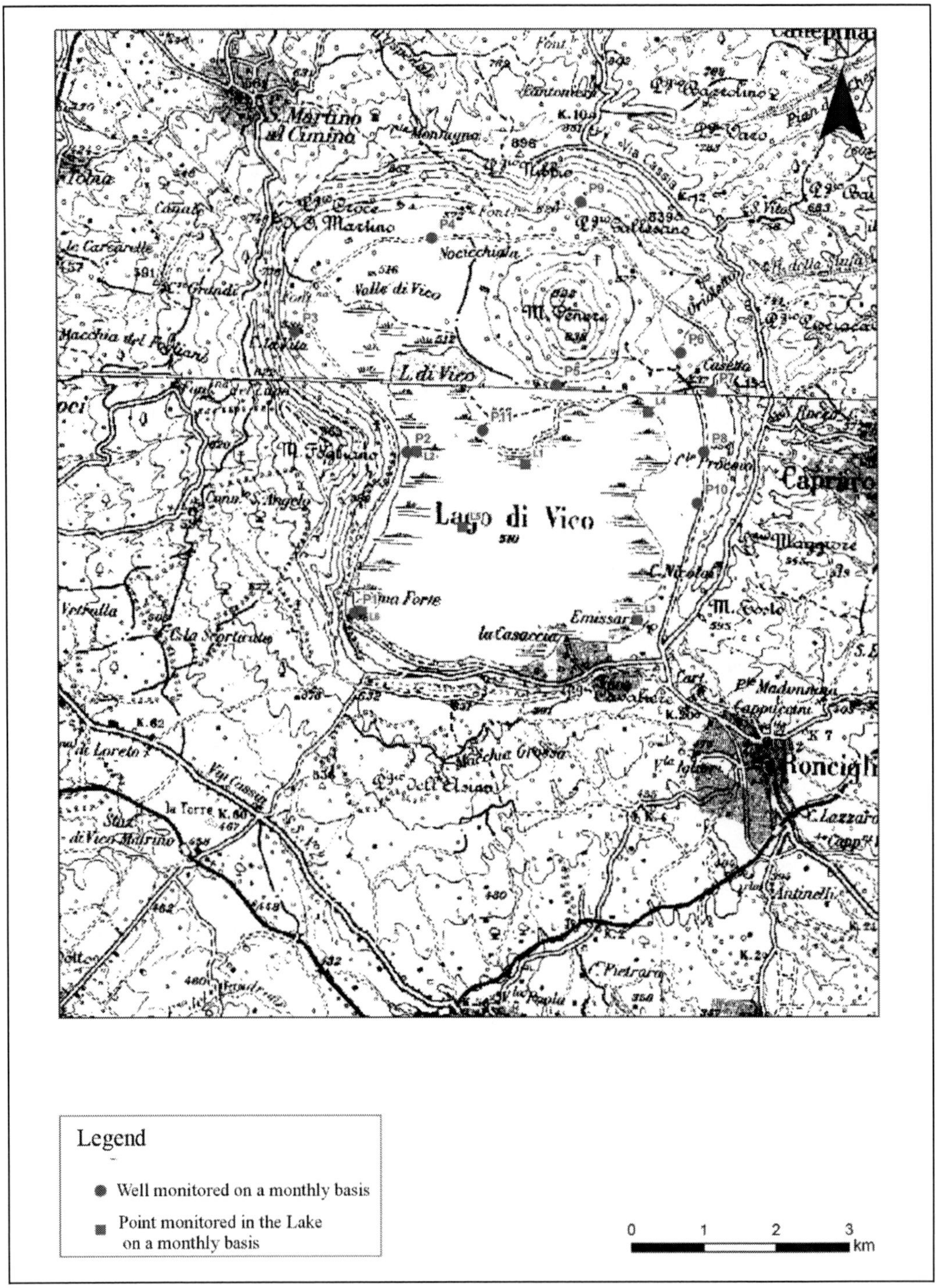

Figure 4.1. Monthly monitored wells and points in the Lake.

5. Groundwater Monitoring

This activity was preceded by a study of the area with a view to assessing bibliographic references, identifying water points to be surveyed and getting the involvement of well owners. The Latium Region informed these owners about the project, so as to secure their consent to sampling and on-site determinations for an entire hydrological year. The monitoring survey (July 2006) was focused on a representative grid of 32 wells, distributed over the study area (Figure 2.3). The screen of these wells lies at an average depth of 470 m above sea level, as shown in the histogram of Figure 5.1.

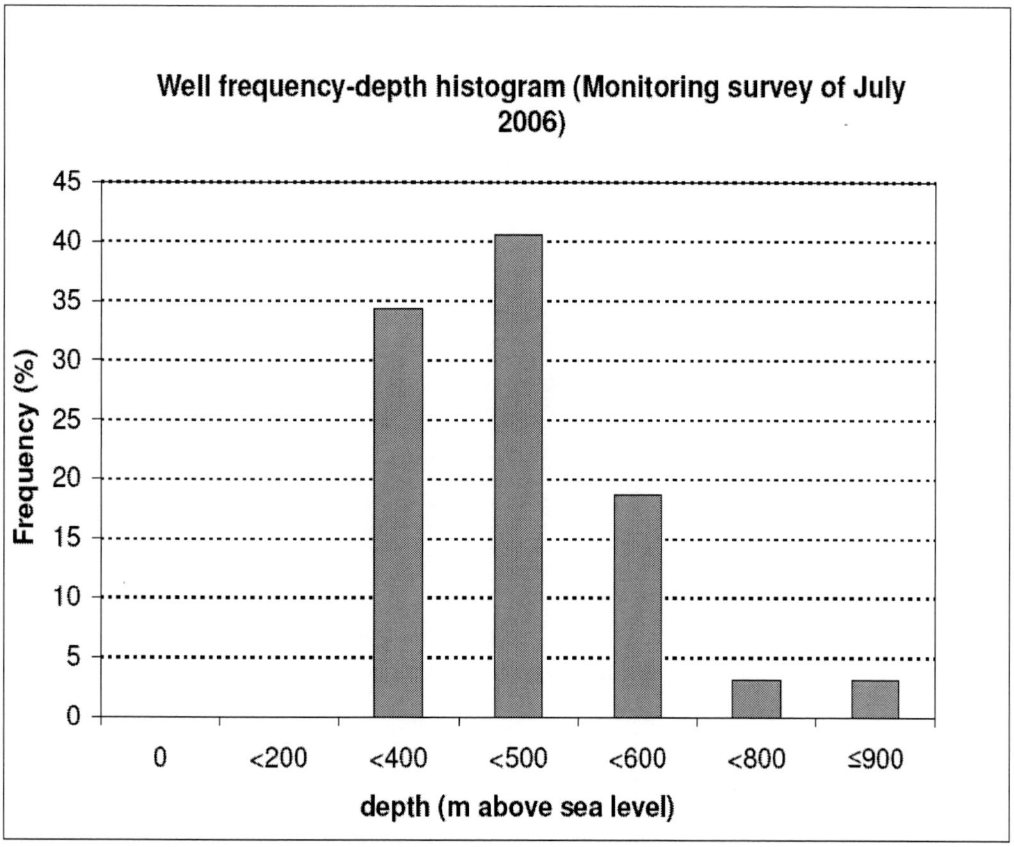

Figure 5.1. Well frequency-depth histogram.

Table 5.1 below summarizes the measures taken in the wells during the July 2006 survey and the characteristics of the monitored wells.

In addition to updating piezometric level data, investigations were conducted to assess the chemical elements contained in the Lake's waters. As previously mentioned, consideration was given not only to physico-chemical parameters (e.g. temperature, pH and conductivity), but also to contents of nitrates, phosphates, total nitrogen and total phosphorus. The survey was made from September 2005 to February 2007 on a monthly basis. These investigations were concentrated on a narrower grid of wells, which are listed below:

Table 5.1.Results of the July 2006 survey (*estimated data)

Well Code	Longitude	Latitude	Well-Head Elevation (m sea-level)	Total Well Depth (m from ground level)	Well Bottom Elevation (m sea-level)	Well Diameter (Filter) (mm)	C.T.R. (Regional Technical Map)	Static Level (m from ground level)	Temperature (°C)	pH	Conductivity (microS/cm)	**Date**
MGV355020P2	271634	4696367	614	90	524	180	355020	73.60	15.2	7.02	123	26/07/06
MGV355020P3	268689	4697570	758		758	250	355020	24.22	14.7	6.74	188	26/07/06
MGV355030P4	270216	4696368	720	130	590	250	355030	97.30	15.9	6.38	162	26/07/06
MGV355070P1	270817	4690426	680	110	570	250	355070	22.60	15.7	6.12	212	27/07/06
MGV355100P1	262290	4686278	470	100	370	180	355100	46.90				27/07/06
MGV355110P1	269737	4685009	420	80	340	300	355110	55.20	15.8	7.34	355	27/07/06
MGV375030P1	269034	4693888	835		835	350	375030	14.65				26/07/06
MGV375060P2	262047	4686940	474		474	180	355060	95.10				27/07/06
S355020P1	260741	4693901	402	114	288	150	355020	98.30	15.2	6.72	293	26/07/06
S355020P3	261136	4697614	308	18	290	250	355020	7.30	15.7	7.92	421	26/07/06
S355020P4	262884	4694323	515		515		355020	97.20	15.2	6.27	236	26/07/06
S355020P5	263022	4697039	440	55	385	180	355020	51.40				26/07/06
S355020P6	265104	4697205	578	75	503	250	355020	47.20	14.3	6.21	203	26/07/06
S355020P8	262564	4697323	418	72	346	250	355020	25.10	15.6	6.23	320	26/07/06
S355020P8	262564	4697323	418	72	346	250	355020	25.10	15.6	6.23	320	26/07/06
S355030P3	272870	4697551	500		500	200	355030	53.20				26/07/06
S355060P2	264284	4692278	534	65	469	200	355060	16.20	15.3	7.34	197	26/07/06
S355070P6	272698	4687503	405	106	299	200	355070	87.20				26/07/06
S355100P2	260020	4687063	412	75	337	200	355100	46.30	15.2	6.89	243	26/07/06
S355100P8	260255	4685209	436	64	372	300	355100	57.30	15.2	6.90	189	26/07/06

Table 5.1 (Continued)

Well Code	Longitude	Latitude	Well-Head Elevation (m sea-level)	Total Well Depth (m from ground level)	Well Bottom Elevation (m sea-level)	Well Diameter (Filter) (mm)	C.T.R. (Regional Technical Map)	Static Level (m from ground level)	Temperature (°C)	pH	Conductivity (microS/cm)	**Date**
S355110P1	271869	4685554	399	74	325	200	355110	63.90	14.3	7.23	356	26/07/06
PT355060P1(P1)	265067	4688005	518	18	500	300	355060	2.87	14.2	6.13	631	13/07/06
PT355060P2(P2)	265794	4690130	518	22	496	200	355060	15.03	24.1	6.85	1330	13/07/06
PT355060P3(P3)	264254	4691720	538	40	498	250	355060	22.60	15.4	7.01	367	13/07/06
PT355020P1(P4)	266139	4692954	531	40	491	250	355020	16.30	15.6	6.90	584	13/07/06
PT355070P1(P5)	267819	4691014	540	40	500	200	355070	35.30	15.3	6.29	563	13/07/06
PT355070P2(P6)	269521	4691439	538	40	498	250	355070	28.70	14.3	6.50	434	13/07/06
PT355070P3(P7)	269939	4690920	538	40*	498	250	355070	29.20	14.8	6.58	396	13/07/06
PT355070P5(P10)	269745	4689446	524	40*	484	250	355070					13/07/06
PT355030P1(P9)	268159	4693433	615	100	515	250	355030		13.60	6.99	186	13/07/06
PT355070P4(P8)	269837	4690126	524	35	489	250	355070	13.51	17.60	6.62	488	13/07/06
PT355070P6(P11)	266825	4690413	513	20	493	250	355070	2.70	16.20	6.35	644	13/07/06

Table 5.2. List of wells monitored from September 2005 to February 2007

Well Code	Longitude	Latitude
PT355060P1(P1)	265067	4688005
PT355060P2(P2)	265794	4690130
PT355060P3(P3)	264254	4691720
PT355020P1(P4)	266139	4692954
PT355070P1(P5)	267819	4691014
PT355070P2(P6)	269521	4691439
PT355070P3(P7)	269939	4690920
PT355070P5(P10)	269745	4689446
PT355030P1(P9)	268159	4693433
PT355070P4(P8)	269837	4690126
PT355070P6(P11)	266825	4690413

The following graphs give the results of analyses made on the samples which were monthly collected from all the wells (monthly tables of all the monitored parameters are reported in the Appendix).

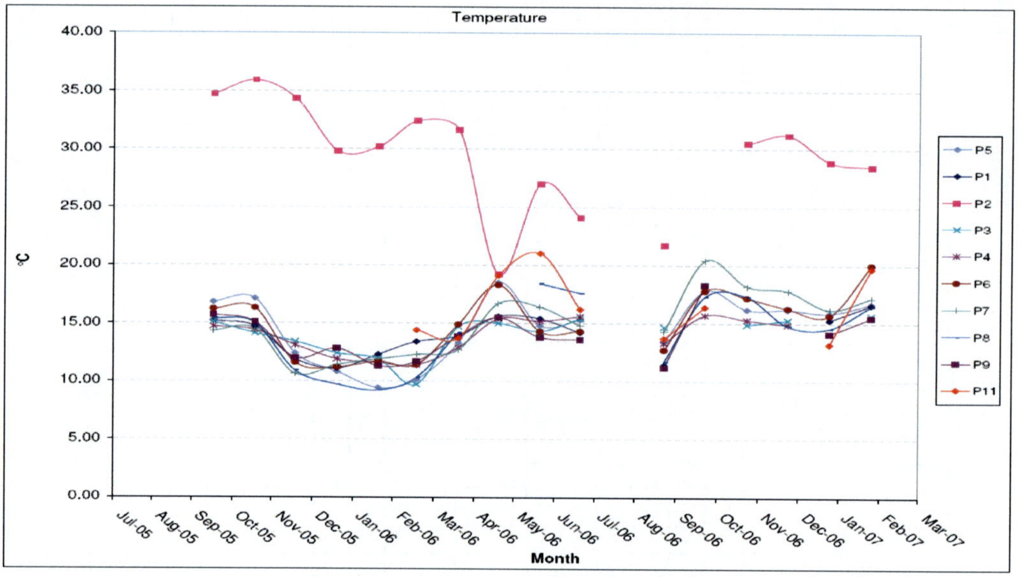

Figure 5.2. Graph of groundwater temperature measured in the wells.

Commentary on the graphs

- *Temperature Graph* (Figure 5.2): the graph immediately shows a groundwater thermal anomaly in well P2 (low-enthalpy thermal fluids, average temperature 30°C). This anomaly, together with visual evidence of rising groundwater convoys along the Lake's shore in front of the well, suggests that previously unreported endogenous fluids rise into the lacustrine caldera. The values measured in the remaining wells have a similar pattern, which is certainly related to seasonal cycles (range: 9.20-21 °C).

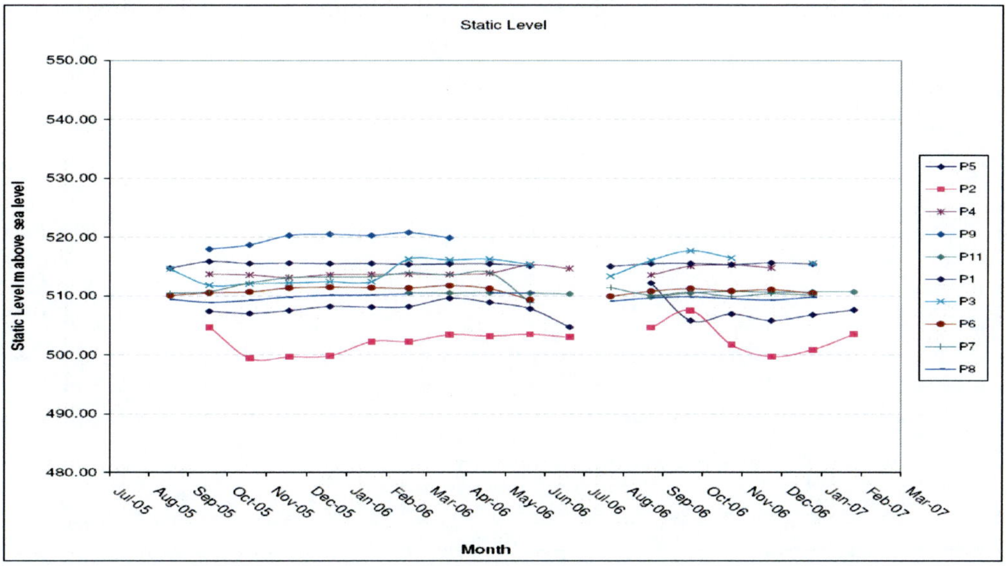

Figure 5.3. Graph of groundwater static level measured in the wells.

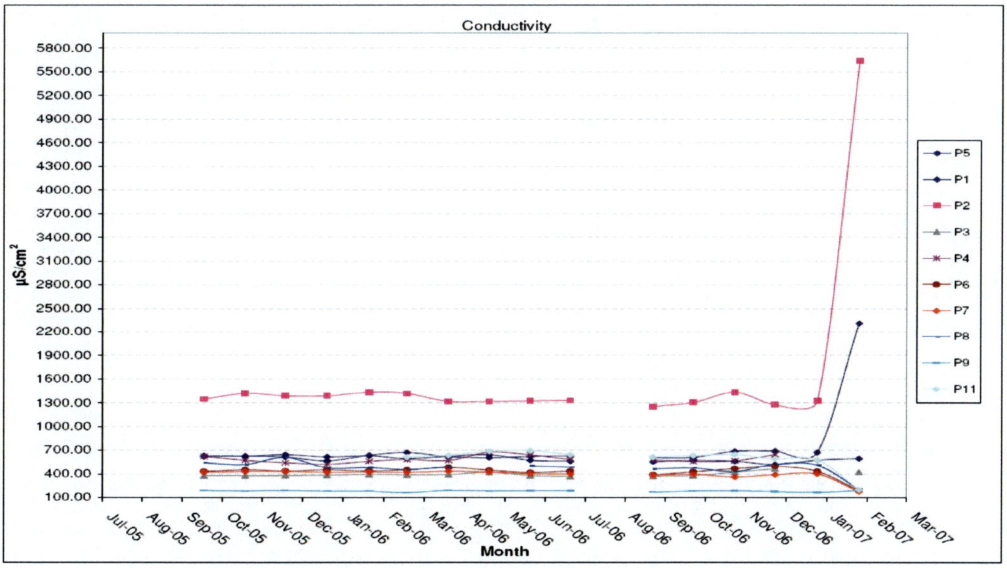

Figure 5.4. Graph of groundwater conductivity measured in the wells.

- *Conductivity Graph* (Figure 5.4): the graph corroborates the isolated nature of well P2, whose average conductivity is 1350 microS/cm, above the one measured in other wells; this finding is certainly associated with rising mineralized endogenous fluids. Conversely, well P9, which is located northwards (Figure4.1), has the lowest conductivity (about 180 microS/cm); this finding may be due to the location of the well. Contrary to the other monitored wells, P9 is the only one which does not lie in the immediate vicinity of the Lake; in fact, it lies at a distance of about 3 km northwards, beyond Mt. Venere, along the Lake's underground recharge path.

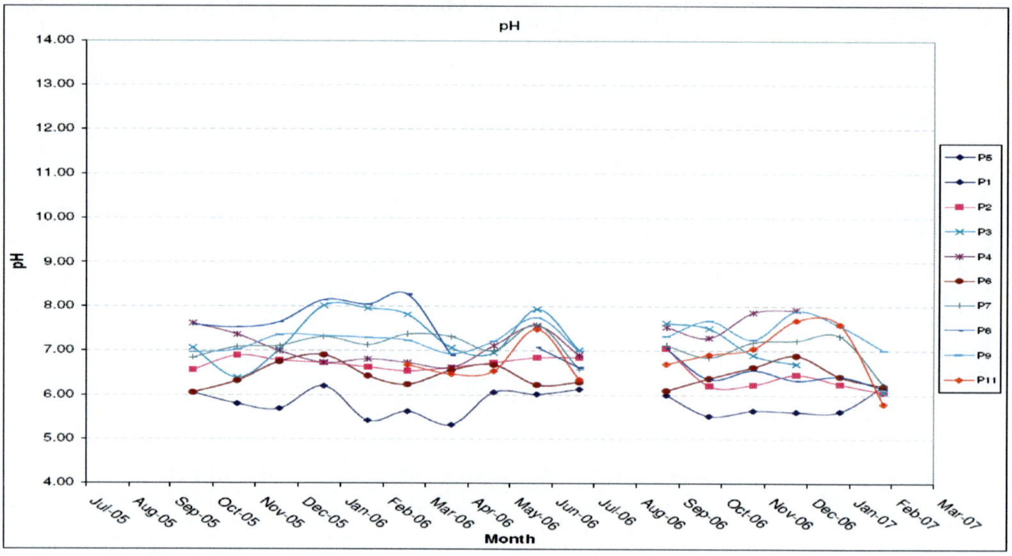

Figure 5.5: Graph of groundwater pH measured in the well

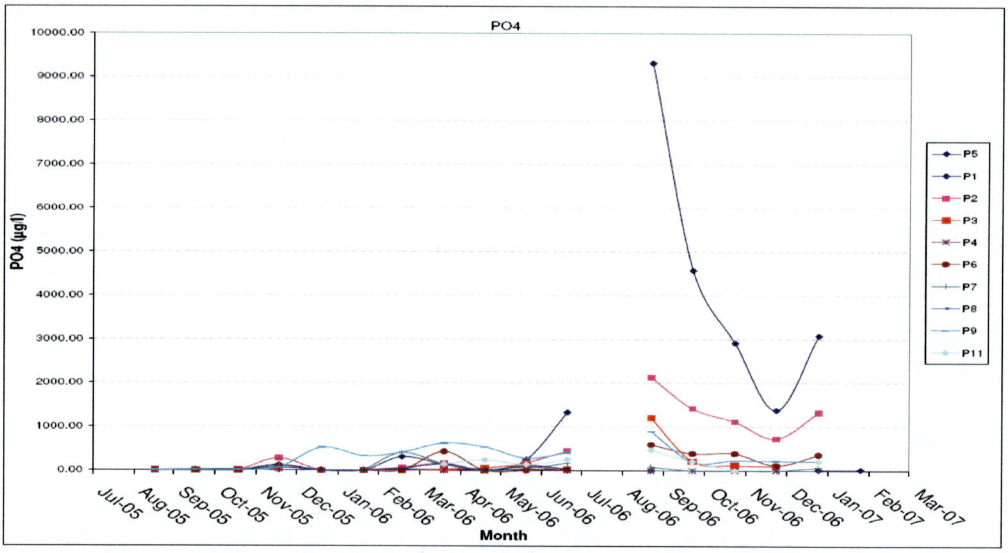

Figure 5.6. Graph of groundwater phosphates measured in the wells.

- *Static Level Graph* (Figure 5.3): the static level measured in the wells validates the uniqueness of well P2, whose water level varies by 1.5 m on average and drops by 7.5 m in the October 2006-February 2007 period, when its maximum fluctuations are recorded. These fluctuations might be related to changes in the pressure of endogenous fluids or in the volume of water abstraction. The groundwater level measured in wells P1, P6, P8, P9 and P11 remains practically constant at about 510 m above sea level. In wells P3, P5, P7, the level fluctuates from 3 to 5 m in periods of poor rainfall (dry period: June-September 2006). As these wells supply farms and tourist resorts, they can reasonably be supposed to have been more intensely used in the dry period. It should be emphasized that, from March 2006 to February 2007, the groundwater level of well P3 climbed to an average value of about 515.9 m above sea level, i.e. up by about 3.3 m from the average of 512.6 m above sea level measured in the September 2005-February 2006 period. Well P4 has negative variations in groundwater level coinciding with the dry period but, in absolute terms, lower (by about 2 m) than those of the above-mentioned wells.

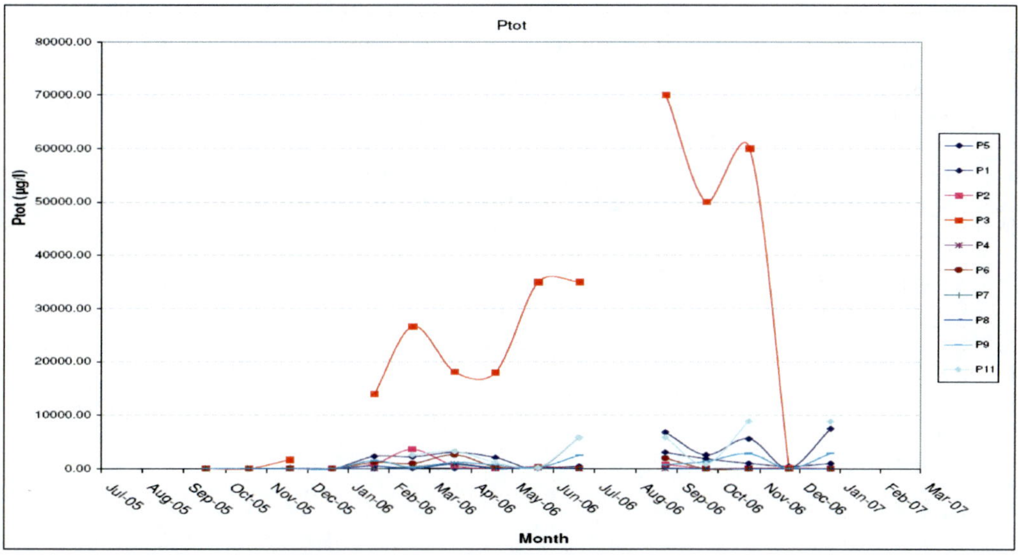

Figure 5.7. Graph of groundwater total phosphorus measured in the wells.

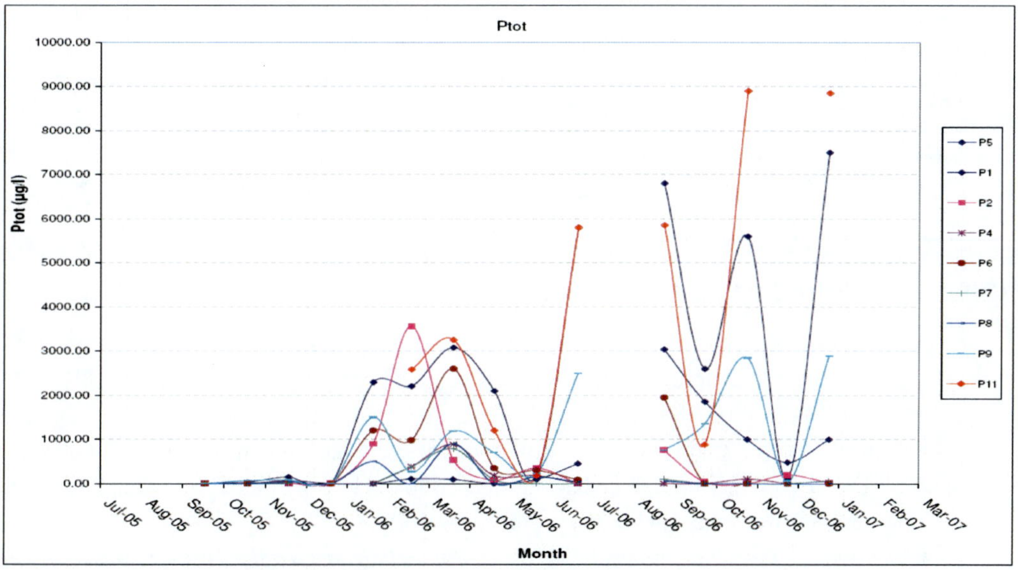

Figure 5.8. Graph of groundwater total phosphorus measured in the wells (excluding P3).

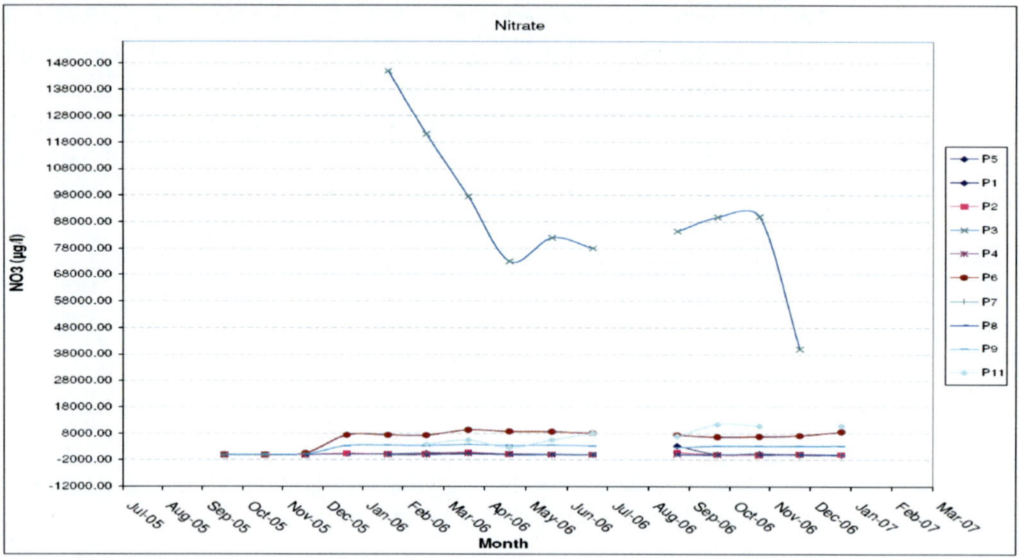

Figure 5.9. Graph of groundwater nitrates in the wells (note the high values of well P3).

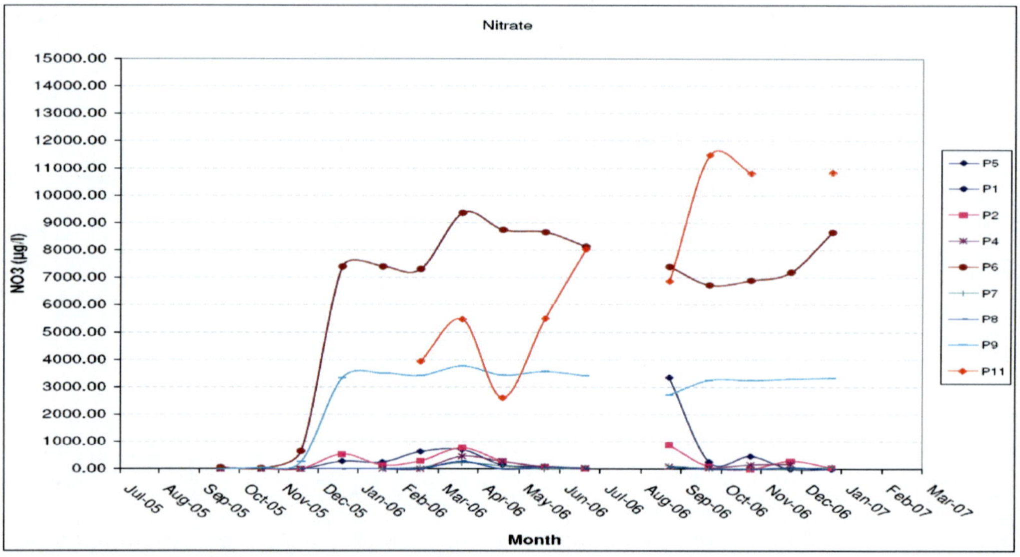

Figure 5.10. Graph of nitrates in the wells (excluding well P3).

- *pH Graph* (Figure 5.5): the average values of groundwater pH measured in the wells are constant, but different in absolute terms. The average pH is 5.81 in well P1, 7.20 in P3, 6.44 in P5, 7.88 in P6 and 7.8 in P7 with minimum values in July and February. Wells P2 and P8 have a downward trend from September 2006 to February 2007. Well P11 has higher peaks in June 2006 (pH 7.48) and December 2006 (pH 7.68) and a lower peak (pH 5.80) in February 2007.
- *Phosphate, Total Phosphorus, Nitrate, Total Nitrogen Graphs* (Figs. 5.6-5.12): the contents of nitrates, total nitrogen and total phosphorus in well P3 groundwater are 1-2 times higher than in the other wells. This may be due to the fact that the well is located near a farm spreading livestock effluents over the soil (as a sewage disposal

technique). This practice favors the transfer of nutrients from the soil to groundwater. Excluding P3, the comparison of nitrate concentrations in the remaining wells highlights two groups of wells: the first (P6, P9 and P11), where groundwater nitrates lie in the 2600-11500 µg/l range; and the second (remaining wells), where nitrate concentrations are below 1000 µg/l. This classification of wells based on their groundwater nitrate contents does not hold for the contents of the other chemical elements (phosphates, NTot, PTot), given their irregular occurrence.

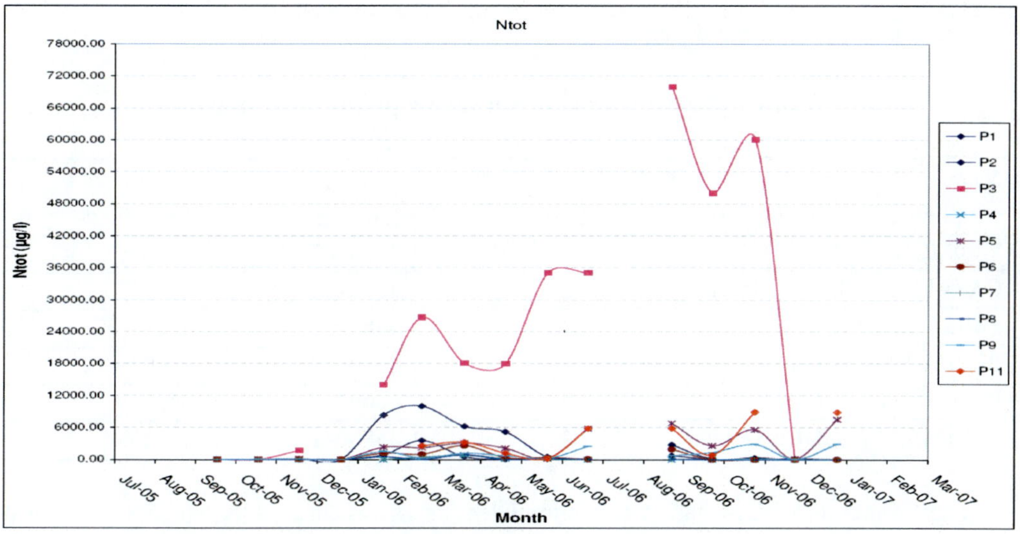

Figure 5.11. Graph of groundwater NTot in the wells.

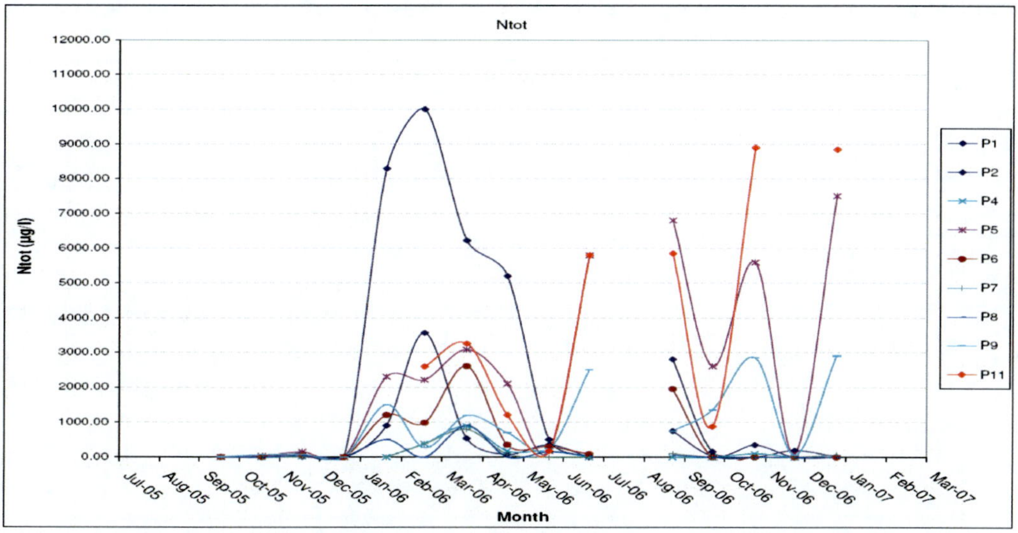

Figure 5.12. Graph of groundwater NTot in the wells (excluding P3).

Lake Monitoring Survey

The first stage of this activity identified the sampling points. The points (Figure 4.1) were selected on the basis of the main targets of the survey, which are summarized below:

- characterizing the pollution of the water body as a whole;
- analyzing and assessing pollution over time and in space;
- pinpointing the areas most exposed to the risk of pollution.

6 monitoring points were chosen:

Table 5.3. Sampling and measuring points in the Lake

Name of sampling and measuring point	Code	Latitude	Longitude
PROMONTORIO (L1)	L1	4689967	267396
CASALE FOGLIANO (L2)	L2	4690141	265922
EMISSARIO (L3)	L3	4687905	268915
BELLA VENERE (L4)	L4	4690652	269070
BOA CENTRO (L5)1	L5	4689133	266549
ULTIMA SPIAGGIA (L6)	L6	4688041	265166

The Promontorio (L1) station was eliminated, as it was located too close to the Casale Fogliano (L2) one. A new station (Ultima spiaggia – L6), located on the northern side and more representative of the geographic coverage of the Lake, was thus identified as an alternative.

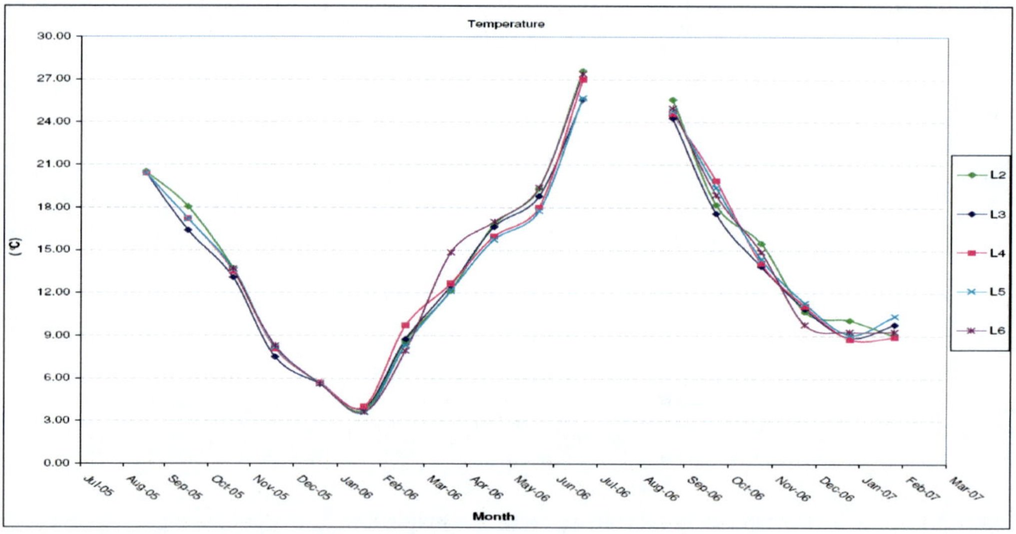

Figure 5.13. Temperature at the Lake's surface.

As pointed out in the previous paragraphs, monthly samples and measures were taken from September 2005 to February 2007 in order to assess and analyze: temperature, pH, conductivity, dissolved oxygen, dissolved salts, nitrates, phosphates, total nitrogen, total phosphorus.

Figure 5.13 displays the temperature curves of surface water samples collected from stations L2, L3, L4, L5, L6 in the Lake. The Lake's surface temperature reflects seasonal changes, with maximum values in summer (May-September) and minimum ones in winter (November-February). Figs. 5.14-5.16 show the temperature curves along the vertical (measured at point L5), which define the stratification of the Lake during the year. No data are available for the periods prior to September 2005; however, stratification appears to occur at 10-15 m below the Lake's surface and to end in December 2005, when the water column has a constant temperature of approximately 8°C. In the following 3 cold months (Jan-Feb-Mar 2006), the Lake has no stratification, which suggests complete mixing of its waters, also thanks to climatic factors.

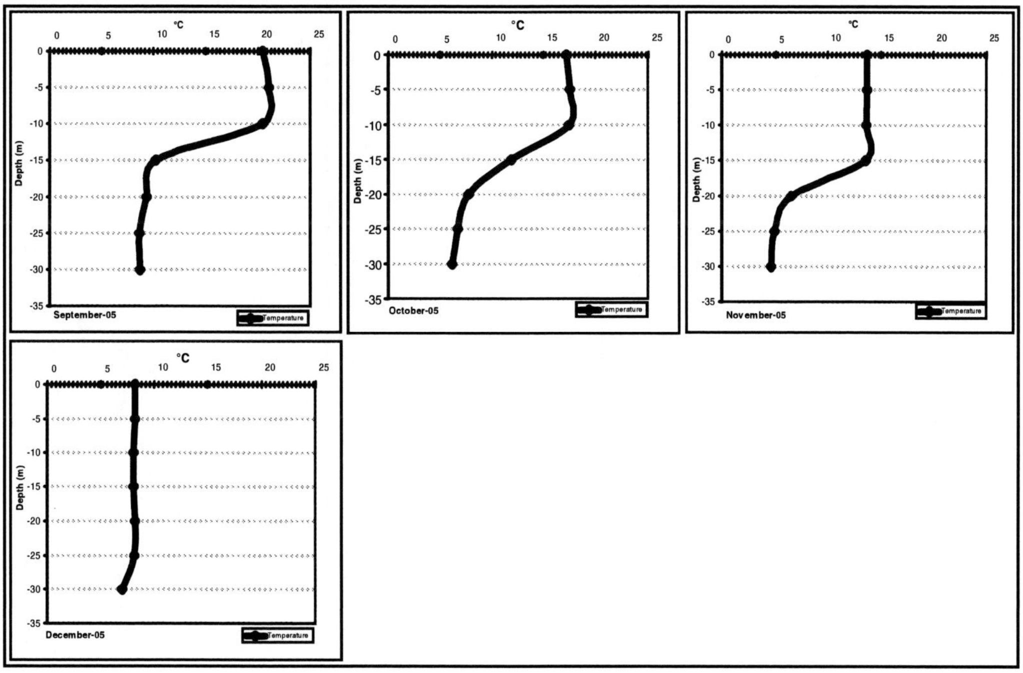

Figure 5.14. Temperature along the vertical, station L5, year 2005.

Although wind intensity and direction data were available, their analysis was not made on specific graphs; however, in the above-mentioned months, permanent breezes with a mean speed of 11.4 km/h were observed (UCEA weather station, Caprarola).

Stratification also reappears in 2006 (April to October) with variable depths of the mesolimnion in the various months. The latest samples collected in January and February 2007 indicate homogenization of the Lake's water mass, as in 2006. A temperature increase is noted along the water column in January 2006 (+3.44 °C on January 2005) and in February 2006 (+5.81°C on February 2005).

Figure 5.17 evidences the variability of pH values measured at the Lake's surface: the mean difference from one station to the other is 1.5; however, at sampling point L2, the pH is 5.28 in November and 5.10 in December.

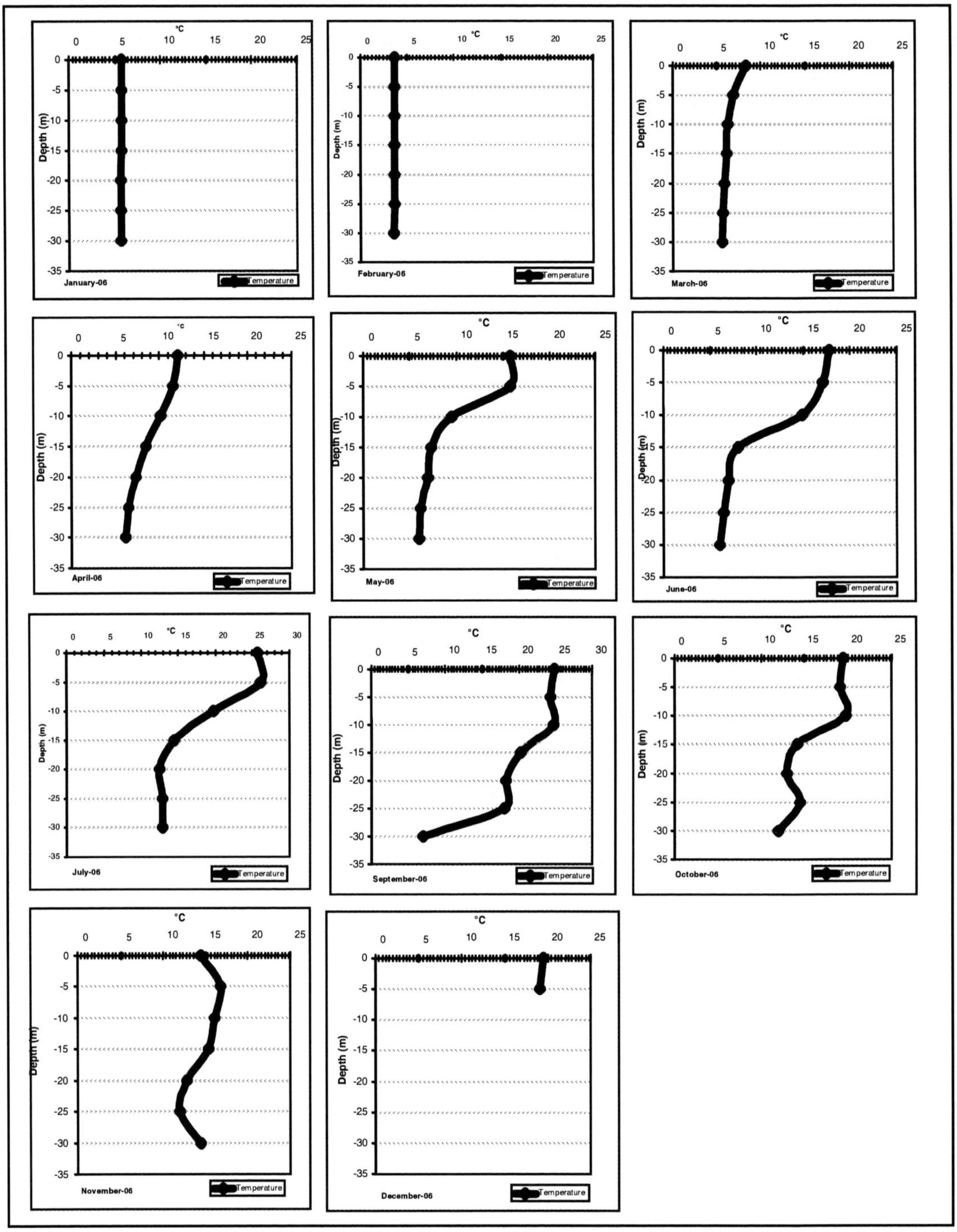

Figure 5.15. Temperature along the vertical, station L5, year 2006

Along the vertical, the pH of the Lake's water (measured at L5) appears to be associated with thermal stratification. Indeed, in January and February 2006 (Figure 5.15), pH is constant along the vertical, just as temperature in the same months, as previously stressed. Throughout the observation period, experimental values stood in the 7.20-9.28 range. In December 2006, a pH value of 6.50 was recorded at a depth of 5 m. This value, not consistent with the scenario of the investigated period, has been neglected in the considerations made until now.

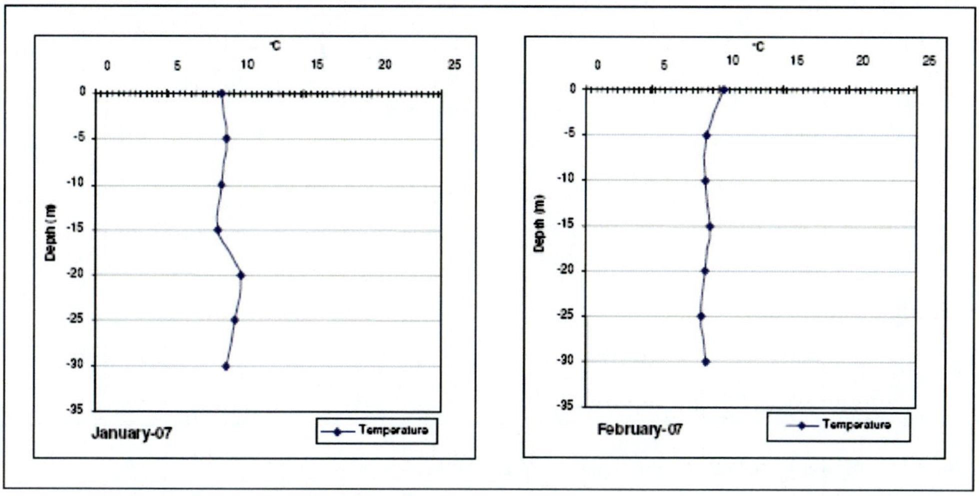

Figure 5.16. Temperature along the vertical, station L5, year 2007.

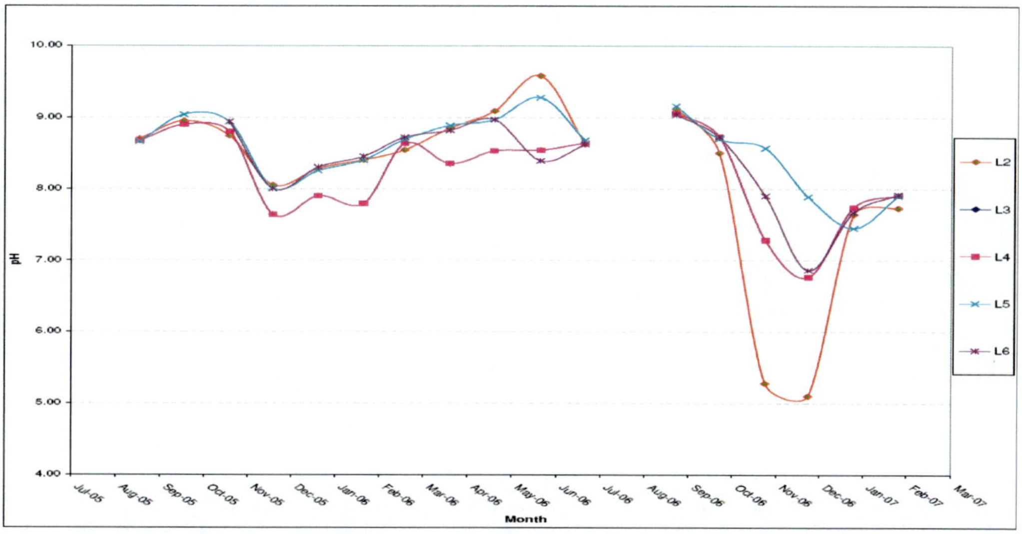

Figure 5.17. pH at the lake' s surface.

Dissolved oxygen at the Lake's surface has a cyclical pattern, in line with seasonal variations. Increases are recorded from November 2005 to March 2006 at all the stations, especially at L3 (peak of 14.5 mg/l). The values of this period may be ascribed to higher seasonal wind intensity and remixing of the Lake's waters. The higher values of station L3, near the outlet, may be justified by a more sustained dynamics of the waters of the Lake in the vicinity of its outlet and of water abstraction points.

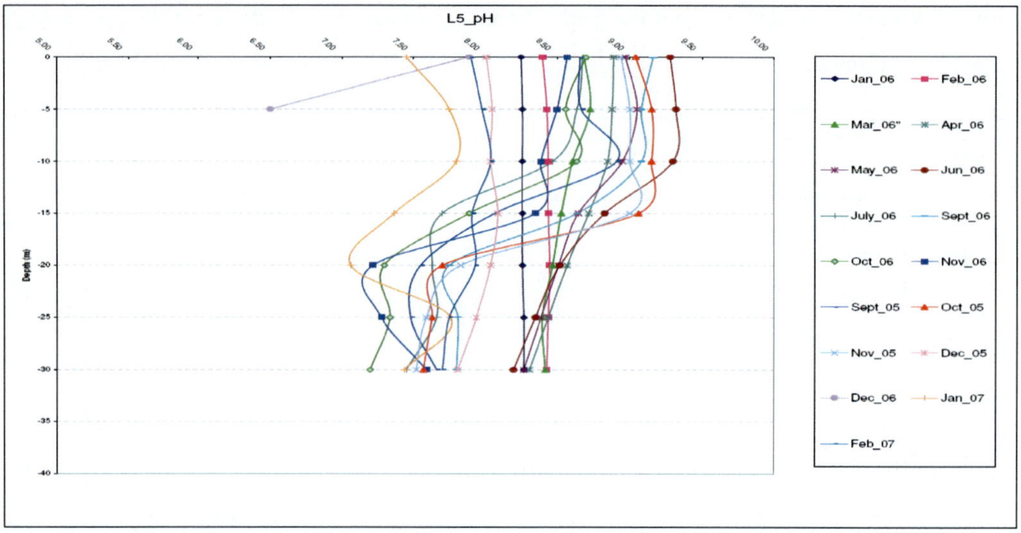

Figure 5.18. pH along the water column sampled in station L5.

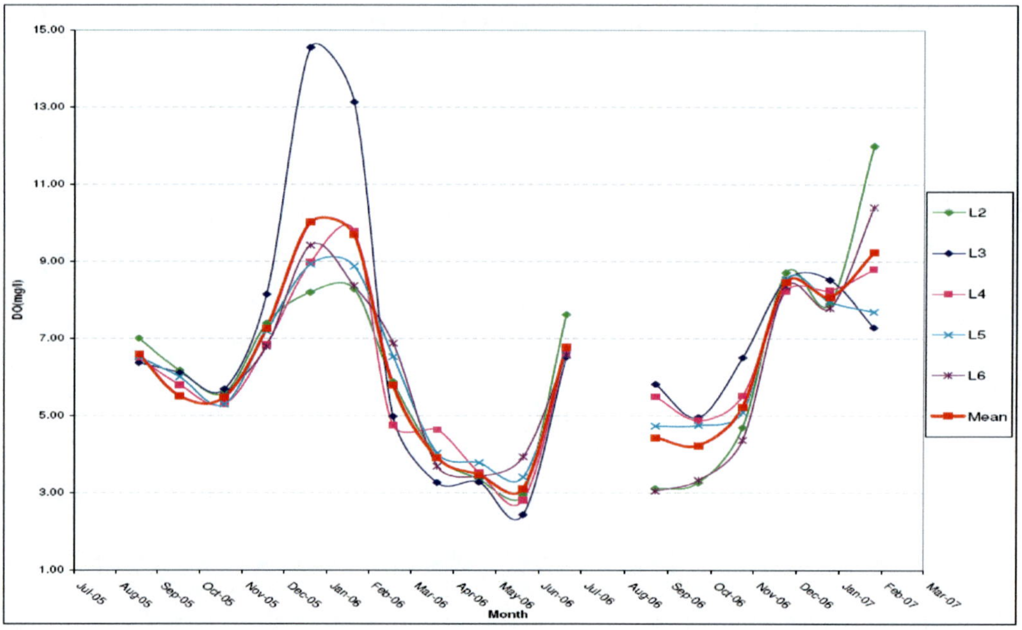

Figure 5.19. Dissolved oxygen at the Lake's surface.

In contrast, in the spring period (April-June 2006), dissolved oxygen decreases until reaching average values close to 2.50 mg/l throughout the lake. This trend is not observed in the central area along the vertical, where minimum values approach 3.50 mg/l. This is due to the different conditions of the riparian zone with respect to the pelagic one (Figure 5.20).

From September 2006 to February 2007, dissolved oxygen has upward trend, reaching an average of 9.24 mg/l at the surface, thus replicating the cycle of the previous fall-winter period.

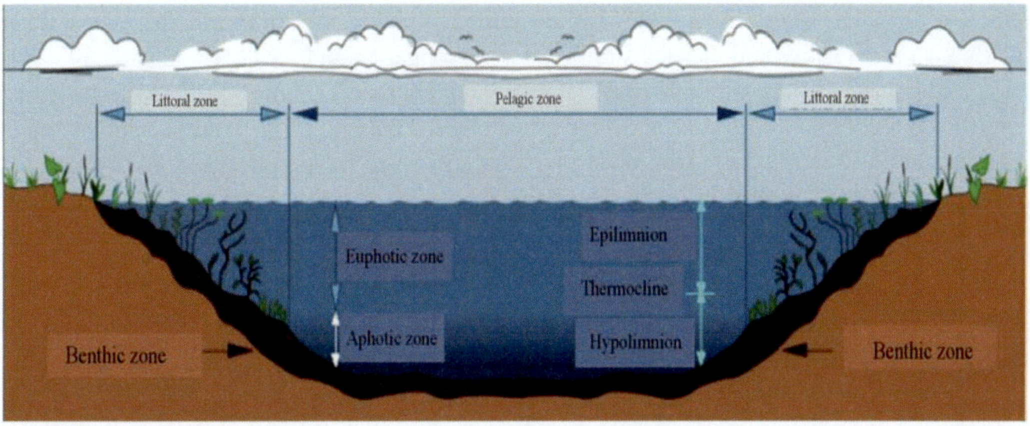

Figure 5.20. Zones of the Lake.

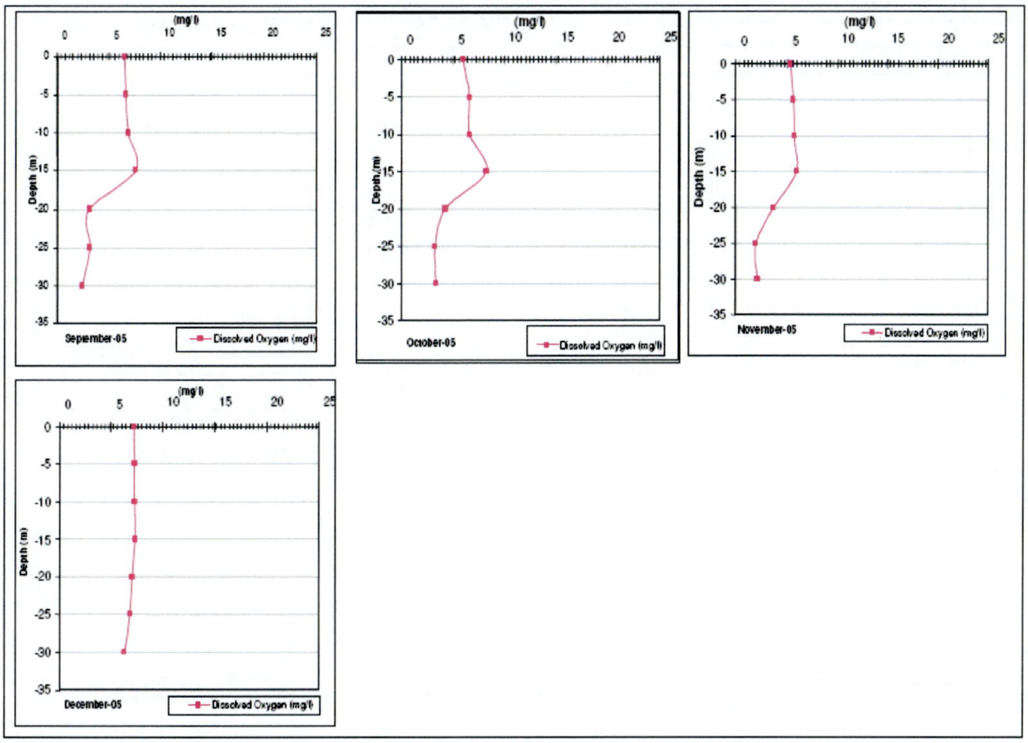

Figure 5.21. Dissolved oxygen along the vertical, station L5, year 2005.

Figure 5.21 shows dissolved oxygen along the vertical of the Lake, measured at point L5 in 2005. A difference between concentrations at the surface and at depth is clear in October, November and December. This difference is related to stratification of the Lake in connection with temperature changes (thermal stratification), as described in the previous pages. Thermal stratification hinders oxygen movement from shallow to deep waters, which tend to become anoxic (difference in dissolved oxygen: 3 to 4 mg/l). The stratification phenomenon tends to disappear in fall months, as evidenced by inflexions in the curves. The inflexion corresponding to the depth of stratification indicates that the interface between anoxic and more oxygenated waters moves downward, until it vanishes in winter months, when water mixing is more intense and thermal stratification no longer occurs.

In the first months of 2006, i.e. January and February (Figure 5.22), when thermal stratification does not occur, dissolved oxygen along the column of the Lake is relatively high on average (9-10 mg/l). In the following months, dissolved oxygen declines and zones with different concentrations (stratification) tend to develop.

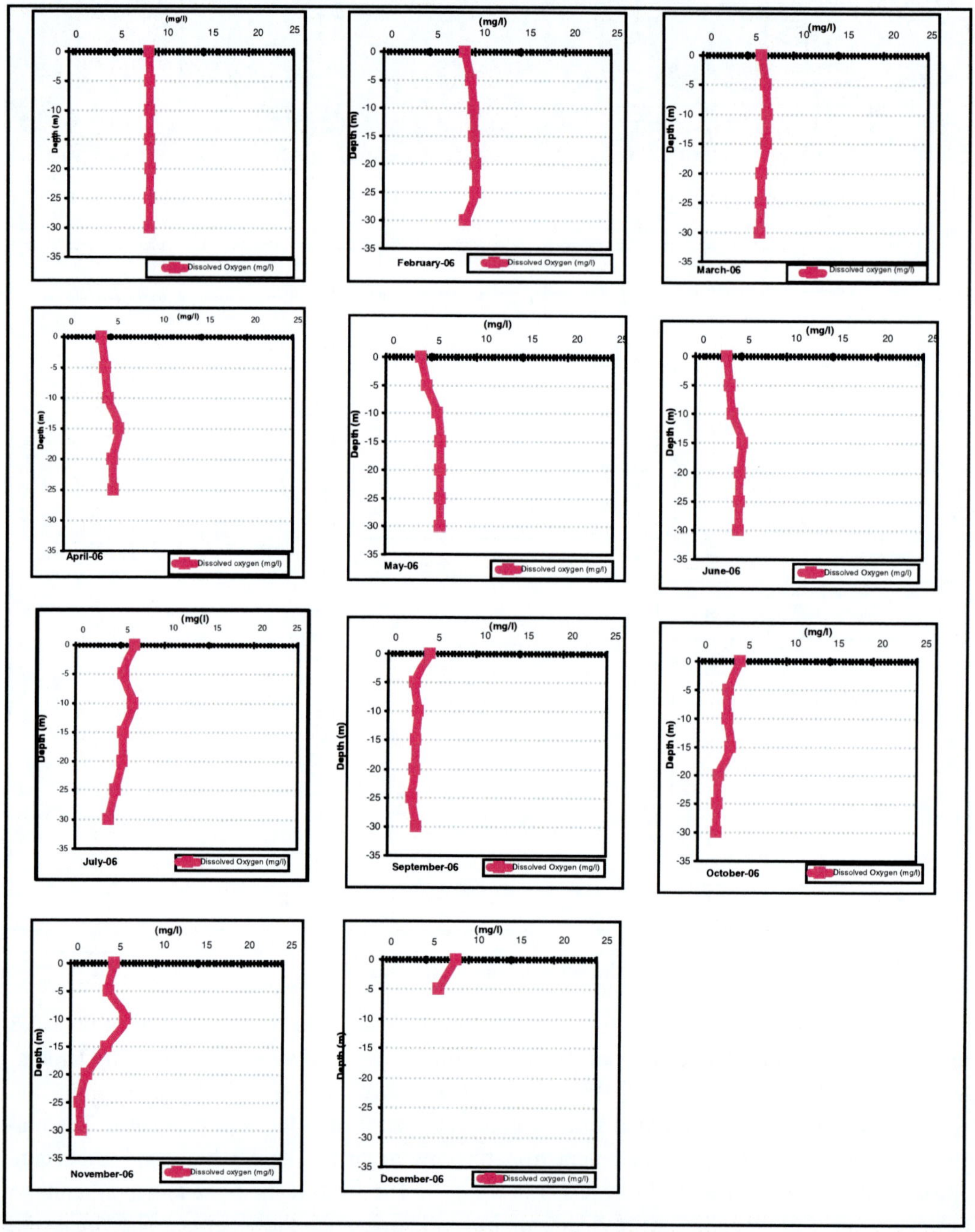

Figure 5.22. Dissolved oxygen along the vertical, station L5, year 2006.

The curve gradients in spring and summer months are generally mild and the difference between the values measured every 5 m does not exceed 1 mg/l.

In the months of 2007 (Figure 5.23), higher temperatures cause the Lake's water oxygen concentrations to drop, in accordance with the law governing gas solubility in water (Figure 5.24).

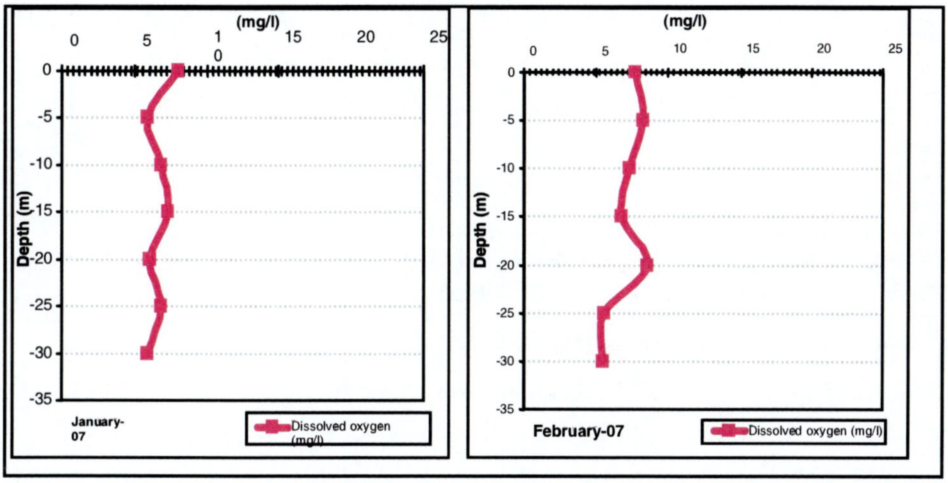

Figure 5.23. Dissolved oxygen along the vertical, station L5, year 2007.

Conductivity varies homogeneously over the months, with a minimum deviation between its values (410-433 μS/cm). In January, the values recorded at the various stations have more significant deviations from one another (of up to 21 μS/cm). However, these deviations lie within the manual measurement error range.

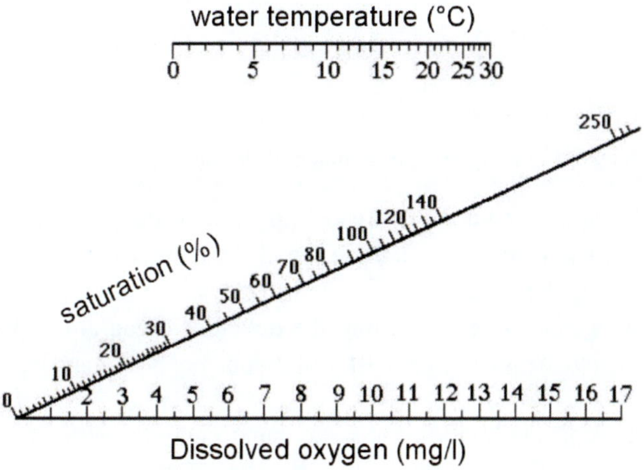

Figure 5.24. Water oxygen saturation vs. temperature.

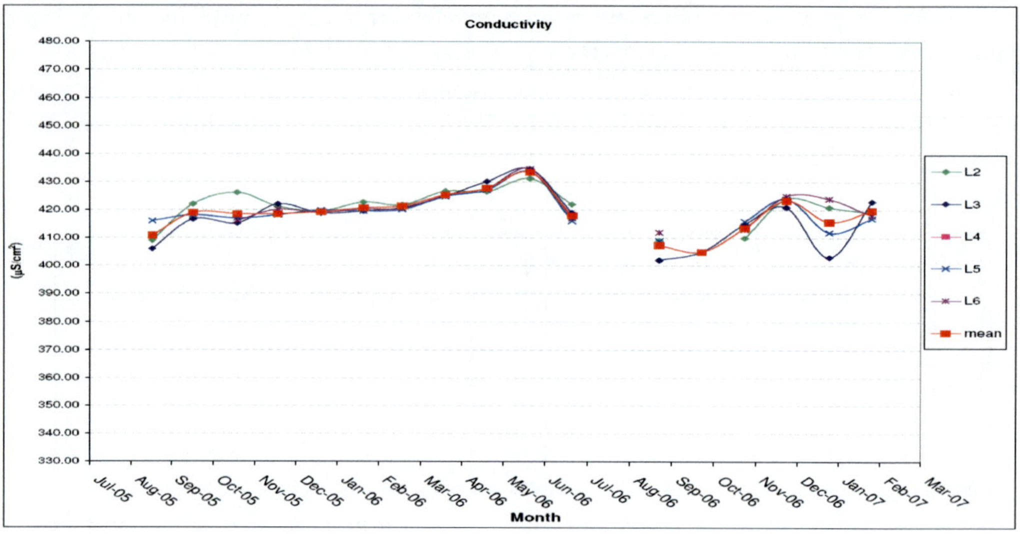

Figure 5.25.Conductivity at the Lake's surface.

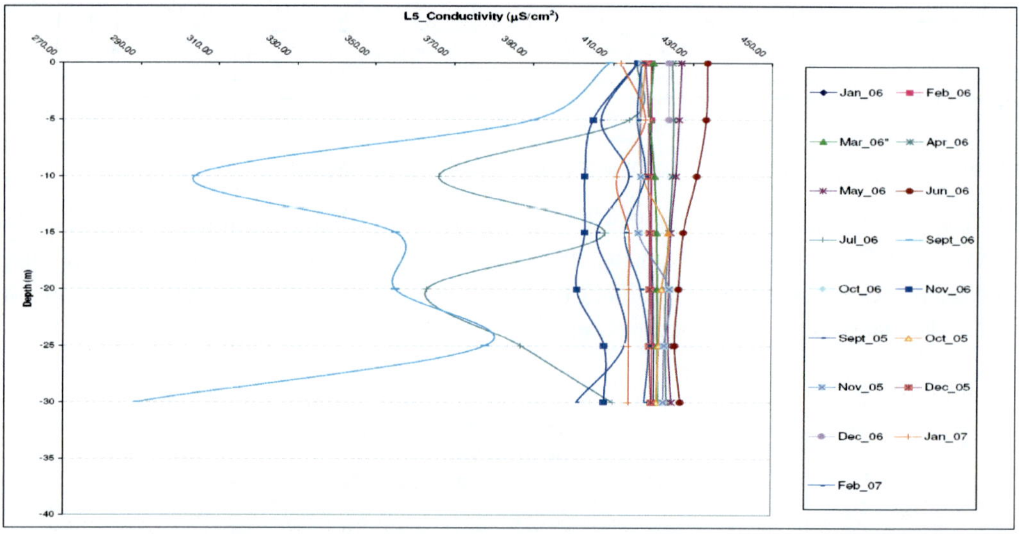

Figure 5.26. Conductivity of the water column sampled in station L5.

Figure 5.26 shows the conductivity read along the water column at L5. During the months of the entire survey period (except July and September 2006), conductivity remains practically constant along the depth.

It is worth pointing out, once again, that the deviations read along the depth for a few months are enhanced by the scale factor of the graph, but that they lie in a range of 100 µS/cm at the most.

Figure 5.27 underscores the stability of salt content in the lacustrine basin (about 0.3 g/l). This value is no longer available for the period after June 2006 (breakdown of the sensor).

The variability of phosphates and PTot, measured at the various stations around the Lake, is consistent with their concentrations throughout the lacustrine basin. The value of station L2 in the month of January is exceptional. For L3, L4,L5 and L6, the phosphate range is 106-<10 μg/l (Figure 5.28), whereas the PTot range is 51-<20 μg/l (Figure 5.29). However, in absolute values, the measured PTot is below 100 μg/l.

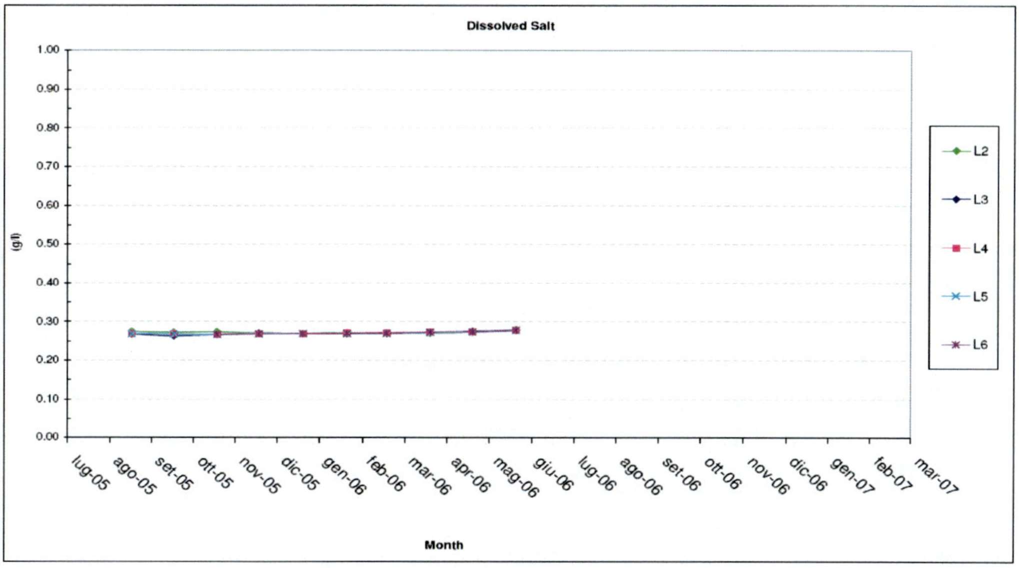

Figure 5.27. Dissolved salts measured at the Lake's surface.

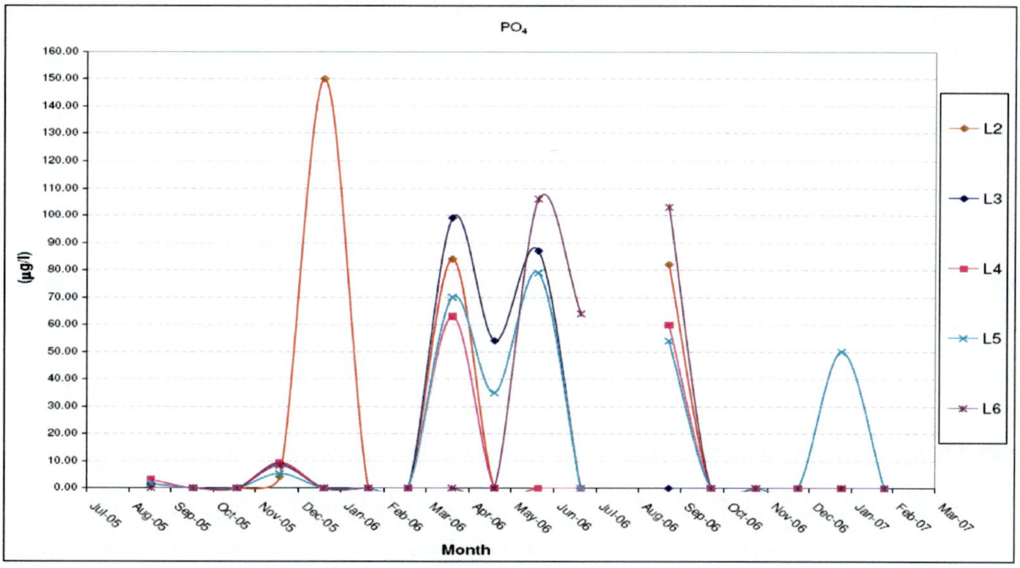

Figure 5.28. Phosphates measured at the Lake's surface.

In some months, the concentration of PTot along the column shows the same stratification as the one observed for other parameters (September 2006); in other months, it shows a more or less marked homogeneity.

The anomaly recorded in January 2007 at a depth of 15 m cannot be easily explained.

Values increase on average with depth, although staying below 150 μg/l.

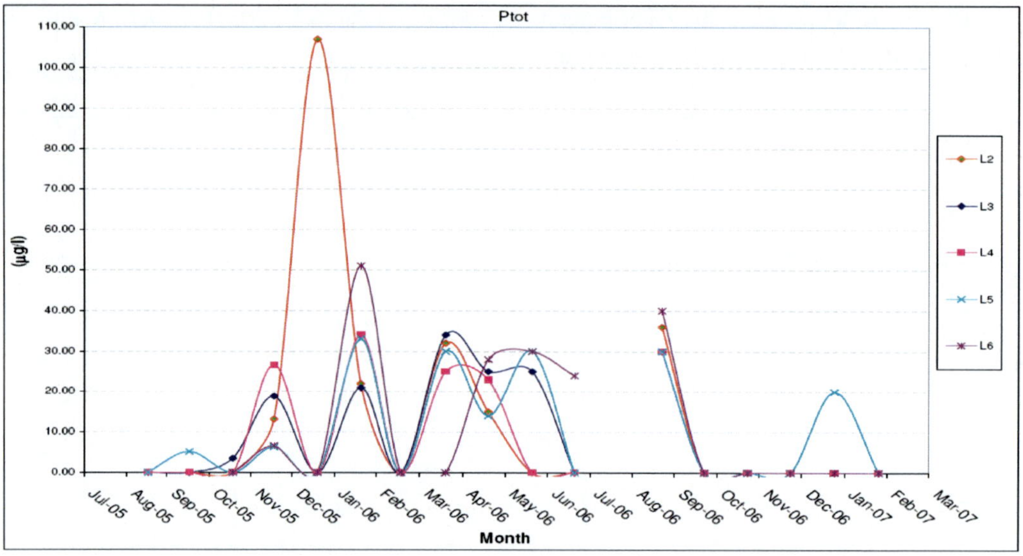

Figure 5.29. PTot measured at the Lake's surface.

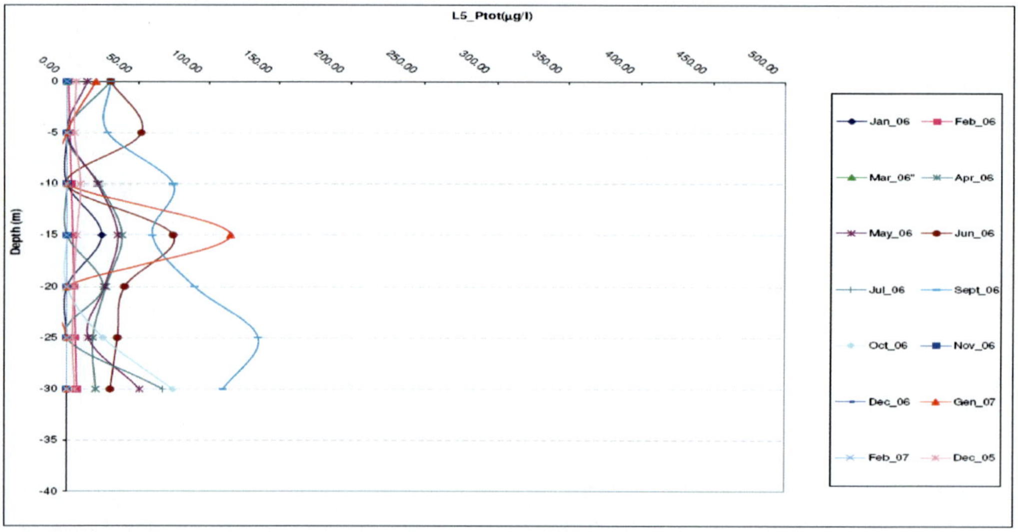

Figure 5.30. Phosphates measured in the water column sampled in station L5.

Just as PTot, nitrates (Figure 5.31) have positive variations in the months elapsing from December 2005 to June 2006. L6 is the station with the peak nitrate values (699 µg/l).

NTot concentrations are consistent at the various stations and in the various months. Peaks are noted from January 2006 to May 2006 (200-1000 µg/l). Figure 5.32 also displays the daily rainfall measured by the station near the Lake (Assofrutti station); its trend is correlated with the peak values of NTot. The correlation is clear in this period only, probably because it coincides with the season in which farmland treatments are concentrated (Figure 5.34).

In Figure 5.33, NTot values are indicative of stratification in warm months and homogeneity of the Lake's waters in cold months. In the spring of 2006, NTot goes up, reaching maximum concentrations of roughly 800 µg/l.

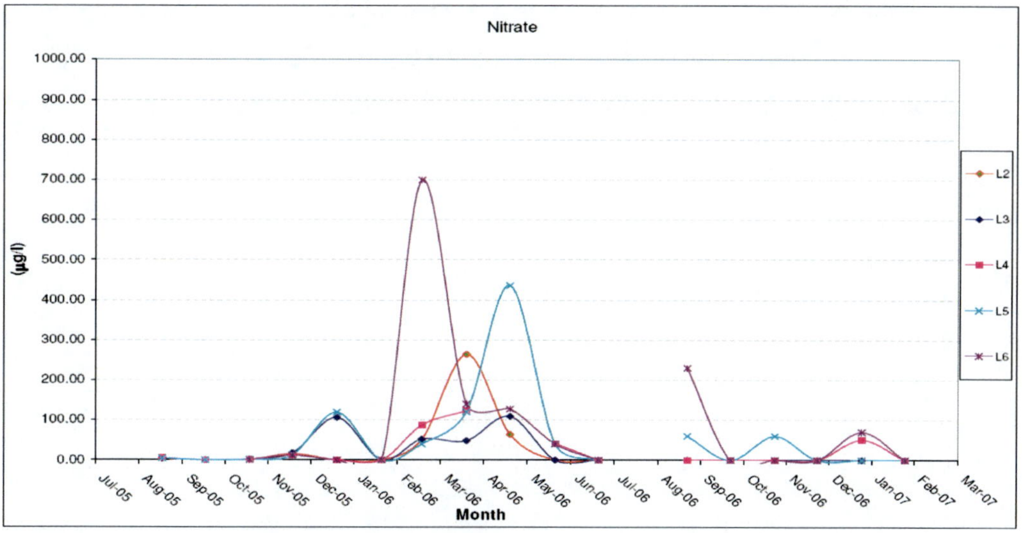

Figure 5.31. Nitrates measured at the Lake's surface.

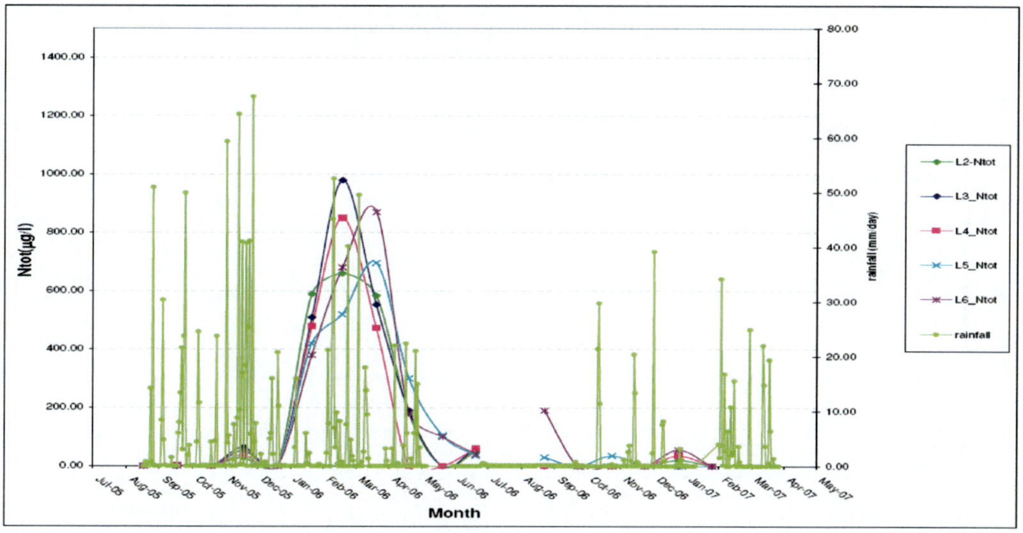

Figure 5.32. NTot measured at the Lake's surface.

6. Statistical Quantification of Eutrophicating Substances Potentially Migrating into Lake Vico from Surface Water, Groundwater and Discharges

The study was expected to determine the loads of eutrophicating substances of both natural and anthropogenic origin released into Lake Vico's recharge area.

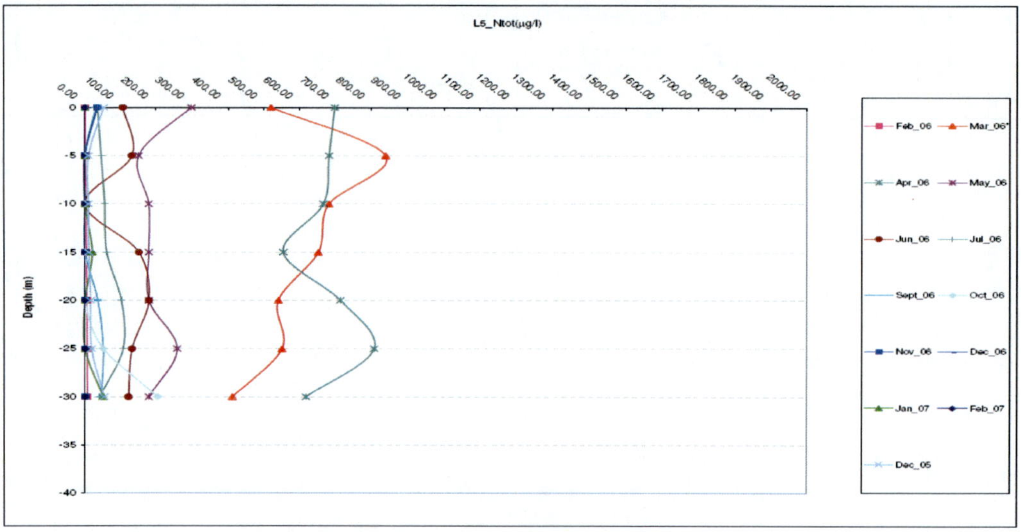

Figure 5.33. NTot measured in the water column sampled in station L5.

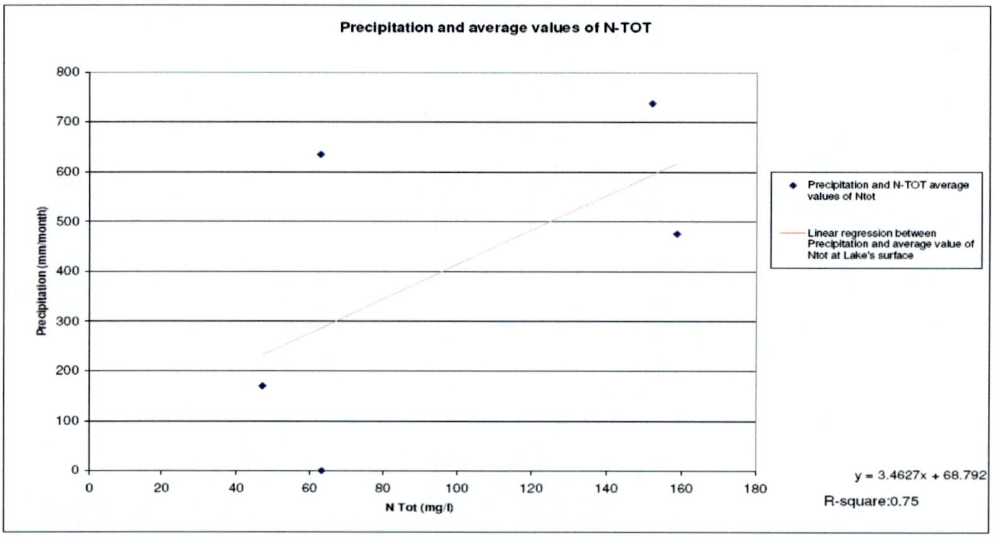

Figure 5.34. Linear regression between values of precipitation and average values of NTot at the Lake's surface (January 2006 – May 2006).

Activities were thus based on the consideration that any land and water monitoring and conservation choice or strategy should be supported by a thorough understanding of: i) the processes that generate polluting loads; and ii) the characteristics of watersheds and receiving water bodies, at least in terms of:

- development and continuity of the watershed;
- size and characteristics of watersheds and their uses;
- characteristics of natural and/or artificial discharge in streams;
- water budget;
- generation and release of polluting loads in the study area.

The following paragraphs will describe the methodologies used and the results obtained in assessing the distribution of polluting load sources and releases into the water environment, by integrating the data available on the human system (data of ISTAT – Istituto Nazionale di Statistica - Italian Statistical Institute, provincial data bases, data on land uses) and on the natural system (weather, climate, soil characteristics, surface water and groundwater circulation).

The study was carried out by:

- searching data bases in order to retrieve the location and characteristics of the reported point sources;
- collecting ISTAT's data to estimate the permanent and temporary population in the census units;
- mapping land uses to a scale of 1:10,000;
- making distributed calculation of water balance parameters, namely of the yearly water distribution between runoff and effective infiltration;
- demarcating the boundaries of hydrological and hydrogeological basins;
- identifying the areas served by public sewerage systems and connected to water purification systems;
- spatially representing the permanent and temporary population on the land use map and assessing the population not connected to the public sewerage system;
- assessing nitrogen and phosphorus releases from residential and productive activities, by quantifying the volumes from the reported point sources and estimating the volumes released into areas not connected to the public sewerage system;
- estimating the amounts of nitrogen released from farmland to surface water and groundwater;
- tentatively quantifying nitrogen and phosphorus carried by surface and groundwater to lakes.

Pollution from eutrophicating substance is chiefly due to human activities. In the worst cases, nitrogen in soil and surface water is so high as to contaminate deep groundwater. An interesting and fairly macroscopic correlation exists between this type of pollution and the morphology of the polluted area: nitrates are mostly concentrated in groundwater that is recharged from flat or low-hill areas, where human activities contribute to accumulating nitrogen substances.

Most of the nitrogen inputs into the environment derive from:

- intensive farming (corn, row crops, vegetables, etc.);
- production of livestock effluents (inappropriate storage of sewage may have a dramatic environmental impact);
- production of urban sewage: generally, rural areas are not served by a sewerage system and residential sewage is treated with the subsurface irrigation method (100 cl of faeces generate 30 cl of free ammonia);
- industrial activities in general.

The following paragraphs make a few concise considerations on the occurrence of phosphorus and nitrates in water.

- *Phosphorus:* phosphorus may be inorganic (from apatite solubilization) and organic (from decomposition of organic matter); its cycle is typically sedimentary (i.e. it takes place in the soil). This phosphorus adds to the one from chemical fertilizers (leached from farmland through rainfall), residential sewage (of metabolic origin and from detergents) and industrial releases. The toxicity of phosphorus in water is given by its elemental form. Italian Legislative Decree 152/99 (Annex 2) on total phosphorus (PTot) specifies a maximum concentration of 0.07 mg/l in salmon breeding waters and 0.14 mg/l in cyprinid breeding waters. Under the same Decree (Annex 5), phosphorus directly released into surface water should not exceed 10 mg/l.
- *Nitrates:* nitrogen (together with carbon) is a fundamental constituent of the biomass of animal and vegetal organisms. In streams, nitrogen substances (together with phosphorus) are the nutrients of algae and the base of their entire food chain; an excessive concentration of these substances may cause eutrophication. Nitrogen substances generally come from vegetal and animal wastes, fertilizers, industrial and residential releases. The nitric form of nitrogen has its peak toxicity when it is reduced to the nitrous form. A substantial amount of nitrates ($NO3^-$) in waters may reflect protracted inorganic pollution. Under Legislative Decree 152/99 (Annex 5), the limit concentration of nitric nitrogen in releases into surface waters is equal to 20 mg/l.

Watersheds and Hydrological Basins

To assess the loads of eutrophicating substances which may reach the lacustrine basin, it is essential to define the boundaries of the watershed and of the hydrogeological basin.

The perimeters of the watershed and of sub-basins were mapped to a scale of 1:10,000 by referring to the Regional Topographic Map (CTR).

This work relied on different topographic data bases and thematic maps, such as:

1) Regional Topographic Map, scale 1:10,000, updated in 1990;
2) Maps of IGM (Istituto Geografico Militare - Italian Military Geographic Institute), scale 1:25,000, updated in the 1950s;
3) Color ortho-photo IT2000, year 1999, nominal scale 1:10,000;
4) Watershed map built by SIRA (Servizio Informativo Regionale Ambientale - Regional Environmental Information System), scale 1:10,000;
5) Digital Elevation Model (DEM), 20 x 20 m mesh;
6) Watershed maps drawn by SIMN, scale 1:25,000;
7) Map of water-demanding areas.

Field surveys were made in case of doubts.

Wherever possible, the sub-basins were mapped by dividing the study area under homogeneous morphology and land use criteria. This activity identified 54 sub-basins (Fig 6.1).

To demarcate the boundaries of the Lake's hydrogeological basin, reference was made to the isopiezometric lines obtained during the 2005-2006 monitoring survey.

Water Budget

This activity was intended to yield data for assessing precipitation, surface runoff and seepage into the aquifer, also in relation to transport of polluting loads towards lakes.

In view of the above, the water budget was calculated by making a distributed analysis of the different parameters, starting with precipitation and evapotranspiration, so as to obtain a meaningful estimation of rainfall and its distribution between runoff (R) and effective infiltration (IE).

The water budget of the Lake was thus determined on the basis of the boundaries of its hydrogeological basin, recharge area and watershed, of data on natural surface discharge (outlet stream) and water abstraction, as well as of the approximate distributed values of IE and R.

For weather parameters (precipitation and temperature), reference was made to the data recorded by the "Caprarola Vico" weather station (managed by Assofrutti) in the two-year period elapsing from March 2005 to February 2007.

For environmental and methodological aspects, reference was made to the studies conducted as part of the Plan for Sustainable Use of Water Resources ("Piano dell'Uso Compatibile delle Risorse Idriche" - PUC") in the volcanic areas of Latium. For a description of these studies, the reader is referred to Capelli *et alii*, 2005.

For estimating water-demanding areas, the available land use maps were updated and integrated.

With regard to the Available Water Capacity (AWC), reliance was made on the maps produced in the above-mentioned studies.

Methodological Approach

The methodology used for water budgeting (summarized in this paragraph) is the one developed within the framework of the Plan for Sustainable Use of Water Resources, which was issued by the Latium and Tiber River Regional Water Authorities (Capelli et. al., 2005).

To compute water budget parameters, the proposed model schematizes the surface as a soil/vegetation system with a single layer coinciding with the rooting height of vegetation. Water discharge from the soil towards the unsaturated zone and then percolation as far as the saturated zone are disregarded.

The water budget simulated in each cell of the soil/vegetation system is based on the following equation:

$$Ie_{\text{year}} = \sum(P_{\text{month}} - EVR_{\text{month}} - R_{\text{month}} + Endo_{\text{month}})$$

where
Ie = effective infiltration (mm)

P_{month} ; EVR_{month} and R_{month} = monthly precipitation, evapotranspiration and runoff
$Endo_{month}$ = contribution of runoff in endorheic or pseudo-endorheic areas.

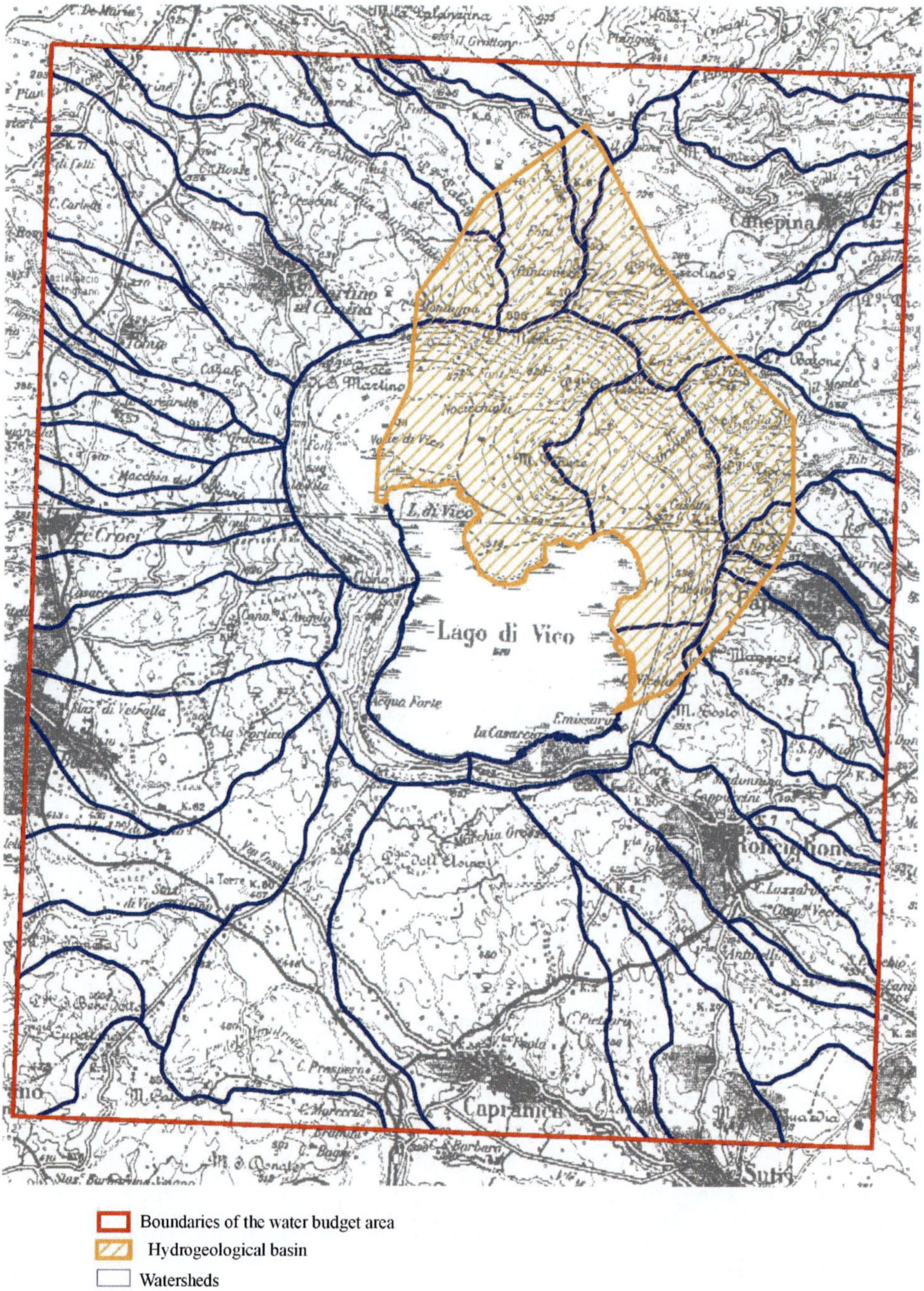

Figure 6.1. Investigated watersheds and hydrogeological basins.

As can be seen, the following simplifications are introduced to analyze the variability of the different factors:

- evaporation in the portion of precipitation intercepted by leaves (in the range of some mm/year) is assumed to be negligible;
- soil evaporation and capillary water rise and evaporation are neglected, as the order of magnitude of these losses is significantly smaller than the discharge measurement and runoff computation error;
- snowfall (possible accumulation-thawing model) is disregarded, as snow falls in the study area for periods that are on average shorter than the temporal scale used (month).

Evapotranspiration is computed as the balance between rainfall, water potentially evapotranspirated from plants (crop evapotranspiration) and actual availability of water in the soil according to the following conceptual model:

$$EVr = ETr \qquad \text{if } P + Ui \geq ETr$$

$$EVr = P + Ui \qquad \text{if } P + Ui \leq ETr$$

where:

EVr = actual evapotranspiration in the period,

ETr = crop evapotranspiration, equivalent to water requirements by the individual crop in the period,

Ui = volume of water usable by plants (fraction of AWC), contained in the soil at the start of the period.

Crop evapotranspiration or crop water requirements are determined by multiplying potential evapotranspiration (ETP) by the crop coefficients (K_C):

$$ETr\ (crop\ evapotranspiration) = ETP * KC$$

where:

ETr = crop evapotranspiration, which is similar to actual evapotranspiration in a given area, if does not exceed the water made available by precipitation and obtainable from the soil (fraction of AWC),
ETP = potential evapotranspiration, i.e. the water evaporated in a given period of time from a wide land area, covered by a homogeneous, dense and low vegetation, in full development, optimally supplied with water and shading the soil completely (e.g. *Festuca Arundinacea*), evaluated through the well-known Hargreaves-Samani formula;

K_c = crop coefficient expressing the proportionality between evapotranspiration of the reference crop and the one of the actual crop.

The AWC denotes the volume of water which can be retained by the soil between -30 and -1500 kPa and corresponding to the volume of water potentially usable by plants.

To estimate runoff, use was made of the well-known Kennessey (1930) method, a relatively straightforward and widely used method giving fairly reliable results.

For a given portion of a study area, the method computes the discharge coefficient. This coefficient is the sum of three components: steepness of slopes, permeability of outcrops, vegetal cover.

Runoff (R) is calculated on a yearly basis as the summation of monthly contributions through the following relation:

$$R_{year} = \sum (P_{month} - EVr_{month}) * CK$$

where

R = runoff (mm),

P_{month} = monthly precipitation,

EVr_{month} = monthly evapotranspiration,

CK = Kennessey's coefficient.

The runoff computation disregards the monthly soil water budget. Indeed, comparison with experimental runoff data (Capelli et al., 2005) indicated that the values of Kennessey's coefficients (CK) are statistically sized for effective rainfall.

Temperature-Rainfall Data

Temperature and rainfall data are those recorded for agronomical purposes by the "Caprarola Vico" weather station in the period going from March 2005 to March 2007.

To determine the water budget, the data were aggregated on a monthly basis. Some gaps in the measurements were observed, especially for January 2007. Therefore, use was made, to a first approximation, of the values of January of the previous year.

As shown in Figure 6.2, the two years have a very different rainfall pattern. Overall precipitation in 2005, though excluding the months of January and February, exceeds 1250 mm; conversely, in 2007, overall precipitation is equal to 707 mm.

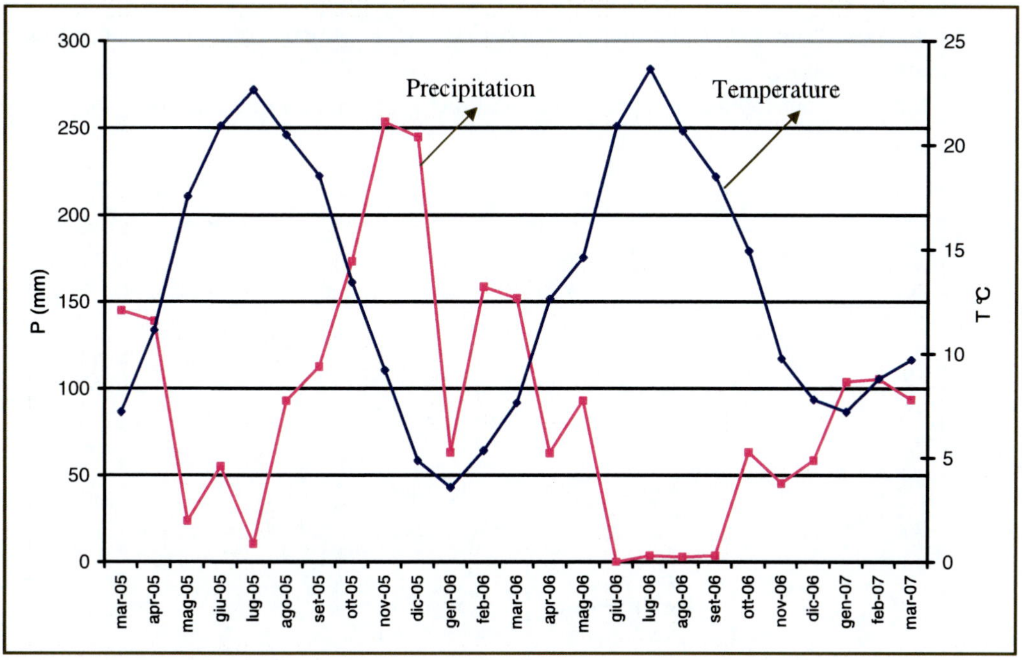

Figure 6.2. Precipitation and temperature from March 2005 to March 2007.

Distributed Water Budget

The computation of the monthly distributed water budget was narrowed to the Lake's recharge area, including the watershed and the hydrogeological basin.

The spatial resolution is given by the size of the mesh (250 m) of the grid, as displayed in Figure 6.3.

The computed evapotranspiration values account for 50% on average of rainfall and they have a very variable pattern in the study area, depending on types of soil and vegetation (Figure 6.4).

Runoff is moderate, i.e. in the range of 10% of precipitation with an average of 80 mm in 2006. Also in this case, runoff has significant local variations (Figure 6.5).

Similar considerations may be made about effective infiltration; however, it is high and accounts for roughly 40% of precipitation (Figure 6.6).

For the purposes of this study, the monthly cumulated data for the hydrogeological basin and the watersheds, albeit approximated, are more meaningful. Table 6.1 below summarizes the values of the various inflows into the Lake, which obey the following model:

- watersheds – inflows into the Lake only consist of runoff water (average discharge 170 l/s), which is extremely variable throughout the year and equal to or close to zero from April to September (extreme weather events are neglected);
- hydrogeological basin – inflows into the Lake are comparable with the ones from seepage waters (in volumetric terms); these inflows are poorly variable during the year owing to the reduction effect exerted by the time of seepage into aquifers. The

average discharge in the study period (Mar. 2005 – Feb. 2007) is supposed to be around 566 l/s;

- the third contribution is represented by direct precipitation into the Lake; its average value in the period was 428 l/s.

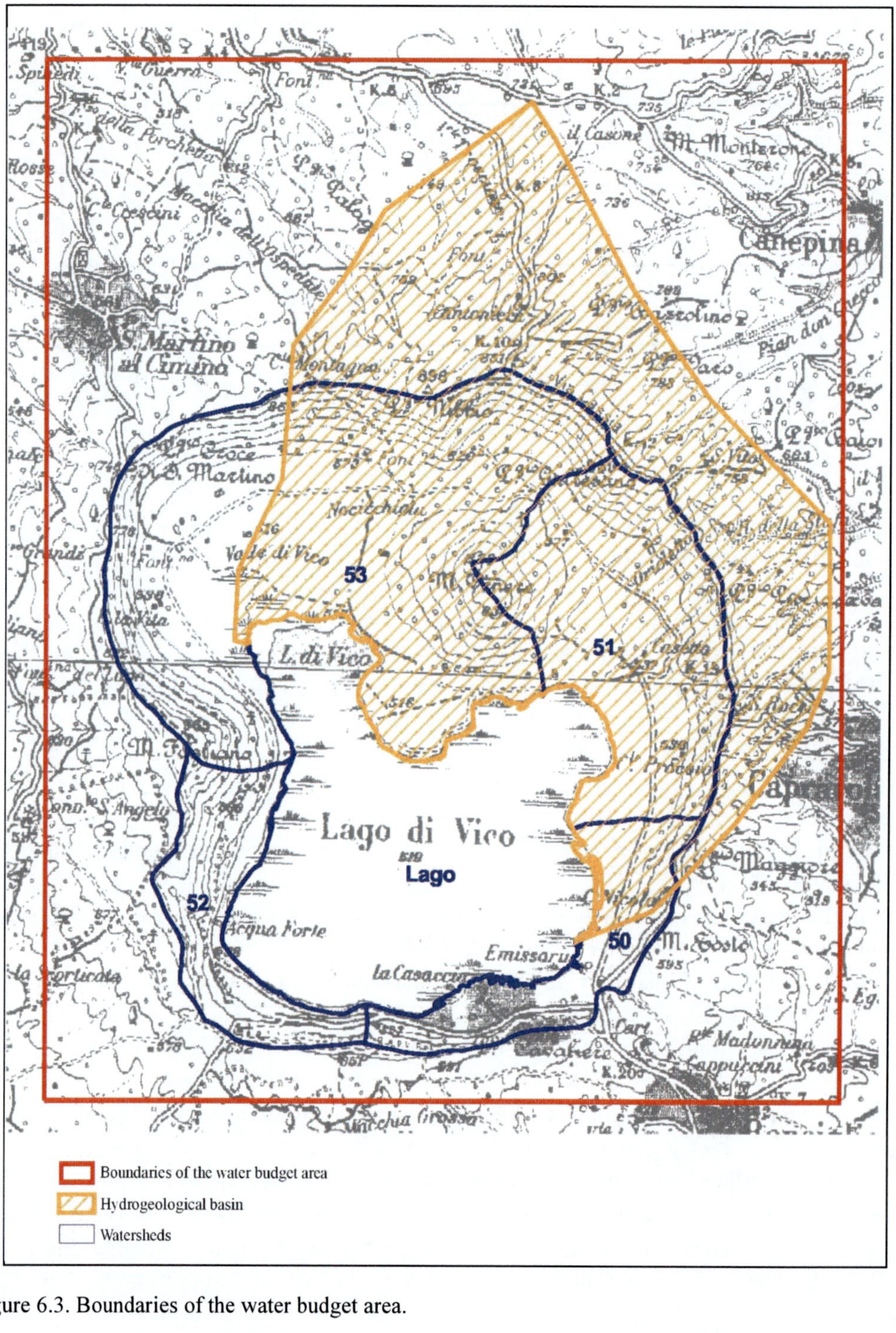

Figure 6.3. Boundaries of the water budget area.

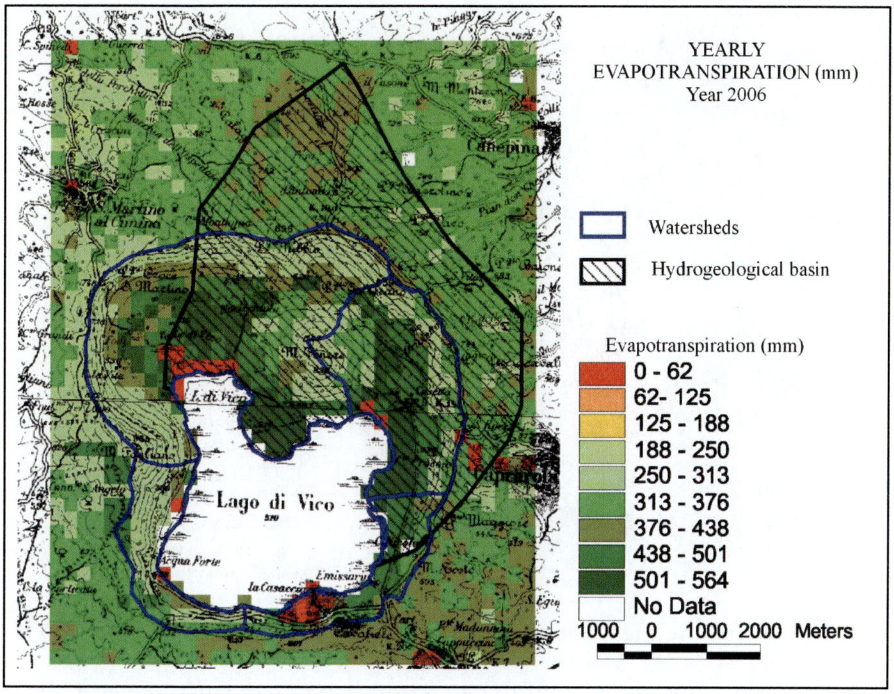

Figure 6.4. Map of yearly average actual evapotranspiration computed for 2006 (average value 354 mm).

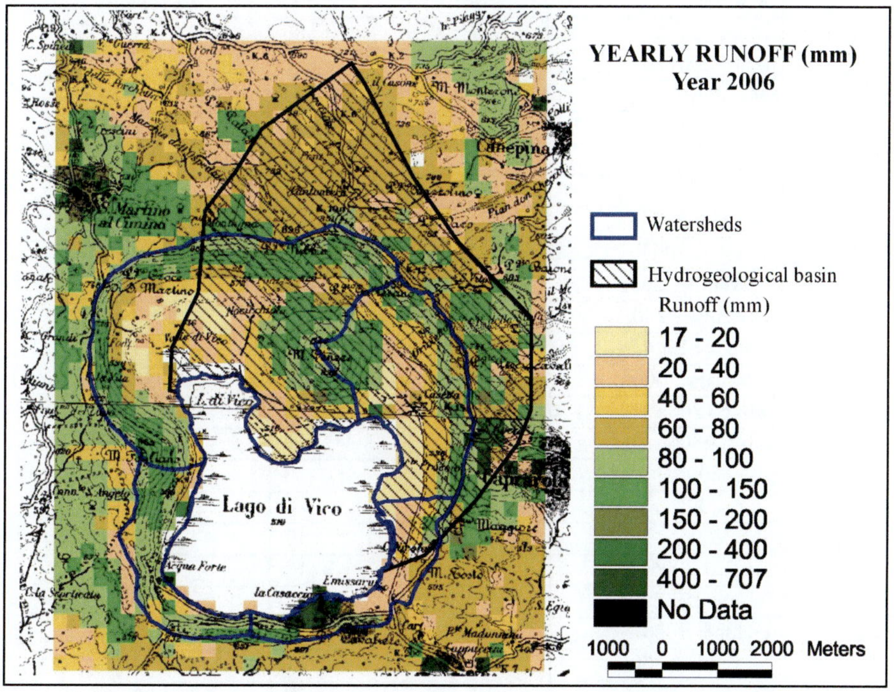

Figure 6.5. Yearly average runoff computed for 2006 (average value 80 mm).

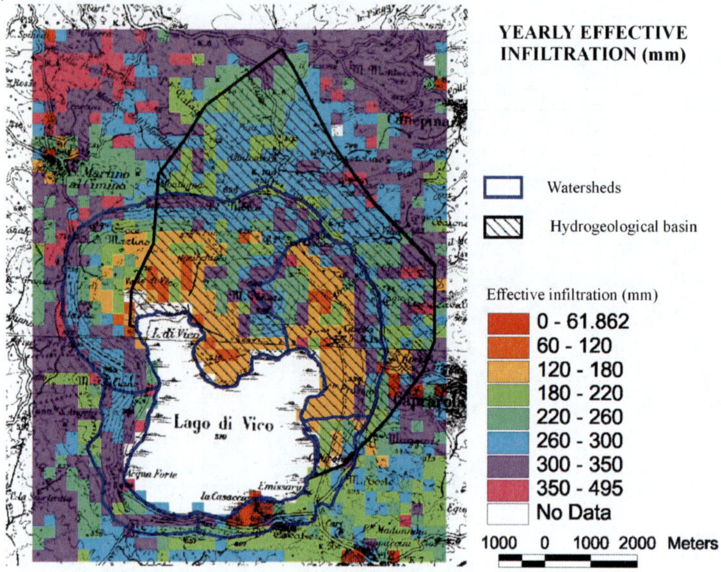

Figure 6.6. Map of yearly average effective infiltration computed for 2006.

Table 6.1. Monthly cumulated runoff (R), precipitation (P) and effective infiltration (IE) in the hydrogeological basin and watersheds.

Month	R (l/s)	P (l/s)	IE (l/s)
March 2005	287	686	1061
April 2005	113	658	374
May 2005	0	113	0
June 2005	0	261	0
July 2005	0	49	0
August 2005	0	440	0
September 2005	24	534	37
October 2005	365	821	1373
November 2005	617	1201	2334
December 2005	622	1159	2357
January 2006	606	299	541
February 2006	386	751	1458
March 2006	319	720	1164
April 2006	0	297	0
May 2006	0	440	0
June 2006	0	0	0
July 2006	0	17	0
August 2006	0	14	0
September 2006	0	17	0
October 2006	72	299	335
November 2006	62	215	230
December 2006	130	277	487
January 2007	243	492	915
February 2007	243	500	917
March 2007	152	443	529
Average	**170**	**428**	**566**

To preliminarily assess the reliability of the computed values of hydrogeological parameters, the computed inflows into Lake Vico were compared with the known and estimated outflows from it.

Total computed inflows: 1164 l/s, broken down as follows:

- precipitation (direct precipitation into the Lake): 428 l/s;
- groundwater: 566 l/s;
- runoff: 170 l/s.

Known outflows:

- outflow from the outlet stream - estimated to be in 30-150 l/s range on the basis of experimental observations;
- abstraction from the Lake (municipal aqueducts and private individuals – 3 water leases) – estimated to be equal to 49 l/s;
- abstraction of groundwater directly connected to the Lake (inside the caldera area) – estimated to be equal to roughly 160 l/s on the basis of average water requirements for irrigation. This estimation is widely validated by the number of wells (reported: 115; used under lease: 35) distributed in the caldera area; for these wells, the water demand specified in the lease exceeds 200 l/s;
- evaporation, estimated to be 403 l/s (Dragoni, 2006).

Subsurface discharge was thus estimated to lie in the range of 400 l/s (difference between inflows and known outflows).

The results of the water budget afford some interesting considerations as to the timescales and modes of transfer of pollutants to the lacustrine basin. The latter has a long period of water deficit from April to September, as evidenced by the absence of runoff and by low precipitation.

Hence, during this period, the amount of eutrophicating substances leached from farmland and entering the watershed and aquifers is likely to negligible and almost exclusively due to groundwater-carried releases from diffuse residential sources.

Sources of Polluting Loads

As previously hinted at, potential sources of polluting loads are:

- farming;
- livestock and fish breeding;
- residential releases;
- industrial releases.

Sources may be divided into point (releases proper) and non-point or diffuse sources (e.g. leaching from farmland).

To identify the location and assess the volumes of residential and industrial releases from point sources, use was also made of the data base developed by the Province of Viterbo. However, this data base is not exhaustive, as it does not include unreported releases.

Diffuse loads may be determined through estimation models: the deeper the understanding of their variables, the more reliable the estimation.

Therefore, two parallel approaches were taken to the determination of polluting loads:

1. analytical computation through the point source data contained in the available data bases;
2. estimation of potential loads via models using indirect data, such as permanent and temporary population or pedological characteristics of farmland soils potentially releasing nitrogen.

Another key factor in the computation of polluting loads is their geographic position (georeferencing) on a scale permitting to analyze and subsequently reaggregate the data into significant geographic subdivisions (sub-basins). Indeed, if the data are spatially represented over wide geographic subdividisions or subdivisions whose boundaries hardly coincide with the physical ones of study areas (e.g. overall load from livestock breeding in each Municipality), no detailed analysis can be made and no effective action can be planned.

Data Base of Releases Developed by the Province of Viterbo

A secure source of reference for point sources of releases into the drainage network of the study area is the Inventory of Releases developed by the Province of Viterbo.

From this data base, the following main data were extracted: types of release (residential or productive), volumes of releases, seasonal trends of releases, category of productive activity, quality of discharged waters (where available), receiving water body and location of releases. For location of releases, a specific activity was carried out to convert and validate the coordinates reported in the individual datasheets.

All the data regarding the Municipalities included or partially included in the study area were collected (Table 6.2).

The study area comprises 28 reported sources of releases, of which 25 residential and the remaining ones from productive activities (industry and accommodation/catering). All the reported sources (in most cases, not served by a purification system) lie outside the Lake's watersheds (Figure 6.7).

Table 6.2. List of Municipalities reported in the Province of Viterbo's data base and whose data were computerized and validated

Barbarano Romano	Canepina
Capranica	Vallerano
Sutri	Soriano nel Cimino
Ronciglione	Vetralla
Caprarola	Viterbo
Carbognano	Vejano

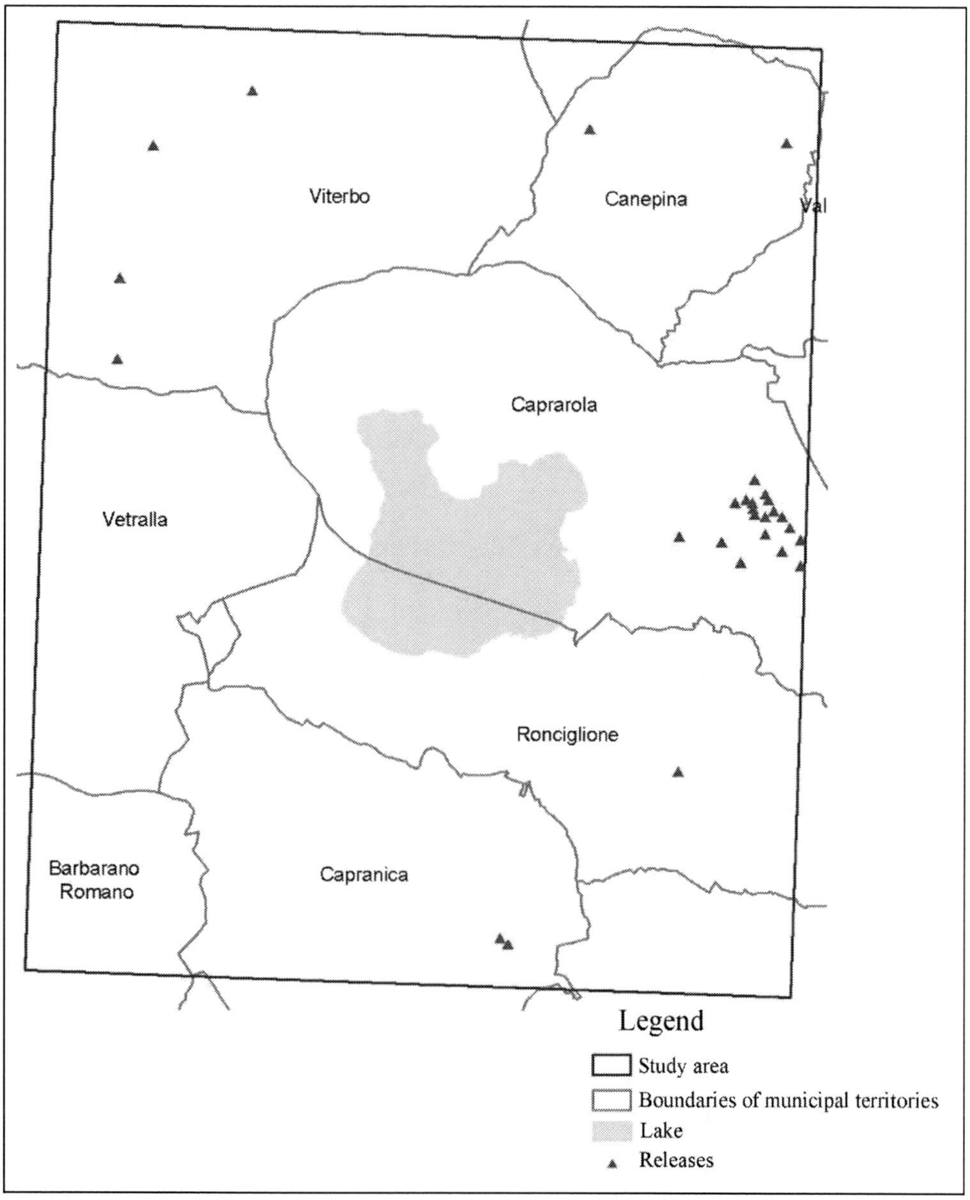

Figure 6.7. Inventoried sources of releases.

Mention should be made, in particular, of the following sources of releases:

- 19 sources in the Municipality of Caprarola: all residential with a flow rate of some l/s and a total load of 4,370 permanent inhabitants and about 700 temporary inhabitants;

- 1 industrial source in the Municipality of Capranica (beverage factory), served by a purification system;
- 1 residential source in the Municipality of Ronciglione, with a load of 5,000 equivalent inhabitants, one part residing in the town and another part temporarily occupying the residential area of Punta Lago.

In conclusion, the main villages or towns located in the study area or outside the basins of the Lake (Ronciglione, Caprarola, Capranica, S. Martino al Cimino, Punta Lago) are served by a public sewerage system (Figure 6.8).

Based on the reports of the Province of Viterbo, the housing units and accommodation/catering facilities located inside the Lake's watersheds have sealed sewage tanks, probably equipped with phytopurification systems, as laid down in the applicable legislation. In particular, the "La Bella Venere" facility (hotel, restaurant, etc.) has a sealed sewage tank, which is periodically emptied via self-purging and feeding of effluents to the municipal pumping system (Municipality of Ronciglione).

Sources of Data for Quantifying Residential and Industrial Releases

Residential and industrial loads from point sources were determined on the basis of the data contained in the Province of Viterbo's data base.

The approach to estimating residential releases from diffuse sources included the following steps:

1) based on the Map of Water-Demanding Areas, spatial representation of ISTAT's estimated permanent and temporary population data;
2) identification of residential areas served by a public sewerage system connected to a purification system vs. residential areas not connected to purification systems.

Approach to Spatial Representation of Permanent Population Data

The data available for spatial representation of the population were:

a) ISTAT's 1991 census with data on permanent population and occupied and unoccupied housing units;
b) ISTAT's 2001 census with population data aggregated by Municipality;
c) Map of Water-Demanding Areas, specially modified to a scale of 1:10,000 for the following classes:

With regard to the Map of Water-Demanding Areas, the data were field-checked, updated and tailored to the requirements of the study. For areas classified as "agricultural-residential", photos were interpreted in order to identify buildings (about 1,000 scattered homes) and crops in areas without buildings.

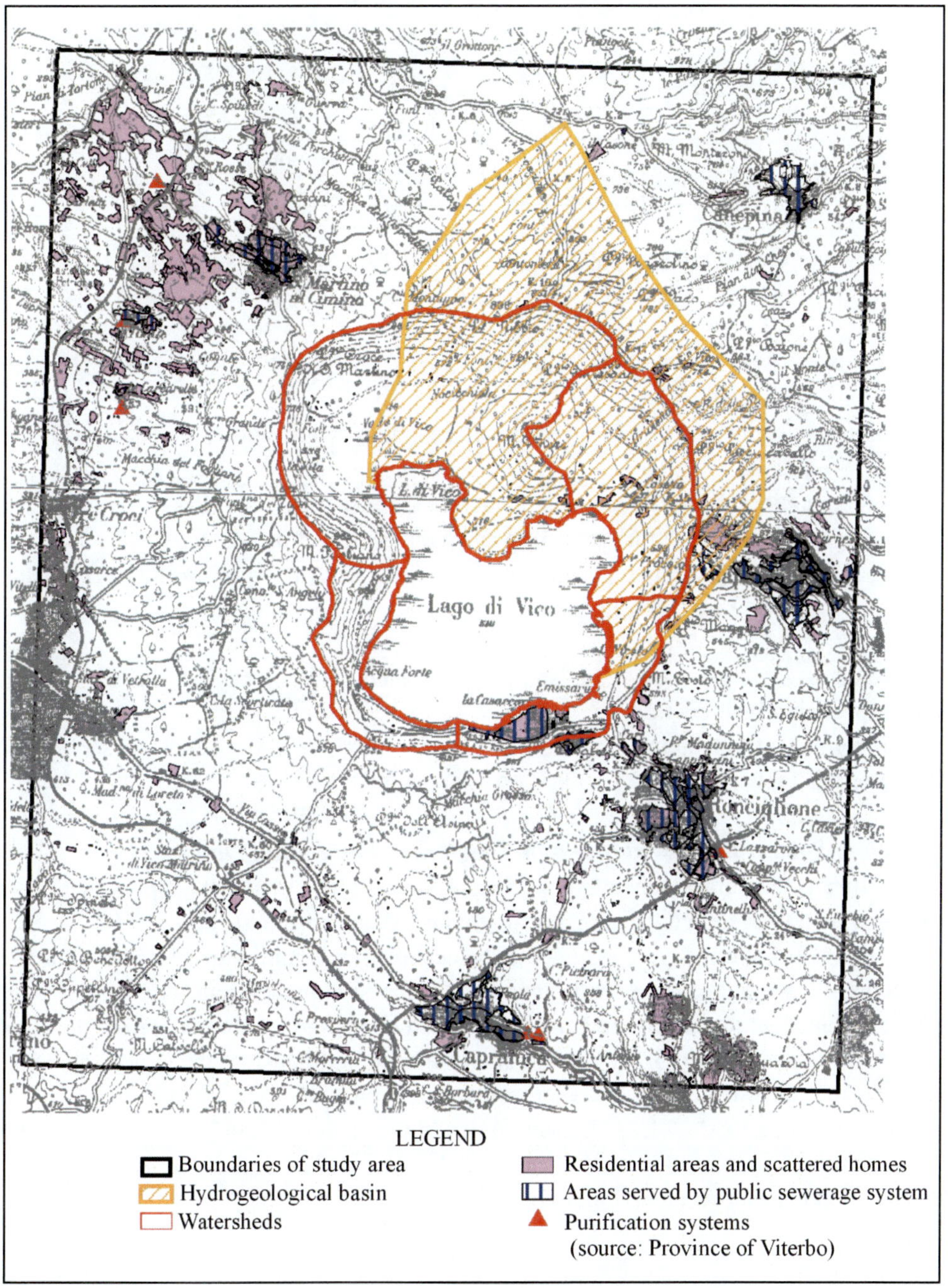

Figure 6.8. Areas served by public sewerage system, urban areas, inhabitantial areas and scattered homes

The "residential-agricultural" category was thus replaced, to the extent possible, by "scattered homes" and farmland of different type, so as to optimize the estimation of polluting loads from residential releases and agricultural practices.

Table 6.3: Corine codes

Corine Code	Description
1.1.1	Permanent settlement – built-up areas
1.1.2	Temporary settlement
1.1.2.1	Scattered homes

To spatially represent the population data from ISTAT's census sections on the Land Use Map, the following steps were taken:

1. Update of permanent population data based on the 1991-2001 trends recorded at municipal level;
2. Intersection between data on Water-Demanding Areas and ISTAT's data in a GIS environment;
3. Computation of the surface area of each of the identified polygons;
4. Computation of the urbanized surface area in each section;
5. Computation of the percentage surface area of each polygon in the total urbanized surface area in each census section.

The population data were then spatially represented by applying the density index obtained for each census section.

Total permanent inhabitants proved to be 28,956, of whom about 8,923 not served by the public sewerage system.

Approach to Spatial Representationo of Temporary Population Data

To estimate temporary inhabitants, reliance was made - to a first approximation - on ISTAT's data on unoccupied vacation housing units in each census section.

Vacation housing units were assumed to be occupied by a minimum of 2 and a maximum of 5 people on average during week-ends and in the summer period. The yearly average number of equivalent temporary inhabitants was then computed by approximating the number to the nearest upper integer: 3 persons for each "unoccupied housing unit" and 2 persons for each "housing unit occupied by non-resident inhabitants". The study area was thus determined to have a total of 4,189 housing units, of which about 2,556 in areas not served by a public sewerage system. This value can thus be regarded as indicative of the maximum weight of temporary inhabitants (Figure 6.9).

Sources of Data for Quantifying Agricultural Loads

Agriculture in the study area represents a significant economic sector, where specialized crops are dominant. These crops almost exclusively consist of hazelnut groves, most of which are irrigated. The remaining part of the study area is dominantly forested.

The chief source of data on type and distribution of farmland is the above-mentioned mentioned Map of Water-Demanding Areas, whose data were detailed and classified for the purposes of the study via photo-interpretation (IT2000 color ortho-photo, year 1999, nominal scale 1:10,000 and AGEA black and white ortho-photo, year 2002, nominal scale 1:10,000) and validated in the field (Figure 6.10).

Table 6.4 shows the complete items of the legend.

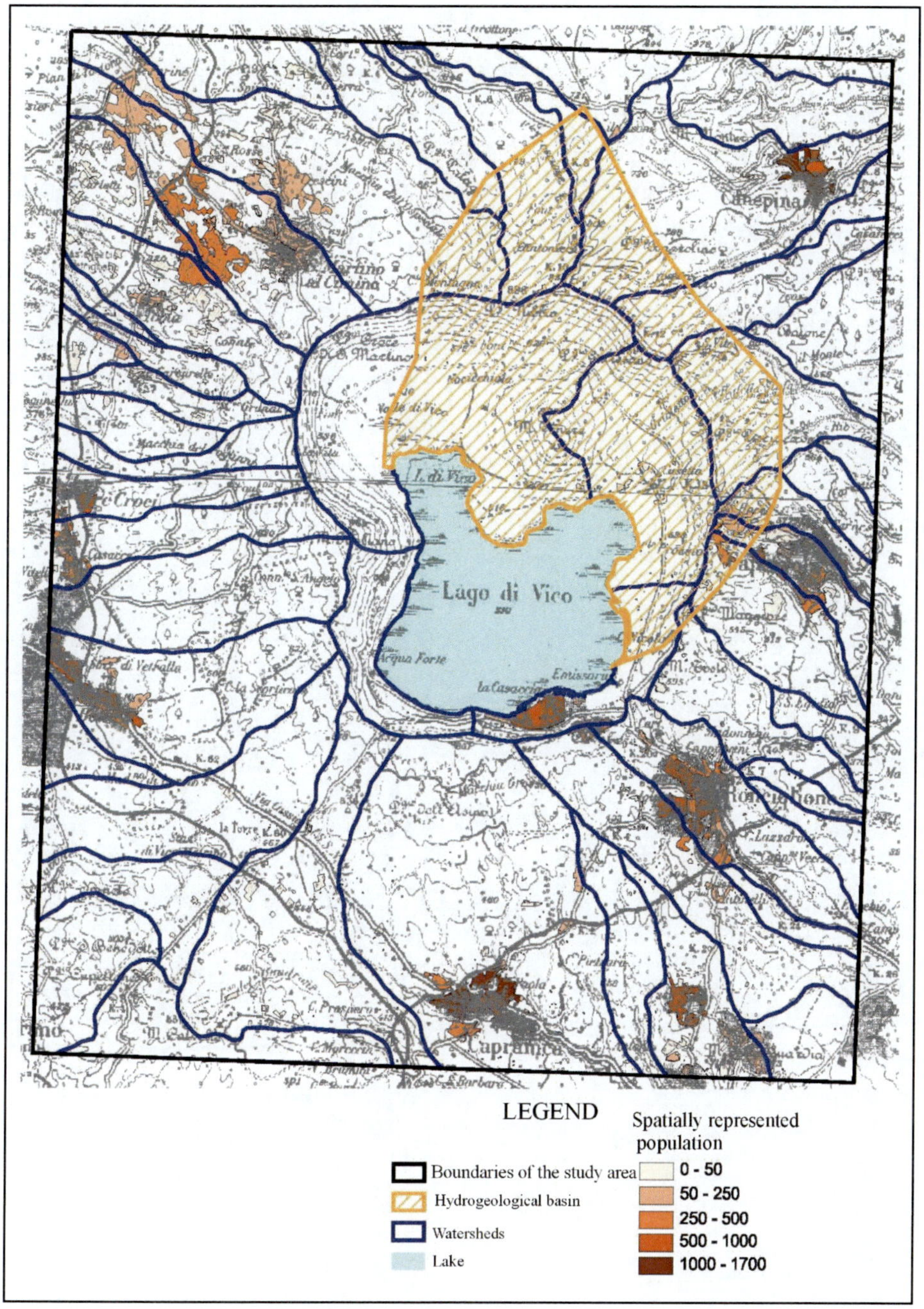

Figure 6.9. Yearly average distribution of equivalent (permanent+temporary) inhabitants.

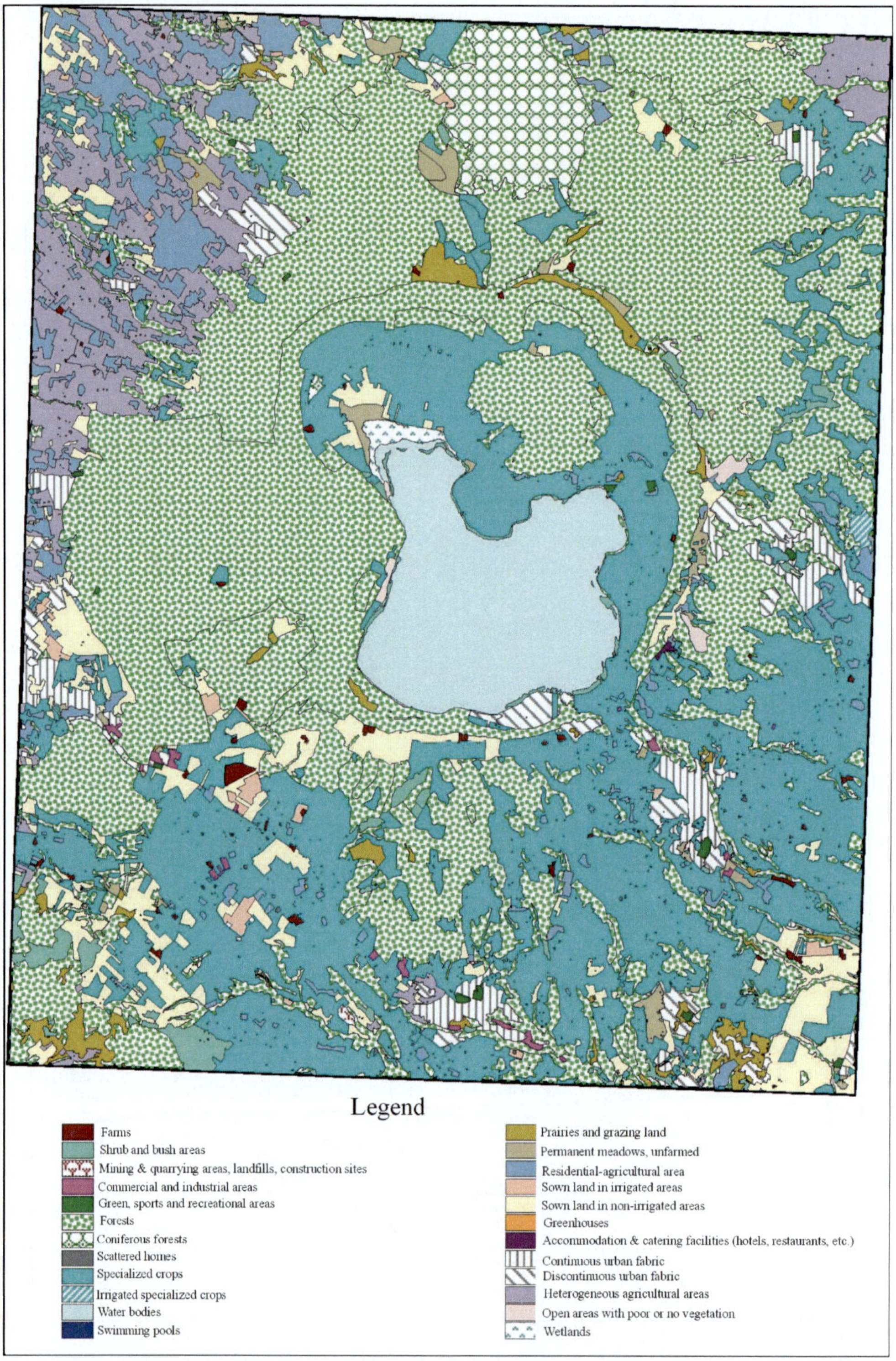

Figure 6.10. Map of Water-Demanding Areas (legend simplified to Level I Corine Land Cover).

Table 6.4. Legend of the Map of Water-Demanding Areas simplified to Level I Corine Land Cover

ARTIFICIALLY MODELED LAND	FARMLAND	NATURAL OR SEMI-NATURAL LAND
111 - Continuous urban fabric	211 - Sown land in non-irrigated areas	3111 - Deciduous forests
112 - Discontinuous urban fabric (residential non-agricultural)	212 - Sown land in irrigated areas	31111 - Chestnut forests
1120 - Residential-agricultural (dominantly sown land)	2121 - Intensive horticultural crops	3113 - Mixed broad-leaved forests
1121 - Residential (homes with gardens)	21 - Vineyards	3114 - Riparian tree vegetation
1123 - Residential-agricultural (dominantly olive groves – vineyards)	222 - Hazelnut groves	3115 - Tall herbaceous and shrub vegetation of wetlands (canebrakes)
11231 - Residential-agricultural (dominantly olive groves/orchards – sown land)	2221 - Kiwi crops	312 - Coniferous forests
11232 - Residential-agricultural (dominantly sown land – olive groves/orchards - vineyards	2222 - Irrigated peach and apricot orchards	313 - Forests, reforested areas (coniferous and broad-leaved)
1124 - Residential-agricultural (dominantly olive groves – gardens)	2223 - Irrigated hazelnut groves	321 - Natural grazing land
1126 - Residential-agricultural (olive groves and/or orchards)	223 - Olive groves	322 - Bush areas
1126bis - Residential-agricultural (dominantly olive groves/orchards)	224 - Abandoned orchards	324 - Areas with evolving shrub and tree vegetation
1127 - Residential in natural vegetation	24 - Heterogeneous agricultural areas	325 - Prairies
1129 - Heterogeneous residential-agricultural	241 - Yearly crops associated with permanent crops	331 - Beaches, dunes and sands
1128 - Cemeteries	245 - Associated permanent crops, irrigated	334 - Areas affected by fires
113 - Scattered homes, shanty homes	23 - Permanent meadows	41 - Inland wetlands (swamps and peat bogs)
1130 - Residential (orchards)	25 - Greenhouses	
114 - Scattered homes with swimming pools	26 - Unfarmed land (abandoned agricultural areas)	
116 - Swimming pool		
119 -Accommodation & catering facilities (prisons, condominiums, hotels, schools, colleges, etc.)		
121 - Industrial areas		
1211 - Sheds (warehouses etc.)		WATER BODIES
125 - Shopping & business centers and adjacent areas, winegrowers' cooperative		512 - Natural water basins
127 - Livestock breeding farms		5121 - Natural water basins
127a - Horse-riding school		5122 - Artificial water basins
128 - Agricultural farm		
129 - Accommodation & catering facilities (hotels, restaurants, etc.)		
131 - Mining & quarrying areas		
133 - Construction sites, areas under construction		
141 - Green urban areas; gardens; watered meadows		

Table 6.4 (Continued)

ARTIFICIALLY MODELED LAND	FARMLAND	NATURAL OR SEMI-NATURAL LAND
1421 - Sports and recreational areas (zoos; sport fishing; race track; thermal spas; shrines, etc		
1422 - Sports and recreational areas with swimming pools		
1424 Beach facilities (bathing)		
1425 - Campsites		
145 - Public and private hospitals		

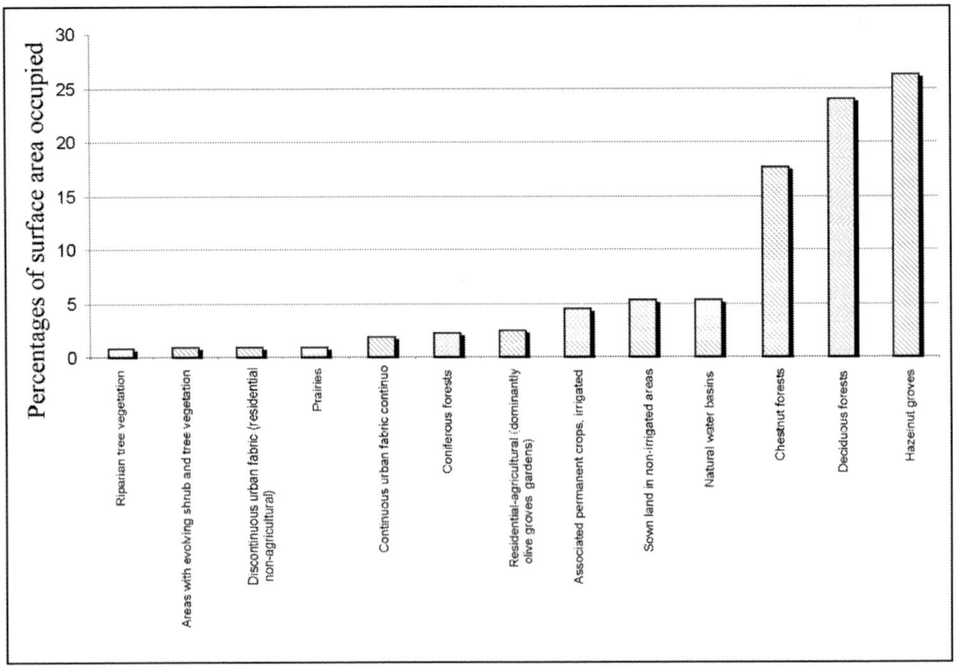

Figure 6.11. Percentages of surface area occupied by the main categories of land use in the study Area.

Figure 6.11 suggests that over 40% of the study area is forested. Farmland consists of: i) hazelnut groves (over 25% of the total surface area), which are dominant and concentrated around the Lake; ii) associated permanent crops (about 4%) and sown land (about 5%). Urban and residential areas account for roughly 2% of the total surface area.

To give a qualitative picture of polluting loads associated with farming, Tables 6.5 and 6.6 shows the results of ISTAT's survey of the distribution of agricultural fertilizers in 2003. The census-based survey involves all the firms operating with their own trademark in the marketing of fertilizers (fertilizers, soil conditioning and amending agents).

In the Tables, each fertilizer is associated with the content of nutrients or fertilizers specified in the applicable legislation (entire content or nutrient content reported by the distributing company).

Table 6.5., Table 1 ISTAT's data

Tavola 1 – Types of fertilizers by region Year 2003 (in quintals)

REGIONS	Mineral Fertilizers							
	Simple				Compound			
	Nitrogen	Phosphate	Potassium	Total	Binary	Ternary	Total	Mesoelement based
Piemonte	1.449.520	27.761	347.601	1.824.882	441.975	980.108	1.422.083	706
Valle d'Aosta	15	0	0	15	1	216	217	1
Lombardia	2.439.510	224.776	386.056	3.050.342	553.869	1.196.219	1.750.088	974
Trentino-Alto Adige	93.040	5.097	15.511	113.648	8.865	216.870	225.735	7.854
Bolzano-Bozen	*78.016*	*4.195*	*11.449*	*93.660*	*5.276*	*154.660*	*159.936*	*4.770*
Trento	*15.024*	*902*	*4.062*	*19.988*	*3.589*	*62.210*	*65.799*	*3.084*
Veneto	2.074.337	279.467	276.511	2.630.315	440.422	1.664.226	2.104.648	6.193
Friuli-Venezia Giulia	671.904	35.861	218.852	926.617	247.200	392.657	639.857	207
Liguria	15.245	636	15.262	31.143	5.361	37.269	42.630	1.483
Emilia-Romagna	2.023.178	485.975	91.220	2.600.373	529.090	539.279	1.068.369	1.956
Toscana	763.088	52.387	36.446	851.921	316.708	343.716	660.424	828
Umbria	570.334	34.062	19.062	623.458	244.385	138.177	382.562	1.920
Marche	657.093	220.075	5.149	882.317	281.720	139.039	420.759	66
Lazio	649.981	39.526	18.462	707.969	251.227	387.744	638.971	5.698
Abruzzo	314.498	94.164	19.092	427.754	213.734	252.614	466.348	432
Molise	154.867	37.720	1.809	194.396	80.489	27.189	107.678	316
Campania	1.007.561	162.057	10.632	1.180.250	319.377	376.358	695.735	4.697
Puglia	2.072.497	361.210	26.368	2.460.075	695.289	703.292	1.398.581	7.080
Basilicata	263.018	30.850	2.457	296.325	150.241	92.785	243.026	406
Calabria	338.921	87.521	5.709	432.151	126.502	291.016	417.518	425
Sicilia	731.987	303.224	57.513	1.092.724	385.566	704.343	1.089.909	14.738
Sardegna	405.316	37.399	15.700	458.415	365.983	194.427	560.410	886
ITALY	**16.695.910**	**2.519.768**	**1.569.412**	**20.785.090**	**5.658.004**	**8.677.544**	**14.335.548**	**56.866**
Northern Italy	***8.766.749***	***1.059.573***	***1.351.013***	***11.177.335***	***2.226.783***	***5.026.844***	***7.253.627***	***19.374***
Central Italy	***2.640.496***	***346.050***	***79.119***	***3.065.665***	***1.094.040***	***1.008.676***	***2.102.716***	***8.512***
Southern Italy	***5.288.665***	***1.114.145***	***139.280***	***6.542.090***	***2.337.181***	***2.642.024***	***4.979.205***	***28.980***

REGIONS	Mineral fertilizers		Organic fertilizers	Organic-mineral fertilizers	Total fertilizers (a)	Amending agents	Corrective agents	Total fertilizers
	Microelement-based	Total						
Piemonte	2.971	3.250.642	308.625	221.464	3.780.731	546.629	47.123	4.374.483
Valle d'Aosta	11	244	834	1.198	2.276	11.734	3	14.013
Lombardia	12.759	4.814.163	433.839	139.841	5.387.843	2.486.336	86.473	7.960.652
Trentino-Alto Adige	4.607	351.844	72.141	7.471	431.456	111.755	11.328	554.539
Bolzano-Bozen	1.542	259.908	49.728	969	310.605	53.330	10.306	*374.241*
Trento	3.065	91.936	22.413	6.502	120.851	58.425	1.022	*180.298*
Veneto	26.832	4.767.988	486.464	273.912	5.528.364	2.255.193	4.260	7.787.817
Friuli-Venezia Giulia	5.357	1.572.038	100.881	109.269	1.782.188	166.452	2.474	1.951.114
Liguria	3.096	78.352	48.688	49.867	176.907	480.922	588	658.417
Emilia-Romagna	26.615	3.697.313	411.831	434.674	4.543.818	744.761	5.149	5.293.728
Toscana	1.846	1.515.019	288.304	440.002	2.243.325	515.658	9.116	2.768.099
Umbria	688	1.008.628	54.770	142.495	1.205.893	105.005	72	1.310.970
Marche	3.880	1.307.022	106.535	211.224	1.624.781	159.832	7.713	1.792.326
Lazio	3.454	1.356.092	198.519	213.911	1.768.522	699.977	8.295	2.476.794
Abruzzo	15.118	909.652	67.924	165.144	1.142.720	93.448	5.026	1.241.194
Molise	508	302.898	8.962	38.443	350.303	16.507	532	367.342
Campania	2.418	1.883.100	75.666	257.524	2.216.290	410.499	2.516	2.629.305
Puglia	23.972	3.889.708	227.088	355.534	4.472.330	245.343	18.511	4.736.184
Basilicata	693	540.450	21.463	47.169	609.082	24.640	3.520	637.242
Calabria	411	850.505	62.917	87.452	1.000.874	79.751	1.379	1.082.004
Sicilia	23.384	2.220.755	265.339	317.008	2.803.102	547.898	16.496	3.367.496
Sardegna	5.227	1.024.938	46.150	40.053	1.111.141	73.406	1.481	1.186.028
ITALY	**163.847**	**35.341.351**	**3.286.940**	**3.553.655**	**42.181.946**	**9.775.746**	**232.055**	**52.189.747**
Northern Italy	***82.248***	***18.532.584***	***1.863.303***	***1.237.696***	***21.633.583***	***6.803.782***	***157.398***	***28.594.763***
Central Italy	***9.868***	***5.186.761***	***648.128***	***1.007.632***	***6.842.521***	***1.480.472***	***25.196***	***8.348.189***
Southern Italy	***71.731***	***11.622.006***	***775.509***	***1.308.327***	***13.705.842***	***1.491.492***	***49.461***	***15.246.795***

(a) including mineral, organic and organic-mineral fertilizers

Table 6.6. Types of fertilizers applied to the soil – Years 2002 and 2003 (in quintals)

Type of fertilizer	2002	2003	Change	
			Absolute	%
FERTILIZERS	42143261	4218194	38685	0.1
MINERAL	35475944	35341351	-134593	-0.4
Simple	20803538	20785090	-18448	-0.1
Nitrogen	16764851	16695910	-68941	-0.4
Calciumcyanamide	154389	138949	-15440	-10
Ammonium nitrate	5653653	5759409	105756	1.9
Calcium nitrate	700736	704255	3519	0.5
Ammonium sulfate	1624940	1277278	-347662	-21.4
Urea	7639300	7704120	64820	0.8
Other	991833	111899	120066	12.1
Phosphate	2515323	2519768	4445	0.2
Simple perphosphate	1660973	1724365	63392	3.8
Triple perphosphate	642074	593213	-48861	-7.6
Other	212276	202190	-10086	-4.8
Potassium	1523364	1569412	46048	3
Potassium sulfate	307890	380876	72986	23.7
Potassium chloride	1065454	1031345	-34109	-3.2
Other	150020	157191	7171	4.8
Compounds	14469135	14335548	-133587	-0.9
Binary	5262640	5658004	395364	7.5
Nitrogen-phosphate	4296497	4692650	396153	9.2
Nitrogen-potassium	618414	651763	33349	5.4
Phosphorus-potassium	347729	313591	-34138	-9.8
Ternary	9206495	8677544	-528951	-5.7
Mesoelement-based	52653	56866	4213	8
With single mesoelement	45990	47843	1853	4
With multiple mesoelements	6663	9023	2360	35.4
Microelement-based	150618	163847	13229	8.8
With single microelement	140089	152602	12513	8.9
With multiple microelements	10529	11245	716	6.8
ORGANIC	3167840	3286940	119100	3.8
Simple nitrogen	1847525	1935178	87653	4.7
Compound	1320315	1351762	31447	2.4
ORGANIC-MINERAL	3499477	3553655	54178	1.5
Simple nitrogen	181253	163655	-17598	-9.7
Compound	3318224	3390000	71776	2.2
AMENDING AGENTS	8080235	9775746	1695511	21
Vegetal	1288405	2200381	911976	70.8
Mixed	2915337	3294227	378890	13
Peaty	1452681	1659058	206377	14.2
Peat	807432	1020661	213229	26.4
Manure	566851	427503	-139349	-24.6
Other	1049529	1173917	124388	11.9
CORRECTIVE AGENTS	229656	232055	2399	1
Lime and limestone	138481	136628	-1853	-1.3
Sulfur	60340	52106	-8234	-13.6
Other	30835	43321	12486	40.5
TOTAL FERTILIZERS	50453152	52189747	1736595	3.4

Figure 6.12 reveals that, in the 1998-2003 period, mineral and oligomineral products decreased while organic products constantly increased.

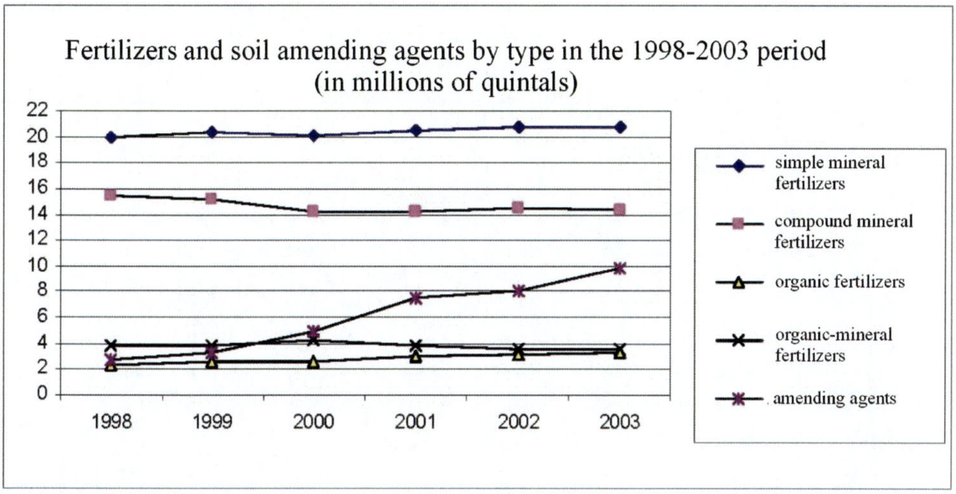

Figure 6.12. Fertilizers and soil amending agents by type in the 1998-2003 period.

Approaches to Quantification of Releases from Point and Diffuse Sources

Determination of Residential and Industrial Loads

Pollutants from point sources (individual residential settlements, purification systems and industrial firms) were computed by applying the following concentration values to the reported flow rates: i) for residential or equivalent releases, the concentration values pertaining to the number of equivalent inhabitants served; and ii) for all other types of releases, the measured or reported concentration values (Figure 6.13).

To compute polluting loads from diffuse residential sources, the values reported in Table 6.7 below were applied only to the share of permanent and temporary population which proved not to be served by a public sewerage system (model described in para. 6.6).

Finally, to estimate actual releases, 200 l/g per equivalent inhabitant were considered.

Figure 6.14 shows no point sources (individual residential settlements, purification systems and industrial firms), but only scattered homes inside the watersheds and the hydrogeological basin of the Lake. The scattered homes were considered to be diffuse sources of pollutants.

The presence of accommodation/catering facilities (hotels, camp sites, restaurants) inside the Lake's recharge area is poorly significant. The only facility which might be a source of pollutants, especially in summer (La Bella Venere), is provided with a sealed sewage tank which is periodically emptied via self-purging and feeding of releases into the pumping system of the Municipality of Ronciglione.

The study area has no hospital facilities or food industries (Figures 6.15-6.16)

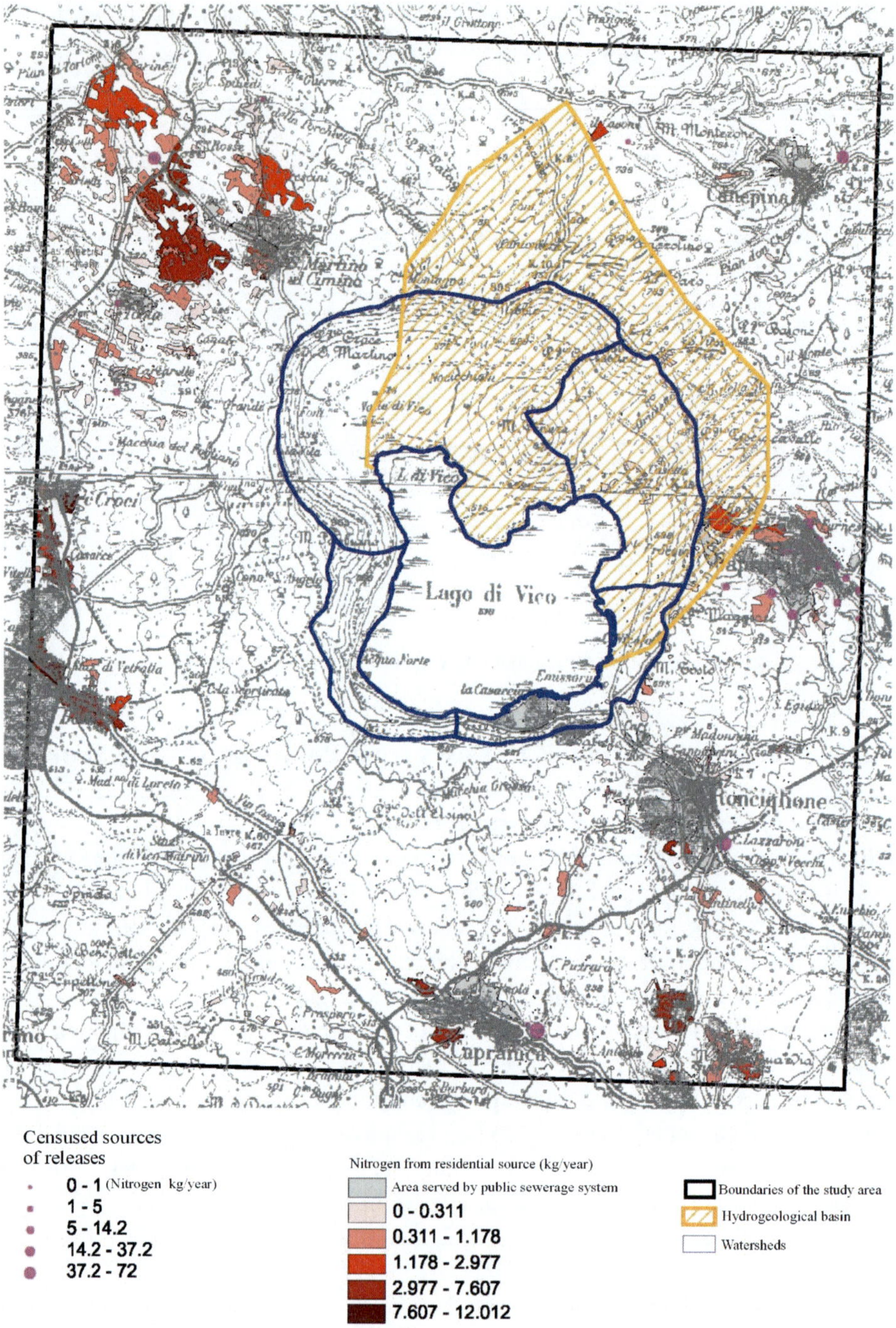

Figure 6.13. Residential loads.

Assessment of Agricultural Loads

Nitrogen

Studies on nitrogen fertilizers used in Italy have been focused on their effects on crop yields rather than on nitrate losses to percolation or runoff waters. These studies have been only recently intensified, driven by the so-called "Nitrates Directive" of the European Community on potential nitrogen contamination and by the Italian Legislative Decree implementing it (152/99). To prevent and monitor water pollution, the Directive (91/676/EEC) provides that vulnerable areas should be defined on the basis of: i) present or predicted nitrogen content in surface water and groundwater; ii) physical and chemical characteristics of waters and soils; and iii) greater understanding of the behavior of nitrogen compounds in the environment.

Table 6.7. Values used for loads from diffuse residential sources

Total nitrogen	12 g/day per equivalent inhabitant
Total phosphorus	2.5 g/day per equivalent inhabitant

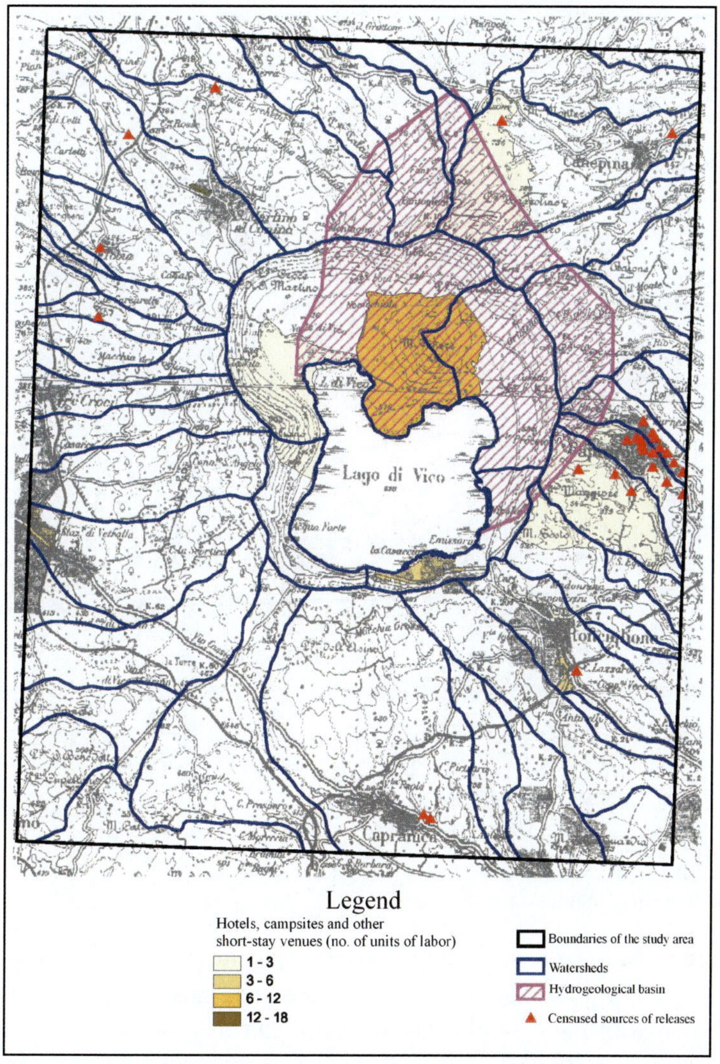

Figure 6.14. Distribution of accommodation/catering facilities (hotels, camp sites, etc.).

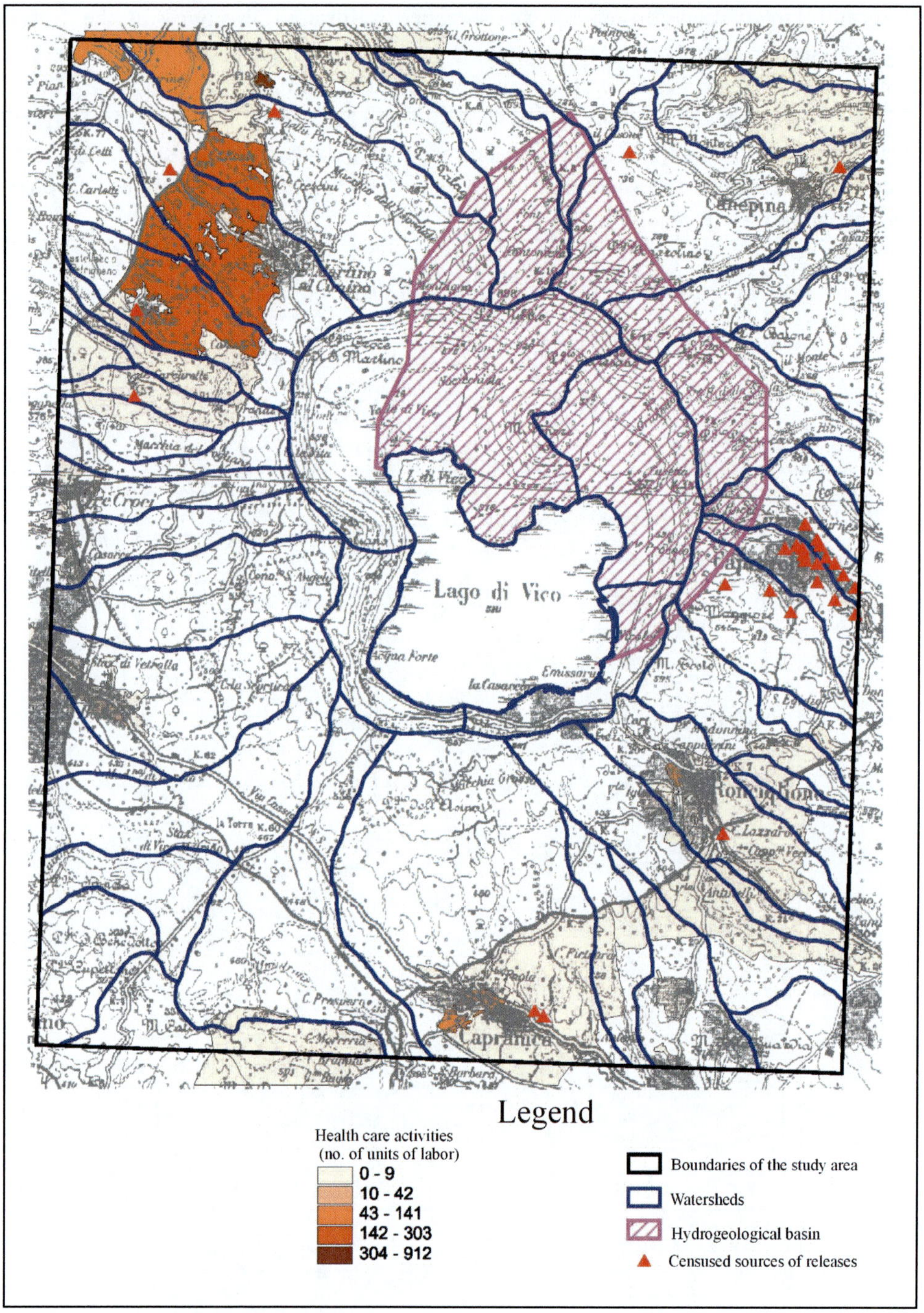

Figure 6.15. Distribution of health care activities.

In some cases, studies on nitrogen compound leaching have given conflicting results, pointing to the need for more thorough investigations. The specificity of all the experiments made in connection with these studies makes it difficult to generalize their results in areas different from those investigated. Furthermore, accurate predictions of nitrogen leaching require in-depth analyses of soils and a huge amount of data on crops and amounts of nitrogen fertilizers released into the cropland.

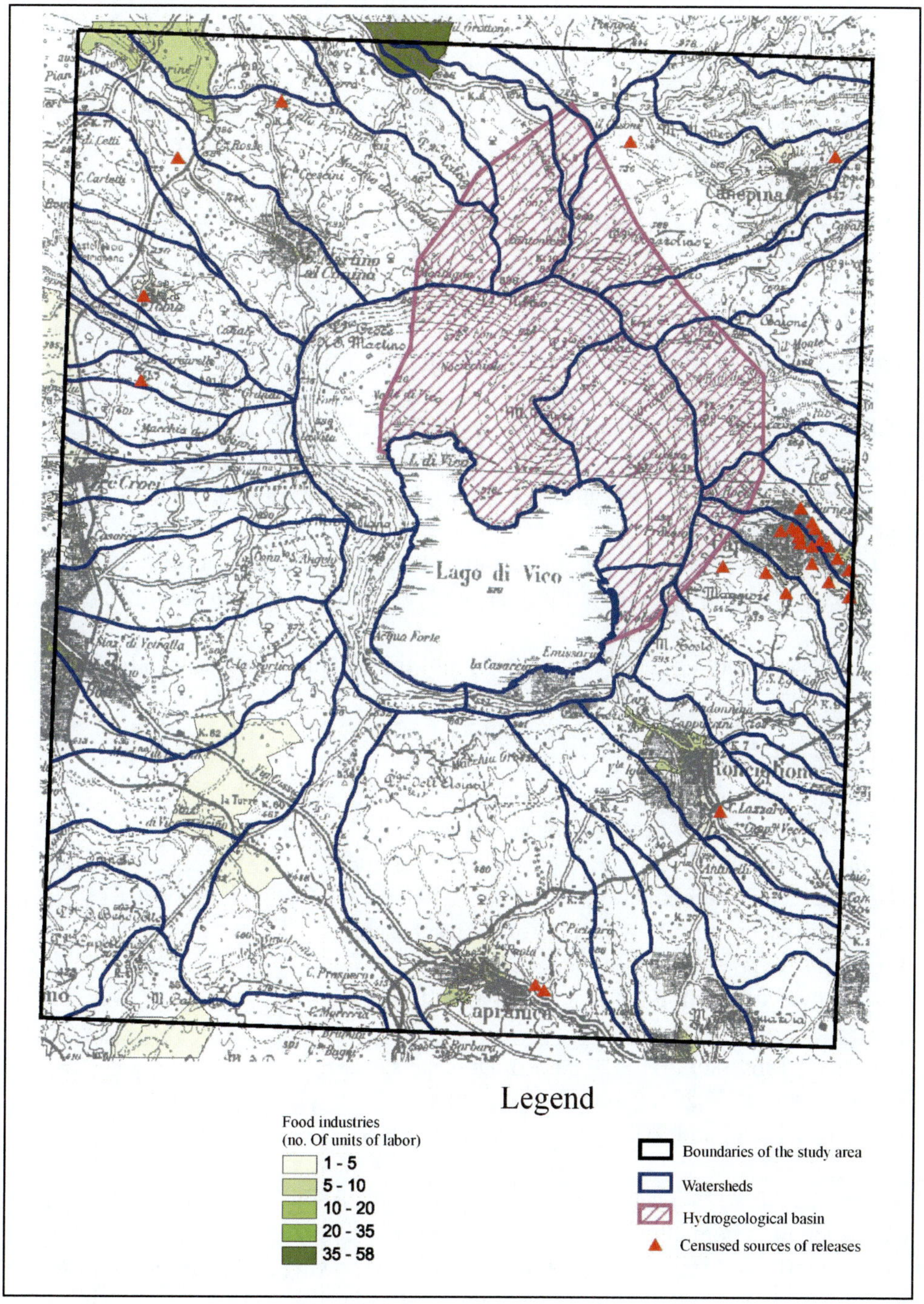

Figure 6.16. Distribution of food industries.

At present, these data are only available in fragmentary form. As a result, to quantify nitrogen leaching from soils, a general approach was taken. This approach derives from the one adopted for the Plan of Agronomical Use of Livestock Effluents ("Piano di utilizzazione agronomica dei liquami zootecnici") issued by the Campania Region.

The original approach applied to the above Plan was expected to produce the limit values of livestock effluents usable as nitrogen fertilizers for the different crops. This nitrogen content depends on 4 factors:

a) crop nitrogen demand;
b) soil nitrogen supply;
c) leaching;
d) immobilization and dispersion.

that is

Nitrogen to be applied as fertilizer = A -B +C +D

If the amount of nitrogen applied to the various crops as a fertilizer is known and if detailed analyses of farmland are available, then the amount of leached nitrogen (**A**) can be easily determined; indeed, crop nitrogen demand is known and reported in an extensive literature.

(B) Soil nitrogen supply can be computed as the sum of nitrogen deriving from mineralization of the organic substance (B1) and of the initial supply of assimilatable nitrogen (B2).

(B1) Nitrogen deriving from mineralization of the organic substance can be computed in the following way:

(B1.1) for a C/N ratio of 9 to 12:
for mostly sandy soil: B1 = 36 x (% of organic substance)
for loam soil: B1 = 24 x (% organic substance)
for mostly clayey soil: B1 = 12 x (% organic substance)
(B1.2) for a C/N ratio below 9:
for mostly sandy soil: B1 = 42 x (% organic substance)
for loam soil: B1 = 26 x (% organic substance)
for mostly clayey soil: B1 = 18 x (% organic substance)
(B1.3) for a C/N ratio above 12
For mostly sandy soil: B1 = 24 x (% organic substance)
For loam soil: B1 = 20 x (% organic substance)
For mostly clayey soil: B1 = 6 x (% organic substance)

The value of B1 so computed refers to the availability of nitrogen deriving from mineralization of the organic substance in one year. If the crop occupies the soil for less than one year, then B1 will be multiplied by the ratio of the growing period to the year (RB1): for instance, if the crop cycle is 4 months, then (RB1 = 4 / 12).

(B2) Nitrogen deriving from the initial supply of assimilatable nitrogen (corresponding to about 1% of total nitrogen) will be quantified as follows:

For mostly sandy soil: B2 = 28.4 x (‰ NTot.)

For loam soil: B2 = 26 x (‰ NTot.)

For mostly clayey soil: B2 = 24.3 x (‰ NTot.)

Hence, soil nitrogen supply (B) will be determined as the sum of B1 and B2.

(C) Soil nitrogen losses by leaching will be computed in accordance with the following Table, taking into account ease of drainage and soil texture.

Nitrogen lost by leaching (kg/ha year):

(D) Nitrogen immobilized through physico-chemical adsorption and by biomass through volatilization and denitrification will be computed as a percentage of soil nitrogen supply (B1 and B2); to do so, use will be made of the formula introducing the following correction factors (Fc):

$$D = (B1 + B2) \times Fc$$

To compute tree crop nitrogen requirements, the units computed according to the previous equation will be multiplied by a correction factor (FcE) taking into account the age of planting.

In view of the above, leaching of nitrates from soil might be calculated with excellent approximation by applying the formula:

Nitrogen to be applied as a fertilizer = A-B+C+D and solving with respect to C

Nonetheless, applying this formula requires data on the applied fertilizers and detailed soil analyses. Since all the required data are not available, the variant of the method applied in this study considers the release of nitrogen to be independent of precipitation and fertilizers spread over the soil, assuming compliance with the Code of Good Agricultural Practice (approved by a Ministerial Decree of Apr. 19, 1994) and thus a nitrogen supply to the plant not exceeding its demand.

Under this assumption, reference may be made to Table 6.8 only. In the Table, correlating soil texture and drainage yields the (estimated) values of leached nitrogen, regardless of types of crop, extent of precipitation and nitrogen applied as a fertilizer.

The map of the textural characteristics of farmland soil was built taking into account the nature of the outcropping lithotypes and slope steepness and making correlations with experimental sampling data.

Table 6.8. Nitrogen lost by leaching (kg/ha/year) (*) these values also account for the negative impact that lack of oxygen causes on mineralization of the organic substance

DRAINAGE	SOIL		
	Mostly sandy	Loam	Mostly clayey
Slow or difficult	50*	40*	50*
Normal	40	30	20
Fast	50	40	30

The amounts of leached nitrogen estimated with this method (Figure 6.17) proved to be consistent with most of the values obtained experimentally and described in the References. Consequently, they may be considered to be valid for this study.

Table 6.9. Correction factors (Fc)

DRAINAGE	SOIL		
	Mostly sandy	Loam	Mostly clayey
Slow or difficult	0.30	0.35	0.40
Normal	0.20	0.25	0.30
Fast	0.15	0.20	0.25

Table 6.10. Correction factor (FcE)

Crop	1st year	2nd year	3rd year	4th year	5th and subsequent years
Actinidia	0.6	0.4	0.8	1	1
Grapevine	0.5	0.4	0.6	0.7	0.85
Pear tree	0.55	0.4	0.7	0.85	1
Apple tree	0.6	0.4	0.8	1	1
Cherry tree	0.5	0.4	0.6	0.7	0.85
Persimmon/Lotus tree	0.55	0.4	0.7	0.85	1
Peach tree	0.75	0.5	1	1	1
Apricot tree	0.6	0.4	0.8	1	1
Plum tree	0.6	0.4	0.8	1	1
Olive tree	0.6	0.4	0.8	1	1
Hazelnut tree	0.6	0.4	0.8	1	1
Lemon tree	0.6	0.4	0.8	1	1

If the above-mentioned data were more detailed, especially in terms of spatial and temporal distribution in the study area, they might certainly improve the estimated values and permit the use of more sophisticated estimation approaches.

Phosphorus

Phosphorus is generally retained in the soil in the form of Fe, Al and Ca phosphates. Therefore, it can hardly be removed from the soil by leaching. In general, leaching of agricultural phosphorus takes place when the supply of fertilizers exceeds by far crop demand.

Livestock breeding

The presence of few small breeding farms and horse-riding grounds does not appear to be a significant source of release of eutrophicating substances into lacustrine basins.

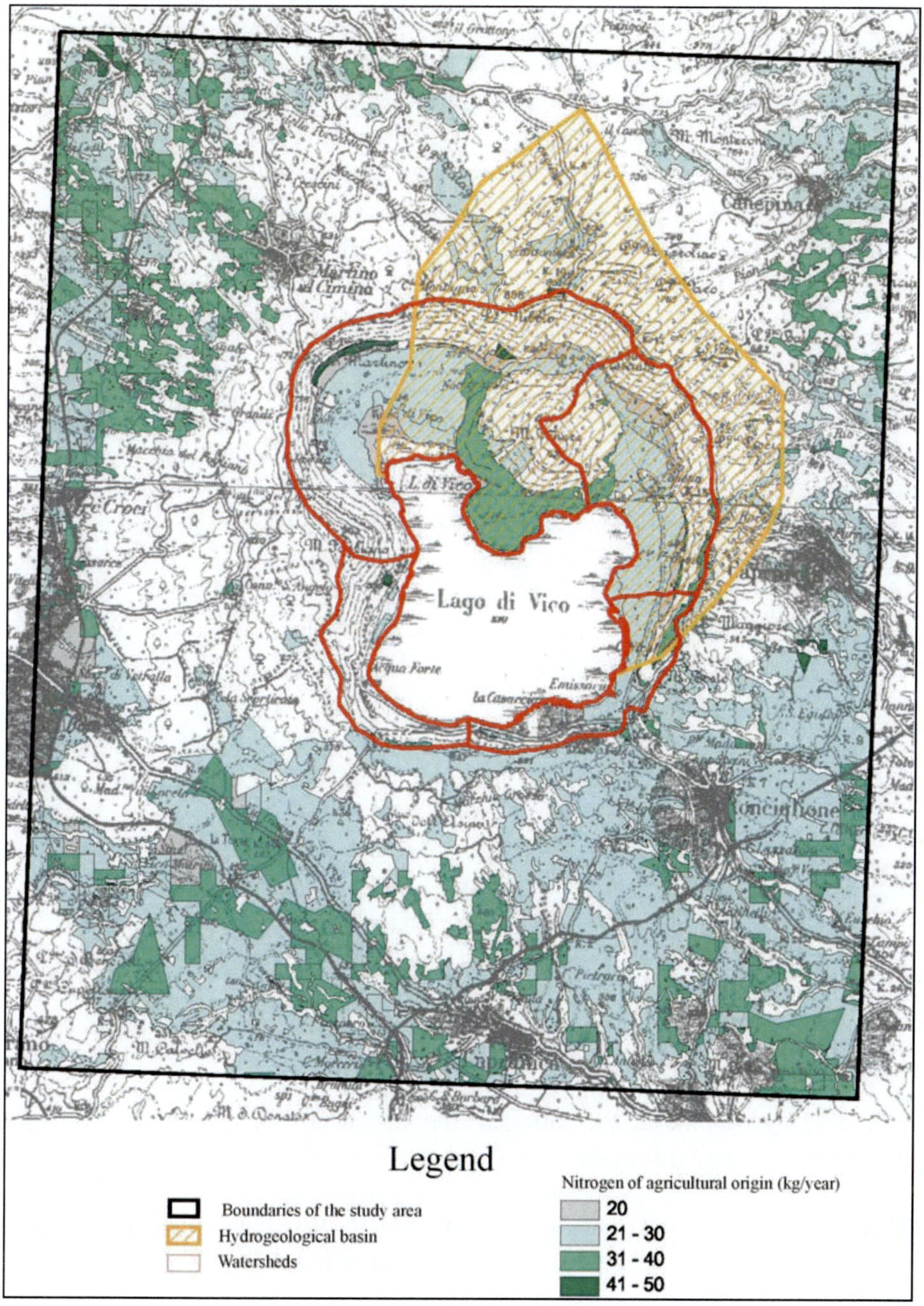

Figure 6.17.Distribution of nitrogen of agricultural origin.

Quantification of Eutrophicating Substances Potentially Migrating into Lakes From Surface Water, Groundwater and Direct Discharges

As to the loads of eutrophicating substances released by residential and industrial activities, the following documents were produced during the study:

- Map of nitrogen loads potentially released into surface water and groundwater (seepage) from residential areas and scattered homes not connected to the public sewerage system and to purification systems (Figure 6.13);
- Location of the main point sources of: releases from purification systems; releases from residential areas served by sewerage systems and not purified; and releases from productive activities (Figure 6.7);
- Map of nitrogen volumes released into surface water and groundwater (Figure 6.17 and enclosed map) from agricultural activities;
- Map of distribution of phosphorus of residential origin (enclosed map).

Quantification of nitrogen input into the Lake on the basis of the above data is an extremely complex problem. This problem lies in determining the quantitative distribution of nitrogen between water runoff towards lakes and water flows into aquifers.

Lack of a dedicated monitoring network and, consequently, unavailability of adequate experimental measures do not permit to develop a reliable model and fine-tune it.

The yearly load that surface water and groundwater may carry into lakes can be roughly estimated by proportionally dividing the estimated volumes of releases into the various sectors (50, 51, 52, 53, hydrogeological system and Lake) by the ratio of monthly runoff and infiltration water to effective rainfall. To do so, the following relations may be used in each sector:

$Vr = Vt*(R/PE)$
$Vi = Vt*(Ie/PE)$
where:
Vt = yearly total volume released
Vr = volume carried by runoff water
Vi = volume carried by infiltration water
R = yearly runoff
Ie = yearly effective infiltration
PE = effective rainfall (PE = P-EVr = R+Ie)

Table 6.11 below shows the nitrogen and phosphorus loads that are obtained by cumulating the polluting loads in the Lake's recharge area only.

Table 6.11. Nitrogen and phosphorus loads shown separately for the investigated watersheds and hydrogeological basin

Basin	Loads by basin (kg/year)				
	N-agr (kg)15.0	N_res (kg)	N-tot (kg)	P-res (kg)	P-tot (kg)
50	4528.0	2.5	15.1	0.5	3.1
51	10127.9	0.2	0.2	0.0	0.0
52	811.1	0.0	0.1	0.0	0.0
53	23496.3	0.1	0.1	0.0	0.0
hydrogeological	33531.6	6.1	7.4	1.3	1.5
Lake (rainfall)	0.0	0.0	0.0	0.0	0.0

If the above-mentioned volume formula is used to multiply the distribution of polluting loads (shown in the above table) by the flows recharging the Lake, then the following nitrogen concentrations (probably the maximum expected nitrogen values) in runoff water, infiltration water and rainfall are obtained (see Table 6.12 below).

Table 6.12. Maximum nitrogen concentrations in waters flowing into the investigated watersheds and hydrogeological basin

Basin	Discarge m^3/year					Maximum concentration of nitrogen in water inflows (mg/l)		
	R	R/PE	IE	IE/PE	P	R mg/l	IE mg/l	P
50	793711.8	0.363				2.071		
51	1224200	0.258				2.134		
52	588645.9	0.26				0.358		
53	2888964	0.259				2.106		
Hydrogeological		0.22	18025000	0.78			1.451	
Lake (Rainfall)	0	0	0	0	13873000			0.000

Table 6.13. Monthly values of nitrogen input into the Lake (IE does not refer to effective infiltration but to groundwater drainage towards the Lake)

Month	Water inflows/ Evaporation			N input into the Lake	
	R m^3/month *1000	P m^3/month *1000	IE m^3/month *1000	Average nitrogen concentration in water inflows (mg/l)	Total nitrogen input into the Lake (kg/month)
mar-05	744	1779	1442	0.888	3522
apr-05	293	1706	1442	0.772	2655
may-05	0		1442	1.207	2093
june-05	0	677	1442	0.988	2093
jul-05	0	128	1442	1.333	2093
agu-05	0	1141	1442	0.810	2093
sept-05	62	1384	1442	0.766	2212
oct-05	947	2128	1442	0.866	3912
nov-05	1600	3112	1442		5166
dec-05	1612	3004	1442	0.857	5189
jan-06	1570	775	1442	1.349	5108
feb-06	999	1946	1442	0.914	4012
mar-06	826	1865	1442	0.890	3679
apr-06	0	771	1442	0.946	2093
may-06	0	1141	1442	0.810	2093
june-06	0	0	1442	1.451	2093
jul-06	0	44	1442	1.408	2093
agu-06	0	37	1442	1.415	2093
sept-06	0	44	1442	1.408	2093
oct-06	185	775	1442	1.019	2449
nov-06	162	557	1442	1.112	2403
dec-06	336	719	1442	1.097	2739
gen-07	629	1276	1442	0.987	3302
feb-07	630	1296	1442	0.981	3303
mar-07	395	1149	1442	0.955	2852

Considering the estimated water flows and nitrogen concentrations, the average nitrogen content in the waters feeding the Lake (Runoff + Rainfall + Infiltration) is 0.98 mg/l. As is obvious, this value does not take into account soil and country rock retention capabilities or biochemical cycles. Therefore, it can be regarded as the maximum expected concentration value for the surveyed polluting loads.

With a similar computation process, nitrogen concentrations and total nitrogen inputs may be distributed among the different months of the study period (see Table 6.13).

Obviously, these values should not be understood in absolute terms, as they are certainly error-biased to a large extent. Nevertheless, they were calculated to highlight the periods in which nitrate input into the Lake is most likely.

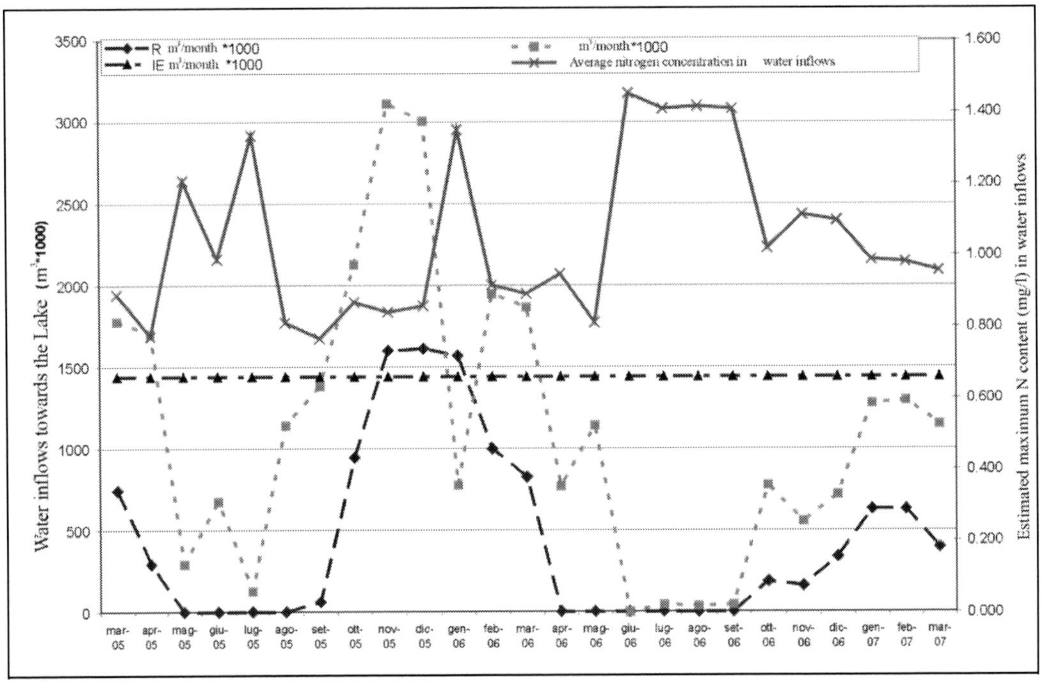

Figure 6.18. Estimated monthly nitrogen concentration in waters flowing towards the Lake vs. estimated flows for the different components.

Plotting (Figure 6.18) the estimated monthly trend of nitrogen concentrations in waters flowing towards the Lake evidences that nitrogen concentrations may: i) significantly change, depending on the extent of rainfall; and ii) reach peak levels in periods of strong runoff or lack of rainfall (when nitrogen-rich groundwater flows become dominant).

The monthly trend of the calculated total nitrogen inputs (Figure 6.19) may be significant, too. Naturally, this trend almost mirrors the one of concentrations. In the study period, namely from October 2005 to March 2006, the estimated nitrogen concentration values are particularly high (Figs. 6.19 and 6.20) and ascribable to the strong contribution of runoff waters.

In this respect, it is worth noting that, in the 2005-2006 fall-winter period, rainfall values are closer to the average ones of the area; by contrast, the corresponding 2006-2007 period appears to be particularly dry.

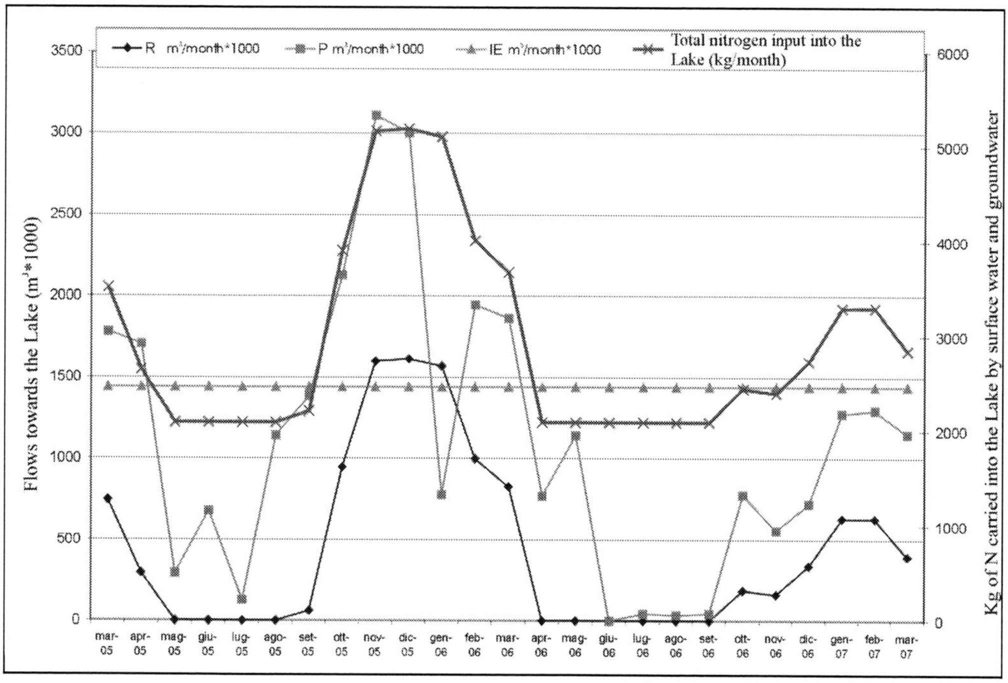

Figure 6.19. Calculated monthly total nitrogen inputs (from surface and groundwater).

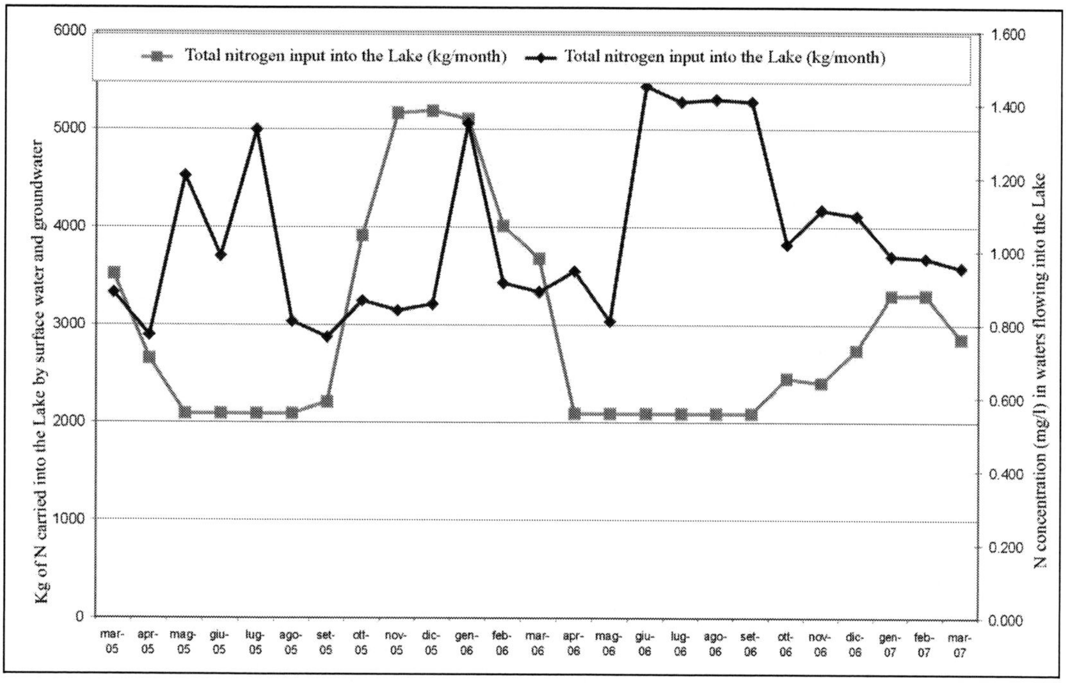

Figure 6.20. Nitrogen concentration in waters flowing into the Lake.

7. Experimentally Measured versus Statisticaly Calculated Concentrations of Eutrophicating Substances

The previous paragraphs have described: i) the results of analyses conducted on water samples from the Lake and from wells; and ii) statistical processes through which the sources of polluting loads and their expected concentrations in the waters flowing into the Lake were identified.

Based on these results, the measured and calculated values were compared.

Table 7.1 below compares the values of the February 2007 analyses, i.e. calculated total nitrogen concentrations and total nitrogen concentrations measured in filtered and non-filtered water samples. The filtered water samples underwent filtration through a 0.45 μm filter prior to analysis.

Table 7.1. Calculated vs. measured values in the Lake's waters

Lake			
Sample	Maximum expected concentration of total nitrogen in the Lake's waters (average value between concentrations of IE, R, P) in February (mg/l)	Average value of total nitrogen measured in non-filtered sample (mg/l)	Average value of total nitrogen measured in filtered sample (mg/l)
L1	0.981		0.12
L2	0.981	2.16	0.13
L3	0.981		0.11
L4	0.981	2.13	0.12
L5/0	0.981	1.88	0.13
L5/5	0.981	2.28	0.11
L5/10	0.981	1.78	0.09
L5/15	0.981	2.31	0.13
L5/20	0.981	1.85	0.16
L5/25	0.981	1.86	0.11
L5/30	0.981		0.14
L6	0.981	3.13	0.00

In non-filtered samples, total nitrogen is higher. This increase is due to the fact that, in these samples, the final total nitrogen value includes the amount of nitrogen that microorganisms capture from the water in which they reproduce. This nitrogen, which is retained and stored inside their structure (biomass), is taken into account only for the non-filtered samples.

This simple comparison evidences that, both in the Lake and in wells, the values of non-filtered samples are higher by one order of magnitude than those measured on filtered samples, which always lie above the "expected values".

Table 7.2. Measured vs. calculated values in well waters

Wells			
Sample	Maximum expected concentration of total nitrogen in the waters flowing into the Lake in February (IE) (mg/l)	Average value of total nitrogen measured in non-filtered sample (mg/l)	Average value of total nitrogen measured in filtered sample (mg/l)
P1	1.45		1.96
P2	1.45	45.18	0.36
P3	1.45	83.69	22.45
P4	1.45		0.10
P5	1.45	10.09	2.52
P6	1.45	4.07	0.42
P7	1.45	0.08	0.08
P8	1.45	2.37	0.10
P9	1.45	4.08	0.96
P10	1.45		0.00
P11	1.45	10.83	2.56

As mentioned in the previous paragraphs, the maximum contribution to the polluting load comes from agricultural sources, as the contribution from residential and industrial sources falls outside the study area. The values obtained through the statistical approach derive from considerations about nitrogen leaching from soil, assuming however that soil fertilization techniques comply with the Code of Good Agricultural Practice (Ministerial Decree of Apr. 19, 1994).

Therefore, this comparison demonstrates that, in the study area, nutrient inputs are due above all to inconsiderate and excessive use of fertilizers and other agents. This conclusion is supported by interviews with many farmers who fail to comply with the provisions of the Ministerial Decree of Apr. 19, 1994.

8. Toxic Eutrophication of Lake Vico

Cyanobacteria and Cyanotoxins

Cyanobacteria are prokaryotic organisms with highly diversified morphology (unicellular, filamentous, ramified or non-ramified) and size (from unicellular with a diameter of 0.2 μm to filamentous with a length of up to 200 μm). These microorganisms are photoautotrophic, but many species use organic compounds as their energy source (Adams, 1997) and colonize a large variety of environments (Bold and Wynne, 1985) including extreme ones (glaciers, deserts, hypersaline and thermal environments). Most of these species occur in fresh water and soil. They may be planktonic or benthic and populate marine or fresh water environments. They can withstand high and low light intensities, high levels of UV radiation and many species can fix atmospheric nitrogen. Hence, they are ubiquitary organisms with significant pioneering capabilities (Van den Hoek et al., 1995; Pupillo et al., 2003).

Under favorable environmental conditions, some cyanobacteria may become dominant in the phytoplankton of a water body and produce blooms.

Many cyanobacterial species generate different types of potent toxins (Chorus and Bartram, 1999). The dominant species in toxic blooms belong to the genera *Microcystis, Anabaena, Aphanizomenon, Nodularia* and *Planktothrix.*

Francis (Francis, 1978), in Australia, was the first to report a toxic bloom causing the death of sheep, bears, pigs and dogs. Subsequently, increasing reports about poisoning of livestock drinking water from lakes and ponds with toxic blooms have focused the attention of breeders and veterinarians on the toxicity of cyanobacteria.

In humans, cyanotoxins may cause skin irritation, gastroenteritis and death.

Human exposure to cyanotoxins may occur through direct ingestion of water containing cyanobacterial blooms (Texeira da Gloria Lima Crux et al., 1993; Annadotter et al., 2001; Falconer, 2005), dialysis with contaminated water (Carmichael et al., 2001) and recreational use of water (Bowling e Baker, 1996; Pilotto et al., 1997; Chorus e Bartram, 1999; Carmichael, 2000; Behm, 2003).

The species affecting Lake Vico is *Planktothrix rubescens*, previously *Oscillatoria rubescens* (D.C. ex Gomont) Komarek and Anagnostidis (Kaebernick and Neilan, 2001; Christiansen et al., 2003), which produces microcystins.

Microcystins

Microcystins are monocyclic heptapeptides of low molecular weight, whose general structure is cyclo-(D-Ala^1-X^2-D-$MeAsp^3$-Z4-$Adda^5$-D-Glu^6-$Mdha^7$-), where X and Z are two variable L-amino acids, D-$MeAsp^3$ is D-erythro-β-methylaspartic acid, Mdha is N-methyldehydroalanine and Adda (2S,3S,8S,9S-3-amino-9-methoxy-2,6,8-trimethyl-10-phenyldeca-4,6-dienoic acid) is the only structural component imparting toxicity to these toxins (Namikoshi et al., 1989). So far, about 70 variants of these toxins have been identified.

Microcystins are genotoxic (Bouaicha et al., 2005), inhibit protein phosphatases (Dawson, 1998) and are responsible for hepatotoxicosis and death in humans (Azevedo et al., 2002), animals, livestock and fish fauna (Carmichael, 1994, 2001; Sivonen and Jones, 1999; Mwaura et al., 2004). They cause acute histological damage to liver (their main target organ), as well as to lungs and kidneys. Their lesions include necrosis of hepatocytes, extensive perilobular bleeding, cell detachment and loss of tissue architecture, rupture of sinusoidal epithelial cells and disappearance of Disse's spaces between hepatocytes and sinusoids.

Evidence supporting the role of microcystins as tumor promoters in humans has been reported by Yu (Yu, 1989) and Zhou (Zhou et al., 2002) in China and by Fleming (Fleming et al., 2002) in Florida.

Microcystins exert their toxic effects on all living organisms: i) zooplankton and fish fauna, as they accumulate in their tissues (Freitas de Magalhes et al., 2001) with severe repercussions on public health; and ii) vegetal organisms, which may become sources of exposure, as these toxins may also accumulate in edible plant tissues (Codd et al., 1999).

The provisional WHO guideline Tolerable Daily Intake (TDI) of microcystin-LR is 0.04 μg/Kg (1 μg/L for a healthy adult weighing 60 kg) (Chorus and Bartram, 1999). Other studies suggested lower limits (0.1 μg/L, Annadotter et al., 2001; 0.01 μg/L, Hernandez et al., 2000; Ueno et al., 1996).

During the WHO Workshop held in Bad Elster (Germany) on April 23-25, 1997, the limits proposed for microcystin-LR were 1 μg/L for acute toxicity and 0.3 μg/L for chronic toxicity and risk of tumor promotion and carcinogenicity (Carmichael W.W., personal communication). Later, only the provisional limit of 1 μg/L (Chorus and Bartram, 1999) was adopted; indeed, at that time, further studies were needed to set an acceptable limit for the risk of carcinogenesis.

In Italy, the limit value in bathing waters is 0.84 μg/L (corresponding to 5 x 10^6 cells/L, (Falconer et al, 1994) and bathing bans are introduced when the cells of toxic cyanobacterial species reach 5,000/ml (Circ. Min. Sal. 31/7/1998 IX.400.4/13.1/3/1447).

Characteristics of Planktothrix rubescens

Planktothrix rubescens is made up of brown red tricomes without visible sheath, mean diameter of 7 μm and variable length reaching few mm. Tricomes consist of adjacent cells, all identical except for apical ones (used for morphological identification), which are smaller, thereby reducing the diameter of the tricome.

Planktothrix rubescens was originally considered to belong to the class of *Cyanophyceae* or blue-green algae, because of its prokaryotic characters, although being capable of photosynthesis. It has now been placed in the taxonomic group of cyanobacteria, which classifies all blue-green algae as true bacteria (Stainer 1977, Ripka et al,1979).

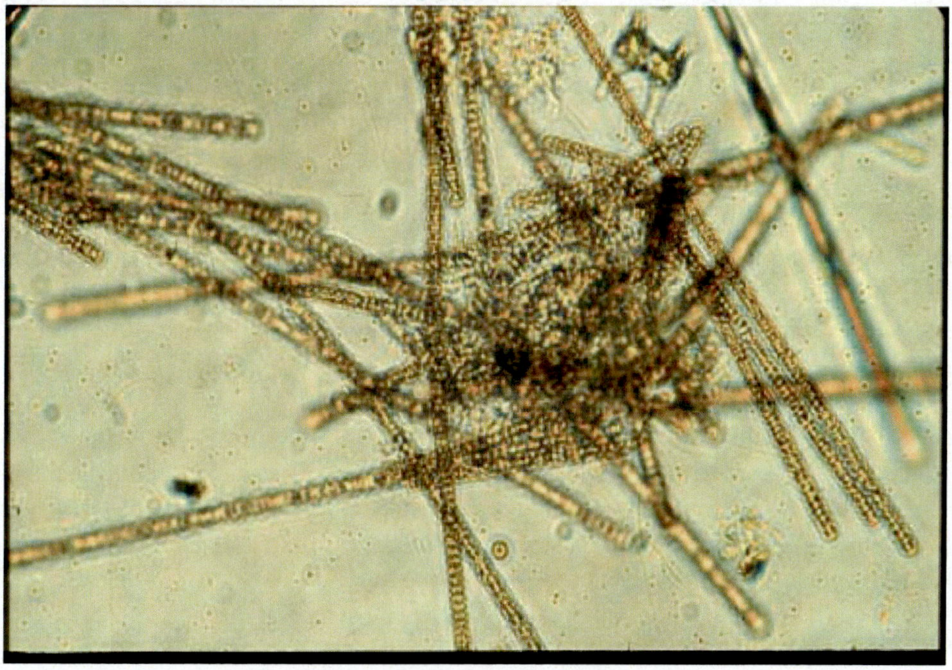

Figure 8.1. Tricomes of Planktothrix rubescens (400x magnification).

Like many other cyanobacteria, *Planktothrix rubescens* is a typical component of lacustrine phytoplankton and its behavior is similar to the one of a true alga. It occurs in many

European lakes (from the regions around the Alps to central Europe, Norway, Sardinia and Sicily). It also inhabited Lake Washington (USA) for some years and then disappeared after the diversion of one of its outlet streams. It was also observed in Lake Crooked (USA, Indiana) (Konopka 1982 a,b; Edmondson 1991).

Ecophysiology

When *Planktothrix rubescens* actively grows in a balanced and nutrient-rich medium, it has a dark red color. Under these conditions, its protein content is similar to the one of Spirulina, which is equivalent to 12% N and 1.3% P (dry weight). Under the electron microscope, it shows many phosphate granules.

Planktothrix rubescens contains chlorophyll *a* only and two typical carotenoids: mixoxanthophyll (typical of all cyanobacteria) and oxillaxanthin (specific for the genus *Oscillatoria).* Unlike chlorophyll *a*, the content of oxillaxanthin is not light-dependent (Feuillade,1972). The light energy that carotenoids absorb has a very low photosynthetic efficiency; these pigments are generally assumed to guard against high levels of irradiation. *Planktothrix rubescens* has three biliproteins: allophycocyanin and C-phycocyanin, giving cyanobacteria their typical blue-green color; and the red C-phycoerythrin. These pigments are water-soluble and propagate outside the cells when cyanobacteria die. They are regarded as the most efficient pigments in light absorption (Stainer,1977). The photosynthetically active radiation (PAR) captured by phycocyanin is used with an efficiency equal to or greater than the one absorbed directly by chlorophyll *a* (Garnier,1974).

Thanks to its biliprotein content, *Planktothrix rubescens* can exploit the entire spectrum of PAR. Abundance of pigments explains why a very low irradiation of 26 $\mu\varepsilon m^{-2}s^{-1}$ was determined to be optimal for the growth of *Planktothrix rubescens* in a long-term experimental project (20 °C, 18 hours, 64 light/dark cycles). Konopka (1982a) found that the optimal irradiation for *Planktothrix rubescens* in Lake Crooked was 180 $\mu\varepsilon m^{-2}s^{-1}$ (25°C, 2 hours). Given their composition (proteins associated with pigments), biliproteins are very sensitive to nitrogen deficiency and may act as nitrogen stores when this nutrient is in adequate supply (Van Liere et al.1975). Consequently, the biliprotein content of cells is correlated with their total nitrogen content and an indicator of deficiency or availability of this nutrient (Feuillade and Feuillade 1981, Feuillade and Davies 1994). Hence, under stable natural conditions, the content of biliproteins in a cyanobacterial population is an indirect measure of nitrogen availability in a lake.

Photosynthetic Metabolism

Planktothrix rubescens does not tolerate even a 1% carbon dioxide enrichment of its culture medium, which would soon become too acidic. During growth, its pH rapidly rises from 8.5 to 9. Water with this pH value has only traces of free carbon dioxide, because dissolved inorganic carbon is present above all as bicarbonate ion.

Feuillade and Feuillade (1981) investigated the photosynthesis of *Planktothrix rubescens*: under suboptimal growth conditions, a typical C3 cycle was observed (Calvin –Benson). Also its PEP carboxylase and carboxykinase are very active enzymes, because PEP is used in prior C3 carboxylation and bicarbonate as a substrate. The prokaryotic structure of the individual cells is an advantage: the entire cell, rich in tylacoids, appears and behaves as a chloroplast; therefore, if enough energy is available, all of the breathed carbon dioxide is immediately reincorporated photosynthetically. The relative proportions of ^{14}C fixed in the intracellular fractions of *Planktothrix rubescens* chiefly depend on the level of prevailing irradiation upon incorporation.

Measurements at various levels of nitrogen limitation demonstrated that the ratio of the sum of the PEP carboxylase+PEP carboxykinase activity to the RuBP carboxylase activity is correlated with the level of nitrogen deficiency and that there is a mathematical relation between this ratio and photosynthetic pigments. Nitrogen deficiency appears to induce a real decrease in the activity of RuBP carboxylase.

The above systems enable the alga to survive upon carbon dioxide starving.

This fact corroborates the assumption that the supremacy of *Planktothrix rubescens* is related to its capability of adaptation to the limiting factors most frequent in lakes: low light intensity, thanks to its pigment systems which capture energy from the light spectrum; low carbon dioxide, thanks to the organization of its cell minimizing the loss of this compound and to its carboxylating enzyme whose substrate is bicarbonate rather than free carbon dioxide; poor availability of inorganic nitrogen, thanks to increasing affinity and choice of metabolic pathways. In effect, Smith (1982) demonstrated that some cyanobacterial species have a heterotrophic potential. Uptake and use of amino acids at natural concentrations was validated by axenic cultures of *Planktothrix rubescens.* Furthermore, *Planktothrix rubescens* can successfully compete with bacteria for uptake of amino acids at environmental concentrations (Feuillade et al.1988; Bourdier et al.1989). Alkaline phosphatase allows *Planktothrix rubescens* to utilize the organic forms of phosphorus when water is phosphate-poor (Feuillade et al.1990).

Floatability

Laboratory and field investigations revealed that Relative Gas Vacuolation (RGV) and floatability of blue-green algae depend on the interaction between light and limiting nutrients, which affect the related rates of photosynthesis, growth and synthesis of gas vesicles

In *Planktothrix rubescens*, RGV declines with nitrogen limitations and rises with transitions to inorganic carbon limitations: when *Planktothrix rubescens* is nitrogen-limited, it loses its floatability and sinks in a stratified lake. In other words, the alga approaches a combination of conditions leading to the build-up of proteins and vesicles. In this way, floatability responds negatively to surface conditions (high light intensity, nutrient deficiency) and positively to deep water conditions (low light intensity, abundance of nutrients), tending to bring *Planktothrix rubescens* towards the bottom of the photic zone.

Storage of carbohydrates plays a key role in controlling the floatability of this species (Utkilen et al.1985): when the alga is exposed to high light intensity, it stores carbohydrates and slowly descends to a depth where the amount of carbohydrates is low, so that the gas vesicle content may increase. An external pressure of 0.7 to 1.2 MPa induces the collapse of both isolated and *in situ* vesicles of *Planktothrix rubescens*. This pressure is by far higher than the one needed to break the vesicles in *Microcystis* (Deacon and Walsby 1990).

Trophic Status of Lake Vico

The trophic status of the waters of the Lake is an important indicator of the problems arising in its environment.

In effect, the Lake is the final destination of all the waters flowing into its watershed and its trophic status is highly conditional not only upon land use, but also and above all upon pedo-climatic factors. These factors impact surface runoff and erosion and, thus, the input of nutrients from the watershed to the Lake. In turn, the trophic status of the Lake influences the uses that can be made of its waters, from abstraction to tourism.

Hence, to achieve a harmonious development of the local socio-ecological system, any policy of management of the watershed should take into account its interaction with the Lake, so as to protect its environment and safeguard economic interests.

In the past, numerous authors investigated water quality in Lake Vico (Barbanti et al., 1971; Gelosi et al., 1985; Franzoi, 1997) and, from 1992 to 1993, Finnish researchers conducted a monitoring program (Dyer, 1995).

The results of these studies are summarized below (for details, the reader is referred to the specific study by Franzoi, 1997):

- in the 1968-1970 period, the Lake was oligomesotrophic (7 μg/L of PO_4 and 30 μg/L of NO_3) with good water oxygenation, except for a small portion of its hypolimnion (at a depth of 35-45 m) in late summer;
- in the 1985-1986 period, the content of nutrients remained more or less unaltered but with a marked deoxygenation of the Lake bottom;
- conversely, the monitoring program conducted in the early 1990s showed a sharp increase in lacustrine trophism, with an incipient eutrophic status; this finding was substantiated by the 1993-1995 survey.

Most of the phosphorus originates from the watershed and reaches the Lake in particulate form. Therefore, seeking a methodology which correlates land use with the trophic status of the Lake is critical to: i) singling out anthropogenic activities with the highest impact on the Lake; and ii) using the resulting data to formulate correct land management policies, which are vital to controlling the above phenomenon.

Phosphorus Balance/Budget at the Watershed Scale

Limnologists unanimously point to nitrogen and phosphorus as the primary culprits of eutrophication in a water body (the so-called "limiting factors"), as confirmed in Lake Vico (IRSA, 1980). As P lacks the gaseous component, it can be modeled in a simpler way. Therefore, phosphorus production in a watershed is a practical parameter to understand anthropogenic pressures on one the most important elements of the environment (Leone e Marini, 1993). In other terms, the extremely diversified factors of environmental stress (urban settlements, tourist activities, farming and animal breeding practices, etc.) can be homogenized and objectively compared in a way that would otherwise be impossible.

Diffuse sources of phosphorus are:

- *precipitation:* P directly reaches the Lake through rainfall, which conveys atmospheric particulate containing P compounds;
- *natural organic refuse:* coming from seeds, pollens, leaves, etc. dropping into the Lake and from wild animal excreta (the related P loads are incorporated in the ones of unfarmed land);

- *unfarmed land:* organic P derives from vegetable and animal refuse, while inorganic P comes from the minerals contained in the soil; both reach the Lake by leaching and runoff of storm water, which carries P compounds or particles contained in or eroded from the soil;
- *farmed land:* P comes from farmland treated with chemical and natural fertilizers through mechanisms that are identical to those mentioned above;
- *urban areas:* P is removed by rainwater, which flows on the (generally impervious) portions of urban surfaces exposed to precipitation.

Point sources are:

- *domestic effluents:* P comes from both human metabolism and detergents;
- *industrial effluents:* industrial processes involving the use of P compounds;
- *animal breeding effluents,* related to metabolism.

Among anthropogenic activities potentially inducing organic contamination of the Lake, mention should be made of:

- tourism (residential and occasional);
- agricultural practices (especially hazelnut groves);
- animal breeding (mostly in the wild state).

P input from precipitation was assumed to be equal to 0.3 kg/ha x year, an intermediate value with respect to the one proposed by Reckhow et al. (Reckhow et al., 1980) for volcanic parent rock soils.

For P input from farmed land, reference was made to the average data obtained from a specific analysis made on the local hazelnut groves (Biondi, Gusman and Miligi, 1988): an average value of 85 kg/ha x year of phosphoric anhydride, corresponding to 37.5 kg/ha x year of P was determined. From this value, the "real load" (actual amount of P supposed to reach the Lake) was calculated by relying on the input coefficients suggested in the literature.

To highlight the impact of the different concentrations of P considered so far, the Table below Table gives the 1985 OECD values used to assess the trophism of a water body.

To define land and water management criteria, a target of 12 μg/L was set, in line with the assumption of mesotrophy under undisturbed conditions. This value epitomizes the resilience of the Lake's system to anthropogenic impact and in particular to agriculture, provided that phosphorus input and erosion risk in its most vulnerable areas are mitigated. In other words, this is an operational definition of the sustainability of agricultural development.

Materials and Methods

During the 18 months of study of Lake Vico, monthly sampling surveys were made with a view to identifying its phytoplanktonic population and detecting possible toxic species, namely *Planktothrix rubescens,* belonging to the *Cyanophyceae* class and responsible for the production of the hepatotoxic microcystins.

Surface Sampling

Samples were collected from: 4 stations located at a distance of 50 m off the shore and arranged in the form of a Greek cross; and 1 station located at the centre of the Lake and corresponding to the point of intersection of the Greek cross. Samples were also taken from 10 wells located in the Lake area (Figure 7.2).

Vertical Sampling

Samples were collected vertically by means of a thin-mesh plankton net (mesh size not exceeding 20 μm), model 20MY KC Maskiner og lab.Udstyr. The net was lowered to the desired depth. The depth of this vertical sampling depends on the type of cyanobacteria to be detected. As *Planktothrix rubescens* was distributed along the entire water column, the net was lowered to a considerable depth.

Sampling at Different Depths

Samples along the water column were taken every 5 m, from the surface of the central station to a depth of 30 m.

Physico-Chemical Parameters

Samples were field-analyzed by using a Mettler Toledo probe (model In Lab 781) for conductivity and pH and a Hanna Instruments HI 9143 probe for temperature and dissolved oxygen.

Table 8.1. Classification of fresh water trophic status

	OLIGOTROPHY	MESOTROPHY	EUTROPHY	HYPERTROPHY
P tot mg/L	8	26.7	84.4	750-1200
N tot mg/L	661	753	1875	
Chl a mg/L	1.7	4.7	14.3	100-150
Secchi Disc (m)	9.9	4.2	2.45	0.4-0.8

Table 8.2. Sampling stations' GPS coordinates

Lake Center	N	42 -19.15' -9"
	E	12 -10.17' -2"
1 - Monte Venere	N	42 -19.33' -0"
	E	12 -10.26' -6"
2 - Fogliano	N	42 -19.36' -3"
	E	12 -9.27' -6"
3 - Emissario	N	42 -18.25' -5"
	E	12 -11.44' -7"
4 - Baia Bella Venere	N	42 -19.54' -0"
	E	12 -11.48' -9"

Table 8.3. Time-scheduled selected reaction monitoring condition for detecting microcystins

Compound	Transition (m/z)	Cone voltage (V)	Collision energy (eV)	Retention window (min)
Microcystin-RR	520>135 520>887	35	35 25	0-6
Desmethyl Microcystin-RR	512.5>135	35	35	-
Nodularin (surrogate)	825>135 825>807	45 45	60 40	
Microcystin-YR	523>135 523>910 1045>135	18 18 70	5 5 70	6-9.0
Desmethyl Microcystin-LR	491>135 981>135	20 70	15 60	-
Microcystin-LR	498>135 498>860 995>135	20 20 70	15 40 60	-
Microcystin-LA	910>402.5 909>776	35	25 25	9-13
Trimethacarb (IS)	194>137	25	10	-
Microcystin -LW	1025>1007 1025>873	30	30	-

Sedimentation and Cell Count

Subsamples were preserved in 1% Lugol's solution and analyzed after 24-48 hours via an inverted microscope (Leitz Labovert FS), according to Utermohl (1931) and Lund et al. (1958), using 25 ml sedimentation chambers for phytoplankton identification and cell density estimation.

Identification and Quantification of Toxins

From July 2005 to February 2007, monthly water samples were collected from Lake Vico (5 surface stations and 6 central stations down to a depth of 30 m) and from 10 wells in the Lake area. Throughout the above period, total (intracellular plus extracellular) toxin content was determined. From February 2006, also the extracellular content was separately determined.

To analyze total microcystin content, the samples were frozen and kept at -18°C ±2 to favor cell lysis. After defreezing, the samples were filtered through black-band filters. Then, they were analyzed according to a methodology developed in our laboratories (Bogialli et al., 2006) and based on liquid chromatography combined with tandem mass spectrometry (LC/MS/MS).

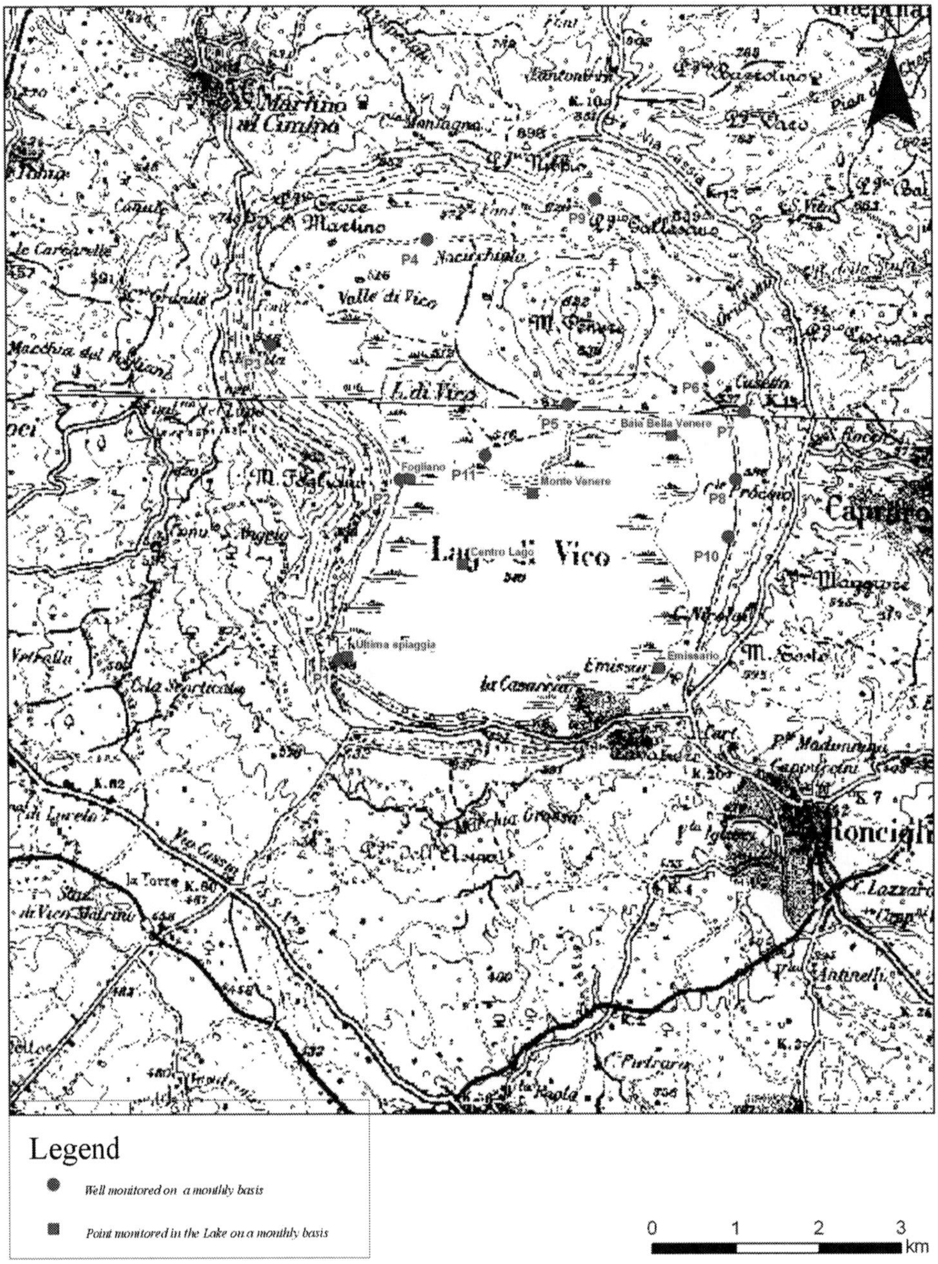

Figure 8.2. Location of sampling stations in Lake Vico and of wells.

For analyzing extracellular content only, the samples were stored in a refrigerator at a temperature of 4 C° ± 4 and extracted within 24 hours with the same procedure. In this case, however, the filters had a smaller pore diameter (0.45 μm) in order to retain the cells.

To carry out the extraction procedure correctly, use was made of a surrogate internal standard. This standard (recommended by the Environmental Protection Agency - EPA) makes possible to detect and offset errors, if any, in sample preparation. Nodularin was chosen as the surrogate standard (SS); this cyanotoxin, occurring in brackish environments and structurally similar to microcystins, was not present in the investigated waters.

In brief, sample preparation consists in extracting the investigated toxins via Solid Phase Extraction (SPE), using Carbograph (a graphitized carbon black) as adsorbent (surface area: 200 m^2/g). 0.5 L of water were contaminated with 50 ng/L of a nodularin (SS) solution and passed through the SPE cartridge at a flow rate of 20 mL/min.

After washing, the cartridge was back-flushed and analytes were eluted with 1 mL of methanol and 4 mL of methylene chloride/methanol (80:20, v/v), both acidified with 10 mmol/L of TFA and gathered in a conical bottom test tube. The eluting phase was gravity percolated to achieve good extraction efficiency.

Before removing the solvents in a thermostated water bath at 50°C under a mild nitrogen flow, 15 ng of internal standard (an obsolete insecticide) were added to the eluate. Solvent evaporation was continued until attaining a volume of 50 μL. The residue was diluted with 200 μL of a water/acetonitrile (70:30 v/v) solution acidified with 10 mM/L of formic acid. 25 μL of this solution were injected into the LC/MS/MS equipment.

Determination Via LC-MS-TANDEM

To separate and determine the investigated compounds, use is made of a liquid chromatography binary pump (Waters 600E), coupled via an electrospray interface with a triple quadruple mass spectrometer (Quattro Micro, Micromass) operating in the "Multiple Reaction Monitoring" (MRM) mode. Table 7.3 displays the instrumental parameters for each cystin.

Analytes were chromatographically separated via a reverse phase column (Altima, 250 mm x 4.6 mm i.d.), packed with chemically modified 5-μm silica particles, and a guard column with the same composition.

To separate analytes chromatographically, the following mobile phases were used:

- *phase A:* acetonitrile 10 mM formic acid
- *phase B:* water 10 mM formic acid

The profile of the elution gradient used for microcystins and nodularin (as a percentage of phase A, where t is the time expressed in minutes) was as follows: t_0, A= 35%; t_5, A= 45%; t_6, A= 57%; t_{11}, A= 67%; t_{12}, A= 100%; t_{15}, A= 100%; t_{16}, A= 35%, t_{25}, A= 35%.

During the Lake Vico monitoring survey, demethylated variants of MC-RR and MC- LR were detected. The conditions for detecting these toxins had been previously optimized as part of a project of monitoring of Lake Albano. As these compounds are not commercially available, no standard can be used for quantitative analysis. However, since their only difference with respect to the corresponding methylated microcystin is the lack of one methyl group, their molar response factor to mass spectrometer is likely to be practically the same. This is why the demethylated forms of MC-RR and MC-LR were quantified vs. the respective analogous methylated forms.

Performance of the Method

The quantification limits of the method ranged from 2 ng/L (MC-RR) to 9 ng/L (MC-YR). In other terms, the method was capable of detecting microcystins in the Lake and well waters at concentrations of a few ppt, i.e. 100 times lower than the WHO guideline value.

Accuracy (trueness and precision) of the method, expressed as relative standard deviation (RSD), was assessed on the surrogate. Accuracy and replicability of the method was estimated on the basis of the inter-day average values recorded for each batch (16 samples); the resulting recovery value was equal to 85% with an RSD always below 28%.

In some cases, the total nitrogen measured in the wells hit very high values (up to 70 mg/L, Figure 7.9, well 3, September 2006).

Total phosphorus in the same wells reached even 3 mg/L (Figure 7.10, well 1, September 2006).

The values of total microcystins (when present) in the wells varied from less than 0.004 μg/L to 0.123 μg/L (Figure 7.6, well 1, September 2005).

The average of orthophosphates measured in the column of the Lake during 12 months (March 2006-February 2007) was 62.2 μg/L, with peaks of 369 μg/L (Figure 7.8, -25 m, September 2006).

Total nitrogen in the Lake peaked at 1000 μg/l (-25 m, February 2006). Total phosphorus hit 115 μg/l in the column (-15 m, January 2007).

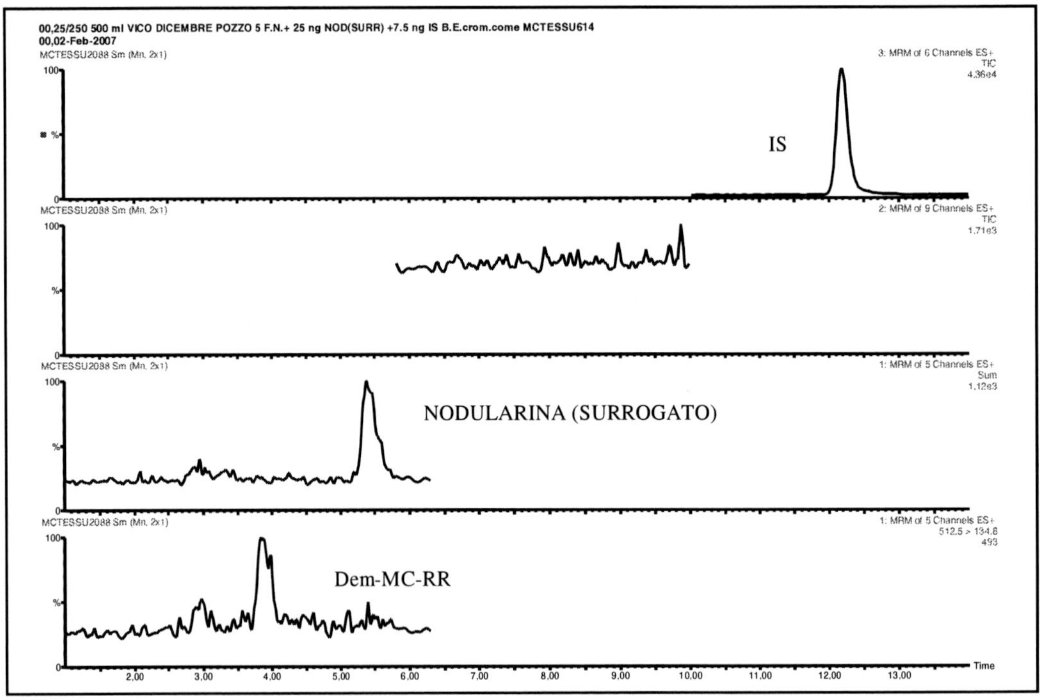

Figure 8.3. Chromatogram of a water sample from a well (well 5) collected in December 2006. The sample is contaminated with dem-MC-RR (~ 4 ng/L). Both the surrogate and internal standard are included.

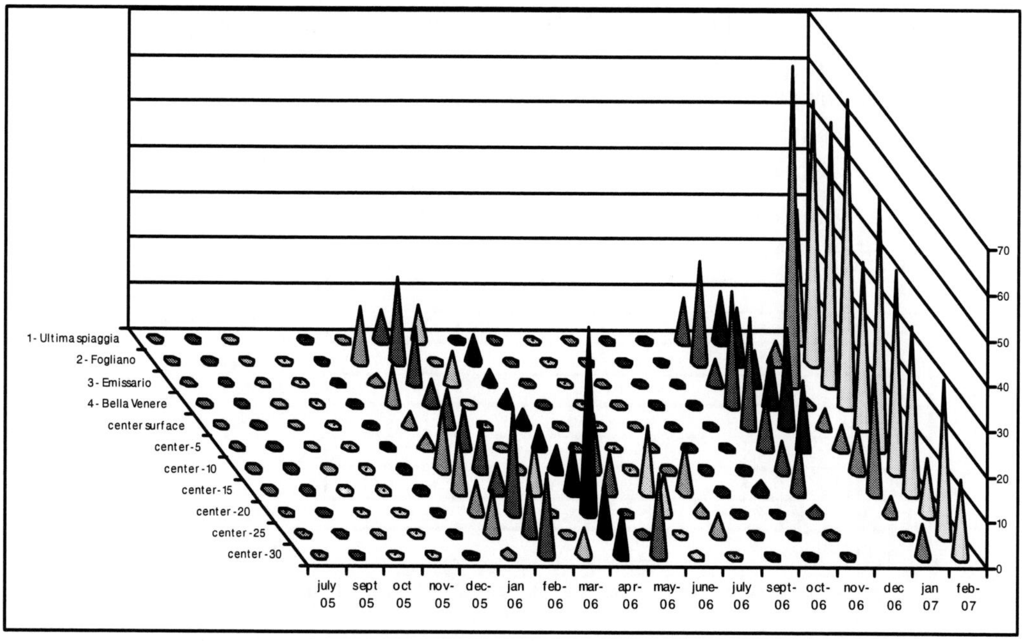

Figure 8.4. Planktothrix rubescens (10^6 cells/l) in Lake Vico.

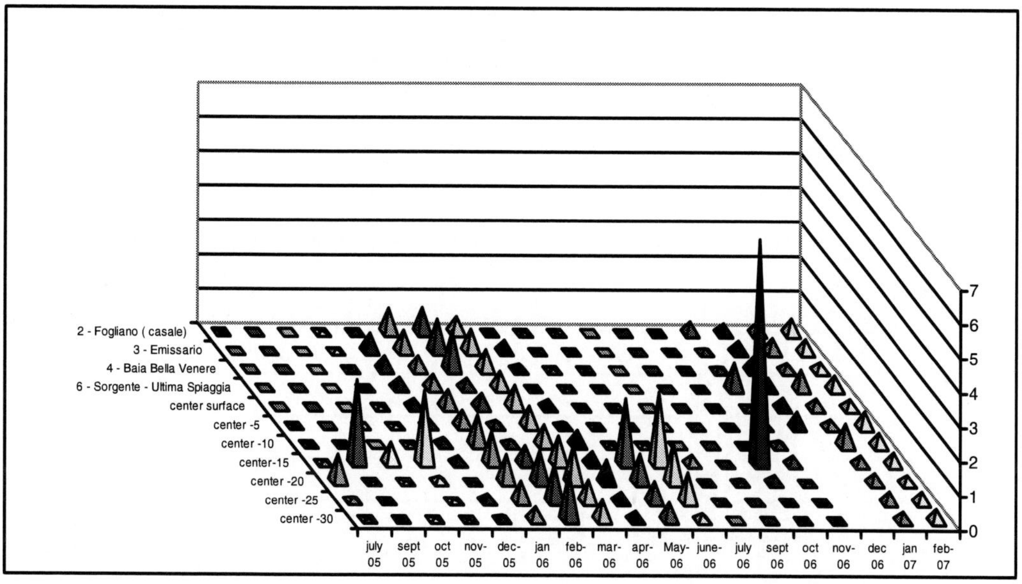

Figure 8.5. Microcystin content of the Lake (mg/l).

The Lake experienced thermal stratification from April to November, with its thermocline at 10-20 m. The densities of *P. rubescens* had their yearly peak in winter, with presence at the surface from January to April and surface blooms of up to 72.5 million cells/liter, i.e. above the Italian limit of 5,000 cells/L (Figure 7.4, "Emissario", January 2007). In the 2006-2007 winter the algal population grew, giving rise to a surface scum from January 07, especially in the bay of the "Ultima spiaggia" station. The scum had a density of 59.5

billion cells/liter, with an intracellular microcystin level of 55.3 μg/L (Italian limit: 0.84 μg/l). The surface scum reappeared in most of the following two months and was visually recorded by the Vico park guards. In the blooming period, the calculated mean TN/TP ratio was 15.

In the Lake, total microcystins varied from 0.002 to 6.5 μg/L (Figure 7.5, October 2006, -15 m).

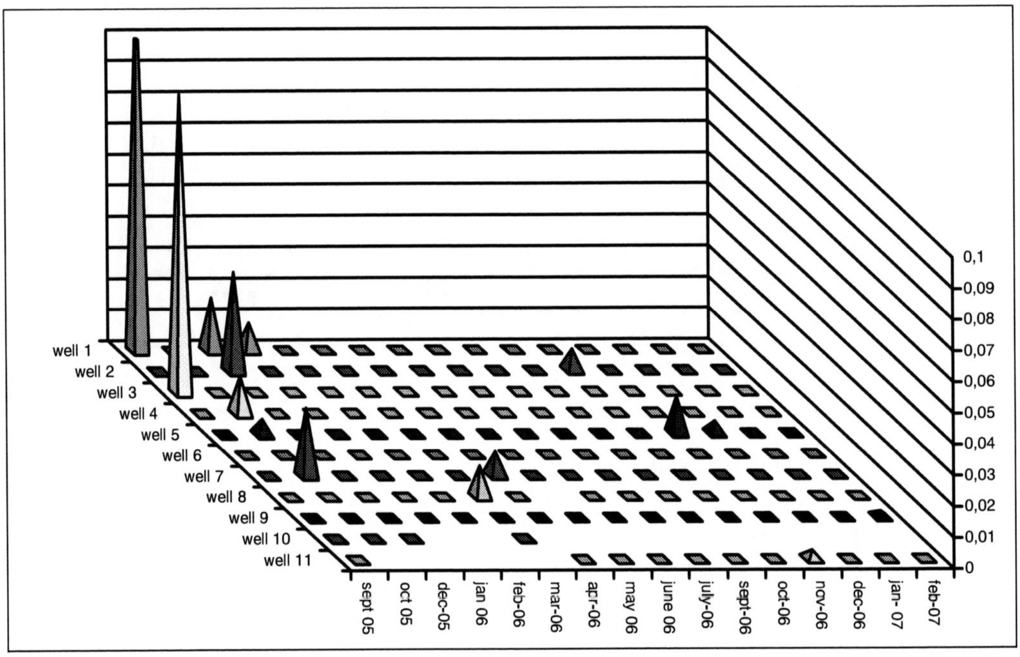

Figure 8.6. Microcystin content in wells (mg/l).

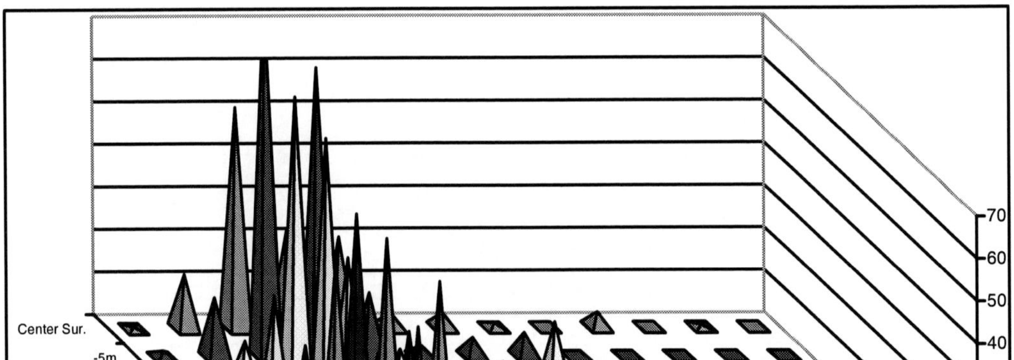

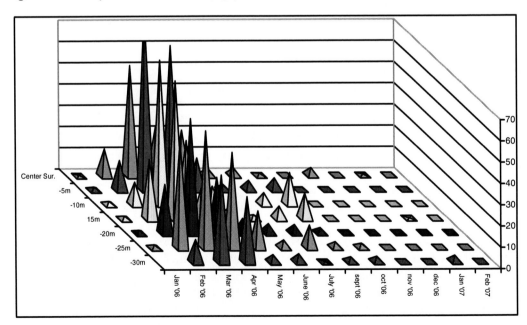

Figure 8.7. N/P ratio in Lake Vico (column).

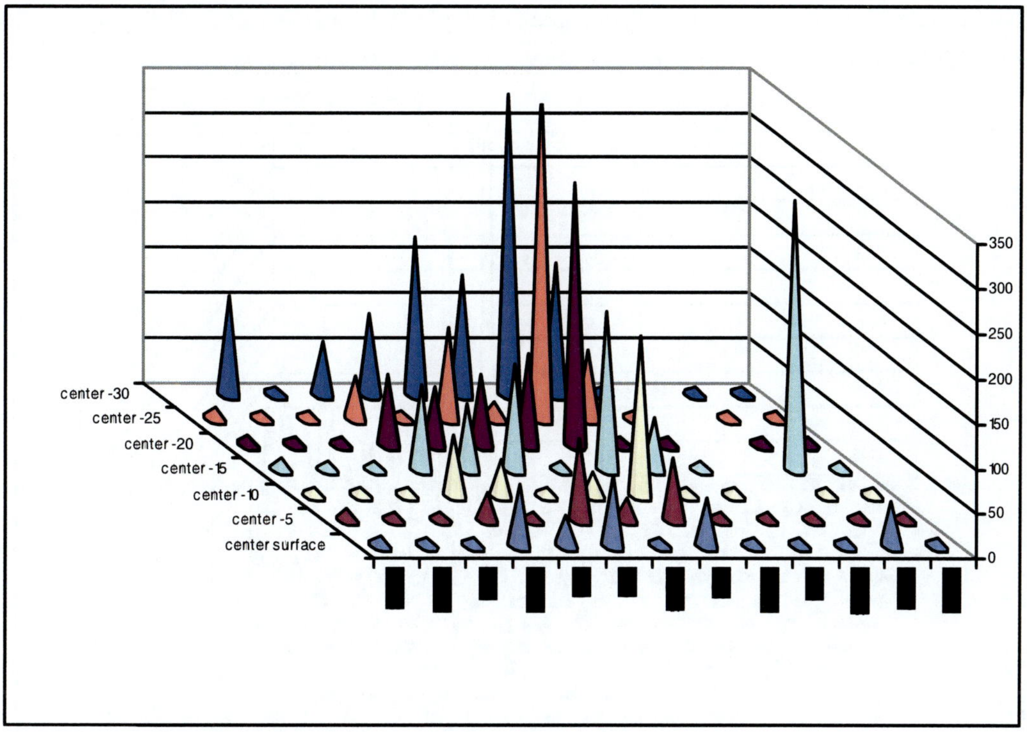

Figure 8.8. Orthophosphates (PO_4) in the Lake (column).

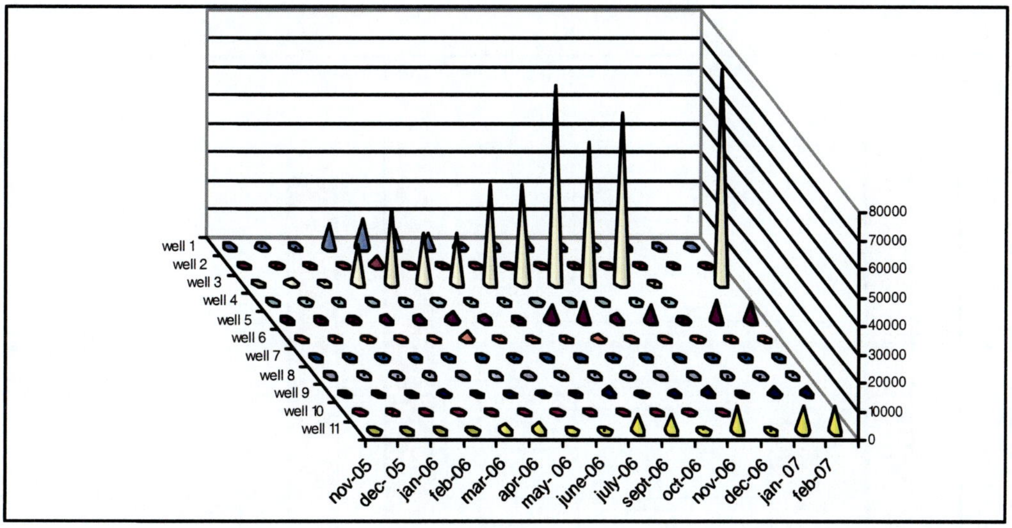

Figure 8.9. Total nitrogen (μg/l) in wells.

Analyses revealed that wells begun to show microcystin contamination when the fluctuating *P. rubescens* populations reached -15 m in the Lake, since the upper part of the lake cuvette is made up of more permeable soil layers with respect to the lavic ones in its lower part.

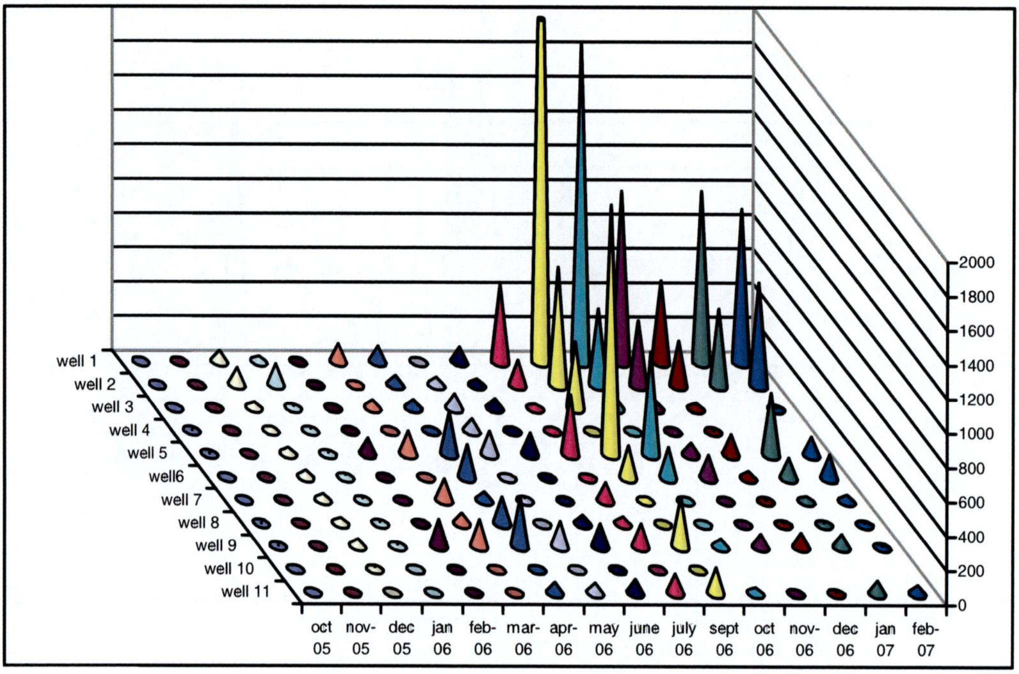

Figure 8.10. Total phosphorus (μg/l) in wells.

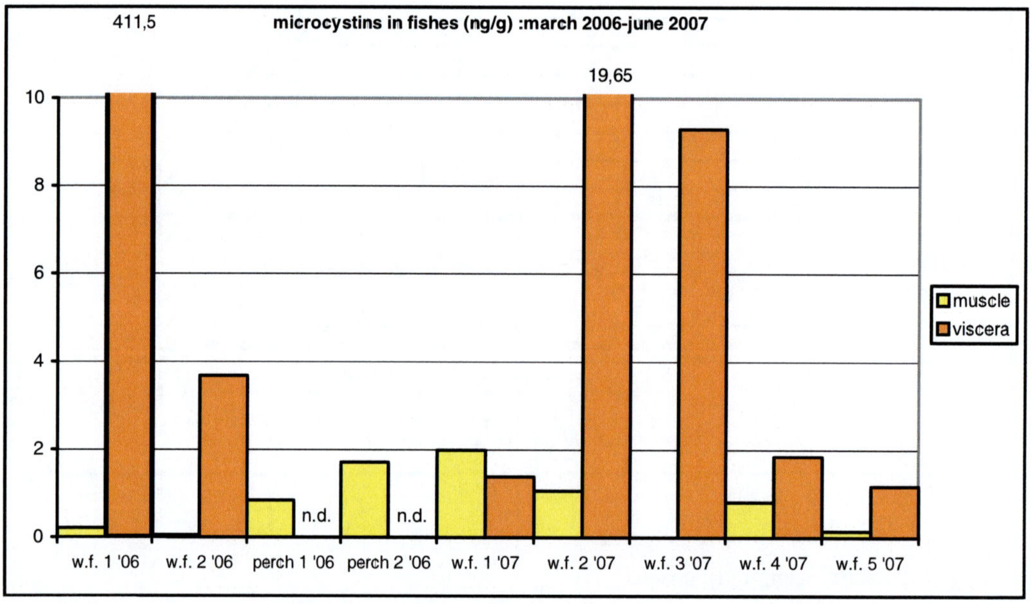

Figure 8.11. Microcystin content (ng/g) in fish samples collected from Lake Vico in March 2006.

Samples of fish species of high commercial value were collected and tested to determine total microcystin content in their viscera and muscular tissue (ELISA immunological method) in March 2006 and June 2007 (Figure 7.11). Toxin content in fish muscle reached 1.99 ng/g, confirming that microcystins can trigger biological magnification in fish tissues.

FINAL CONSIDERATIONS

The following considerations can be drawn from the analysis of the data collected by both the Department of Chemistry, University of Rome "La Sapienza" and the Department of Geological Sciences of the University of Rome "Roma Tre".

1) The rocks that are exposed in the study area chiefly consist of volcanites, originating from the activity of the Cimino and Vico volcanic complex and including pozzolans, lavas, laccoliths and intracalderic sediments. These materials, which have fairly good permeability values, host groundwater and overlie terrigenous sediments of Cenozoic and Plio-Pleistocene age with poor or zero permeability values. These sedimentary terrains lie at an elevation of about 100 m to about 300 m above sea level in the Vetralla area (outcrops) and to about 500 m above sea level under Mt. Cimino and in the Mt. Calvello area (outcrops). A lacustrine sediment cover with a thickness of some tens of meters and poor permeability is located at the bottom of Lake Vico. This cover has an impact on the Lake's water level.
2) Lake Vico's hydrogeological system represents one of the most elevated portions of the regional aquifer extending in the entire peri-Tyrrhenian volcanic domain.
3) In particular, the Lake is recharged by rainfall, runoff in the internal area of the caldera and groundwater in its north-eastern sector. Water losses from the Lake are due to: evaporation; abstraction for drinking water uses; overflow into its outlet stream; groundwater flow mainly towards south, south-east and south-west, as well as towards the bottom of the Lake via drainage (albeit poor) of the lacustrine bottom sediments. Most the groundwater flows through the volcanic complex underlying the Lake, from Mt. Cimino towards south-west.
4) Inputs of eutrophicating substances in terms of equivalent nitrogen and phosphorus were assessed on the basis of actual land uses and inventories of releases made available by the relevant authorities. The assessment showed that Lake Vico receives on average waters with a maximum potential nitrogen concentration of 0.98 mg/l. In particular, the Lake receives on average a maximum concentration of nitrogen of: i) 1.451 mg/l from groundwater; ii) of 0.358 mg/l from runoff water coming from its dominantly forested portion; and iii) over 2 mg/l from the other sectors inside the caldera. Therefore, *in terms of concentration*, the input of eutrophicating substances into Lake Vico is maximum in: i) late winter-spring periods, when rainfall is abundant and fertilizers are applied to the soil; and ii) summer period, when the absence of rainfall and runoff does not permit the dilution of the eutrophicating substances carried by groundwater. Conversely, if eutrophicating substances are considered *in terms of weight* (kg/month), then the highest inputs into the Lake mostly occur in the months during which fertilizers are applied to the soil.

5) Experimental analyses conducted on groundwater and lacustrine water confirmed that the groundwater recharging the Lake is highly polluted, whereas the water of the Lake appears to be diluted with respect to its hydrogeological system. The load of eutrophicating substances is higher than the maximum potentially expected one. This fact clearly implies that inputs of nutrients into the area are well above the level specified in the Code of Good Agricultural Practice (approved by Ministerial Decree of Apr. 19, 1994). Hence, if this trend is not reverted, the hydrogeological system will become increasingly oversaturated in polluting substances.
6) The study revealed that the *Planktothrix rubescens* species has become firmly settled in the Lake's phytoplankton and that its density typically increases in winter, causing blooms of varying extent and surface scums (2006-2007 winter). The total concentration of these toxins in the Lake have often exceeded the Italian limit value in bathing waters (0.84 μg/l). By travelling through the geological strata neighboring the water body, these toxins have largely migrated into groundwaters from which drinking water is abstracted. Their values, measured in wells, have often exceeded the international precautionary values recommended to prevent the risk of carcinogenesis (0.01 μg/l, Hernandez et al., 2000; Ueno et al., 1996). However, their values as a whole are variable and unforeseeable. Consequently, during the period in which this study was conducted, no reliable prediction could be made as to the periodical risk posed by these toxins. Although no quantitative prediction can be made, microcystins were actually detected in the Lake's waters, thereby making them unsuitable for human consumption (Regulation (EC) No 178/2002, Italian Law Decrees 31/2001 and 27/2002).
7) Water with microcystins is withdrawn from the Lake and fed to the potable water systems of the Municipalities of Ronciglione and Caprarola. Given the periodically high intensity of *P. rubescens* blooms, we recommend the use, in both systems, of activated carbon filters throughout the year, in order to protect the health of the local population. The filters will remove both algal biomass and toxins.
8) Contamination of aquatic organisms and of the entire fish fauna with microcystins is instead associated with bloom intensity. As a result, during blooms and one month after their disappearance, we recommend the introduction of a fishing ban and periodical monitoring of microcystin content in the fish fauna to be marketed.

Remediation

P. rubescens is a species typical of open lakes, where it shows outstanding resistance to water mixing and cold temperature. However, it always needs a high supply of nitrogen, as it is unable to produce it. Its phosphorus requirements (with clear preference for orthophosphate) are instead much lower.

Previous studies on the lakes of Latium (Bruno et al., 2004; Messineo et al., 2006) proposed interesting remediation models based on the change of the environmental conditions which favor the dominance of *P. rubescens.* In particular, a project of diversion of a municipal sewerage system in Lake Nemi halved the average surface orthophosphate (from 20 μg/l to 10 μg/l) in 6 months (September 2002 – February 2003), causing the toxic populations to lose their dominance in the phytoplankton in 5 months (December 2002 – April 2003).

Remediation of Lake Vico's waters may follow the above example and be focused on minimization of orthophosphate levels, whose yearly average measured in the water column during the study is very high (62,2 μg/l). This objective may be pursued by changing agronomical practices and issuing stricter regulations on waste water releases. The need thus arises for involving the relevant authorities in efforts to raise eutrophication awareness and encourage farmers to adopt methods and practices cutting down the amounts of eutrophicating substances.

REFERENCES

Adams, D.G. (1997). Cyanobacteria. In: J.A.Shapiro and M.Dworkin eds. Bacteria as multicellular organisms: pp. 109-148. New York, Oxford University Press.

Amministrazione Provinciale di Siena (1992-1994) – Il piano acquedottistico della Provincia di Siena.

Annadotter, H; Cronberg, G; Lawton, L.A; Hansson, H.B; Gothe, U., Skulberg, O.M. (2001). An extensive outbreak of gastroenteritis associated with the toxic cyanobacterium Planktothrix aghardii (Oscillatoriales, Cyanophyceae) in Scania, South Sweden. In: Ingrid Chorus ed. Springer Verlag. Cyanotoxins - occurrence, causes, consequences: pp. 200-208. Berlin,.

ARSIA LIVORNO: "Progetto vulnerabilità da nitrati". La vulnerabilità da nitrati della pianura costiera. Provincia di Livorno – settore 10 -"difesa del suolo"

ARSIA LIVORNO: "Progetto vulnerabilità da nitrati". Studio degli apporti azotati derivanti da attività agricola che danno luogo a lisciviazione nel territorio di Vada-S.Pietro in Palazzi

Azevedo, S.M.F.O.; Carmichael, W.W.; Jochimsen, E.M.; Rinehart, K.L.; Lau, S.; Shaw, G..R. and Eaglesham, G.K. (2002). Human intoxication by microcystins during renal dialysis treatment in Caruaru - Brazil. *Toxicology*, 181, pp. 441-446.

Barbanti, L.; Carolla, A. and Libera, V. (1968). Carta batimetria del Lago di Vico. In: Quaderni dell'Istituto di Ricerca sulle Acque. CNR, 1974. Indagini limnologiche sui Laghi di Bolsena, Bracciano, Vico e Trasimeno, 17.

Behm, D. (2003). Coroner cites algae in teen's death. In: *Milwakee Journal Sentinel*, Milwakee.

Benedini, M. (1993). Esigenze dell'agricoltura e competizione con le industrie e gli insediamenti civili. Agricoltura e Innovazione, 26-27.

Benvenuti, L.; Bodo, G.; Cingolani, L.; Proietti, L. Attività di Arpa Umbria per la gestione e l'utilizzazione agronomica dei reflui.

Bogialli, S.;Bruno, M.; Curini, R.; Di Corcia, A.; Fanali, C.; Laganà, A. (2006). Monitoring Algal Toxins in Lake Water by Liquid Chromatography Tandem Mass Spectrometry. Environ Sci Technol, 40, pp. 2917-2923.

Bold, H.C. and Wynne, M.J. (1985). Introduction to the algae. Prentice-Hall, Inc., Englewood Cliffs, N.J.

Boni, C.; Bono, P.; Capelli, G. (1986). Schema idrogeologico dell'Italia centrale. *Mem. Soc. Geol. It.*, 35, pp. 991-1012.

Boni, C.; Bono, P.; Capelli, G. (1988). Carta Idrogeologica del Territorio della Regione Lazio Scala 1:250.000. Regione Lazio, Assessorato alla Programmazione, Ufficio Parchi e Riserve Naturali, Università degli Studi di Roma "La Sapienza" Dipartimento Scienze della Terra.

Bouaicha, N.; Maatouk, I.; Plessis, M.J.; Perin, F. (2005). Genotoxic potential of microcystin-LR and nodularin in vitro in primary cultured rat hepatocytes and in vivo in rat liver. *Environ. Toxicol*, 20, pp. 341-347.

Bourdier et al. (1989). Amino acids incorporation by a natural population of Oscillatoria rubescens: a microautoradiographic study. *FEMS Microbiol. Ecol.*, 62, pp.185-190.

Bowling, L.C. and Baker, P.D. (1996). Mayor cyanobacterial bloom in the Barwon-Darling River, Australia, 1991 and underlying limnological conditions. *Mar. Freshwat. Res.*, 47, pp. 643-657.

Bruno, M.; Messineo, V.; Mattei, D. and Melchiorre, S. (2004). Dinamica di specie algali tossiche nei laghi di Albano e di Nemi. Rapporti Istisan 04/32.

Bruno, R. and Raspa, G. (1994). La Pratica della Geostatistica non Lineare. In: Guerini Ed. Il Trattamento dei Dati Spaziali: p.170.

Capelli, G.; Mazza, R. and Gazzetti, C. (2005) . Strumenti e strategie per la tutela e l'uso compatibile della risorsa idrica nel Lazio. Gli acquiferi vulcanici. Quaderni di Tecniche di Protezione Ambientale n°78, Pitagora Bologna.

Carmichael, W.W. (1994). The toxins of cyanobacteria. *Sci. Am.*, 270, pp. 64-70.

Carmichael, W.W. (2000). Health effects of toxin producing cyanobacteria: "the cyanoHABS". *Hum. Ecol. Risk Assess.*, 7, pp. 1393-1407.

Carollo, A., Barbanti, L., Merletti, M., Chiaudani, G., Ferrari, I., Nocenti, A. M., Bonomi, G. Roggia, D., Tonolli L. (1974). Indagini limnologiche sui Laghi di Bolsena, Bracciano, Vico e Trasimeno. CNR, Quaderni dell'Istituto di Ricerca sulle Acque, 17.

Carollo, A.; Barbanti, L.; Bonomi, G.; Chiaudani, G.; Ferrari, I. (1971). Limnologia ed ecologia dei Laghi di Bolsena, Bracciano, Trasimeno e Vico: situazione attuale e prevedibili conseguenze derivanti da una loro utilizzazione multipla. Rapporto finale. Istituto Italiano di Idrobiologia, Pallanza.

Cassa per il Mezzogiorno (1983). Progetti speciali per gli schemi idrici del nel Mezzogiorno – Idrogeologia dell'Italia Centro-meridionale" Quaderni della Cassa del Mezzogiorno, Roma.

Chiaudani, G, Gerletti, M, Marchetti, R, Provini, A, Vighi, M. (1978). Il problema dell'eutrofizzazione in Italia. Quaderni Irsa 42.

Chilès, J.P. and Delfiner, P. (1999). Geostatistics: modelling of spatial uncertainty. Wiley series in Probability and Statistics, 695 pp.

Chorus, I. and Bartram, J. (1999). Toxic cyanobacteria in water. World Health Organisation, London, E and FN Spon..

Christiansen, G., Fastner, J., Erhard, T., Borner, T., Dittmann, E. (2003). Microcystin biosynthesis in Planktothrix: genes, evolution and manipulation. *J. Bacteriol,* 185, pp.564-572.

Cimarelli, C., and de Rita, D. (2006). Relatively rapid emplacement of dome-forming magma inferred from strain analyses: The case of acid Latian dome complexes (Central Italy). *Journal of volcanology and geothermal research*, pp. 106-116.

Codd, G.A., Metcalf, J.S., Beattie, K.A. (1999). Retention of Microcystis aeruginosa and microcystin by salad lettuce (Latuca sativa) after spray irrigation with water containing cyanobacteria. *Toxicon*, 37, pp. 1181-1185.

Dawson, R.M. (1998). The toxicology of microcystins. Toxicon, 36, pp. 953-962.

de Rita, D., Bertagnini, A. and Landi, P. (1993b). Itinerario n°13. Da Roma a Civita di Bagnoregio via Viterbo. Società Geologica Italiana, Guide Geologiche Regionali, 5, pp. 301-322.

de Rita, D., Di Filippo, M. and Sposata, A. (1993). Carta geologica del Complesso Vulcanico Sabatino. In Michele Di Filippo (Eds.)"Sabatini Volcanic Complex" (Carta fuori testo alla scala 1:50.000). Quaderni de"La Ricerca Scientifica", CNR 114,.

de Rita, D., Faccenna, C., Rosa, C. and Zarlenga, F. (1993c). Itinerario n°12. Da Civitavecchia a Torrita Tiberina. Società Geologica Italiana, Guide Geologiche Regionali, 5, pp. 285-300.

Deacon and Walsby (1990). Gas vesicle formation in the dark, and in light of different irradiances, by the cyanobacterium Microcystis sp. *Br. Phycol. J.,* 25, pp. 133-139.

Dragoni, W.; Lotti, F.; Piscopo, V.; Sibi, A. (2002). Bilancio Idrogeologico del Lago di Vico (Lazio-Italia). International Conference "Residence Times in Lakes: Science, Management, Education, 29 settembre - 3 ottobre 2002, Bolsena (Viterbo –Italia).

Dyer, M. (1995). The water quality at Lago di Vico during 1992-1993. Science of the total environment, 171, no 1-3, pp. 77-83.

Edmondson (1991).The uses of ecology, Lake Washington and beyond. Seattle, University of Washington Press,.

ENEL (1994). Carta del letto delle Vulcaniti. Joint Venture ENEL-AGIP, unpublished document.

ENEL VDAG-URM (1994). Aggiornamento delle caratteristiche geologiche di superficie e profonde del Lazio settentrionale. ENEL, Università degli Studi "La Sapienza" Dipartimento di Scienze della Terra, 110 pp., unpublished document.

Falconer, I.R. (1994). Health problems from exposure to cyanobacteria and proposed safety guidelines for drinking and recreational water. In: Codd, G.A., Jefferies, T.M., Keevil, C.W. and Potter, E., Eds. Detection Methods for Cyanobacterial Toxins, The Royal Society for Chemistry, Cambridge, pp. 3–10.

Falconer, I.R. (2004). Cyanobacterial toxins of drinking water supplies: cylindrospermopsins and microcystins. Boca Raton: CRC Press.

Feuillade and Davies (1994). Studies on Lake Nantua. Seasonal variations and long-term trends in phytoplankton pigments. Arch. Hydrobiol. *Beih. Ergebn. Limnol.*, 41, pp. 95-111.

Feuillade and Feuillade (1981). Le métabolisme photosynthétique d' Oscillatoria rubescens D.C.(Cyanophyceae). I-Carboxylations initiales. *Arch.Hydrobiol.*, 90, pp. 410-426.

Feuillade (1972). Croissance d' Oscillatoria rubescens et variations quantitatives de la chlorophylle et des différents caroténoïdes en fonction de l'éclairement. *Ann Hydrobiol.,* 3, pp. 21-31.

Feuillade et al. (1988). Amino acids uptake by a natural population of Oscillatoria rubescens in relation to uptake by bacterioplankton. *Arch. Hydrobiol.*, 113, pp. 345-358.

Feuillade et al. (1990). Alkaline phosphatase activities fluctuations and associated factors, in an eutrophic lake dominated by Oscillatoria rubescens. Hydrobiologia, 207, pp. 233-240.

Fleming, L.E., Rivero, C., Burns, J., Williams, C., Bean, J.A., Shea, K.A., and Stinn, J. (2002) Blue green algal (cyanobacterial) toxins, surface drinking water, and liver cancer in Florida. Harmful Algae 157.

Francaviglia, R, Donatelli, M, Stöckle, C, Marchetti, A. Applicazione del sistema Arcview-Cropsyst nella valutazione della percolazione di acqua e della lisciviazione di nitrati.

Francis, G. (1978). Poisonous Australian lake. *Nature,* 18, pp. 11-12.

Franzoi, P., F. Scialanca and G. Castaldelli. (1997). Lago di Vico (Italia Centrale): analisi delle principali variabili fisiche e chimiche delle acque in relazione all'evoluzione trofica. Atti 8° Congresso Naz. S.It.E., Parma, 10-12 settembre 1997. S.It.E./Atti, 18, pp. 159-160.

Freitas De Magalhaes, V, Moraes Soares, R, Azevedo, S. (2001). Microcystin contamination in fish from the Jacarepagua Lagoon (Rio de Janeiro, Brazil): Ecological implication and human health risk. *Toxicon*, 39, pp. 1077-1085.

Garnier (1974). Influence de la température sur l'accumulation, le renouvellement et l'efficacité photosynthétique des pigments d' Oscillatoria subbrevis Schmidle (Cyanophyceae)-Physiol. Veg., 2, pp. 273-318.

Gazzetti, C. and Ismail, H. (2000). Un approccio per la gestione delle risorse idriche sotterranee. Atti del convegno: La Gestione delle risorse Territoriali: il Supporto allo Sviluppo e le Interazioni con l'Ambiente. Tirana 17-18 ottobre 2000.

Gelosi, E. et al. (1985). Physio-chemical and biological characteristics of Lake Vico, Central Italy. Preliminary results. International Symposium Environ. Biogeochemic., 42.

Giardini L. (1982). Agronomia generale, Bologna: Patron.

Hernandez, M., Macia, M., Padilla, C., Del Campo, F. (2000). Modulation of human polymorphonuclear leukocyte adherence by cyanopeptide toxins. Environ. Res. Section A, 84, pp. 64-68.

Kaebernick, M. and Neilan, B.A. (2001). Ecological and molecular investigations of cyanotoxin production. FEMS Microbiology Ecology, 35.

Konopka (1982 a). Physiological ecology of a metalimnic Oscillatoria rubescens population.-Limnol.Oceanogr., 27, pp. 1154-1161.

Konopka (1982 b). Buoyancy regulation and vertical migration by Oscillatoria rubescens in Crooked Lake, Indiana.*Br. Phytol. J.,* 17, pp. 427-442.

La Torre, P., Nannini, R. and Sollevanti, F. (1981). Geothermal exploration in Central Italy: geophysical surveys in Cimini range, area 43°. Meeting European Association of Exploration Geophysicists, Venezia Lido, 26-29 maggio 1981.

Leone, A. and R.Marini, (1993). Assessment and Mitigation of the Effects of Land Use in a Lake Basin (Lake Vico in Central Italy). *Journal of Environmental Management*, 39, pp. 39-50.

Locardi, E. (1965). Tipi di ignimbriti del Vulcano di Vico. Atti Soc. Tosc. Sc. Nat., 72, ser. A, pp. 55-173.

Lund, J.W.G., Kipling, C., Le Cren, E. (1958). The inverted microscope method of estimating algal numbers and the statistical basis of estimations by counting. *Hydrobiology* 11, pp.143–170.

Mattias, P. P. and Ventriglia, U. (1970). La Regione Vulcanica dei Monti Sabatini e Cimini. *Mem. Soc. Geol. It.*, 9, pp. 331-384.

Messineo V., Mattei D., Melchiorre S., Salvatore G., Bogialli S., Salzano R., Mazza R., Capelli G., Bruno M. (2006). Microcystin diversity in a Planktothrix rubescens population from Lake Albano (Central Italy). Toxicon 48, pp. 160-174.

Michelle, A., Ferree and Robert, D. (2001). Shannon Evaluation of a second derivative uv/visibile spectroscopy technique for nitrate and total nitrogen analysis of wastewater *samples. Wat. Res.* 35 (1), pp. 327-332

Ministero dell'Agricoltura (1994). Codice di buona pratica agricola. Decreto Ministeriale del 19/04/1994. Gazzetta Ufficiale Serie Generale n. 102 del 04/05/99.

Murphy, J., Riley, J.P. (1962). A modified single solution method for the determination of phosphate in natural waters. Analytica Chim. Acta, 27, pp. 31-36.

Mwaura, F., Koyo, A.O., Zech, B. (2004). Cyanobacterial blooms and the presence of cyanotoxins in small high altitude tropical headwater reservoirs in Kenya. J. Wat. and Health, 2(1), pp. 49-57.

Namikoshi, M., Rinehart, K.L., Dahlem, A.M., Beasley, V.R., Carmichael, W.W. (1989). Total synthesis of Adda, the unique C20 amino acid of cyanobacterial hepatotoxins. Tetrahedron Lett., 30, pp. 4349-4352.

OECD (1995). Eutrophication of waters. Monitoring, assessment and control. Paris: Organisation for Economic Co-Operation and Development, 154 pp.

Ottaviani M. and Bonadonna L. (2000). Metodi analitici per le acque destinate al consumo umano. Rapporti ISTISAN 00/14, parte 2, Roma: Istituto Superiore di Sanità.

Pilotto, L.S., Douglas, R.M., Burch, M.D., Cameron, S., Beers, M., Rouch, G.J., Robinson, P., Kirk, M., Cowie, C.T., Hardiman, S., Moore, C., Attewell, R.G. (1997). Health effects of exposure to cyanobacteria (blue-green algae) during recreational water-related activities. Australian and New Zealand. *J. of Public Health*, 21(6), pp. 562-566.

Raspa, G. Il ruolo della Geostatistica nella modellizzazione ambientale. In "Aspetti Applicativi delle tecniche geostatistiche alle acque sotterranee", Quad. Ist. Ric. Acque, 114 Roma (Pubblicazione n.2035 del Gndci-Cnr).

Raspa, G., Bruno R. (1993). Integration between Geostatistical Methodologies and GIS Environment Geo-data: the External Drift. Conference Proceedings of Fourth European and Exhibition on Geographical Information Systems, Genova, pp. 1067 -1075.

Raspa, G., Bruno, R., Tucci, M. (1997). Reconstruction of rainfall field by combining ground raingauges data with radar maps using external drift method. Geostatics Wollongong '96, Kluver Academic Publisher.

Regione Campania (2003) – Piano Di Utilizzazione Agronomica Dei Liquami Zootecnici, D.G.R. - Campania n. 2382 del 25/7/03 in attuazione della DGR-Campania n. del 610 del 14 febbraio 2003.

Regione Lazio (1997) – Studi e progettazioni per l'aggiornamento del P.R.G.A. del Lazio settentrionale. Attività di prima fase. Regione Lazio Assessorato Opere e Reti Di Servizi Mobilità, Domanda Civile.

Regione Lazio (1998) – Studi e progettazioni per l'aggiornamento del P.R.G.A. del Lazio settentrionale. Attività di prima fase. Regione Lazio Assessorato Opere e Reti Di Servizi Mobilità, Domanda Industriale.

Reynolds (1987). Cyanobacterial waterblooms. *Advances in Botanical Research,* 13, pp. 67-143.

Ripka et al (1979). Generic assignments, strains Histories and properties of pure cultures of cyanobacteria. *J. Gen. Microbiol.*, 3, pp. 1-61.

Sempere, A., Oliver, J. and Ramos, C. (1993). Simple determination of nitrate in soils by second-derivative spectroscopy. *Journal of Soil Science* 44, pp. 633-639.

Sivonen, K. and Jones, G. (1999). Toxic Cyanobacteria in Water. In: I. Chorus, *J. Bartram* (Eds.) World Health Organisation, London: E and FN Spon.

Smith M. (1990). Guidelines for prediction of crop water requirements. Report on the Expert Consultation on procedures for revision of FAO. Roma: FAO.

Smith, A. J. (1982). Modes of cyanobacterial carbon metabolism. In: N. G. Carr and B. A. Whitton Eds. The Biology of Cyanobacteria (pp. 47-85). Oxford: Blackwell Scientific Publications.

Spallacci P. (1986). Concimazione azotata del terreno e rilascio di nitrati nelle acque di percolazione e ruscellamento. Rassegna degli esperimenti italiani- Le Acque a cura di Zavatti A, Vol 1, Pitagora Ed.

Stainer (1977). The position of cyanobacteria in the world or phototrophs. *Carlsb. Res. Commun.*, 42, pp. 77-98.

Teixeira da Gloria Lima Crux, M., Da Conceicao Nascimento Costa ,M., Lucia Pires de Carvalho, V., Dos Santos Pereira, M. and Hage, E., (1993). Gastroenteritis epidemic in the area of the Itaparica dam-Bahia *Brazil. Bull. Pan Am. Health Organ.*, 27, pp. 244-253.

Ueno, Y., Nagata, S., Tsutsumi, T., Hasegawa, A., Watanabe, M., Park, H.D., Chen, G.C., Chen, G., and Yu, S.Z. (1996). Detection of microcystins, a blue-green algal hepatotoxin, in drinking water sampled in Haimen and Fusui, endemic areas of primary liver cancer in China, by highly sensitive immunoassay. *Carcinogenesis* 17(6), pp. 1317-1321.

Utermohl, H. (1931). Neue Wege in der quantitativen Earfassung des Planktons (mit besonderer Berucksichtigung des Ultraplanktons). Verh. int. Ver. theor. angew. Limnol. 5, pp.567–596.

Utkilen et al. (1985). Buoyancy regulation in a red Oscillatoria unable to collapse gas vacuoles by turgor pressure. *Arch. Hydrobiol.* 102, pp. 319-329.

Van den Hoek, C., Mann, D.G., Jahns, H.M. (1995). Algae: An Introduction to Phycology. Cambridge (Great Britain): Cambridge University Press.

Van Liere et al. (1975). Growth of Oscillatoria agardhii Gom. Hydrobiol. Bull. 9, pp. 62-70.

Various Authors (1990) – Proceeding of the Conference on “La conoscenza dei consumi per una migliore gestione delle infrastrutture acquedottistiche”.

Yu, S.Z. (1989). Drinking water and primary liver cancer. In: Tung Z.G., Wu M.C., Xia S.S. Eds. Primary liver cancer. Berlin, Germany: Springer Verlag. p. 30.

Wetzel Robert G. (2001). – Limnology Lake and River ecosystems, Academic Press

Zhou, L., Yu, H., and Chen, K. (2002). Relationship between microcystins in drinking water and colorectal cancer. *Biomedical and Environmental Sciences* 15, pp. 166.

In: Drinking Water: Contamination, Toxicity and Treatment ISBN: 978-1-60456-747-2
Editors: J. D. Romero and P. S. Molina

Chapter 2

DRINKING WATER CONTAMINATION WITH METALS

M.S. Gimenez[1,*], S.M. Alvarez**, E. Larregle and A.M. Calderoni*

[*]Department of Biochemistry and Biological Sciences. University National of San Luis, IMIBIO, CONICET, San Luis, Argentine

[**]Dept. of Biochemistry and Molecular Biology, Virginia Commonwealth University, Richmond, Virginia, USA

ABSTRCT

It is estimated that contamination of drinking water with heavy metals has considerable impact on the health of the world population. The supply of high quality drinking water of is, therefore, necessary to develop and apply suitable processes which allow the reduction of hazardous metals - arsenic, cadmium, lead - below the international standards set for drinking water. Due to the discharge of untreated or insufficiently treated industrial waste waters or waste disposal, many water resources exhibit increased concentrations of heavy metals. The contamination of drinking water with metals represents a serious problem in human health that results in direct toxic effects causing nephropathies, hepatic necrosis, pulmonary emphysema, osteoporosis, hormone alterations, exacerbation of autoimmune disease, and its well-known carcinogenic effect. The adverse consequences of exposure to water contamination on reproductive organs have been widely considered; some of them can be the long-term impairment of neurobehavioral status, and alteration in complex behaviors, such as learning. Particular attention has to be paid to the fact that the treatment of drinking water - due the large volumes needed - must be selective when eliminating heavy metals, and must not eliminate other components which should remain in the water.

Keywords: drinking water, chemical contamination, metabolic effect, metals

---

[1] Corresponding author. Chacabuco y Pedernera. (5700). San Luis. Argentina, Tel.: + 54-2652-423789 int (107/152); fax: + 54 –2652-431301 , E-mail adress: mgimenez@unsl.edu.ar (M. S. Giménez).

INTRODUCTION

Access to safe drinking-water is essential to health. It has been shown that investments in water supply and sanitation can yield a net economic benefit, since the reductions adverse health effects and health care costs outweigh the costs of undertaking those interventions.

The acceptability of drinking-water by consumers is subjective and can be influenced by many different factors. The concentration at which certain constituents are objectionable to consumers is variable and dependent on individual and local factors, including the quality of the water to which the community is accustomed and a variety of social, environmental and cultural considerations.

Pathogenic contamination of drinking water is the most significant health risk to humans. There have been countless disease outbreaks and poisonings throughout history resulting from exposure to untreated or poorly treated drinking water. Significant risks to human health may also result from exposure to nonpathogenic, toxic contaminants that are often globally ubiquitous in waters from which drinking water is derived. Drinking water is the result of a large pathway through which its components travel from primary sources. The concentration of these components, particularly the toxic metal and their structure along the different transport flow paths is modified. Understanding the sources, fate, and concentrations of chemicals in water, in conjunction with assessment of effects, not only forms the basis of risk characterization, but also provides critical information required to render decisions regarding regulatory initiatives, remediation, monitoring, and management.

It is known that the major sources of contaminants come from anthropogenic activities and migrate to aquatic surface as well as to groundwater. The contaminants move to become incorporated into drinking water supplies. Loading of contaminants to surface waters, groundwater, sediments, and drinking water occurs via two primary sources: the first one originates from discrete sources whose inputs into aquatic systems can often be defined in a spatially explicit manner. Examples of this type of pollution include industrial effluents (pulp and paper mills, steel plants, food processing plants), municipal sewage treatment plants and combined sewage-storm-water overflows, resource extraction (mining), and land disposal sites (landfill sites, industrial impoundments); the other source of contamination originates from poorly defined, diffuse sources that typically occur over broad geographical scales. Examples of this include agricultural runoff (pesticides, pathogens, and fertilizers), storm-water and urban runoff, and atmospheric deposition (wet and dry deposition of persistent organic pollutants, such as polychlorinated biphenyls [PCBs] and mercury). The substances that may be present in source waters include microbiological organisms, such as bacteria and viruses - which may come from sewage treatment plants, septic systems, agricultural livestock operations and wildlife -, inorganic substances, such as salts and metals - which can be naturally occurring substances, or result from urban storm water runoff, industrial or domestic wastewater discharges, oil and gas production, mining, farming, or domestic plumbing. Other substances that can also be found in source waters are synthetic and volatile organic substances, by-products of industrial processes and petroleum production - which can come from gas stations, urban storm water runoff, old landfill sites and septic systems –, pesticides and herbicides - which may come from a variety of sources such as agriculture, storm water runoff and residential use -, radioactive materials - which can occur naturally or result from nuclear power production and mining activities - , and disinfection by-products that are not found in source waters, but that are produced as a function of the disinfection treatment process.

The process of water treatment is designed to reduce the levels of some of these parameters. Some metals, such as copper, zinc, nickel and lead, may leach into the drinking water from the distribution system and from domestic plumbing.

Fluoride is a chemical substance that may be added to municipal water during the treatment process to promote strong teeth. Fluoride can also be present in the source water as a result of erosion of natural deposits or discharge of fertilizers and aluminium factories. Nitrate is present in source water as a result of run-off of fertilizer use, leaching from septic tanks, sewage and erosion of natural deposits.

Lead can occur in source water as a result of erosion of natural deposits. The most common source of lead is corrosion of household plumbing. First flush water at the consumer's tap may contain higher concentrations of lead than water that has been flushed for several minutes.

Selenium occurs naturally in water at trace levels as a result of geochemical processes, such as weathering of rocks and soil erosion. It is difficult to establish levels of selenium that can be considered toxic because of the complex interrelationships between selenium and dietary constituents, such as protein, vitamin E, and other trace elements. Selenium is an essential trace element in the human diet. [1, 2].

Due to rapid industrialization and urbanization during the last two decades, contamination of drinking water by heavy metals is on global increase. [3]. All countries in the world have considered this problem, and they have agreed on the maximum permissible levels of metals in drinking water in order to protect the public health. In relation to this problem the World Health Organization (WHO) has published the corresponding guideline (See Table 1). Removal of trace amounts of heavy metals can be achieved by means of selective sorption processes. One of the possibilities is the application of weak base anion exchangers [4].

The goal of this chapter will thus be to give information about the metabolic effects of different metals that can contaminate drinking water and their consequences on health.

CADMIUM

Cadmium (Cd) has no known physiological function. Cd is an environmental toxic metal involved in human diseases. Cd is one of the most important toxic chemicals due to its increasing level in the environment as a result of tobacco smoking, industrial and agricultural practices It has a very long biological half-life (10–30 years) in humans, and its toxicity depends on the route, dose and duration of exposure. Acute cadmium intoxication primarily induces hepatic and testicular damage, whereas chronic exposure results in renal injury and osteotoxicity.

Most of the Cd pollution incidents, which occur particularly in eastern countries, result from the waste-water discharge of Cd batteries [5]. Cd induces hypertension in both humans and animals; however, its mechanism has not been clearly elucidated. Rats have been exposed to Cd via drinking water (5, 10 and 50ppm) for 3 months.

Table 1. Guideline of permitted values of metals concentration in drinking water

Chemical	Guideline value[a] (mg/litre)	Chemical	Guideline value[a] (mg/litre)
Antimony	0.02	Arsenic	0.01 (P)
Barium	0.7	Boron	0.5 (T)
Cadmium	0.003	Chlorine	5 (C)
Chlorite	0.7 (D)	Chromium	0.05 (P)
Copper	2	Cyanide	0.07
Fluoride	1.5	Hydrogen sulfide	0.05-0.1
Lead	0.01	Manganese	0.4 (C)
Mercury	0.006	Molybdenum	0.07
Nickel	0.07	Nitrate (as NO_3^-)	50
Nitrite (as NO_2^-)	3	Selenium	0.01
Uranium	0.015 (P,T)		

a P = provisional guideline value, as there is evidence of a hazard, but the available information on health effects is limited; T = provisional guideline value because calculated guideline value is below the level that can be achieved through practical treatment methods, source protection.

D = provisional guideline value because disinfection is likely to result in the guideline value being exceeded

C = concentrations of the substance at or below the health based guideline value may affect the appearance, taste or odor of the water, leading to consumer complaints.

Information adapted and obtained from http://www.who.int/water_sanitation_ health /dwq/ gdwq3rev/en/index.html

Guidelines for drinking-water quality, third edition, incorporating first addendum

Volume 1 - Recommendations

Publishing and ordering information World Health Organization 2006.

Exposure to Cd 10 and 50ppm has been shown to cause significant decreases in the sensitivity of vascular muscarinic receptors to Ach. Treatment with Cd has been proved to decrease endothelial nitric oxide synthase (eNOS) protein level in blood vessels. These results suggest that Cd suppresses ACh-induced vascular relaxation by interfering with the muscarinic receptor function, and its downstream signaling pathway may be one of the contributing factors for the development of hypertension. Cd intakes that had induced hypertension were associated with mean renal cadmium concentrations ranging from 5 to 50 mug/g in kidney [6,7]. It has also been examined the possibility that the subchronic exposure to low doses of a mixture of metals (arsenic, cadmium, lead, mercury, chromium, manganese, iron, and nickel) via drinking water can alter systemic physiology of male rats. The above mentioned metals have been found as contaminants in various water sources of India, and in concentrations equivalent to WHO's Maximum Permissible Limits (MPL) in drinking water for each metal. Male Wistar rats were exposed to a mixture at 0, 1, 10, and 100 times the mode concentrations (the most frequently occurring concentration) of the individual metals via drinking water for 90 days. Another group of rats was exposed to a mixture at a concentration equivalent to the WHO's MPL (in drinking water for individual metals). Subchronic exposure to the metal mixture affected general health of male rats by altering the functional and structural integrity of kidney, liver, and brain at 10 and 100 times the mode concentrations of the individual metals in Indian water sources [8]. Lead acetate (300 mg

Pb/L) and/or Cd acetate (10 mg Cd/L) in blood and liver were administrated as drinking water to pregnant Wistar rats from day 1 of pregnancy to parturition (day 0) or until weaning (day 21), to investigate the toxic effects in blood and in the liver. Both metals produced mycrocitic anaemia in the pups, as well as oxidative damage in the liver. The toxikinetics is different for Pb and Cd. While Cd is a hepatotoxic from day 0, Pb is not until day 21 [9].The main target organ for Cd is the kidney, and more specifically, the proximal tubular cells. Long-term Cd exposure has been shown to modify the ultrastructure of the kidney and cause tubulointerstitial fibrosis. Mice exposed to Cd concentrations varying from 10 to 500 mg CdCl (2)/l in drinking water during 4, 16 and 23 weeks, presented ultra structural changes in the kidney. The morphologic changes in the kidney became more pronounced by exposure to higher Cd concentrations [10]. Oxidative stress is believed to participate in the early processes of Cd-induced proximal tubular kidney damage. Low cadmium exposure triggers a biphasic oxidative stress response in mice kidneys. Mice chronically exposed to low Cd concentrations (10 and 100 mg CdCl (2)/l) up to 23 weeks via drinking water showed alteration in the pro- and antioxidant gene expression levels, glutathione, ascorbate and lipid peroxidation. [11].On the other hand, chronic exposure of mice to low doses of Cd was shown to lead to early renal damage, not predicted by blood or urine cadmium levels. Mice were exposed to Cd concentrations ranging from 0 to 100mg CdCl(2)/l in the drinking water for 1, 4, 8, 16 and 23 weeks. Urine samples were taken regularly; Cd content was determined in blood, liver, kidney and urine and histological analyses of the kidney were performed. Urinary Cd levels were not correlated with the kidney Cd content. Chronic exposure to low doses of Cd induced functional and histological signs of early damage at concentrations in or below the ones generally accepted as safe [12]. On the other hand Cd inhibits the antioxidant enzyme activities in liver and kidney. This was observed in rats receiving Cd in drinking water (250 mg/l $CdCl_2$, in tap water) that were killed after 10, 20 and 30 days of treatment [13].

Chronic Cd administration affects the circadian release of pituitary hormones. Rats that received Cd as CdCl (2) (5ppm in drinking water) with or without melatonin (3 microg/mL drinking water) presented modified expression of two major clock genes period: (Per) 1 and Per 2, in the hypothalamic-pituitary unit. Melatonin prevented the changes produced by Cd. [14]. Cd decreased dopamine content in the median eminence, while it increased its content in the posterior pituitary. Cd exposure (at the dose of 25 ppm of CdCl2 in the drinking water for 1 month) during adulthood affects the daily pattern of prolactin secretion and dopamine and serotonin content in the median eminence and pituitary [15].

The dose of 25 ppm of CdCl2 is approximately 835 times higher than provisional tolerable weekly intake (PTWI) for this metal (WHO, 2000.) [16].

The neurotoxic properties of Cd during development have rendered increasing interest. The immature blood–brain barrier permits transport of Cd to the developing brain, as has been shown in rodents. At lower doses, repeated oral Cd exposure resulted in increased locomotor activity and electrophysiological effects. Sprague–Dawley lactating rats were exposed to 0, 5 or 25 ppm cadmium as cadmium chloride in drinking water to day 17 of lactation. A battery of neurobehavioral tests was applied to the male offsprings after weaning at 5 weeks until 4 months of age. It was observed an increased spontaneous motor activity in offsprings after maternal Cd exposure during lactation [17].

This report confirms previous results obtained by other authors. Cd exposure leads to motor hyperactivity, increased aggressive behavior, impaired social memory processes, and altered drinking behavior [18].

Perinatal Cd exposure has been shown to alter behaviors and reduce offspring learning ability. Mice were administered 10ppm Cd (from gestational day 1 to postnatal day 10) and/or 0.025% methimazole (MMI; anti-thyroid drug) (from gestational day 12 to postnatal day 10) in drinking water. Also, 0.1% MMI was administered as a positive control (high MMI group). It was found that Cd increased the expression of specific genes, neurogranin (RC3) and myelin basic protein (BMP). This protein is regulated by TH through TH receptors (TR). It has been suggested that RC3 may play roles in memory and learning lobuloalveolar development and ductal branching. Simultaneously, perinatal exposure to Cd disrupted the gene expressions of sex hormone receptors [19].

It was found that Cd is a heavy metal classified as a human carcinogen.

Female mice of different ages and estrogenic status (prepuberal, adult intact) were treated with CdCl (2) (2-3 mg/kg, i.p.), melatonin (10 microg/mL in drinking water), CdCl (2) + melatonin, or diluents. Melatonin has antioxidant and antiestrogenic properties. Cd exerts estrogenic effects on mammary glands, increasing lobuloalveolar development and ductal branching; uterine weight also increased as a result of Cd treatment. These Cd effects are attenuated by melatonin [20].

Cd damages cellular metabolism at the level of various enzymatic systems of the cell, which may disturb functioning of the salivary glands. Mutual interactions of cadmium and zinc suggest a protective role of zinc through the induction of metallothionein which inactivates cadmium effect. Male Wistar rats were exposed to Cd and/or zinc for 6 months. CD was received in aqueous solutions of Cd chloride with drinking water at a concentration of 5 mg Cd/dm3 or 50 mg Cd/dm3. Zinc was also given in aqueous solutions of zinc chloride ad libitum at concentrations of 30 mg Zn/dm3 and 60 mg Zn/dm3. It was observed that exposure to Cd induces ultrastructural changes in the submandibular gland and zinc did not affect significantly the ultrastructural picture of cells of the submandibular gland, though when administered together with Cd, it reduces the intensity of ultrastructural changes in the submandibular gland [21]. Other authors have communicated alteration of rat submandibular gland secretory function by cadmium. Sprague–Dawley rats received cadmium (10 mg /l) as chloride in drinking water for 14 days. This caused significant alterations on the salivary function. Salivary flow rate, total protein concentration and amylase activity of saliva were decreased, while secretion of calcium was increased by cadmium [22].

At our laboratory, we found that Cd altered the prostatic lipid profile.Wistar male rats exposed to 15 ppm in drinking water during eight weeks showed that triglycerides and esterified cholesterol decreased, while free cholesterol and phospholipids increased, and total cholesterol did not change. In addition, the activity and expression of lipids synthesis were modified. Ultrastructural analysis showed a decrease in lipid droplets and signs of cellular damage in the Cd group.Cadmium exposure induced important changes in prostatic lipid profile and metabolism, confirmed by the morphology analyses, which also showed signs of cellular damage [23].

On the other hand, Cd and Zinc induced preneoplastic changes in ventral rat prostate.

Sprague-Dawley rats were separated into 3 groups according to treatment. The first group received Cd chloride in drinking water at a concentration of 60 ppm, 24 months. The second group received 50 ppm of zinc plus 60ppm Cd chloride in drinking water. The third group was used as control and received drinking water that was shown to be free of these metals. Some authors have found that cell proliferation is enhanced in rats exposed to cadmium for a long time. Similarly, the exposure to zinc combined with Cd has no effect on Bcl-2, p53 and Caspase-3 expression. [24], and this protein participates in cancer and apoptosis balance. There are more than 10 million new cancer cases each year worldwide, and cancer is the cause of approximately 12% of all deaths. Given this, a large number of epidemiologic studies have been undertaken to identify potential risk factors for cancer, amongst which the association with trace elements has received considerable attention. Trace elements, such as selenium, zinc, arsenic, cadmium, and nickel are found naturally in the environment, and human exposure derives from a variety of sources, including drinking water [25]

It is well known that long-term Cd exposure has carcinogenic effects on the male reproductive organ. [26] Mice exposed to low doses of Cd showed quantitative changes in the testicular structure. The animals received 0.0015 g/l of Cd in drinking water for 1, 3, 6 and 12 months. It was found that Cd exposure alters the morphology of testes. They showed a marked testicular hypoplasia in the Cd-administered mice [27].

The exposure to low levels of Cd through maternal milk alters the development and immune functions of juvenile and adult rats [20]. Others authors have reported that the exposure to 3, 30, 3000 or 10 000 parts per billion (ppb) of cadmium in drinking water for 2, 4, 28, or 31 weeks, determines the exacerbation of autoimmune disease in mice [28].

At our laboratory, we observed that Cd exposure through drinking water alters the lipid composition in resident peritoneal macrophages of Balb/c mice. After 2 months, adult male Balb/c mice that had drunk water with 15 ppm of Cd, showed tissue damage mediated by oxidative stress, as assessed by serum measuring of tissue damage and lipoperoxidation indicators [29].

CHROMIUM

Chromium (VI) (CrVI) compounds are genotoxic in a variety of cellular systems. Their potential carcinogenicity is affected by toxicokinetic patterns restricting bioavailability to certain targets, and by metabolic pathways affecting interaction of chromate-derived reactive species with DNA. Epidemiological data indicate that CrVI can be carcinogenic to the human respiratory tract following inhalation at doses that are only achieved in certain occupational settings. Swiss mice of both genders and different ages received sodium dichromate dihydrate administered with drinking water, up to a concentration of 500 mg/ l CrVI for up to 210 consecutive days. No increment of the micronucleus frequency was observed in either bone marrow or peripheral blood erythrocytes. Simultaneously, due to the hypothesis that susceptibility may be increased during the period of embryogenesis, authors treated pregnant mice, up to a concentration of 10mg CrVI / l in drinking water. There was no effect on the numbers of fetuses/dam and on body weight of fetuses. No toxic or genotoxic effect was observed either in bone marrow of pregnant mice or in liver and peripheral blood of their fetuses [30].

Others authors had communicated that sodium dichromate dihydrate administered to mice for 9 consecutive months, at doses corresponding to 5 and 20mg CrVI/l, which exceed drinking water standards by 100 and 400 times, respectively, failed to enhance the frequency of DNA-protein crosslinks and did not cause oxidative DNA damage, measured in terms of 8-oxo-2'-deoxyguanosine, in the forestomach, glandular stomach and duodenum [31].

The lack of carcinogenicity of oral chromium (VI) is explained by the high efficiency of chromium (VI) detoxification processes in the gastrointestinal tract. However investigators in Liaoning Province, China, reported that mortality rates for all cancer, stomach cancer, and lung cancer in 1970-1978 were higher in villages with hexavalent chromium (Cr+6)-contaminated drinking water than in the general population [32].

Recent analyses have revealed that 38% of municipal sources of drinking water in California have detectable levels of hexavalent chromium. This observation provided new impetus to characterize the carcinogenic risk associated with oral exposure to hexavalent chromium in drinking water. The marked reduction of hexavalent chromium to trivalent chromium in the stomach suggests that exposure to hexavalent chromium in drinking water may not pose a carcinogenic risk. Diverse studies suggest that the potential for a carcinogenic response is present only if hexavalent chromium enters in the cells. Both toxicokinetic and genotoxicity studies indicate that a portion of an orally administered dose of hexavalent chromium is absorbed and gets into cells of several tissues, causing DNA damage. The increases of stomach tumors in both human and animal studies, suggests that oral exposure to chromium (VI) has a carcinogenic risk [33].

LEAD

Lead (Pb) can occur in source water as a result of erosion of natural deposits. The most common source of Pb is corrosion of household plumbing. First flush water at the consumer's tap may contain higher concentrations of Pb than water that has been flushed for several minutes.

Low level exposure to Pb increases blood pressure in human and rats. Perinatal Pb exposure causes hypertension. In an experiment, Wistar dams received 1000 ppm of Pb or sodium acetate (control) in drinking water during pregnancy and lactation. Thoracic aortas of weaned (23-day old) male pups showed alteration in production of Nitric Oxide and Cyclooxygenase activity [34, 35]. Pb seems to be involved in the etiology of psychological pathologies. Swiss male and female mice (21 days) were exposed to 0, 50, 100 or 500 ppm of Pb, as Pb acetate, in drinking water for 70 days and were performed different tests. The authors observed induced an anti-depressant-like effect in both males and females, whereas exposure to 500 ppm induced anxiogenic effect only in males. Interruption of exposure was able to reverse the behavioral alterations in females, but not in males exposed to the highest concentration (500 ppm) [36].

Neurotoxicity has been associated with Pb exposure and this may be the result of a series of small perturbations in brain metabolism, in particular, of oxidative stress. Some studies have suggested a lead-induced enhancement on lipid peroxidation as a possible mechanism for some of its effects. Pregnant rats were treated with 0.4% lead acetate through drinking water from 6th day of gestation. This treatment continued until day 21 post-natal. The lipid peroxidation products catalase and superoxide dismutase activities of rats exposed to lead were increased in hippocampus, cerebellum and frontal cortex [37]. Similar results were obtained by the authors using Wistar male rats treated with lead acetate (500 ppm) through drinking water for a period of 8 weeks [38, 39].

It is known that Pb exposure leads to impaired learning and memorizing abilities in children, and it has also been observed that rats exposed to low levels of lead in drinking water containing 0.025%, 0.05% and 0.075% Pb acetate, during the developing period, presented inhibited activity of nitric oxide synthase activity in the hippocampus, the cerebral cortex and the cerebellum. The degree of the inhibitory effect depends on the time span of exposure and the lead concentration. [40]. Pb affects the developing central nervous system and may potentially induce apoptotic cell death. The ascorbic acid may potentially be beneficial in treating lead-induced brain injury in the developing rat brain. Female Sprague-Dawley rats that received Pb (0.2% Pb acetate) during pregnancy and lactation showed a decreased level of Bax protein and increased levels of Bcl-2 protein in the hippocampus of 21-day-old male pups. This situation was significantly attenuated by the simultaneous oral administration of 100 mg/kg/day ascorbic acid [41, 42].

COPPER

Copper (Cu) is an essential element, being a vital component in several enzyme systems. Some intake, therefore, is necessary for human health. At high doses, however, it can have toxic effects [43]. Despite the fact that Cu is an essential element for human health, elevated concentrations may result in serious liver and/or kidney damage, gastrointestinal distress and, in a lesser scale, vomiting and nausea [44]. Chronic toxicity was principally derived from patients with Wilson disease and cases of infantile cirrhosis that were related to excessive copper intakes [45]. There is a lot of evidence suggesting copper may influence the progression of Alzheimer's disease by reducing clearance of the amyloid beta protein (Abeta) from the brain. Previous experiments have shown that addition of only 0.12 PPM of copper (one-tenth the Environmental Protection Agency Human consumption limits) to distilled water was sufficient to precipitate the accumulation of Abeta in the brains of cholesterol-fed rabbits [46].

The quality of water has a pronounced effect on cholesterol-induced accumulation of Alzheimer amyloid beta (Abeta) in rabbit brain [47].

ALUMINUM

Aluminum (Al) is a neurotoxic metal and Al exposure may be a factor in the etiology of various neurodegenerative diseases such as Alzheimer's disease (AD). Chronic exposure over 12 months to Al sulphate in drinking water resulted in deposition of Abeta similar to that seen in congophilic amyloid angiopathy (CAA) in humans, and a reduction in neuronal expression of GRP78 similar to what has previously been observed in AD [48]. Al is neurotoxic for both experimental animals and certain human diseases. However, the results are contradictory

Low quantities injected intracerebrally into rabbits will induce severe neurological symptoms and neuropathological features of neurodegeneration. Hyper-aluminemia often develops in patients with renal failure being treated with intermittent hemodialysis on a chronic basis, and in severe cases results in an encephalopathy. Uremic adults and premature infants who are not under dialysis treatment can also develop encephalopathy due to Al toxicity, as is the case when large amounts of Al are used as a urinary bladder irrigate [49].

Al added to distilled water (0.36 PPM) administered to drink exacerbated cholesterol-induced hepatic pathology but not splenic pathology, and addition of 0.36 PPM zinc to the distilled drinking water failed to affect the pathology of either the liver or spleen [50].

Epidemiological studies tested the relationship between Al in drinking water and AD, showing a significant correlation between elevated levels of monomeric Al in water and AD, This has been confirmed by in vitro studies showing that Al mediated cell toxicity [51].

Others authors consider that Al in drinking water under different forms of speciation is not a risk for AD, after studying an elderly population in Canada [52]. It has not been demonstrated that the concentration of Al in the brains of AD patients is significantly higher than that of control subjects, however, detectable levels of Al have been observed

in senile plaques and neurofibrillary tangles of subjects with AD. For this reason Al was nominated as possible agent that could provoke AD [53].

ARSENIC

Arsenic is a ubiquitous metalloid found in various chemical forms in soil, ground water and foods. Because arsenic in the bedrock is easily dissolved to surrounding water, inorganic arsenic is frequently present at elevated concentrations in ground water [54].

There is some data suggesting that Arsenic could be an essential mineral. It has been suggested that arsenic has a biological role that affects formation of various metabolites from methionine, including taurine and polyamines [55]. The arsenic requirement for growing chicks and rats has been suggested to be near 25 ng/g diet. A possible human arsenic requirement is 12 μg/day. The reported arsenic content of diets from various parts of the world indicates that the average daily intake of arsenic is in the range of 12-40 μg [56].

On the other hand, Arsenic is a well-documented potent human carcinogen causing cancer of the bladder, lung, skin, and possibly also kidney and liver [1 IARC]. A large number of reports show associations between arsenic exposure and multiple non cancer health effects, e.g., diabetes, skin diseases, chronic cough, and toxic effects on liver, kidney, cardiovascular system, and peripheral and central nervous systems [57].

Millions of people worldwide, mainly in the developing countries, are exposed to arsenic because of emissions from mining activities, industrial or pesticide use, or contaminated well water. Arsenic in the bedrock or soil is easily dissolved in the surrounding ground water, and elevated concentrations of arsenic, i.e., above the World Health Organization (WHO) guideline level of 10 ug.L^{-1} [WHO], are present in most countries, although the prevalence and concentrations vary considerably.

The liver has long been identified as a target organ of arsenic exposure. Non-malignant hepatic abnormalities include hepatomegaly, non-cirrhotic portal fibrosis, and portal hypertension [58, 59, 60]. Furthermore, arsenic exposure has been linked to hepatic malignancies, namely hepatic angiosarcoma and hepatocellular carcinoma in both humans and in animal models [61, 62]. Straub et al. (2007) [63] recently demonstrated that mouse liver is also sensitive to more subtle hepatic changes (e.g., SEC capillarization and vessel remodeling) at lower arsenic exposure levels (250 ppb) without any gross pathologic effects.

Arsenic exposure has also been associated with impaired lung function and bronchiectasis [64, 65]. Arsenic has also been shown to synergize with cigarette smoking and other risk factors to induce mutations, DNA adducts, and cancer risk [66, 67]. Andrew et al. assessed the long-term effects of increasing doses of drinking water arsenic on expression levels in the mouse lung by using microarrays [68]. Mice were exposed to 0, 0.1, 1 and 50 ppb of arsenic for 5 weeks. They observed statistically significant expression changes for transcripts involved in lipid metabolism, oxygen transport, apoptosis, cell cycle, and immune response. For example, they found that the decreased expression of transcripts involved in lipid metabolic processes was strongest in the animals exposed to the lowest dose, 0.1 ppb. Among those was Angptl4, which inhibits lipoprotein lipase, an enzyme that regulates triglyceride clearance and lipid homeostasis [69]. This low dose of arsenic was also associated with decreased Acyl-CoA thioesterases (Cte1), enzymes that regulate the levels of free fatty acid and coenzyme A by catalyzing the hydrolysis of acyl-CoAs [70]. Likewise, expression of carnitine palmitoyltransferase 1A (Cpt1a), the enzyme that catalyzes the primary rate controlling step of fatty acid oxidation, was decreased in arsenic-exposed animals. Arsenic-exposed animals had also decreased expression of a number of transcripts involved in oxygen transport (Alas2, Hbb-b1, Hba-a1, Bpgm). The effects on multiple hemoglobin complex and oxygen transporter transcripts were particularly notable at the lower doses. Alas2 catalyzes the first step in the heme biosynthetic pathway, the hemoglobin complex members Hbb-b1 and Hba-a1 actually bind and transport oxygen and Bpgm is an enzyme that controls the levels of an allosteric effector of hemoglobin and the dissociation of oxygen [71].

They also observed differential expression of several apoptotic transcripts, including decreases in Ahr, Stk17B, Nr4a1, Egr1, and Angptl4 and increases in Hspa1b/ Hsp70 and ZFP145/ZBTB16 (above 1 ppb). Nr4a1 is a nuclear orphan steroid receptor that induces apoptosis. Interestingly, the protein levels of Nr4a1 were higher in the arsenic-exposed mice, indicating a possible role for posttranslational modification. They also observed increased levels of Hspa1b/Hsp70 with arsenic exposure levels 1 ppb and higher, as observed previously in the lymphocytes of arsenic exposed individuals in Bangladesh [72]. Immunoblot analysis confirmed increased protein levels of Hsp70 with arsenic exposure. Hsp70 is induced in response to stress and is important for DNA repair and maintaining genomic stability. The most dramatic effect of arsenic exposure in their experiment was a decrease in transcripts involved in immune response, including Igh-VJ558, Igj, Igk-V28, Igk-V8, Cd79b, Cxcl7, Ian6, s100a9, and Ahr. Igh-VJ558 is involved in B-cell antigen recognition and Igj, Igk-V28, and Igk-V8 in humoral immune response and antigen binding. Decreases in these transcripts were evident mainly at 50 ppb. Expression of the precursor plasma chemokine Cxcl7 that may affect tumor development by attracting immunocompetent cells was also decreased in the arsenic-exposed groups, particularly at 0.1 and 1 ppb. Similarly, the chemokines Cxcl2, Ccl3, Ccrl2, Cxcl3, Ccl4, and Ccl20 were decreased in the lymphocytes of arsenic-exposed individuals from Bangladesh [73]. A decreased expression of Calgranulin B/s100a9 with arsenic exposure, particularly at the lowest dose, 0.1 ppb was also observed. Calgranulin B is involved in the inflammatory response to stimuli, including the release of neutrophils and is regulated by Ahr signaling. Ahr, which was also decreased at the transcript and protein level in arsenic-exposed mice, is involved in the generation of regulatory T cells.

On the other hand, many epidemiological studies have revealed that people are more likely to have vascular diseases when living for many years in areas where aquifers or well water are contaminated with inorganic arsenic (as arsenite and arsenate) [74]. High prevalence of carotid atherosclerosis, ischemic heart disease, and cerebrovascular disease have been reported for chronic exposure to arsenic in water [75, 76].

Yang et al administered 50 ppm arsenic (as arsenite and arsenate) in drinking water to Wistar rats for 200 successive days and studied blood pressure and antioxidative enzyme system. They successfully induced hypertension in rats exposed to 50 ppm arsenic for 200 days and demonstrated the time-sequential changes in antioxidative enzyme systems. Their data provided a picture of how chronic arsenic exposure affects the antioxidative system leading to hypertension. Moreover, arsenicals were verified to induce CYP4A expression which is involved in 20- HETE metabolism, and which may be more important than ACE in contributing to arsenic-induced hypertension [77].

Regarding the association between Arsenic and cancer, several studies have been made. In cell and animal models, arsenite and perhaps some methylated metabolites may activate signal transduction pathways, which enhance cell proliferation, reduce antiproliferative signaling or inhibit differentiation, and override checkpoints controlling cell division after genotoxic insult [78, 79]. C57BL/6 male mice were provided 0.01% arsenic, as sodium arsenite in their drinking water for up to 16 weeks and mild hyperplasia was seen in the urinary bladder epithelium [80].

Four-week-old male A/J mice were exposed to water containing 0, 1, 10, and, 100 ppm of sodium arsenate for 18 months. With this model Cui et al showed that the As exposure significantly increased additional lung tumor numbers and tumor size and that high doses of arsenic seemed to be more carcinogenic to the mice because the tumor size in 100 ppm As–exposed mice was larger than in the 1 and 10 ppm As–exposed mice. They also showed that chronic exposure to As leads to arsenic accumulation, alters expression of p16INK4a and RASSF1A through influencing methylation patterns of the genes, and induced additional lung tumors in A/J mice [81]. This reduced or lost expression was the result of hypermethylation of these genes. Significant DNA hypermethylation of the promoter region of p53 gene was observed in the DNA of arsenic-exposed humans compared to that in control subjects [82]. This hypermethylation showed a dose–response relationship. Furthermore, hypermethylation of the p53 gene was also observed in arsenic-induced skin cancer patients compared to that in subjects having skin cancer unrelated to arsenic, though not at a significant level. However, a small subgroup of cases showed hypomethylation with high arsenic exposure. Significant hypermethylation of the gene p16 was also observed in cases of arsenicosis caused by high levels of arsenic. Thus, the ability of arsenic to alter DNA methylation in humans may be important in carcinogenesis.

Recently Waalkes et al studied carcinogenic effects of in utero arsenic exposure in mice. Pregnant C3H mice received drinking water with arsenite (0, 42.5 or 85 ppm) from gestation days (GD) 8 to 18. They found that brief in utero arsenic exposure in mice induces or initiates tumors or preneoplastic lesions in the liver, lung, urinary bladder, adrenal, kidney, ovaries, uterus, oviduct and vagina in the offspring as adults [83 84, 85]. Therefore they hypothesized that fetal arsenic exposure may induce aberrant genetic reprogramming as part of its carcinogenic mechanism.

In recent years, a few reports on adverse effects of arsenic on fetal growth and development in populations exposed to arsenic from drinking water have appeared [86 87, 88, 89]. Because several of the studies are ecological in design or include few subjects, more research is needed for firm conclusions on dose-response relations. However, it is quite likely that arsenic has adverse effects on the fetus because it readily crosses the placenta [90], possibly by Glut1, which has been shown to catalyze the cellular uptake of both arsenite and its monomethylated metabolite [91], and to be the main transplacental glucose transporter. Arsenic also accumulates in the placenta, possibly producing toxic effects in placental tissues, mediated via oxidative stress, and interfering with nutrient transport to the fetus, thereby affecting fetal growth. In contrast to the extensive fetal exposure in women exposed to arsenic during pregnancy, the breast-fed infant is protected against arsenic exposure because the excretion of arsenic in breast milk is limited [92]. Still, fetal exposure may give rise to long-lasting effects [93].

Conclusion

Drinking water could be free of toxic chemicals or these could be present below the concentrations where science suggests they are not of health concerns. By this way the health risks associated with drinking water could be reduced.The means of attaining and maintaining clean drinking water sources requires effective policies that identify, document, and reduce watershed risks. These risks are defined by their potential impact to human health. Governments and agencies, including, the US Environmental Protection Agency (EPA) and the World Health Organization (WHO) have established guidelines that specify acceptable concentrations and limits (e.g., MCL: maximum contaminant levels; MAC: maximum acceptable concentrations), for chemical/physical parameters,radiological amounts and many microbiological components. Looking to the future: Only with such directed policies can the future availability of clean drinking water sources be ensured. All levels of governments (local, provincial/state, and federal) bear the responsibility for setting policies to ensure the protection of our water resources and for providing instruments for the attainment of these policies.

References

[1] Moe, CL; Rheingans RD Global challenges in water, sanitation and health. *J Water Health.* 2006;4 Suppl 1:41-57.
[2] Onishchenko GG, Urgent problems in the implementation of the resolutions of the United Nations Organization on declaration of the decade 2005-2015 as the international decade "Water for Life" *Gig Sanit.* 2005 Jul-Aug;(4):3-5
[3] Kaur, R; Rani, R. Spatial characterization and prioritization of heavy metal contaminated soil-water resources in peri-urban areas of National Capital Territory (NCT), Delhi. *Environ Monit Assess,* 2006 Dec, 123(1-3):233-47.
[4] Zhao, X; Höll, WH; Yun G. Elimination of cadmium trace contaminations from drinking water. *Water Res,* 2002 Feb, 36(4):851-8.

[5] Lu, LT; Chang, IC; Hsiao, TY; Yu,YH; Ma, HW. Identification of pollution source of cadmium in soil: application of material flow analysis and a case study in Taiwan. *Environ Sci Pollut Res Int*, 2007 Jan, 14(1):49-59.

[6] Yoopan, N; Watcharasit, P; Wongsawatkul, O; Piyachaturawat, P; Satayavivad, J. Attenuation of eNOS expression in cadmium-induced hypertensive rats. *Toxicol Lett*, 2008 Jan 30,176(2):157-61.

[7] Perry, HM Jr; Erlanger, M; Perry, EF. Elevated systolic pressure following chronic low-level cadmiun feeding. *Am J Physiol,* 1977 Feb, 232(2):H114-21.

[8] Jadhav, SH; Sarkar, SN; Patil, RD; Tripathi, HC. Effects of subchronic exposure via drinking water to a mixture of eight wa ter-contaminating metals: a biochemical and histopathological study in male rats. *Arch Environ Contam Toxicol*, 2007 Nov, 53(4):667-77.

[9] Massó, EL; Corredor, L; Antonio, MT. Oxidative damage in liver after perinatal intoxication with lead and/or cadmium. *J Trace Elem Med Biol*, 2007 Sep 10, 21(3):210-216

[10] Thijssen, S; Lambrichts, I; Maringwa, J; Van Kerkhove E. Changes in expression of fibrotic markers and histopathological alterations in kidneys of mice chronically exposed to low and high Cd doses. *Toxicology,* 2007 Sep 5, 238(2-3):200-10.

[11] Thijssen, S; Cuypers, A; Maringwa, J; Smeets, K; Horemans, N; Lambrichts, I; Van Kerkhove, E. Low cadmium exposure triggers a biphasic oxidative stress response in mice kidneys. *Toxicology*, 2007 Jul 1, 236(1-2):29-41.

[12] Thijssen, S; Maringwa, J; Faes, C; Lambrichts, I; Van Kerkhove, E. Chronic exposure of mice to environmentally relevant, low doses of cadmium leads to early renal damage, not predicted by blood or urine cadmium levels. *Toxicology*, 2007 Jan 5, 229(1-2):145-56.

[13] Casalino, E; Calzaretti, G; Sblano, C; Landriscina,C. Molecular inhibitory mechanisms of antioxidant enzymes in rat liver and kidney by cadmium. *Toxicology,* 2002, 179, 37–50

[14] Cano, P; Poliandri, AH; Jiménez,V;Cardinali, DP; Esquifino, AI. Cadmium- induced changes in Per 1 and Per 2 gene expression in rat hypothalamus and anterior pituitary: effect of melatonin. *Toxicol Lett*, 2007 Aug, 172(3):131-6.

[15] Lafuente, A;González-Carracedo, A; Romero, A;Cabaleiro, T; Esquifino, AI.Toxic effects of cadmium on the regulatory mechanism of dopamine and serotonin on prolactin secretion in adult male rats. *Toxicol Lett,* 2005, 155, 87–96.

[16] World Health Organisation, 2000. Evaluation of certain food additives and contaminants. In: 55th Report of the Joint FAO/WHO Expert Committee on Food Additives, Geneva, Switzerland

[17] Petersson Grawéa, K; Teiling-Gårdlund, A; Jalkesten, E; Oskarsson, A. Increased spontaneous motor activity in offspring aftermaternal cadmium exposure during lactation. *Environ Toxicol Pharmacol*, 2004, 17, 35–43.

[18] Desi, I; Nagyrnajtenyi, L; Schulz, H. Behavioural and neurotoxicological changes caused by cadmium treatment of rats during development. *J Appl Toxicol,* 1998, 18, 63-/70

[19] Ishitobi, H; Mori, K; Yoshida, K; Watanabe, C. Effects of perinatal exposure to low-dose cadmium on thyroid hormone-related and sex hormone receptor gene expressions in brain of offspring. *Neurotoxicology*, 2007 Jul, 28(4):790-7.

[20] Alonso-González, C; González, A; Mazarrasa, O; Güezmes, A;Sanchez- Mateos, S; Martínez-Campa, C; Cos, S; Sánchez-Barceló, EJ; Mediavilla, MD. Melatonin prevents the estrogenic effects of sub-chronic administration of cadmium on mice mammary glands and uterus. *J Pineal Res*, 2007 Apr, 42(4):403-10.

[21] Dabrowska, E; Szynaka, B; Kulikowska-Karpińska, E. Ultrastructural study of the submandibular gland of the rat after 6-month exposure to cadmium and zinc in drinking water. *Adv Med Sci*, 2006, 51:245-9.

[22] Abdollahia, M; Dehpourb, A; Kazemiana,P. Alteration by cadmium of rat submandibular gland secretory function and the role of the l-arginine/nitric oxide pathway. *Pharmacol Res, 2000, vol. 42, no. 6.*

[23] Alvarez, SM; Gómez, NN; Scardapane, L; Fornés, MW; Giménez, MS. Effects of chronic exposure to cadmium on prostate lipids and morphology. *Biometals,* 2007 Oct, 20(5):727-41

[24] Arriazu, R; Pozuelo, JM; Henriques-Gil, N; Perucho, T; Martín, R; Rodríguez, R; Santamaría, L. Immunohistochemical Study of Cell Proliferation, Bcl-2, p53 and Caspase-3 Expression on Preneoplastic Changes Induced by Cadmium and Zinc Chloride in the Ventral Rat Prostate *J Histochem Cytochem*. 2006 Sep;54(9):981-90.

[25] Navarro Silvera, SA; Rohan, TE. Trace elements and cancer risk: a review of the epidemiologic evidence. *Cancer Causes Control,* 2007 Feb, 18(1):7-27.

[26] Haffor, AS; Abou-Tarboush FM. Testicular cellular toxicity of cadmium : transmission electron microscopy examination. *J Environ Biol*. 2004 Jul;25(3):251-8.

[27] Pillet, S; Rooney, AA; Bouquegneau, JM; Cyr, DG; Fournier, M. Sex-specific effects of neonatal exposures to low levels of cadmium through maternal milk on development and immune functions of juvenile and adult rats. *Toxicology*, 2005, 209, 289–301.

[28] Leffel, EK; Wolf, C; Poklis, A; White, KL Jr. Drinking water exposure to cadmium, an environmental contaminant, results in the exacerbation of autoimmune disease in the murine model. *Toxicology*, 2003, 188, 233-250.

[29] Ramirez, DC; Gimenez MS. Lipid modification in mouse peritoneal macrophages after chronic cadmium exposure. *Toxicology*. 2002 Mar 5;172(1):1-12

[30] De Flora, S; Iltcheva, M; Balansky, RM.Oral chromium(VI) does not affect the frequency of micronuclei in hematopoietic cells of adult mice and of transplacentally exposed fetuses. *Mutat Res*, 2006 Nov 7, 610(1-2):38-47.

[31] Salnikow, K; Zhitkovich, A. Genetic and epigenetic mechanisms in metal carcinogenesis and cocarcinogenesis: nickel, arsenic, and chromium. *Chem. Res.Toxicol*. 2008 Jan 21, 21(1):28-44.

[32] Beaumont, JJ; Sedman, RM; Reynolds, SD; Sherman, CD; Li, LH; Howd, RA; Sandy, MS; Zeise, L; Alexeeff, GV. Cancer mortality in a chinese population exposed to hexavalent chromium in drinking water. *Epidemiology,* 2008 Jan, 19(1):12-23.

[33] Sedman, RM; Beaumont, J; McDonald, TA; Reynolds, S; Krowech, G; Howd, R. Review of the evidence regarding the carcinogenicity of hexavalent chromium in drinking water. *J Environ Sci Health C Environ Carcinog Ecotoxicol Rev*, 2006 Apr, 24(1):155-82.

[34] Grizzo, LT; Cordellini, S. Perinatal Lead Exposure Affects Nitric Oxide and Cyclooxygenase Pathways in Aorta of Weaned Rats. *Toxicol Sci,* 2008 Jan 30 [Epub ahead of print].

[35] Karimi, G; Khoshbaten, A; Abdollahi, M; Sharifzadeh, M; Namiranian, K; Dehpour, AR. Effects of subacute lead acetate administration on nitric oxide and cyclooxygenase pathways in rat isolated aortic ring. *Pharmacol Res*, 2002 Jul, 46(1), 31-7.

[36] Soeiro, AC; Gouvêa, TS; Moreira, EG. Behavioral effects induced by subchronic exposure to Pb and their reversion are concentration and gender dependent. *Hum. Exp. Toxicol.,* 2007 Sep, 26(9), 733-9.

[37] Bokara, KK; Brown, E; McCormick, R; Yallapragada, PR; Rajanna, S; Bettaiya, R. Lead-induced increase in antioxidant enzymes and lipid peroxidation products in developing rat brain. *Biometals*, 2008 Feb, 21(1), 9-16.

[38] Bennet, C; Bettaiya, R; Rajanna, S; Baker, L; Yallapragada, PR; Brice, JJ; White, SL, Bokara, KK. Region specific increase in the antioxidant enzymes and lipid peroxidation products in the brain of rats exposed to lead. *Free Radic Res,* 2007 Mar, 41(3), 267-73.
[39] Villeda-Hernández, J; Barroso-Moguel, R; Méndez-Armenta, M; Nava-Ruíz, C; Huerta-Romero, R; Ríos, C.Enhanced brain regional lipid peroxidation in developing rats exposed to low level lead acetate. *Brain Res Bull,* 2001 May 15, 55(2), 247-51.
[40] Zhu, ZW; Yang, RL; Dong, GJ; Zhao, ZY. Study on the neurotoxic effects of low- level lead exposure in rats. *J- Zhejiang Univ Sci B*, 2005 Jul, 6(7):686-92.
[41] Han, JM; Chang, BJ; Li, TZ; Choe, NH; Quan, FS; Jang, BJ; Cho, IH; Hong, HN; Lee. JH. Protective effects of ascorbic acid against lead-induced apoptotic neurodegeneration in the developing rat hippocampus in vivo. *Brain Res,* 2007 Dec 14, 1185,68-74.
[42] Chao, SL; Moss, JM; Harry,GJ. Lead-induced alterations of apoptosis and neurotrophic factor mRNA in the developing rat cortex, hippocampus, and cerebellum. *J Biochem Mol Toxicol,* 2007, 21(5), 265-72.
[43] Fewtrell, L; Kay, D; MacGill, S. A review of the science behind drinking water standards for copper. *Int J Environ Health Res*, 2001 Jun, 11(2), 161-167.
[44] Pyatt, FB; Pyatt, AJ; Walker, C; Sheen, T; Grattan, JP. The heavy metal content of skeletons from an ancient metalliferous polluted area in southern Jordan with particular reference to bioaccumulation and human health, *Ecotoxicol Environ Safety,* 2005, 60 (3), 295–300.
[45] Olivares, M; Uauy, R. Limits of metabolic tolerance to copper and biological basis for present recommendations and regulations. *Am J Clin Nutr*, 1996 May, 63(5), 846S-52S.
[46] Sparks, DL; Lochhead, J; Horstman, D; Wagoner, T; Martin, T. Water quality has a pronounced effect on cholesterol-induced accumulation of Alzheimer amyloid beta (Abeta) in rabbit brain. *J Alzheimers Dis*, 2002 Dec, 4(6), 523-529.
[47] Sparks, DL; Friedland, R; Petanceska, S; Schreurs, BG; Shi, J; Perry, G; Smith, MA, Sharma, A; Derosa, S; Ziolkowski, C; Stankovic, G. Trace copper levels in the drinking water, but not zinc or aluminum influence CNS Alzheimer-like pathology. *J Nutr Health Aging*, 2006 Jul-Aug, 10(4), 247-54.
[48] Rodella, LF; Ricci, F; Borsani, E; Stacchiotti, A; Foglio, E; Favero, G; Rezzani,R, Mariani, C; Bianchi, A. Aluminium exposure induces Alzheimer's disease-like histopathological alterations in mouse brain. *Histol Histopathol*, 2008 Apr, 23(4), 433-439.
[49] Savory, J; Exley, C; Forbes, WF; Huang ,Y; Joshi, JG; Kruck, T;McLachlan ,DR; Wakayama I. The controversy of the role of aluminum in Alzheimer's disease be resolved? What are the suggested approaches to this controversy and methodological issues to be considered? *J Toxicol Environ Health*. 1996 Aug 30;48(6):615-35
[50] Sparks, DL; Martin, T; Stankovic, G; Wagoner, T; Van Andel, R. Influence of water quality on cholesterol induced systemic pathology. *J Nutr Health Aging*, 2007 Mar-Apr, 11(2), 189-193.
[51] Prolo, P; Chiappelli, F; Grasso, E; Rosso, MG; Neagos, N; Dovio, A; Sartori, ML; Perotti, P; Fantò, F; Civita, M; Fiorucci, A; Villanueva, P; Angeli, A. Aluminium blunts the proliferative response and increases apoptosis of cultured human cells: putative relationship to alzheimer's disease. *Bioinformation*, 2007 Jun 4, 2(1), 24-27.
[52] Gauthier, E; Fortier, I; Courchesne, F; Pepin, P; Mortimer, J; Gauvreau, D. Aluminum forms in drinking water and risk of Alzheimer's disease. *Environ Res,* 2000 Nov, 84(3), 234-246.

[53] Itzhaki RF. Possible factors in the etiology of Alzheimer's disease. *Mol Neurobiol.* 1994 Aug-Dec;9(1-3):1-13.
[54] IARC. Volume 84. Some drinking water disinfectants and contaminants, including arsenic. *International Agency for Research on Cancer, Lyon, France, 2004*].
[55] Uthus, E0. Effects of arsenic deprivation in hamsters. *Magnesium Tr. Elem.*, 1990, *9*, 227-232.
[56] Nielsen, FH. Other trace elements. In Brown M (Ed). *Present Knowledge in Nutrition.* Washington, D.C.: International Life Sciences Institute; 1990; pp. 294-307 -57 WHO. Arsenic and arsenic compounds. Geneva: International Programme on Chemical Safety, World Health Organization; 2001.
[58] Santra, A ; Das, GJ ; De, BK ; Roy, B and Guha Mazumder, DN. Hepatic manifestations in chronic arsenic toxicity. *Indian J. Gastroenterol.* 1999, 18, 152–155.
[59] Santra, A; Maiti, A; Das, S; Lahiri, S; Charkaborty, SK; Mazumder, DN. Hepatic damage caused by chronic arsenic toxicity in experimental animals. *J. Toxicol. Clin. Toxicol.* 2000, 38, 395–405.
[60] Mazumder, DN. Effect of chronic intake of arsenic-contaminated water on liver. *Toxicol. Appl. Pharmacol.* 2005, 206, 169–175.
[61] Smith, AH; Hopenhayn-Rich, C; Bates, MN; Goeden, HM; Hertz-Picciotto, I; Duggan, HM; Wood, R; Kosnett, MJ; Smith, MT. Cancer risks from arsenic in drinking water. *Environ. Health Perspect.* 1992, 97, 259–267.
[62] Waalkes, MP; Liu, J; Ward, JM; Diwan, BA. Enhanced urinary bladder and liver carcinogenesis in male CD1 mice exposed to transplacental inorganic arsenic and postnatal diethylstilbestrol or tamoxifen. *Toxicol. Appl. Pharmacol.* 2006, 215, 295–305.
[63] Straub, AC; Stolz, DB; Ross, MA; Hernandez-Zavala, A; Soucy, NV; Klei, LR ; Barchowsky, A. Arsenic stimulates sinusoidal endothelial cell capillarization and vessel remodeling in mouse liver. *Hepatology* 2007, 45, 205–212.
[64] De, BK; Majumdar, D; Sen, S; Guru, S; Kundu, S. Pulmonary involvement in chronic arsenic poisoning from drinking contaminated ground-water. *J. Assoc. Physicians India* 2004, 52, 395–400.
[65] Smith, AH; Marshall, G; Yuan, Y; Ferreccio, C; Liaw, J; von Ehrenstein, O; Steinmaus, C; Bates, MN; Selvin, S. Increased mortality from lung cancer and bronchiectasis in young adults after exposure to arsenic in utero and in early childhood. *Environ. Health Perspect.* 2006, 114, 1293–1296.
[66] Ahsan, H; Thomas, DC. Lung cancer etiology: Independent and joint effects of genetics, tobacco, and arsenic. *JAMA* 2004, 292, 3026–3029.
[67] Chen, CL; Hsu, LI; Chiou, HY; Hsueh, YM; Chen, SY; Wu, MM and Chen, CJ. Ingested arsenic, cigarette smoking, and lung cancer risk: A follow-up study in arseniasis-endemic areas in Taiwan. *JAMA* 2004, 292, 2984–2990.
[68] Andrew, AS; Bernardo, V; Warnke, LA; Davey, JC; Hampton, T; Mason, RA; Thorpe, JE; Ihnat, MA; Hamilton JW. Exposure to Arsenic at Levels Found in U.S. Drinking Water Modifies Expression in the Mouse Lung, *Toxicological Sciences* 2007, 100(1), 75–87.
[69] Koster, A; Chao, YB; Mosior, M; Ford, A; Gonzalez-Dewhitt, PA; Hale, JE; Li, D; Qiu, Y; Fraser, CC; Yang, DD; Heuer, JG; Jaskunas, SR ; Eacho P. Transgenic Angptl4 overexpression and targeted disruption of Angptl4 and Angptl3: Regulation of triglyceride metabolism. *Endocrinology* 2005, 146, 4943–4950.
[70] Hunt, MC; Rautanen, A; Westin, MA; Svensson, LT; Alexson, SE. Analysis of the mouse and human acyl-CoA thioesterase (ACOT) gene clusters shows that convergent,

functional evolution results in a reduced number of human peroxisomal ACOTs. *FASEB J.* 2006, 20, 1855–1864.

[71] Garel, MC; Lemarchandel, V; Prehu, MO; Calvin, MC; Arous, N; Rosa, R; Rosa, J; Cohen-Solal, M. Natural and artificial mutants of the human 2,3-bisphosphoglycerate as a tool for the evaluation of structure-function relationships. *Biomed. Biochim. Acta* 1990, 49, S166–S1171.

[72] Argos, M; Kibriya, MG; Parvez, F; Jasmine, F; Rakibuz-Zaman, M; Ahsan, H. Gene expression profiles in peripheral lymphocytes by arsenic exposure and skin lesion status in a Bangladeshi population. *Cancer Epidemiol. Biomarkers Prev.* 2006, 15, 1367–1375.

[73] Argos, M; Kibriya, MG; Parvez, F; Jasmine, F; Rakibuz-Zaman, M; Ahsan, H. Gene expression profiles in peripheral lymphocytes by arsenic exposure and skin lesion status in a Bangladeshi population. *Cancer Epidemiol. Biomarkers Prev.* 2006, 15, 1367–1375.

[74] Lilienfeld, DE. Arsenic, geographical isolates, environmental epidemiology, and arteriosclerosis. *Arteriosclerosis* 1998, 8, 449–451

[75] Chen, CJ; Chiou, HY; Chiang, MH; Lin, LJ; Tai, TY. Dose–response relationship between ischemic heart disease mortality and long-term arsenic exposure. *Arterioscler. Thromb. Vasc. Biol.* 1996, 16, 504–510.

[76] Wang, CH; Jeng, JS; Yip, PK; Chen, CL; Hsu, LI; Hsueh, YM; Chiou, HY; Wu, MM; Chen, CJ. Biological gradient between long-term arsenic exposure and carotid atherosclerosis. *Circulation* 2002, 105, 1804–1809.

[77] Yang, H-T; Chou, H-J; Han, B-C; Huang, S-Y. Lifelong inorganic arsenic compounds consumption affected blood pressure in rats. *Food and Chemical Toxicology* 2007, 45, 2479–2487.

[78] Simeonova, PP, Luster, MI. Mechanisms of arsenic carcinogenicity: genetic or epigenetic mechanisms? J. EnViron. Pathol. Toxicol. Oncol. 2000, 9, 281–286.

[79] Salnikow, K ; Cohen, MD. Backing into Cancer: Effects of Arsenic on Cell Differentiation. *Toxicol. Sci.* 2002, 65, 161–163.

[80] Simeonova, PP; Wang, S; Toriuma, W; Kommineni, V; Matheson, J; Unimye, N; Kayama, F; Harki, D; Ding, M; Vallyathan, V; Luster, MI. Arsenic mediates cell proliferation and gene expression in the bladder epithelium: association with activating protein-1 transactivation. *Cancer Res.* 2000, 60, 3445–3453.

[81] Cui, X; Wakai, T; Shirai, Y; Hatakeyama, K ; Hirano. S. Chronic Oral Exposure to Inorganic Arsenate Interferes with Methylation Status of p16INK4a and RASSF1A and Induces Lung Cancer in A/J Mice. *Toxicological Sciences* 2006, 91(2), 372–381.

[8]2 Chanda, S; Dasgupta, UB; GuhaMazumder, D; Gupta, M; Chaudhuri, U; Lahiri, S; Das, S; Ghosh, N; Chatterjee, D. DNA hypermethylation of promoter of gene p53 and p16 in arsenic-exposed people with and without malignancy. *Toxicol. Sci.* 2006, 89, 431–437.

[83] Waalkes, MP; Ward, JM., Liu, J; Diwan, BA. Transplacental carcinogenicity of inorganic arsenic in the drinking water: Induction of hepatic, ovarian, pulmonary and adrenal tumors in mice. *Toxicol. Appl. Pharmacol.* 2003, 86, 7–17.

[84] Waalkes, MP; Ward, JM; Diwan, BA. Induction of tumors of the liver, lung, ovary and adrenal in adult mice after brief maternal gestational exposure to inorganic arsenic: Promotional effects of postnatal phorbol ester exposure on hepatic and pulmonary, but not dermal cancers. *Carcinogenesis* 2004, 25, 133–141.

[85] Waalkes, MP; Liu, J; Ward, JM; Diwan, BA. Enhanced urinary bladder and liver carcinogenesis in male CD1 mice exposed to transplacental inorganic arsenic and postnatal diethylstilbestrol or tamoxifen. *Toxicol. Appl. Pharmacol.* 2006, 215, 295–303.

[86] Ahmad, SA; Sayed, MH; Barua, S; Khan, MH; Faruquee, MH; Jalil, A; Hadi, SA; Talukder HK. Arsenic in drinking water and pregnancy outcomes. *Environ Health Perspect.* 2001,109: 629–631.

[87] Milton, AH; Smith, W; Rahman, B; Hasan, Z; Kulsum, U; Dear, K; Rakibuddin, M; Ali, A. Chronic arsenic exposure and adverse pregnancy outcomes in Bangladesh. *Epidemiology* 2005, 16:82–86.

[88] Pehayn-Rich, C; Browning, SR; Hertz-Picciotto, I; Ferreccio, C; Peralta, C; Gibb, H. Chronic arsenic exposure and risk of infant mortality in two areas of Chile. *Environ Health Perspect.* 2000, 108: 667–673.

[89] Hopenhayn, C; Ferreccio, C; Browning, SR; Huang, B; Peralta, C; Gibb, H; Hertz-Picciotto, I. Arsenic exposure from drinking water and birth weight. *Epidemiology* 2003, 14: 593–602.

[90] Concha, G; Vogler, G; Lezcano, D; Nermell, B; Vahter, M. Exposure to inorganic arsenic metabolites during early human development. *Toxicol Sci.* 1998, 44, 185–190.

[91 Liu, Z; Sanchez, MA; Jiang, X; Boles, E; Landfear, SM; Rosen, BP. Mammalian glucose permease GLUT1 facilitates transport of arsenic trioxide and methylarsonous acid. *Biochem Biophys Res Commun.* 2006,351,424–430.

[92] Concha, G; Vogler, G; Nermell, B; Vahter, M. Low-level arsenic excretion in breast milk of native Andean women exposed to high levels of arsenic in the drinking water. *Int Arch Occup Environ Health.* 1998, 71, 42–46.

[93] Vahter, ME. Interactions between Arsenic-Induced Toxicity and Nutrition in Early Life. *J. Nutr.*2007, 137, 2798–2804.

In: Drinking Water: Contamination, Toxicity and Treatment ISBN: 978-1-60456-747-2
Editors: J. D. Romero and P. S. Molina

Chapter 3

CURRENT AND EMERGING MICROBIOLOGY ISSUES OF POTABLE WATER IN DEVELOPED COUNTRIES

William J. Snelling[1,*], Roy D. Sleator[2,*], Catherine D. Carrillo[3], Colm J. Lowery[1], John E. Moore[4], John P. Pezacki[5] and James S.G. Dooley[1]

[1]Centre for Molecular Biosciences, School of Biomedical Sciences, University of Ulster, Cromore Road, Coleraine, Co., Londonderry, Northern Ireland, United Kingdom, BT52 1SA

[2]Alimentary Pharmabiotic Centre, University College Cork, Cork, Ireland

[3]Bureau of Microbial Hazards, Health Canada, 251 Sir Frederick Banting Driveway, Ottawa, ON, Canada, K1A 0L2

[4]Department of Bacteriology, Northern Ireland Public Health Laboratory, Belfast City Hospital, Belfast, United Kingdom, BT9 7AD

[5]The Steacie Institute for Molecular Sciences, National Research Council of Canada 100 Sussex Drive, Ottawa, Ontario, Canada K1A OR6

ABSTRACT

Water is vital for life, for commercial and industrial purposes and for leisure activities in the daily lives of the world's population. Diarrhoeal disease associated with consumption of poor quality water is one of the leading causes of morbidity and mortality in developing countries (especially in children <5 years old). In developed countries, whilst potable water is not a leading cause of death, it still can still pose a significant health risk. Water quality is assessed using a number of criteria, e.g., microbial load and nutrient content which affects microbial survival, as well as aesthetic factors such as odor. In water systems, the presence of disinfectant, low temperatures, flow regimes and low organic carbon sources do not appear to be conducive to microbial persistence. However, frequently this is not the case. A variety of human pathogens can be

* These authors contributed equally

transmitted orally by water and in the developed world water quality regulations require that potable water contains no microbial pathogens.

Chlorine dioxide is a safe, relatively effective biocide that has been widely used for drinking water disinfection for 40 years. Providing the water is of low turbidity, standard chlorination procedures are sufficient to prevent the spread of planktonic bacteria along water mains. However, despite this, bacterial contamination of water distribution systems is well documented, with growth typically occuring on surfaces, including pipe walls and sediments. Rivers, streams and lakes are all important sources of drinking water and are used routinely for recreational purposes. However, due to fouling by farm and wild animals, these sources can be contaminated with microbes, e.g., chlorine-resistant *Cryptosporidium* oocysts, no matter how pristine the source or well maintained the water delivery system. The high incidence of *Cryptosporidium* in surface water sources underlines the need for frequent monitoring of the parasite in drinking water. The use of coliforms as indicator organisms, although considered relevant to most cases, is not without limitations, and is thus not a completely reliable parameter of water safety, e.g., *Campylobacter* contamination cannot be accurately predicted by coliform enumeration. Furthermore, the presence of biofilms and bacterial interactions with protozoa in water facilitate increased resistance to antimicrobial agents and procedures such as disinfectants and heating, e.g., Legionnaires' disease caused by *Legionella pneumophila*.

The high cost of waterborne disease outbreaks should be considered in decisions regarding water utility improvement and treatment plant construction. The control of human illnesses associated with water would be aided by a greater understanding of the interactions between water-borne protozoa and bacterial pathogens, which until relatively recently have been overlooked.

INTRODUCTION

Currently, over 1 billion people worldwide have no access to safe drinking water (CDC, 2007a). In the developed world water quality regulations require that potable water does not contain any microbial pathogens (Percival and Walker, 1999). Residents of affluent nations are remarkably lucky to have high-quality, safe drinking water supplies that most residents of modem cities enjoy, particularly when considered in contrast to the toll of death and misery that unsafe drinking water causes for most of the world's population (Hrudey and Hrudey, 2007). The supply of clean drinking water is a major, and relatively recent, public health milestone (Berry et al., 2006). However, some may presume that drinking-water disease outbreaks are a thing of the past, but complacency can easily arise (Hrudey and Hrudey, 2007). Increasing human populations and urbanisation have placed burdens on water sources, i.e. rivers, streams and lakes, used to provide potable water to most metropolitan areas (Calderon et al., 2006). Excreta from humans, pets, livestock and wildlife, e.g. foxes, present in these source waters have the potential to harbour hundreds of pathogenic microorganisms of public health concern (Leclerc et al., 2002). In developed countries, major outbreaks of waterborne infections have fuelled widespread public concern regarding the microbiological quality of potable water (De Paula et al., 2007; Pankhurst and Coulter, 2007).

Potable water can be a source of various potentially infectious microorganisms, which in the majority of cases are contracted by ingestion, but can also be contracted *via* inhalation of aerosol droplets or by direct dermal exposure (Stojek and Dutkiewicz, 2006). Rivers, streams and lakes, which are not only important sources of drinking water but are also routinely used in the pursuit of human leisure interests, are often contaminated with microbes, e.g. chlorine resistant *Cryptosporidium* oocysts, no matter how pristine the source, or well maintained the water delivery system (Snelling et al., 2006). Fecal contamination of surface waters can occur *via* wastewater discharge, farming activities and fouling by wild animals. Perhaps the best known potential pathogens are certain strains of *Escherichia coli* and related Gram-negative species of fecal origin belonging to the Enterobacteriaceae family, usually referred to as "coliforms", which are commonly used as sanitary indicators of potable water quality (Stojek and Dutkiewicz, 2006). Important agents of waterborne infections include various genera of Gram-negative bacteria such as *Campylobacter, Escherichia coli* O157:H7, *Legionella, Salmonella, Shigella, Yersinia, Vibrio cholerae*, mycobacteria (Gram positive), enteroviruses and intestinal protozoa (*Giardia, Cryptosporidium*) (Stojek and Dutkiewicz, 2006).

Potable water distribution systems generally present a hostile environment for the growth of microorganisms due to low nutrient levels as well as the presence of disinfection residuals (Långmark et al., 2007; Momba et al., 2007). Yet despite this, biofilms form ubiquitously in environments that have been subjected to a range of disinfection processes including chlorination and ultra-violet (UV)-treatment (Momba et al., 1998). Most of the bacteria in drinking water distribution systems, e.g. *M. avium, L. pneumophila*, and *E. coli*, are associated with biofilms. In biofilms, their nutrient supply is generally improved and biofilms can provide shelter against disinfection (Lehtola et al., 2007). Pathogenic bacteria and viruses entering water distribution systems can survive in biofilms for at least several weeks, even under conditions of high-shear turbulent flow, and may be a risk to water consumers (Lehtola et al., 2007). Also, considering the low number of virus particles needed to result in an infection, their extended survival in biofilms must be taken into account as a risk for the consumer (Lehtola et al., 2007). Recent inquiries into the microbial ecology of distribution systems have found that pathogen resistance to chlorination is affected by microbial community diversity and interspecies relationships (Berry, et al., 2006). Research indicates that multispecies biofilms are generally more resistant to disinfection than single-species biofilms (Berry, et al., 2006).

Control of microbial growth in drinking water distribution systems, often achieved through the addition of disinfectants, is essential to limiting waterborne illness, particularly in immunocompromised subpopulations (Berry et al., 2006). Whilst disinfection residuals within distribution systems reduce the growth of autochthonous and allochthonous pathogens and reduce the potential re-growth of heterotrophic bacteria, they do not insure the sterility of distribution waters (Payment et al., 1993). The omnipresence of heterotrophs does not in itself constitute a potential health hazard, they can be used to evaluate the microbiological quality of the water and the efficacy of water treatment and distribution processes, as well as the biological stability of distributed waters (Långmark et al., 2007).

Waterborne diseases occur worldwide, and outbreaks caused by the contamination of community water systems have the potential to cause disease in large numbers of consumers (Table 1) (Karanis et al., 2007). These cases create a lack of confidence in potable water quality and in the water industry in general; waterborne outbreaks have economic consequences far beyond the cost of health care for affected patients, their families and contacts. In addition to outbreaks caused by contaminated potable water, there is also a risk associated with the accidental ingestion of recreational (or other) waters. National statistics on outbreaks linked to contaminated water have been available in the USA since 1920, and since 1971, the Centers for Disease Control (CDC), the US Environmental Protection Agency (USEPA), and the Council of State and Territorial Epidemiologists have maintained a collaborative surveillance system for collecting data pertaining to the occurrence and causes of outbreaks of waterborne disease (Karanis et al., 2007). In Europe during 1986–96, 277 outbreaks associated with drinking and recreational water were reported from 16 European countries (Kramer et al. 2001).

VIRUSES

Viruses are possibly the most hazardous of all enteric pathogens and have relatively low infectious doses (Santamaría and Toranzos 2003). Viruses are found in very low concentrations in treated water due to the effect of dilution and to the potabilization process (Gutiérrez et al., 2007). For this reason, it has been proposed that large amounts of water need to be collected in order to detect them. A negative result might not be meaningful, while a positive result is important, particularly when viruses are detected in small water samples (Gutiérrez et al., 2007).Thus, the paucity of data regarding waterborne viruses makes it difficult to determine the true risk they represent and precludes the development of plans to prevent viral transmission through contact with environmental water (De Paula et al., 2007).

It has been proposed that more than 140 different types of viruses, apparently not eliminated by massive purification treatments, can be found in drinking water (Gutiérrez et al., 2007). The main source for water contamination is human excreta (Gutiérrez et al., 2007). In fact, viruses infiltrate the ground, penetrating to depths greater than 67 m and can remain latent there for several months, as long as the temperature remains low and the environment humid (Gutiérrez et al., 2007). Under these conditions they can easily reach aquifers. Enteric pathologies due to the presence of Rotavirus (RV), Norovirus (NV), Astrovirus (HAstV), Adenovirus (Ad), Hepatitis A (HAV), polio, Coxsakie, and echo type Enteroviruses, among others, have been reported to be associated with consumption of fresh water (Gutiérrez et al., 2007).

Enteroviruses are a group of viruses including the polioviruses, coxsackieviruses, echoviruses, and others (CDC, 2007c). In addition to the three different polioviruses, there are 62 non-polio enteroviruses that can cause disease in humans: 23 Coxsackie A viruses, 6 Coxsackie B viruses, 28 echoviruses, and 5 other enteroviruses (CDC, 2007c). Non-polio enteroviruses are second only to rhinoviruses - causative agent of the "common cold" as the most common viral infectious agents in humans (CDC, 2007c). The enteroviruses cause an estimated 10-15 million or more symptomatic infections per year in the United States (CDC, 2007c). All three types of polioviruses have been eliminated from the Western Hemisphere, as well as Western Pacific and European regions, by the widespread use of vaccines (CDC, 2007c). Most people who are infected with an enterovirus exhibit no clinical manifestations of disease. Infected individuals who do exhibit clinical signs of disease usually develop either mild upper respiratory symptoms (a "summer cold"), a flu-like illness with fever and muscle aches, or illness with an associated rash (CDC, 2007c). A less common, though none the less possible, manifestation of the illness is "aseptic" or viral meningitis (CDC, 2007c). Rarely, infected individuals may develop an illness that affects the heart (myocarditis) or the brain (encephalitis) or causes paralysis (CDC, 2007c).

Human noroviruses (NoVs) are a significant cause of non-bacterial gastroenteritis worldwide with contaminated drinking water a potential transmission route (Bae and Schwab, 2008). As members of the *Caliciviridae* family, NoVs (previously known as "Norwalk-like viruses") are small (27 nm), icosahedral, non-enveloped human enteric viruses that cause acute gastroenteritis (Bae and Schwab, 2008). Due to their non-enveloped structure which is similar to other human enteric viruses such as poliovirus (PV), coxsackievirus, and echovirus, NoVs are presumed to be as resistant to environmental degradation and chemical inactivation as the aforementioned viruses (Bae and Schwab, 2008).

The most common vehicles for Hepatitis A virus (HAV) transmission are ingestion of contaminated water, consumption of contaminated foods and contact with infected individuals (De Paula et al., 2007). Hepatitis E virus (HEV) is an emerging food borne pathogen in developing countries, e.g. Asia, Africa and Latin America (Skovgaard, 2007). Transmitted by the fecal–oral route, infection which results in liver inflammation, is linked to the consumption of swine; the primary source of the virus for humans. The incidence of sporadic cases has begun to increase recently in developed countries, but it is uncertain whether these cases were food- or water borne (Skovgaard, 2007). The incidence of hepatitis E infection is highest in adults between the ages of 15 and 40. Children are also succeptable to infection but are less likely to become symptomatic. Mortality rates are generally low as Hepatitis E is a "self-limiting" disease; symptoms generally dissipate by themselves and the patient usually recovers. It is spread mainly through fecal contamination of water supplies or food; person-to-person transmission is uncommon. Outbreaks of epidemic Hepatitis E most commonly occur following disruption to water supplies caused by heavy rainfalls and monsoons.

Viruses can be inactivated in the environment by the breakage of their capsid and the liberation of the contained nucleic acid, which can be easily degraded when deprived of the capsid protection (Gutiérrez et al., 2007). Environmental degradation of viruses can result from extremes in pH, thermal inactivation, sunlight and predation or release of virucidial agents from endogenous microorganism in environmental water (Hurst 1988; Yates etal., 1985). Chlorine, the most commonly used drinking water disinfectant, can also inactivate enteric viruses if sufficient dose and contact time are provided (Ellis, 1991). Despite some opinions to the contrary, this means that detection of viral protein (VP) in laboratory assays is not a sufficient criterion as to the infectious nature of the virus (Gassilloud et al., 2003).

TOXIGENICITY

Microcystis aeruginosa is a gas vacuolate, bloom-forming cyanobacterium that is of interest from a water quality perspective because of its ability to produce a range of toxic compounds (Saker et al., 2005). The cyclic peptides, also known as microcystins, are produced by several species of cyanobacteria including *M. aeruginosa*, and have been implicated in the death of humans as well as domestic and wild animals (Saker et al., 2005). These compounds inhibit protein phosphatases 1 and 2A in a similar manner to okadaic acid, and have been linked to liver cancer in humans (Saker et al., 2005). To date, more than 60 microcystin variants have been identified and chemically characterized (Saker et al., 2005). In recognition of their toxic properties, the World Health Organization has adopted a maximum allowable concentration in drinking water of $1 \mu g l^{-1}$ (WHO 1998). With appropriate water treatment, maximum exposure to total microcystins is probably less than $1 \mu g l^{-1}$, based on the above data. While average exposure is generally well below this level (WHO, 1998), not all water supplies are treated by filtration or adsorption; many are untreated or simply chlorinated (WHO, 1998). In addition to microcystins, many strains of Microcystis are known to produce other peptides including aeruginosins, anabaenopeptilides, cyanopeptolins, anabaenopeptins and microginins, which show a diverse range of bioactivities (Saker et al., 2005).

FOODBORNE 'CONTAMINATING' BACTERIA

Several foodborne pathogens, e.g. *S. typhimurium*, *C. jejuni*, *Y. enterocolitica*, *Mycobacteria*, *E. coli* O157:H7 and *Cryptosporidium*, are common intestinal contaminants frequently found in both farmed animals and wildlife (Doyle and Erickson 2006), and in several cases are carried asymptomatically (Snelling et al., 2005; Thompson et al., 2007). These pathogens are generally shed in feces in large populations and can be transmitted to surface water, both directly and indirectly (Doyle and Erickson 2006). Soil as a recipient of solid wastes is able to contain enteric pathogens in high concentrations, which can then be spread to surface water (Figure 1) (Santamaría and Toranzos, 2003).

Although most *E. coli* strains are harmless, *E. coli* O157:H7 produces a powerful toxin that can cause severe human illness (CDC, 2007b). *E. coli* O157:H7 has been found in the intestines of livestock (CDC, 2007b), is part of the natural microbiota of the soil, and is therefore also associated with manure, crops, minimally processed ready to eat foods and surface water (Selma et al., 2007; Himathongkham et al., 2007). Infections often lead to bloody diarrhea, and in some individuals, particularly children under 5 years of age and the elderly, the infection can cause haemolytic uremic syndrome (HUS), in which the red blood cells are destroyed and the kidneys fail (CDC, 2007b). *Escherichia coli* O157:H7 is a major global cause of foodborne illness (CDC, 2007b). Based on a 1999 estimate, 73,000 cases of infection and 61 deaths occur in the United States each year. People can become infected with *E.coli* O157:H7 in a variety of ways (CDC, 2007b). Though most illness has been associated with eating undercooked, contaminated ground beef, people have also become ill from eating contaminated bean sprouts or fresh leafy vegetables such as lettuce and spinach (CDC, 2007b). In addition, infection can occur after drinking raw unpasturized milk and after swimming in or drinking sewage-contaminated water (CDC, 2007b).

Salmonella are enteric bacterial pathogens, of which mainly *S. enterica* and *S. typhimurium* cause a variety of food and water-borne diseases ranging from gastroenteritis to typhoid fever (Ly and Casanova 2007). Previous investigations have found that sewage effluent regularly contain *Salmonella* and *Campylobacter* (Kinde et al., 1997). Globally, *C. jejuni* is the major cause of human bacterial diarrhoeal illness and is by far the most common *Campylobacter* species associated with human illness. The major *C. jejuni* reservoir is poultry (Snelling et al., 2005). However, *C. jejuni* has been linked to several drinking water-related epidemics in Finland (Lehtola et al., 2006). *C. jejuni* is not normally able to multiply in drinking water or in biofilms, although it may survive in biofilms and/or within protozoa (Lehtola et al., 2006; Snelling et al., 2006).

Shigellosis is a global human diarrhoeal disease leading to dysentery and is caused by *S. dysenteriae*, *S. flexneri*, *S. boydii* and *S. sonnei* (Niyogi et al., 2005). Shigella dysenteriae type

1 produces severe disease and may be associated with life-threatening complications (Niyogi et al., 2005).

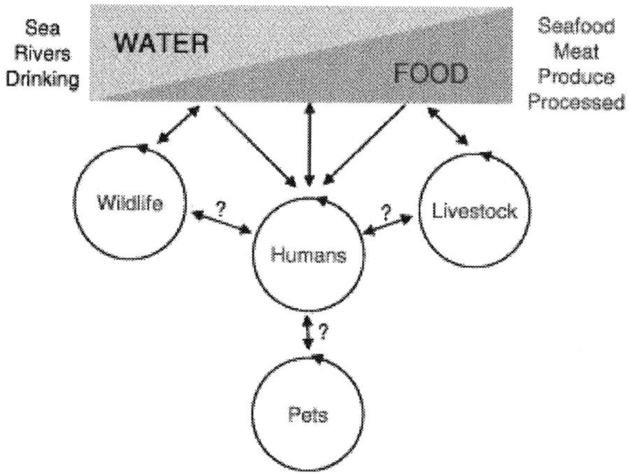

Figure 1. Diagram showing the different, most important cycles of transmission for maintaining *Giardia* and *Cryptosporidium*. As well as direct transmission, water and food may also play a role in transmission. Question marks indicate uncertainty regarding the frequency of interaction between cycles (Hunter and Thompson, 2005).

The symptoms of shigellosis include diarrhoea and/or dysentery with frequent mucoid bloody stools, abdominal cramps and tenesmus (Niyogi et al., 2005). Transmission usually occurs *via* contaminated food, e.g. ready to eat salads (Ghosh et al., 2007), water, livestock manure, or through person-to-person contact (Niyogi et al., 2005). Of the estimated 165 million cases of *Shigella* diarrhoea that occur annually 99% occur in developing countries while in developing countries 69% of episodes occur in children under 5 years of age (Kotloff et al.,1999).

Mycobacterium is a genus of *Actinobacteria*, given its own family. Mycobacteria are aerobic and generally non-motile bacteria, that are characteristically acid-alcohol fast (Ryan and Ray, 2004). Environmental mycobacteria are common heterotrophic bacteria in soils and natural waters, especially in boreal drinking water systems (Torvinen et al., 2004). Mycobacteria are classified as an acid-fast Gram-positive bacterium due to their lack of an outer cell membrane (Ryan and Ray, 2004). All *Mycobacterium* species share a characteristic cell wall, thicker than in many other bacteria, making a substantial contribution to the hardiness of this genus (Ryan and Ray, 2004). Therefore, compared to most other bacterial species, mycobacteria are exceptionally resistant to chlorination and heating (Torvinen et al., 2004). Some thermotolerant species, including the important mycobacterial pathogens *M. avium* complex (MAC) and *M. xenopi*, tolerate heating and may even survive in hot water (>60°C) (Torvinen et al., 2004). Thus, they may survive chemical treatments at waterworks and enter the distributed water and finally tap water, where their occurrence has been known since the beginning of the 20th century.

Tuberculosis (TB) affects one-third of the world's population, claiming a life every 10 seconds and global mortality rates are increasing, especially in developing countries (Sacchettini et al., 2008; Tabbara, 2007). The incidence of tuberculosis is increasing with the increase in the HIV infected population and increased strain drug resistance (Tabbara, 2007).

Complex interactions involving humans, domestic animals, and wildlife create environments favorable to the emergence of new diseases (Palmer, 2007). Today, reservoirs of *M. bovis* subsp. *paratuberculosis* (MAP), the causative agent of tuberculosis in animals and a serious zoonosis, exist in wildlife (Palmer, 2007). The presence of these wildlife reservoirs is the direct result of spillover from domestic livestock, especially cows, in combination with anthropogenic factors such as translocation of wildlife, supplemental feeding of wildlife and wildlife populations reaching densities beyond normal habitat carrying capacities (Palmer, 2007).

Toxigenic *Vibrio cholerae*, the causative agent of cholera, is a native inhabitant of the aquatic environment (rivers, estuaries, and coastal waters) which is transmitted through drinking water and still remains a leading cause of morbidity and mortality in many developing countries, especially in Asia and Africa (Chomvarin et al., 2007). In aquatic environments *V. cholerae* associates with the chitinous exoskeletons of copepod molts, which serves as a surface of nutrients thus facilitating biofilm formation and induces competence for natural transformation (Blokesch and Schoolnik, 2007). Despite more than a century of investigation, much remains to be discovered about how pathogenic strains of *V. cholerae* interact with the human host and how the biology of disseminating stool *V. cholerae* drives devastating cholera outbreaks (Nelson et al., 2007). The *V. cholerae* species encompasses more than 200 serogroups (Blokesch and Schoolnik, 2007). Cholera is an ancient secretory diarrheal disease caused by the O1 and O139 serogroups of *V. cholerae* (Nelson et al., 2007). Despite the dramatic reduction of mortality rates due to the development of oral rehydration solution, the emergence of multiple drug-resistant *V. cholerae* may reduce the efficacy of antimicrobial treatment and alter the dynamics of outbreaks (Nelson et al., 2007). *V. cholerae* is a bacterial pathogen of the gastrointestinal tract and the secreted toxin is largely responsible for the massive fluid loss that may reach between 0.5 and 1.0 liter per hour (Nelson et al., 2007).

Yersinia (family Enterobacteriaceae) are faculatative anerboic bacteria, of which *Y. enterocolitica* are foodborne pathogens which are mostly associated with human disease. *Y. enterocolitica* is responsibel for outbreaks of acute human gastroenteritis and chronic sequela, e.g. reactive arthritis, and livestock morbidity, e.g. mastitis (Meusburger et al., 2007; Rudwaleit et al., 2000; Shwimmer et al., 2007). *Y. pseudotuberculosis* (food-borne route) and *Y. pestis* (flea-borne zoonotic disease) can also cause disease in humans. Rodents are the major reservoirs of *Y. enterocolitica*; and while humans sre the primary hosts other mammals may also be infected. Infection may occur either through blood (in the case of *Y. pestis*) or occasionally *via* consumption of food products (especially vegetables, milk-derived products and meat) contaminated with infected urine or feces. An important property of *Yersinia* is its ability to multiply at low temperatures (4°C) (Skovgaard, 2007). The minimum infection dose of *Yersinia* is relatively high (>100,000), and in many countries, e.g. Denmark, *Yersinia* interest and concern have declined (Skovgaard, 2007).

PROTOZOAN PARASITES

At least 325 water-associated outbreaks of parasitic protozoan disease have been reported (Karanis et al., 2007). North American and European outbreaks accounted for 93% of all reports and nearly two-thirds of outbreaks occurred in North America (Karanis et al., 2007). Over 30% of all outbreaks were documented from Europe, with the UK accounting for 24% of outbreaks, worldwide (Karanis et al., 2007). *Giardia duodenalis* and *C. parvum* account for the majority of outbreaks (132; 40.6% and 165; 50.8%, respectively), *Entamoeba histolytica* and *Cyclospora cayetanensis* have been the aetiological agents in nine (2.8%) and six (1.8%) outbreaks respectively, while *Toxoplasma gondii* and *Isospora belli* have been responsible for three outbreaks each (0.9%) and *Blastocystis hominis* for two outbreaks (0.6%) (Karanis et al., 2007). *Balantidium coli*, the *microsporidia*, *Acanthamoeba* and *Naegleria fowleri* were responsible for one outbreak each (0.3%) (Karanis et al., 2007).

Waterborne parasites produce transmission stages which are highly resistant to external environmental conditions, and to many physical and chemical disinfection methods routinely used as bactericides in drinking water plants, swimming pools or irrigation systems (Gajadhar and Allen, 2004). Resistant stages include cysts of amoebae, *Balantidium*, and *Giardia*, spores of *Blastocystis* and microsporidia, oocysts of *Toxoplasma gondii*, *Isospora*, *Cyclospora* and *Cryptosporidium* and eggs of nematodes, trematodes, and cestode(Gajadhar and Allen, 2004). These exogenous transmission stages are microscopic in size and of low specific gravity, which facilitate their easy dissemination in fresh water, or seawater (Gajadhar and Allen, 2004). The exogenous stages of waterborne parasites possess outer surfaces capable of withstanding a variety of physical and chemical treatments (Gajadhar and Allen, 2004). The resistant surfaces are comprized of multiple polymeric layers of lipids, polysaccharide, proteins or chitin (Gajadhar and Allen, 2004). Examples of these are the two protein layers of coccidian oocysts derived from the coalescence of wall-forming bodies, the chitinous wall of microsporidian spores, the multi-layered (inner lipid/protein-, middle protein/chitin-, outer protein/mucopolysaccharide) shell of *Ascaris* eggs, and the impermeable embryophore of the *Echinococcus* egg which is constructed of polygonal blocks of keratin-like protein held together by a cement substance (Gajadhar and Allen, 2004).

The intestinal protozoan parasites *Cryptosporidium* (Apicomplexan) and *Giardia* (*G. duodenalis*) are major global causes of diarrhoeal disease in humans (Smith et al., 2007). Significantly, normal concentrations of chlorine and ozone used in mass water treatment are not adequate to kill these microbes (Smith et al., 2007), which have life cycles suited to both waterborne and foodborne transmission (Smith et al., 2007). *Giardia* causes intestinal malabsorption and diarrhoea (giardiasis) in humans and other mammals worldwide (Smith et al., 2007). *Giardia* is one of the most prevalent pathogens that should be removed from drinking water (Smith et al., 2007). In developing countries, the prevalence of human giardiasis is on average 20% (4–43%), compared with 5% (3–7%) in developed countries, where it is associated mainly with travel and waterborne outbreaks (Smith et al., 2007). *G. duodenalis* assemblages A and B have been found in humans and most mammalian orders (Smith et al., 2007). Giardiasis is a disease with economic ramifications due to the large impact on domestic animals such as cattle and sheep (Smith et al., 2007). The role of animal transmission of human giardiasis is unclear, but the greatest risk of zoonotic transmission seems to be from companion animals such as dogs and cats. Interestingly, the capacity of *Acanthamoeba* to predate *Cryptosporidium* oocysts has recently been demonstrated (Gómez-Couso et al., 2007). Free-living-amoeba (FLA) may act as environmentally resistant carriers of *Cryptosporidium* oocysts and, thus, may play an important role in the transmission of cryptosporidiosis (Gómez-Couso et al., 2007).

Zoonotic *Cryptosporidium parvum* and anthroponotic *C. hominis* are the major cause of human cryptosporidiosis, although other species including *C. meleagridis*, *C. felis*, *C. canis*, *C. suis*, *C. muris* and two corvine genotypes of *Cryptosporidium* have been associated with human gastroenteritis (Xiao and Ryan 2004; Caccio et al. 2005). *Cryptosporidium* can survive for months in a latent form outside hosts, as its oocysts retain their infectivity for several months in both salt and fresh water (Fayer et al. 1998; Sunnotel *et al.*, 2006a). *Cryptosporidium* causes self-limited watery diarrhoea in immunocompetent subjects, but has far more devastating effects in the immunocompromized and in some cases can be life-threatening due to dehydration caused by chronic diarrhoea (Caccio 2005, Chen et al. 2005). Cryptosporidiosis is responsible for significant neonatal morbidity in farmed livestock and causes weight loss and growth retardation, leading to significant economic losses (McDonald 2000). Over the last two decades, increasing numbers of cryptosporidiosis (in particular water related) outbreaks have been recorded in developed countries (Craun et al. 2005). In 1993, the largest *Cryptosporidium* outbreak was registered in Milwaukee, Wisconsin, USA where 403,000 people were infected through contaminated drinking water (MacKenzie et al. 1994). This outbreak was caused by *C. hominis* (Peng et al. 1997), and the total cost of outbreak-associated illness was estimated at >$96 million in both medical costs and productivity losses (Corso et al. 2003).

Cryptosporidium has also been associated with treated water in swimming and wading pools (Craun et al. 2005). Shellfish have the ability to filter large amounts of water and concentrate oocysts within their gills (Gomez-Couso et al. 2006). Thus, despite prevention measures, e.g. standard UV depuration treatment, the consumption of raw or undercooked shellfish, may still be a potential health risk (Sunnotel et al., 2007). Also, the usage of surface water for irrigation can indirectly cause human infection *via* the consumption of contaminated fresh produce, e.g. lettuce (Shigematsu et al., 2007). Outbreaks have been reported in healthcare facilities and day-care centres, within households, among bathers and water sports enthusiasts in lakes and swimming pools, and in municipalities with contaminated public water supplies or people served by private water supplies. In 2005 the European Basic Surveillance Network (BSN) recorded 7,960 human cases of cryptosporidiosis cases from 16 countries.

Microsporidial gastroenteritis; a serious disease of immunocompromized people, can have a waterborne etiology (Graczyk et al., 2007). Microsporidia are obligate intracellular eukaryotes parasitizing a wide range of invertebrates and vertebrates with over 1,200 species, of which 14 are opportunistic human pathogens, with *Encephalitozoon intestinalis*, *E. hellem*, *E. cuniculi*, and *Enterocytozoon bieneusi* being the most common (Graczyk et al., 2007). Currently, *Microsporidia* are on the Contaminant Candidate List of the U.S. Environmental Protection Agency due to their unknown transmission routes, technologically challenging identification and the difficult treatment of human infections (Graczyk et al., 2007). Considerable evidence indicates involvement of water in the epidemiology of microsporidiosis, however, this link has not been conclusively substantiated (Graczyk et al., 2007). Risk factor analysis for encephalito zoonosis has previously suggested groundwater as a source of infection, and a massive outbreak of microsporidiosis was epidemiologically linked to a drinking water distribution system (Cotte, et al., 1999). More accurate Microsporidial epidemiological data is urgently required to accurately assess the importance of this emerging pathogen.

Emerging and Opportunistic Pathogens

Opportunistic microbes are a subset of the emerging pathogens and include species of both fecal and environmental origin. Treated potable water contains a variety of microbes that are not well characterized (Stelma et al., 2004). Many of these organisms grow slowly and require nutrient-poor media for culturing (Stelma et al., 2004). Although there is evidence that these microbes are generally not hazardous to the general healthy population, there is a possibility that some of them may be opportunistic pathogens and may be capable of causing adverse health effects in individuals with impaired body defences (Stelma et al., 2004). Examples of opportunistic bacterial pathogens of potable water origin include *Aeromonas hydrophila*, *L. pneumophila*, *M. avium*, and *Pseudomonas aeruginosa*. There is no reason to assume that the currently known opportunistic pathogens are the only opportunistic pathogens indigenous to potable water (Stelma et al., 2004). Many cases of respiratory infections and digestive system infections are still of unknown etiology and it is possible that some of them could be due to pathogens that are currently unknown (Stelma et al., 2004).

Opportunistic bacterial pathogens have the potential to reproduce in the natural environment, particularly in the presence of increased temperatures and nutrients, often when associated with free-living protozoa (Långmark et al., 2007). Free-living protozoa are in most cases non-parasitic, however, some species are known to cause human illness (Långmark et al., 2007). The potential significance of free-living protozoa, e.g. amoeba, as potential environmental reservoirs for aquatic pathogens has been recognized for more than twenty years (King et al., 1988), as has the often complex interspecies relationships between mixed biofilm populations, e.g. bacteria and protozoa (Snelling et al., 2006; Långmark et al., 2007). Yet despite this, with the exception of *L. pneumophila*, relatively little attention has been payed to studying and understanding interactions between free-living protozoa and other smaller human pathogens (Brown and Barker, 1999; Snelling et al., 2006). Some microorganisms have evolved to become resistant to protozoa digestion and these amoeba-resistant microorganisms include established bacterial pathogens, such as *Legionella* spp., *Chlamydophila pneumoniae*, *M. avium*, *P. aeruginosa*, and *C. jejuni*, and emerging pathogens such *Simkania negevensis*, *Parachlamydia acanthamoebae*, and *Legionella*-like amoebal pathogens (Snelling et al., 2006).

The fate of internalized bacteria can be divided into three main outcomes;

i) Bacteria which multiply and cause lysis of amoebal cells (FLA), e.g. *Legionella* and *Listeria monocytogenes*.
ii) Bacteria which multiply without causing cell lysis, e.g. *Vibrio cholera*.
iii) Bacteria which survive without multiplication, e.g. mycobacteria (Snelling et al., 2006).

FLA feed on bacteria, fungi and algae and as such are ubiquitous predators that control microbial communities while at the same time acting as reservoirs for many human pathogens. FLAs are thus often regarded as the "Trojan horses" of the microbial world (Chang and Jung, 2004; Greub and Raoult 2004; Molmeret et al., 2005). *Dictyostelium discoideum* is an established host model for several pathogens including, *P. aeruginosa*, *Mycobacterium* spp., and *L. pneumophila* (Snelling et al., 2006). Significantly, numerous bacterial species survive disinfection treatments when internalized within protozoa, compared to the exposed and thus more sensitive planktonic state which forms the basis of current disinfection policies (Snelling et al., 2005b; King et al., 1998; Whan et al., 2006). Interactions such as these are unaccounted for in current disinfection models (Berry, et al., 2006).

FLA provide habitats for the environmental survival of *L. pneumophila*, which have been observed undergoing binary fusion within intracellular vacuoles of amoeba (Atlas, 1999; Colbourne et al., 1984). To date it has never been shown that *Legionella* can multiply in its own natural environment outside protozoa (Atlas, 1999; Colbourne et al., 1984). *Legionella pneumophila* are fastidious bacteria which develop mostly in water, causing legionnaires'disease [(LD), legionellosis, or atypical pneumonia/flu-like illness] in humans (Stout *et al.* 1992; Stojek and Dutkiewicz, 2006). In environmental water bodies *L. pneumophila* proliferates intracellularly in more than 15 protozoan genera, e.g. *Acanthamoeba*, *Hartmannella*, *Vahlkamphia* and *Ehinamoeba* (Steinert et al., 1994). Infection of humans, generally the elderly and immunocompromized, occurs after inhalation of aerosolized bacteria from contaminated water sources (Steinert et al., 1994). The infectious particle is unknown, but may be excreted as legionellae-filled vesicles, intact legionellae-filled amoebae, or free legionellae that have lysed their host cell, resulting in *L. pneuomphila*, not amoeba, persisting within the respiratory tract (Steinert et al., 1994). The processing of *L. pneumophila* by *A. castellanii* shows many similarities to monocyte phagocytosis, e.g. the uptake of *L. pneumophila* by coiling phagocytosis (Steinert et al., 1994). Phagosome–lysosome fusion is prevented based upon *L. pneumophila* expressing *dot*/*icm* genes that code for a putative large membrane complex which forms a type IV secretion system that is used to alter the endocytic pathway (Steinert et al., 1994). Genes such as *hfq* appear to play a major role in exponential phase regulatory cascades of *L. pneumophila* (McNealy et al., 2005; Solomon and Isberg, 2000). Globally, potable water supplies which harbour *L. pneumophila* are important sources of community acquired LD (Stout *et al.* 1992). LD accounts for an estimated 8000 to 18,000 cases of hospitalized community-acquired pneumonia in the United States annually (Phares et al., 2007). Individuals are most often infected with *Legionella* by inhaling bacteria-laden aerosol droplets, e.g. *via* bathing, but can also become infected by the oral route through drinking water or *via* traumatized skin or mucous membranes (Stojek and Dutkiewicz, 2006). Approximately 35% of all LD cases reported to the Centers for Disease Control and Prevention (CDC) are acquired in health care facilities (Phares et al., 2007).

M. avium is a common cause of systemic bacterial infection in patients with AIDS (Miltner and Bermudez, 2000; Snelling et al., 2006). Infection with *M. avium*, commonly occurs through the gastrointestinal tract and has been linked to bacterial colonization of domestic water supplies, where *A. castellanii* may serve as an environmental host for *M. avium* AIDS (Miltner and Bermudez, 2000). *M. avium* can also survive within *A. polyphaga* cyst outer walls and bacterial growth occurs in cocultures (Steinert et al., 1998; Snelling et al., 2006). *M. avium* enters and replicates in *A. castellanii*, and similar to mycobacteria within human macrophages, inhibits lysosomal fusion and replicates in vacuoles that are tightly juxtaposed to the bacterial surfaces within amoebae (Cirillo et al., 1997). Growing *M. avium* in amoebae enhances invasion and intracellular replication of the bacterium in human macrophages, the intestinal epithelial cell line HT-29, as well as in mice (Miltner and Bermudez, 2000), and also increases resistance to rifabutin, azithromycin, and clarithromycin, which might have significant implications for prophylaxis of *M. avium* infection in AIDS patients (Miltner and Bermudez, 2000).

INDICATOR METHODS

For more than thirty years the measurement of bacteria of the coliform group has been extensively relied on as an indicator of water quality (Yáñez et al., 2006). Among the coliforms, the specific determination of *Escherichia coli* contamination can be performed as one of the best means of estimating the degree of recent fecal pollution (Edberg et al., 2000). The European Drinking Water Directive 98/83/EC (Anon, 1998) defines that the isolation of coliforms using Lactose TTC agar with Tergitol (Tergitol-7 agar) by membrane filtration (Anon, 2000) be the reference method for the enumeration of total coliforms, including *E. coli* in drinking water (Yáñez et al., 2006). However, numerous outbreaks, e.g. *Crytosporidium*, *Campylobacter*, and viruses, have made it clear that the presence of bacterial indicators of fecal contamination does not consistently correlate with pathogen levels (De Paula et al., 2007).

Alternatively, a number of instrumental methods for the rapid determination of microorganisms based on their metabolic activity have been developed, but not really utilized, for example, impedance (Madden and Gilmour, 1995), conductance (Gibson, 1987), chemiluminescence and fluorescence (Van Poucke and Nelis, 2000). Different chromogenic and/or fluorogenic culture media have also been developed in recent years (Manafi, 2000). One is TBX agar medium, a modification of Tryptone bile agar medium where the substrate 5-bromo-4-chloro-3-indolyl-b-d-glucuronide (BCIG) is added. This substrate is cleaved by the enzyme β-d-glucuronidase (GUD) that is produced by approximately 95% of the *E. coli* strains investigated (Adams et al., 1990), and the released chromophore produces easy to read blue-green coloured *E. coli* colonies. Furthermore, the TBX agar method complies with the ISO/DIS Standard 16649 for the enumeration of *E. coli* in food and animal foodstuffs (Anon, 2001). In addition, different chromogenic media such as Chromocult Coliform agar (CC agar), and coli ID agar have been developed for the simultaneous detection of total coliforms and *E. coli*, based on the presence of chromogenic substrates for the enzymes β–galactosidase (Lac) and β-D-glucuronidase (GUD) (Geissler et al., 2000). Results from the evaluation of these two media have been reported previously (Finney et al., 2003). In spite of their advantages, the main drawback of chromogenic media is the relatively high cost for routine analysis laboratories where a great number of analysis are performed (Yáñez et al., 2006). The results obtained with the combination of the two media, Tergitol agar and TBX agar, for the enumeration of total coliforms and *E. coli* have been comparable to those obtained with the individual chromogenic media assayed, CC agar and coli ID (Yáñez et al., 2006). However, the combined method has the advantage of reduced analytical cost since the TBX agar is used only for total coliform-positive samples (Yáñez et al., 2006). In addition, using this small modification, the enumeration of *E. coli* in total coliforms-positive samples can be performed in only two hours, maintaining the advantages of rapid turn-around time and the specificity of the individual chromogenic media (Yáñez et al., 2006). The combination of Tergitol agar with TBX fulfils the European Drinking Water Directive where the ISO Standard 9308 is indicated for the analysis of total coliforms and *E. coli* (Yáñez et al., 2006). Moreover, this small modification provides a simple and comparatively cost-effective method that is as specific and rapid as the chromogenic media, and which can be used easily by laboratories dedicated to routine analysis of water (Yáñez et al., 2006).

Rapid Detection Methods

Many bacterial pathogens are routinely grown on selective/differential agar or broth, e.g. *Shigella* (Niyogi et al., 2005). Infection with *E. coli* O157:H7 is diagnosed by detecting the bacterium from stool samples (CDC, 2007b). About one-third of laboratories that culture stool still do not test for *E. coli* O157:H7, so it is important to request that the stool specimen be tested on sorbitol-MacConkey (SMAC) agar for this organism (CDC, 2007b). All individuals presenting with bloody diarrhoea should have their stool tested for *E. coli* O157:H7 (CDC, 2007b). However, a major limitation of enrichment methods is the length of time required to complete the testing, since at least two days are needed for the complete identification of bacterial typical colonies (Yáñez et al., 2006). To overcome this problem, new techniques are continually being developed, and in particular, molecular biology methods appear to be interesting alternatives for rapidly detecting pathogens in water samples (Yáñez et al., 2006). To maximise the useful impact of epidemiological data, it is important to be able to rapidly detect and identify low numbers of microbes from different sample types, including water, and if possible to ascertain if they are viable (Sunnotel et al., 2006a; Sunnotel et al., 2006b). It is important to be able to accurately differentiate between non-pathogenic and pathogenic species of *Cryptosporidium*, and between different *Cryptosporidium* isolates of the same species (Sunnotel et al., 2006a; Sunnotel et al., 2006b). Also, the rapid and accurate detection and identification of *E. coli* O157:H7 and *V. cholerae* O139 pathogens is crucial for the diagnosis, treatment and eventual control of the contagious disease outbreak (Jin et al., 2007).

Accurate and reliable epidemiological and molecular typing techniques are crucial for tracing sources of pathogen infection, the recognition of which is important for the implementation of preventive measures (Fendukly et al., 2007). Methods include ribotyping, amplified fragment length polymorphism analysis (AFLP), pulsed field gel electrophoresis (PFGE), restriction fragment length polymorphism analysis (RFLP), restriction endonuclease analysis (REA), and arbitrary primed PCR (Fendukly et al., 2007). Nevertheless, molecular techniques are unfortunately still incompatible with most routine water laboratories because of the expense and the need for trained personnel (Angles d'Auriac et al., 2000). Disappointingly, despite the steady improvement in modern molecular biology techniques, the epidemiology of many infections remains unclear (Snelling et al., 2005b). This confusing epidemiological evidence is partly because of the lack of standard global typing methods and communication between laboratories (Wassenaar and Newell 2000). However, these problems are being addressed *via* initiatives like CAMPYNET (Wassenaar and Newell 2000) and PulseNet for standard molecular typing.

Nucleic acid based detection methods such as PCR and real-time PCR are more rapid and sensitive than traditional culturing techniques. However, they are limited by the number of targets that can be detected in a single assay. PCR products can be multiplexed, but typically not more than six targets can be assayed at a single time, and optimizing such a complex reaction can be challenging. For example, primers must not interact with one another to form primer-dimers, and amplification of each target must proceed with equal efficiency. Analysis of the products of the reaction requires size separation by electrophoresis, thus amplicons must be of significantly different size in order to be easily distinguished. Real-time PCR (qPCR) is more rapid as it monitors the amount of PCR products as they are being produced and does not require size separation (Kubista et al. 2006; Zhang and Fang, 2006). In addition qPCR is more sensitive than traditional PCR methods and can estimate the initial concentration of nucleic acids. As with standard PCR methods, multiplexing qPCR is limited by the difficulty in optimizing reactions so that all targets are amplified with similar efficiencies. Moreover, the number of assays that can be multiplexed is limited by the number of fluorescent dyes that can be distinguished in a single reaction (typically no more than 4, depending on the specifications of the instrument being used).

Saker et al., (2007) found that PCR can be used to detect inocula for cyanobacterial populations and therefore provides a useful tool for assessing which conditions particular species can grow into bloom populations. PCR methods were particularly useful when the concentration of the target organism was very low compared with other organisms (Saker et al., 2007).

The European Working Group for *Legionella* Infections (EWGLI) has implemented the AFLP and more recently the sequence-based typing (SBT) to define its collection of *L. pneumophila* serogroup 1 strains (Fendukly et al., 2007). However, the SBT has not been widely used for genotyping non-serogroup 1 *L. pneumophila* isolates (Fendukly et al., 2007). Of the many phenotypical, immunological and molecular typing methods which have been implemented for the characterization and epidemiological typing of *L. pneumophila*, only the AFLP and the SBT techniques have gained acceptance by the EWGLI. Both methods were shown to have good reproducibility and accuracy and obviated the need to submit *Legionella* strains between microbiological laboratories in different countries (Fendukly et al., 2007). The AFLP gel images were, however, more difficult to interpret (Fendukly et al., 2007). This method allows for comparison of the allelic profile of an isolate to previously assigned allele numbers of 6 genes in the following given order: *flaA*, *pilE*, *asd*, *mip*, *mompS* and *proA* (Fendukly et al., 2007).

Some enteric viruses grow poorly in cell culture, which constitutes a problem when investigating strategies for virus control and prevention (Atmar and Estes, 2001; De Paula et al., 2007). Current methods of detecting such viruses in environmental water samples rely on genome amplification using molecular techniques such as qualitative and quantitative real-time reverse-transcription polymerase chain reaction (RT–PCR) (De Paula et al., 2007). Worldwide, virus detection in environmental and potable water samples is becoming an important strategy for preventing outbreaks of infection with waterborne viruses, e.g. HAV (De Paula et al., 2007).

To evaluate the public health threat posed by *V. cholerae* occurring in water, a rapid and accurate method for detection of toxigenic *V. cholerae* is essential (Chomvarin et al., 2007). The culture method (CM), which is routinely used for assessing water quality, has not proven as efficient as molecular methods because the notorious pathogen survives in water mostly in the viable but non-culturable state (VBNC) (Chomvarin et al., 2007). The isolation and identification of *V. cholerae* by CM is expensive, time-consuming, labour intensive, and is unable to precisely distinguish between toxigenic *V. cholerae* and non-toxigenic *V. cholerae* (Chomvarin et al., 2007). Fluorescent monoclonal antibody combined with molecular genetic based methods can demonstrate the presence of toxigenic *V. cholerae* in the aquatic environment (Chomvarin et al., 2007).

The last three decades have witnessed exponential progress in our understanding of the biology of pathogens. The complete sequencing of the human genome and of many microbial pathogens associated with potable water have rapidly fuelled the development of gene array technology, improved pathogen detection systems and analysis methods. A host of different molecular techniques for studying the transmission dynamics and detection of drug resistant pathogens have been developed (Katoch et al., 2007).

DNA microarrays (or gene arrays) consist of segments of DNA called probes or reporters arrayed on a solid surface (Call, 2005; Lemarchand, 2004). DNA spots are typically very small, on the micrometer length scale, and can be arrayed in high density with hundreds to thousands of spots present on a single array. Detection using microarrays is based on the hybridization of fluorescently labelled nucleic acids (RNA or DNA) isolated from the test matrix to these probes, and subsequent measurement of fluorescence at each spot. Microarrays allow the simultaneous detection of thousands of targets in a single assay. Thus, a single array could be used for the detection of a wide range of pathogens, virulent strains of a particular pathogen, antimicrobial resistant strains, and even molecular typing. Microarrays have been applied to the microbial quality monitoring of water (Lee et al., 2006; Lemarchand et al., 2004; Maynard et al., 2005). However, methods involving direct hybridization onto microarrays are inherently lower in sensitivity relative to PCR because of the lack of oligonucleotide amplification. For example, Lee et al. (2006) were only able to detect pathogens in wastewater at levels of 2×10^8 copies of the target gene, using direct hybridization of DNA isolated from wastewater. Straub et al. (2005) have developed a microarray-based system for analysis of RNA isolated from microbial communities. This system shows great promise, as detection of RNA could be linked to the presence of viable cells; however, the sensitivity of this system is currently too low to be useful for pathogen detection.

Sensitivity can be greatly improved by PCR amplification of the target, before hybridization to the array. However, production of multiple PCR products can be labour intensive and slow. Amplification of universal genes (i.e. rRNA genes) gets around this problem as a single primer set can be used to amplify a segment from a gene that is conserved among the pathogens of interest. Species-specific sequence variability within the amplicon is detected by hybridization of the PCR products to a microarray. Using this method, with the 23s rRNA gene, Lee et al. (2006) demonstrated an exponential increase in the sensitivity of their array to 20 copies of the target gene. Maynard et al. (2005) used a similar strategy, but improved discriminatory power of their array by the inclusion of multiple genes (*cpn60*, 16s rRNA and *wecE*). They were able to detect 100 copies of a target gene, but this detection level was decreased when the genomic DNA of interest was in a background of complex DNA mixtures. PCR amplification favors the amplification of the most abundant bacterial species and low-level targets may be underrepresented, or undetectable on the array. This is an important problem, as pathogens are likely to represent a small proportion of the microbial communities in water supplies. Kostic et al. (2007) were able to overcome this problem to a degree by using a unique labelling method (sequence-specific end-labelling of oligonucleotides) and inclusion of competitive oligonucleotides during labelling. They were able to detect pathogens which comprised 0.1% of the total microbial community analyzed. Wu et al (2006) used a method of total DNA amplification (multiple displacement amplification) to non-specifically amplify total genomic DNA isolated from contaminated groundwater prior to microarray hybridization. This method seems promising, as it appears to eliminate the bias of PCR based amplification, while increasing the sensitivity of the microarray-based assay.

Microarrays have proven to be useful tools for monitoring microbial communities in water and environmental samples. However, there are practical limitations in the use of microarray-based approaches including the requirement of specialized equipment and trained personnel. Automation of array systems would contribute significantly to the ease of deployment of these systems. Suspension arrays, such as the Luminex® 100™ system, are particularly amenable to automation (Baums et al., 2007; Gilbride et al., 2006). These arrays work on similar principles to planar microarrays, but they are rapid, cost effective, flexible and provide high reproducibility. In these systems, probes are immobilized onto solid surfaces of microspheres that are labelled with fluorophores (up to 100 colours are available) to facilitate their identification. After hybridization, the beads are analyzed in a flow cytometer. A red laser identifies the bead, and a green laser registers if a target has been captured. This system was found to be effective for detecting six different fecal indicating bacteria in environmental samples (Baums et al., 2007). Straub et al. (2005) have developed a comprehensive system for sample preparation and pathogen detection called "BEADS" (biodetection enabling analyte delivery system). This system incorporates immunomagnetic separation to capture cells for concentration and purification of the analytes, a flow through thermal cycler and detection of PCR products on a suspension array. Using this system with river water, Straub et al. were able to detect 10 cfu of *E. coli*. When multiplexed with *Shigella* and *Salmonella*, sensitivity was reduced, but 100 cfu of each organism could be detected.

Biosensor technologies offer the promise of portable devices delivering reliable results within short periods of time. Biosensors can be defined as analytical devices incorporating a biological material such as specific antibodies or DNA probes to confer specificity, integrated within a transducing microsystem (optical, electrochemical, thermometric, piezoelectric, magnetic or micromechanical) (Lazcka et al., 2007; Gilbride et al., 2006). Biosensors can be incorporated into miniaturized microfluidic devices commonly referred to as "lab-on-a-chip" devices. Typical "lab-on-a-chip" devices contain wells for samples and storage of reagents, and microchannels for distribution of the samples and reagents to reaction chambers. They are connected to an instrument for control and detection of reactions by biosensors and to a computer for data analysis. The use of such devices reduces reagent costs, increases speed of analysis, minimizes handling steps, and provides portability, capability for parallel operations and the automation of complex assays. Several biological assays have been miniaturized and automated, including PCR, qPCR, DNA electrophoresis and microarrays, and immunoassays (Chen et al., 2007). A hand held device comprising electrochemical biosensors was successfully used to screen environmental water samples for 8 pathogens, within 3-5 h (LaGier et al., 2007). The detection limit of the device was equivalent to 10 cells of *Karenia brevis* when purified genomic DNA was assayed with the device. The scale of such devices, in terms of the number of targets that can be analyzed at a time, is still limited. The incorporation of high-density microarrays into integrated microfluidic devices containing biosensors has the potential to address the important issues of sensitivity and assay automation while maintaining the high information content of a microarray. Such systems may enable the reliable real-time monitoring of microbial communities, and detection of pathogens in water systems. While current devices show promise, there is a need to improve sensitivity, to improve specificity when dealing with complex samples, to provide portability and low cost in order to gain widespread use at water treatment facilities.

Conclusion

In most developed countries, the removal or inactivation of the majority of pathogens, e.g. *Cryptosporidium* oocysts, can be accomplished by conventional water treatment technology, which generally includes flocculation, coagulation, sedimentation, filtration and chlorination (Betancourt and Rose 2004). Conventional treatment consists of coagulation/flocculation, sedimentation, and filtration (Sunnotel et al., 2006a). Implementation of multiple barriers to safeguard drinking water is recommended as pathogen loads vary during the season, with peak loads during early spring and late autumn (Ferguson et al. 2003). The high occurrence of many important human pathogens in surface water sources underlines the need for frequent monitoring of the parasite in drinking water (Frost et al. 2002). The high cost of waterborne disease outbreaks should be considered in decisions regarding water utility improvements and the construction of treatment plants (Morgan-Ryan et al. 2002). The integration of watershed and source water management and protection, scientific management of agricultural discharge and run-off, pathogen or indicator organism monitoring for source and treated water, outbreak and waterborne disease surveillance is needed to reduce waterborne transmission of human diseases (Ferguson et al. 2003). Water treatment facilities employing second-line treatment practices such as UV irradiation and ozone treatment can alleviate the danger imposed by chlorine resistant pathogens, e.g. *Cryptosporidium* (Keegan et al. 2003).

However, outbreaks of waterborne illness can still occur in developed countries, because of malfunction or mismanagement of water treatment facilities (Ferguson et al. 2003). In these cases, hazard analyses protocols (microbiological hazards based on fecal coliforms (FC) and turbidity (TBY) as indicators) for critical control points (CCPs) within each facility may help to minimise the risk of contaminated water distribution in cases of system component failure, where CCPs include raw resource water, sedimentation, filtration and chlorine-disinfection (Jagals and Jagals 2004).

Without knowing the occurrence of pathogens in water it is difficult to determine what risk they present to consumers of contaminated potable water (Karanis et al., 2007). Standardized methods are required to determine the occurrence of both established and emerging pathogens in raw untreated water abstracted for potable water, water treatment systems, potable water, and also recreational waters (Karanis et al., 2007). Also, because of the potential for pathogens to interact within protozoa and/or biofilms, the total populations of potable water must be assessed and monitored, e.g. how do populations vary with seasonality and from one country to another? An improved understanding of the microbial ecology of distribution systems is necessary to design innovative and effective control strategies that will ensure safe and high-quality drinking water (Berry, et al., 2006).

Better education and increased awareness by the general public and pool operators could potentially reduce the number and impact of swimming pool and other recreational water-related outbreaks (Robertson et al. 2002a). Improved detection methods, with the ability to differentiate species, can be useful in the assessment of infection and identification of contamination sources. These will also provide vital data on the levels of disease burden due to zoonotic transmission (Fayer 2004; Sunnotel et al., 2006a). The roles of humans, livestock and wildlife in the transmission of microbial pathogens remain largely unclear for many different areas. The continued and improved monitoring (using appropriate molecular methods) of pathogens in surface water, livestock, wild life and humans will increase our knowledge of infection patterns and transmission of pathogens in potable water (Fayer 2004; Sunnotel et al., 2006a).

For zoonotic pathogens which are spread both directly and indirectly to potable water, a wide array of interventions have been developed to reduce the carriage of foodborne pathogens in poultry and livestock, including genetic selection of animals resistant to colonisation, treatments to prevent vertical transmission of enteric pathogens, sanitation practices to prevent contamination on the farm and during transportation, elimination of pathogens from feed and water, additives that create an adverse environment for colonization by the pathogen, and biological treatments that directly or indirectly inactivate the pathogen within the host (Doyle and Erickson, 2006; Sunnotel et al., 2006a). To successfully reduce the carriage of foodborne pathogens, it is likely that a combination of intervention strategies will be required (Doyle and Erickson, 2006). Collaborative efforts are needed to decrease environmental contamination and improve the safety of produce. There is a very real need for studies that monitor the quality of the water as well as for policies and investments that focus on sanitation (De Paula et al., 2007).

Unfortunately, in developed countries, the currently adopted short-term and blinkered strategy of simply detecting pathogens, e.g. a positive/negative *Cryptosporidium* result, without further species and isolate information impedes any significant or meaningful epidemiological analysis and/or effective implementation of intervention strategies. Because of the current necessity to minimise costs, current molecular methods (outlined previously) and future nanotechnology based methods are not being routinely employed, and thus may never fulfil their maximum potential.

This situation demands significant and immediate attention and collaborative input from governments and water industries world-wide – resulting in a more concerted effort from both scientists and policy makers to standardise the global epidemiological response. Currently, due the significant amount of work involved in sample collection and statistical analysis, there is a time lag of approximately 2-3 years in attaining up to date pathogen data (see CDC table 1). A global data base and collective goals for biological contaminant loading are vital for achieving safe potable water. More funding is needed to speed up this process, and to better educate the general public of the advantages of effective intervention strategies. Globally, rapid and accurate monitoring and species typing must be carried out routinely, not just when outbreaks occur. There is also a need for much more accurate and conclusive information on the overall microbial populations present in water and their relationships to one another symbiotic, parasitic or otherwise (Snelling et al., 2006). Once successfully introduced in developed countries, then and only then, can this template be effectively applied to the developing world.

If successful the substantial economic cost (clinical costs and lost working hours) of water borne pathogens might, in time, be reimbursed through improved epidemiological data collection which, with proactive intervention, will dramatically reduce the health and economic burden of waterborne disease.

Table 1. Summaries of Notifiable Diseases in the United States in 2005 (McNabb et al., 2007)

TABLE 1. Reported cases of notifiable diseases,* by month — United States, 2005

Disease	Jan	Feb	Mar	Apr	May	Jun	Jul	Aug	Sep	Oct	Nov	Dec	Total
AIDS†	2,905	2,695	4,274	3,130	3,255	3,877	3,631	3,219	3,353	3,963	2,943	3,875	**41,120**
Botulism													
foodborne	1	—	—	—	—	1	2	9	1	—	2	3	**19**
infant	4	4	5	4	10	10	9	8	8	10	3	10	**85**
other (wound & unspecified)	1	1	1	3	3	1	1	4	3	5	2	6	**31**
Brucellosis	3	8	8	11	12	8	13	14	7	12	9	15	**120**
Chancroid§	2	2	2	2	1	1	—	—	—	1	2	4	**17**
Chlamydia§¶	67,989	76,735	76,283	91,530	75,649	72,200	91,765	75,576	71,290	94,206	70,134	113,088	**976,445**
Cholera	—	—	—	1	—	1	—	2	—	2	1	1	**8**
Coccidioidomycosis	360	335	251	304	326	295	328	510	319	584	565	2,365	**6,542**
Cryptosporidiosis	129	138	147	212	175	179	394	947	1,495	874	354	615	**5,659**
Cyclosporiasis	2	2	6	44	229	123	79	20	7	2	13	16	**543**
Domestic arboviral diseases**													
California serogroup													
neuroinvasive	—	—	—	1	1	5	15	20	20	11	—	—	**73**
nonneuroinvasive	—	—	—	—	—	—	1	5	1	—	—	—	**7**
eastern equine, neuroinvasive	—	—	—	—	—	2	4	11	3	1	—	—	**21**
Powassan, neuroinvasive	—	—	—	—	—	—	—	1	—	—	—	—	**1**
St. Louis													
neuroinvasive	—	—	—	—	—	—	—	1	5	—	1	—	**7**
nonneuroinvasive	—	—	—	—	—	1	1	1	3	—	—	—	**6**
West Nile													
neuroinvasive	—	—	1	—	1	21	191	590	407	91	6	1	**1,309**
nonneuroinvasive	1	1	1	1	10	39	326	849	402	54	7	—	**1,691**
Ehrlichiosis													
human granulocytic	—	4	7	29	36	97	175	96	96	68	32	146	**786**
human onocytic	4	5	10	8	16	35	87	66	72	59	34	110	**506**
human (other & unspecified)	2	2	2	1	5	23	38	10	9	10	2	8	**112**
Enterohemorrhagic *Escherichia coli* infection													
O157:H7	58	73	87	127	116	190	317	338	367	451	181	316	**2,621**
Shiga toxin-positive													
non-O157	13	17	14	18	22	29	58	53	55	68	31	123	**501**
not serogrouped	14	11	8	22	19	12	29	62	56	61	26	87	**407**
Giardiasis	1,047	1,179	1,284	1,579	1,242	1,261	1,899	1,916	2,096	2,464	1,365	2,401	**19,733**
Gonorrhea§	25,339	24,520	24,706	29,739	23,995	24,610	33,106	27,189	26,335	33,221	25,012	41,821	**339,593**
Haemophilus influenzae, invasive disease													
all ages, serotypes	182	205	220	255	208	192	188	113	146	158	129	308	**2,304**
age <5 yrs													
serotype b	—	—	—	1	1	1	—	—	2	—	1	3	**9**
nonserotype b	3	15	19	9	10	9	10	13	11	16	3	17	**135**
unknown serotype	14	24	24	22	17	16	13	20	14	13	13	27	**217**
Hansen disease (leprosy)	2	2	4	6	5	19	6	6	3	7	3	24	**87**
Hantavirus pulmonary syndrome	—	2	—	1	4	5	5	2	3	1	1	2	**26**
Hemolytic uremic syndrome, postdiarrheal	4	11	9	17	10	17	17	33	22	24	10	47	**221**
Hepatitis, viral, acute													
A	267	331	278	337	262	276	344	362	482	498	272	779	**4,488**
B	331	382	341	469	343	337	468	352	367	454	334	941	**5,119**
C	33	32	39	47	38	50	63	45	58	70	42	135	**652**

* No cases of anthrax; diphtheria; domestic arbovial disease, western equine encephalitis virus, neuroinvasive and nonneuroinvasive, eastern equine nonneuroinvasive, and Powassen nonneuroinvasive; severe acute respiratory syndrome–associated coronavirus (SARS-CoV) disease; smallpox; or yellow fever were reported in 2005. Data on chronic hepatitis B and hepatitis C virus infection (past or present) are not included because they are undergoing data quality review. Data on human immunodeficiency virus (HIV) infections are not included because HIV infection reporting has been implemented on different dates and using different methods than for acquired immunodeficiency syndrome (AIDS) case reporting.

† Total number of AIDS cases reported to the Division of HIV/AIDS Prevention, National Center for HIV/AIDS, Viral Hepatitis, STD, and TB Prevention (NCHHSTP) (proposed), through December 31, 2005.

§ Totals reported to the Division of STD Prevention, NCHHSTP (proposed), as of May 5, 2006.

¶ Chlamydia refers to genital infections caused by *Chlamydia trachomatis*.

** Totals reported to the Division of Vector-Borne Infectious Diseases, National Center for Zoonotic, Vector-Borne, and Enteric Diseases (NCZVED) (proposed) (ArboNET Surveillance), as of June 23, 2006.

TABLE 1. (*Continued*) Reported cases of notifiable diseases,* by month — United States, 2005

Disease	Jan	Feb	Mar	Apr	May	Jun	Jul	Aug	Sep	Oct	Nov	Dec	Total
Influenza-associated													
pediatric mortality††	4	10	10	4	4	3	3	1	1	—	1	4	**45**
Legionellosis	95	76	78	88	96	141	348	250	284	319	222	304	**2,301**
Listeriosis	40	34	42	47	38	54	114	109	98	130	79	111	**896**
Lyme disease	448	377	470	562	784	2,293	5,929	3,965	2,124	1,957	1,211	3,185	**23,305**
Malaria	105	79	80	99	90	118	173	150	146	127	96	231	**1,494**
Measles	3	4	3	5	1	2	33	2	4	—	3	6	**66**
Meningococcal disease, invasive													
all serogroups	102	121	142	129	108	115	82	55	59	81	78	173	**1,245**
serogroup A, C, Y, & W-135	26	31	39	34	29	30	15	14	14	17	16	32	**297**
serogroup B	12	12	16	16	10	16	11	4	8	12	6	33	**156**
other erogroups	5	4	3	4	2	2	1	2	1	—	1	2	**27**
serogroup unknown	59	74	84	75	67	67	55	35	36	52	55	106	**765**
Mumps	18	28	19	25	26	26	27	52	15	19	23	36	**314**
Pertussis	1,724	1,630	1,196	1,598	1,816	1,818	2,508	2,137	1,974	2,584	1,879	4,752	**25,616**
Plague	—	—	—	—	2	—	2	1	2	—	—	1	**8**
Poliomyelitis, paralytic§§	—	—	—	—	—	—	—	—	—	1	—	—	**1**
Psittacosis	—	1	—	5	—	1	4	1	1	1	—	2	**16**
Q fever	6	2	6	10	14	24	19	14	12	14	3	12	**136**
Rabies													
animal	485	291	464	732	551	466	565	582	550	525	332	372	**5,915**
human	1	—	—	—	—	—	—	—	1	—	—	—	**2**
Rocky Mountain spotted fever	41	40	35	57	81	185	243	290	234	192	168	370	**1,936**
Rubella	—	1	1	2	1	1	2	1	—	—	—	2	**11**
Rubella, congenital syndrome	—	—	1	—	—	—	—	—	—	—	—	—	**1**
Salmonellosis	1,745	1,730	2,009	2,731	3,154	3,777	5,585	5,149	5,016	5,589	3,384	5,453	**45,322**
Shigellosis	655	790	918	1,071	1,092	1,195	1,574	1,485	1,641	2,060	1,322	2,365	**16,168**
Streptococcal disease, invasive, group A	345	421	469	600	436	362	378	260	215	294	265	670	**4,715**
Streptococcal toxic-shock syndrome	13	14	22	31	12	9	6	3	2	2	5	10	**129**
Streptococcus pneumoniae, invasive disease													
drug resistant, all ages	223	268	335	371	263	207	161	93	99	161	194	621	**2,996**
age <5 yrs	94	112	167	164	155	118	80	48	45	103	117	292	**1,495**
Syphilis¶¶													
all stages***	2,056	2,370	2,489	3,392	2,660	2,662	3,156	2,631	2,326	3,268	2,429	3,839	**33,278**
congenital (age <1 yr)	25	32	25	26	27	36	28	24	28	21	20	37	**329**
primary & secondary	532	612	562	880	698	675	830	716	592	916	672	1,039	**8,724**
Tetanus	—	2	3	1	3	3	3	1	2	2	1	6	**27**
Toxic-shock syndrome	8	6	7	6	8	10	9	6	8	1	3	18	**90**
Trichinellosis	1	—	—	1	1	2	5	2	2	—	—	2	**16**
Tuberculosis†††	589	799	1,116	1,036	1,103	1,334	1,110	1,174	1,231	1,146	1,150	2,309	**14,097**
Tularemia	1	—	2	3	7	31	24	26	18	20	4	18	**154**
Typhoid fever	20	10	19	25	17	24	32	29	39	51	14	44	**324**
Vancomycin-intermediate *Staphylococcus aureus*	—	—	—	—	—	—	—	—	1	—	1	—	**2**
Vancomycin-resistant *Staphylococcus aureus*	—	—	—	2	—	—	—	—	—	1	—	—	**3**
Varicella (chickenpox)	1,869	2,261	2,851	3,180	2,813	2,401	1,776	1,211	1,363	3,167	2,924	6,426	**32,242**
Varicella (deaths)§§§	—	—	—	1	1	—	—	—	—	1	—	—	**3**

†† Totals reported to the Influenza Division, National Center for Immunization and Respiratory Diseases (NCIRD) (proposed), as of December 31, 2005.

§§ Cases of vaccine-associated paralytic polio (VAPP) caused by polio vaccine virus.

¶¶ Totals reported to the Division of STD Prevention, NCHHSTP (proposed), as of May 5, 2006.

*** Includes the following categories: primary, secondary, latent (including early latent, late latent, and latent syphilis of unknown duration), neurosyphilis, late (including late syphilis with clinical manifestations other than neurosyphilis), and congenital syphilis.

††† Totals reported to the Division of TB Elimination, NCHHSTP (proposed), as of May 12, 2006.

§§§ Death counts provided by the Division of Viral Diseases, NCIRD (proposed), as of December 31, 2005.

References

Adams, M.R., Grubb, S.M., Hamer, A., and Clifford, M.N. (1990) Colorimetric enumeration of Escherichia coli based on β-glucuronidase activity. *Applied and Environmental Microbiology. 56*, 2021–2024.

Angles d'Auriac, M.B., Roberts, H., Shaw, T, Sirevag, R., Hermansen, L.F., and Berg, J.D. (2000) Field evaluation of a semiautomated method for rapid and simple analysis of recreational water microbiological quality. Applied and Environmental Microbiology. 66, 4401–4407.

Arrowood, M.J. (1997) Diagnosis. In Cryptosporidium and Cryptosporidiosis ed. Fayer, R. p. 45. Boca Raton, FL: CRC Press.

Atlas, R.M. (1999) Legionella: from environmental habitats to disease pathology, detection and control. Environmental Microbiology. 1, 283–293.

Atmar, R.L., and Estes, M.K. (2001) Diagnosis of noncultivatable gastroenteritis viruses, the human caliciviruses. Clinical Microbiology Reviews. 14, 15–23.

Bae, J., and Schwab, K.J. (2008) Evaluation of murine norovirus, feline calicivirus, poliovirus, and MS2 as surrogates for human norovirus persistence in surface and groundwater. Applied and Environmental Microbiology. 74, 477-484.

Baums, I.B., Goodwin, K.D., Kiesling, T., Wanless, D., Diaz, M.R., and Fell, J.W. (2007) Luminex detection of fecal indicators in river samples, marine recreational water, and beach sand. Marine Pollution Bulletin. 54, 521-536.

Berry, D., Xi, C.W., and Raskin, L. (2006) Microbial ecology of drinking water distribution systems. Current Opinion in Biotechnology. 17, 297-302.

Betancourt, W.Q. and Rose, J.B. (2004) Drinking water treatment processes for removal of Cryptosporidium and Giardia. Veterinary Parasitology. 126, 219–234.

Blokesch, M., Schoolnik, G.K. (2007) Serogroup conversion of Vibrio cholerae in aquatic reservoirs. *PLoS Pathogens. 3, e81.*

Brown, M.R.W., and Barker, J. (1999) Unexplored reservoirs of pathogenic bacteria: protozoa and biofilms. Trends in Microbiology. 7, 46–50.

Caccio, S.M. (2005) Molecular epidemiology of human cryptosporidiosis. Parassitologia. 47, 185–192.

Caccio, S.M., Thompson, R.C., McLauchlin, J. and Smith, H.V. (2005) Unravelling Cryptosporidium and Giardia epidemiology. Trends in Parasitology. 21, 430–437.

Calderon, R. L., and Craun, G. F. (2006) Estimates of endemic waterborne risks from community-intervention studies. *Journal of Water and Health.* 4, 89-99.

Call, D.R. (2005) Challenges and opportunities for pathogen detection using DNA microarrays. Critical Reviews in Microbiology. 31, 91-99.

CDC, (2007a) Healthy Drinking Water, Department of Health and Human Services, Centers for Disease Control and Prevention, Parasitic Diseases (last accessed on 12/01/2008), http://www.cdc.gov/ncidod/dpd/healthywater/index.htm

CDC, (2007b) Centres for Disease Control and Prevention, Department of Health and Human Services, Division of Bacterial and Mycotic Diseases, Escherichia coli O157:H7, http://www.cdc.gov/ncidod/dbmd/diseaseinfo/escherichiacoli_g.htm

CDC, (2007c) National Centre for Infectious Disease-Respiratory and Enteric Virus Branch,Non-Polio Enterovirus Infections, http://www.cdc.gov/ncidod /dvrd/revb /enterovirus/non-polio_entero.htm

Council Directive 98/83/EC of 3 November 1998 on the quality of water intended for human consumption, Off. J. Eur. Communities (1998), pp. L330/32–L330/53 5.12.98

Chang, Y. C., and Jung, K. (2004) Effect of distribution system materials and water quality on heterotrophic plate counts and biofilm proliferation. *Journal of Microbiology and Biotechnology*. 14, 1114–1119.

Charnock, C., and Kjonno, O. (2000) Assimilable organic carbon and biodegradable dissolved organic carbon in Norwegian raw and drinking waters. Water Research. 34 2629–2642.

Chen, X.M., O'Hara, S.P., Huang, B.Q., Splinter, P.L., Nelson, J.B., and LaRusso, N.F. (2005) Localized glucose and water influx facilitates Cryptosporidium parvum cellular invasion by means of modulation of host-cell membrane protrusion. *Proceedings of the National Academy of Sciences* of the United States of America. 102, 6338–6343.

Chen, L., Manz, A., and Day, P.J.R. (2007) Total nucleic acid analysis integrated on microfluidic devices. Lab Chip. 7, 1413-1423.

Chomvarin, C., Namwat, W., Wongwajana, S., Alam, M., Thaew-Nonngiew, K., Sinchaturus, A., and Engchanil, C. (2007) Application of duplex-PCR in rapid and reliable detection of toxigenic Vibrio cholerae in water samples in Thailand. *The Journal of General and Applied Microbiology*. 53, 229-237.

Cirillo, J.D., Falkow, S., Tompkins, L.S., and Bermudez, L.E. (1997) Interaction of Mycobacterium avium with environmental amoebae enhances virulence, Infection and Immunity. 65, 3759–3767.

Colbourne, J.S., Pratt, D.J., Smith, M.G., Fisher-Hoch, S.P., and Harper, D. (1984) Water fittings as sources of Legionella pneumophila in a hospital plumbing system, Lancet. 1, 210–213.

Corso, P.S., Kramer, M.H., Blair, K.A., Addiss, D.G., Davis, J.P., and Haddix, A.C. (2003) Cost of illness in the 1993 waterborne Cryptosporidium outbreak, Milwaukee, Wisconsin. Emerging Infectious Diseases. 9, 426–431.

Cotte, L., Rabodonirina, M., Chapuis F., Bailly, F., Bissuel, F., Raynal C., Gelas, P., Persat, F., Piens, M.A., and Treppo, C. (1999) Waterborne outbreak of intestinal microsporidiosis in persons with and without human immunodeficiency virus infection. *Journal of Infectious Diseases*. 180, 2003–2008.

Craun, G.F., Calderon, R.L. and Craun, M.F. (2005) Outbreaks associated with recreational water in the United States. *International Journal of Environmental Research and Public Health*.15, 243–262.

De Paula, V.S., Diniz-Mendes, L., Villar, L.M., Luz, S.L., Silva, L.A., Jesus, M.S., da Silva, N.M., and Gaspar, A.M. (2007) Hepatitis A virus in environmental water samples from the Amazon Basin. Water Research. 41, 1169-1176.

Doganci, T., Araz, E., Ensari, A., Tanyuksel, M. and Doganci, L. (2002) Detection of Cryptosporidium parvum infection in childhood using various techniques. Medical Science Monitor. 8, 223–226.

Doyle, M.P., and Erickson, M.C. (2006) Reducing the carriage of foodborne pathogens in livestock and poultry. Poultry Science. 85, 960-973.

Edberg, S.C., Rice, E.W., Karlin, R.J., and Allen, M.J. (2000) Escherichia coli: the best biological drinking water indicator for public health protection. Symposium series (Society for Applied Microbiology). 29, 106S–116S.

Ellis, K.V. (1991) Water disinfection: A review with some consideration of the
508 requirements of the third world. Critical Reviews in Environmental Control
20, 341-393.

Fayer, R., Speer, C.A. and Dubey, J.P. (1997) The general biology of Cryptosporidium. In Cryptosporidium and Cryptosporidiosis ed. Fayer, R. pp. 1–41. Boca Raton, FL: CRC Press

Fayer, R., Trout, J.M. and Jenkins, M.C. (1998) Infectivity of Cryptosporidium parvum oocysts stored in water at environmental temperatures. *Journal of Parasitology*. 84, 1165–1169.

Fayer, R. (2004) Cryptosporidium: a water-borne zoonotic parasite. Veterinary Parasitology. 126, 37–56.

Fendukly, F., Bernander, S., and Hanson, H.S. (2007) Nosocomial Legionnaires' disease caused by Legionella pneumophila serogroup 6: implication of the sequence-based typing method (SBT). *Scandinavian Journal of Infectious Diseases*. 39, 213-216.

Ferguson, C., Medema, G., Teunis, P., Davidson, A., and Deere, D. (2003) Microbial health criteria for Cryptosporidium. In Cryptosporidium: From to Disease, ed. Thompson, R.C.A., Armson, A. and Ryan, U.M. pp. 295–301. Amsterdam: Elsevier BV.

Finney, M., Smullen, J., Foster, H.A., Brokx, S., and Storey D.M. (2003) Evaluation of Chromocult coliform agar for the detection and enumeration of Enterobacteriaceae from faecal samples from healthy subjects. *Journal of Microbiological Methods*. 54, 353–358.

Frost, F.J., Muller, T., Craun, G.F., Lockwood, W.B. and Calderon, R.L. (2002) Serological evidence of endemic waterborne Cryptosporidium infections. Annals of Epidemiology. 12, 222–227.

Gajadhar, A.A., and Allen, J.R. (2004) Factors contributing to the public health and economic importance of waterborne zoonotic parasites. Veterinary Parasitology. 126, 3-14.

Gassilloud, B., Schwartzbrod, L., and Gantzer, C. (2003) Presence of viral genomes in mineral water: a sufficient condition to assume infectious risk? Applied and Environmental Microbiology. 69, 3965–3969.

Geissler, K., Manafi, M., Amorós, I., and Alonso, J.L. (2000) Quantitative determination of total coliforms and Escherichia coli in marine waters with chromogenic and fluorogenic media. *Journal of Applied Microbiology*. 88, 280–282.

Ghosh, M., Wahi, S., Kumar, M., and Ganguli, A. (2007) Prevalence of enterotoxigenic Staphylococcus aureus and Shigella spp. in some raw street vended Indian foods. *International Journal of Environmental Health Research.* 17, 151-156.

Gibson, D.M. (1987) Some modification to the media for rapid automated detection of salmonellas by conductance measurement. *Journal of Applied Bacteriology*. 63, 299–304.

Gilbride, K.A., Lee, D.Y., and Beaudette, L.A. (2006) Molecular techniques in wastewater: Understanding microbial communities, detecting pathogens, and real-time process control. *Journal of Microbiological Methods*. 66, 1-20.

Gómez-Couso, H., Paniagua-Crespo, E., and Ares-Mazás, E. (2007) Acanthamoeba as a temporal vehicle of Cryptosporidium. Parasitology Research.100, 1151-1154.

Gomez-Couso, H., Mendez-Hermida, F., Castro-Hermida, J.A., Ares-Mazas, and E. (2006) Cryptosporidium contamination in harvesting areas of bivalve molluscs. *Journal of Food Protection*. 69, 185–190.

Governal, R.A., Yahya, M.T., Gerba. C.P., Shadman, F. (1992) Comparison of assimilable organic carbon and UV-oxidizable carbon for evaluation of ultrapure-water systems. Applied and Environmental Microbiology. 58, 724–726.

Graczyk, T.K., Sunderland, D., Tamang, L., Shields, T.M., Lucy, F.E., and Breysse, P.N. (2007) Quantitative evaluation of the impact of bather density on levels of human-virulent microsporidian spores in recreational water. Applied and Environmental Microbiology. 73, 4095-4099.

Greub, G., and Raoult, D. (2004) Microorganisms resistant to free-living amoebae, Clinical Microbiology Reviews. 17, 413–433.

Harvey, P.A., and Drouin, T. (2006) The case for the rope-pump in Africa: a comparative performance analysis. *Journal of Water and Health.* 4, 499-510.

Hill, V.R., Polaczyk, A.L., Hahn, D., Narayanan, J., Cromeans, T.L., Roberts, J.M. and Amburgey, J.E. (2005) Development of a rapid method for simultaneous recovery of diverse microbes in drinking water by ultrafiltration with sodium polyphosphate and surfactants. Applied and Environmental Microbiology. 71, 6878–6884.

Himathongkham, S., Dodd, M.L., Yee, J.K., Lau, D.K., Bryant, R.G., Badoiu, A.S., Lau, H.K., Guthertz, L.S., Crawford-Miksza, L., and Soliman, M.A. (2007) Recirculating immunomagnetic separation and optimal enrichment conditions for enhanced detection and recovery of low levels of Escherichia coli O157:H7 from fresh leafy produce and surface water. *Journal of Food Protection. 70, 2717-2724.*

Hrudey, S E., and Hrudey, EJ (2007) Published case studies of waterborne disease outbreaks - Evidence of a recurrent threat. Water Environment Research. 79, 233-245.

Hunter, P.R., and Thompson, R.C.A. (2005) The zoonotic transmission of Giardia and Cryptosporidium. *International Journal for Parasitology*. 35, 1181-1190.

Hurst, C.J. (1988) Effect of environmental variables on enteric virus survival in
536 surface freshwaters. Water Science and Technology. 20, 473-476.

ISO/DIS Standard 9308, Water Quality–Detection and Enumeration of Escherichia coli and Coliform Bacteria: Part 1. Membrane Filtration Method (2000).

ISO/DIS Standard 16649, Microbiology of Food and Animal Feeding Stuffs—Horizontal Method for the Enumeration of Beta-glucuronidase-positive Escherichia coli: Part 1. Colony-count Technique at 44 Degrees C Using Membranes and 5-bromo-4-chloro-3-indolyl beta-d-glucoronide (2001).

Jagals, C., and Jagals, P.. (2004) Application of HACCP principles as a management tool for monitoring and controlling microbiological hazards in water treatment facilities. Water Science and Technology 50, 69-76.

Jin, D.Z., Xu, X.J., Chen, S.H., Wen, S.Y., Ma, X.E., Zhang, Z., Lin, F., and Wang, S.Q. (2007) Detection and identification of enterohemorrhagic Escherichia coli O157:H7 and Vibrio cholerae O139 using oligonucleotide microarray. Infectious Agents and Cancer. 2, 23.

Kapley, A. and Purohit, H.J. (2001) Detection of etiological agent for cholera by PCR protocol. Medical Science Monitor. 7, 242–245.

Karanis, P., Kourenti, C., and Smith, H. (2007) Waterborne transmission of protozoan parasites: a worldwide review of outbreaks and lessons learnt. *Journal of Water and Health.* 5, 1-38.

Keegan, A.R., Fanok, S., Monis, P.T. and Saint, C.P. (2003) Cell culture-Taqman PCR assay for evaluation of Cryptosporidium parvum disinfection. Applied and Environmental Microbiology. 69, 2505–2511.

Kinde, H., Adelson, M., Ardans, A., Little, E.H., Willoughby, D., Berchtold, D., Read, D.H., Breitmeyer, R., Kerr, D., Tarbell, R., and Hughes E. (1997) Prevalence of Salmonella in municipal sewage treatment plant effluents in southern California. Avian Diseases. 41, 392-398.

King, C.H., Shotts, E.B.Jr., Wooley, R.E., and Porter, K.G. (1988) Survival of coliforms and bacterial pathogens within protozoa during chlorination. Applied and Environmental Microbiology. 54, 3023-3033.

Kostić, T., Weilharter, A., Rubino, S., Delogu, G., Uzzau, S., Rudi, K., Sessitsch, A., and Bodrossy, L. (2007) A microbial diagnostic microarray technique for the sensitive detection and identification of pathogenic bacteria in a background of nonpathogens, Analytical Biochemistry. 360, 244-254.

Kotloff, K.L. (1999) Bacterial diarrheal pathogens. Advances in Pediatric Infectious Diseases. 14, 219-267.

Kramer, M. H., Quade, G., Hartemann, P. and Exner, M. 2001 Waterborne diseases in Europe: 1986–96. *Journal of the American Water Works Association.* 93, 48–53.

Kubista, M., Andrade, J.M., Bengtsson, M., Forootan, A., Jonák, J., Lind, K., Sindelka, R., Sjöback, R., Sjögreen, B., Strömbom, L., Ståhlberg, A., and Zoric, N. (2006) The real-time polymerase chain reaction. Molecular Aspects of Medicine. 27, 95-125.

LaGier MJ, Fell JW, Goodwin KD (2007) Electrochemical detection of harmful algae and other microbial contaminants in coastal waters using hand-held biosensors. Marine Pollution Bulletin. 54, 757-770.

Långmark, J., Storey, M.V., Ashbolt, N.J., Stenström, T.A. (2007) The effects of UV disinfection on distribution pipe biofilm growth and pathogen incidence within the greater Stockholm area, Sweden. Water Research. 41, 3327-3336.

Lazcka, O., Del Campo, F.J., and Muñoz, F.X. (2007) Pathogen detection: a perspective of traditional methods and biosensors. Biosensors and Bioelectronics. 22, 1205-1217.

Lee, D.Y., Shannon, K., and Beaudette, L.A. (2006) Detection of bacterial pathogens in municipal wastewater using an oligonucleotide microarray and real-time quantitative PCR. *Journal of Microbiological Methods.* 65, 453-467.

Leclerc, H., Schwartzbrod, L., and Dei-Cas, E. (2002) Microbial agents associated with waterborne diseases. Critical Reviews in Microbiology. 28, 371-409.

Lehtola, M.J., Pitkänen, T., Miebach, L., and Miettinen, I.T. (2006) Survival of Campylobacter jejuni in potable water biofilms: a comparative study with different detection methods. Water Science and Technology 54, 57-61.

Lehtola, M.J., Torvinen, E., Kusnetsov, J., Pitkanen, T., Maunula, L., von Bonsdorff, C.H., Carl-Martikainen, P.J., Wilks, S.A., Keevil, C.W., and Miettinen, I.T. (2007) Survival of Mycobacterium avium, Legionella pneumophila, Escherichia coli, and caliciviruses in drinking water-associated biofilms grown under high-shear turbulent flow. Applied and Environmental Microbiology. 73, 2854-2859.

Lemarchand, K., Masson, L., and Brousseau, R. (2004) Molecular biology and DNA microarray technology for microbial quality monitoring of water. Critical Reviews in Microbiology. 30, 145-172.

Ly, K.T., and Casanova, J.E. (2007) Mechanisms of Salmonella entry into host cells. Cellular Microbiology. 9, 2103-11.

Mackenzie, W., Hoxie, N., Proctor, M., Gradus, M.S., Blair, K.A., and Peterson, D.E.. (1994) A massive outbreak in Milwaukee of cryptosporidium infection transmitted through the public water supply. *New England Journal of Medicine.* 331, 161–167.

Madden, R.H. and Gilmour, A. (1995) Impedance as an alternative to MPN enumeration of coliforms in pasteurized milks. Letters in Applied Microbiology. 21, 387–388.

Manafi, M. (2000) New developments in chromogenic and fluorogenic culture media. *International Journal of Food Microbiology*. 60, 205–218.

Maynard, C., Berthiaume, F., Lemarchand, K., Harel, J., Payment, P., Bayardelle, P., Masson, L., and Brousseau, R. (2005) Waterborne pathogen detection by use of oligonucleotide-based microarrays. Applied and Environmental Microbiology. 71, 8548-8557.

McDonald, V. (2000) Host cell-mediated responses to infection with Cryptosporidium. Parasite Immunology. 22, 597–604.

McNabb, S.J.N., Jajosky, R.A., Hall-Baker, P.A., Adams, D.A., Sharp, P., Anderson, W.J., Aponte, J.J., Jones, G.F., Nitschke, D.A., Worsham, C.A., and Richard, R.A. Jr. (2007) Morbidity and Mortality Weekly Report. March 30, 2007 / 54, 2-92. http://www.cdc.gov/mmwr/preview/mmwrhtml/mm5453a1.htm

McNealy, T.L., Forsbach-Birk, V., Shi, C., and Marre, R. (2005) The Hfq homolog in Legionella pneumophila demonstrates regulation by LetA and RpoS and interacts with the global regulator CsrA. *Journal of Bacteriology.* 187, 1527–1532.

Meusburger, S., Reichart, S., Kapfer, S., Schableger, K., Fretz, R., and Allerberger, F. 2007. Outbreak of acute gastroenteritis of unknown etiology caused by contaminated drinking water in a rural village in Austria, August 2006. Wien Klin Wochenschr. 119, 717-721.

Miltner, E.C., and Bermudez, L.E. (2000) Mycobacterium avium grown in Acanthamoeba castellanii is protected from the effects of antimicrobials. Antimicrobial Agents and Chemotherapy. 44, 1990–1994.

Molmeret, M., Horn, M., Wagner, M., Santic, M., and Abu Kwaik,Y. (2005) Amoebae as training grounds for intracellular bacterial pathogens, Applied and Environmental Microbiology. 71, 20–28.

Momba, M.N.B., Cloete, T.E., Venter, S.N., and Kfir, R. (2007) The effects of UV disinfection on distribution pipe biofilm growth and pathogen incidence within the greater Stockholm area, Sweden. Water Research. 41, 3327-3336.

Morgan-Ryan, U.M., Fall, A., Ward, L.A., Hijjawi, N., Sulaiman, I., Fayer, R., Thompson, R.C., Olson, M., Lal, A., and Xiao, L. (2002) Cryptosporidium hominis n. sp. (Apicomplexa: Cryptosporidiidae) from Homo sapiens. *Journal of Eukaryotic Microbiology.* 49, 433–440.

Nainan, O.V., Xia, G., Vaughan, G., and Margolis, H.S. (2006) Diagnosis of Hepatitis A virus infection: a molecular approach. Clinical Microbiology Reviews. 19, 63–79.

Nelson, E.J., Chowdhury, A., Harris, J.B., Begum, Y.A., Chowdhury, F., Khan, A.I., Larocque, R.C., Bishop, A.L., Ryan, E.T., Camilli, A., Qadri, F., and Calderwood, S.B. (2007) Complexity of rice-water stool from patients with Vibrio cholerae plays a role in the transmission of infectious diarrhea. *Proceedings of the National Academy of Sciences* of the United States of America.*104,19091-19096.*

Niyogi, S.K. (2005) Shigellosis. *Journal of Microbiology.* 43, 133-143.

Palmer, M.V. (2007) Tuberculosis: a reemerging disease at the interface of domestic animals and wildlife. Current Topics in Microbiology and Immunology. 315, 195-215.

Pankhurst, C.L.' and Coulter, W.A. (2007) Do contaminated dental unit waterlines pose a risk of infection? *Journal of Dentistry.* 35, 712–720.

Papapetropoulou, M., Tsintzou, A., and Vantarakis, A. (1997) Environmental mycobacteria in bottled table waters in Greece. *Canadian Journal of Microbiology* 43, 499–502.

Payment, P., Franco, E., and Siemiatycki. J. (1993) Absence of relationship between health effects due to tap water consumption and drinking water quality parameters, Water Science and Technology. 27,137–143.

Peng, M.M., Xiao, L., Freeman, A.R., Arrowood, M.J., Escalante, A.A., Weltman, A.C., Ong, C.S., MacKenzie, W.R., Lal AA, Beard CB. (1997) Genetic polymorphism among Cryptosporidium parvum isolates: evidence of two distinct human transmission cycles. Emerging Infectious Diseases. 3, 567–573.

Percival, S.L., and Walker, J.T. (1999) Potable water and biofilms: a review of the public health implications. Biofouling 14, 99–115.

Phares, C.R., Russell, E., Thigpen, M.C., Service, W., Crist, M.B., Salyers, M., Engel, J., Benson, R.F., Fields. B., and Moore, M.R. (2007) Legionnaires' disease among residents of a long-term care facility: The sentinel event in a community outbreak. *American Journal of Infection Control.* 35, 319-323.

Robertson, B., Sinclair, M.I., Forbes, A.B., Veitch, M., Kirk, M., Cunliffe, D., Willis, J. and Fairley, C.K. (2002a) Case-control studies of sporadic cryptosporidiosis in Melbourne and Adelaide, Australia. Epidemiology and Infection. 128,419–431.

Rudwaleit, M., Braun, J., and Sieper, J. (2000) Treatment of reactive arthritis: a practical guide. BioDrugs. 13, :21-28.

Ryan, K.J., and Ray, C.G. (editors) (2004). Sherris Medical Microbiology, 4th ed., McGraw Hill. ISBN 0-8385-8529-9.

Sacchettini, J.C., Rubin, E.J., and Freundlich, J.S. (2008) Drugs versus bugs: in pursuit of the persistent predator Mycobacterium tuberculosis. Nature Reviews Microbiology. 6, 41-52.

Saker, M.L., Fastner, J., Dittmann, E., Christiansen, G., and Vasconcelos, V.M. (2005) Variation between strains of the cyanobacterium Microcystis aeruginosa isolated from a Portuguese river. *Journal of Applied Microbiology* 99, 749-757.

Saker, M.L., Vale, M., Kramer, D., and Vasconcelos, V.M. (2007) Molecular techniques for the early warning of toxic cyanobacteria blooms in freshwater lakes and rivers. Applied Microbiology and Biotechnology. 75, 441-449.

Santamaría, J., and Toranzos, G.A. (2003) Enteric pathogens and soil: a short review. International Microbiology. 6, 5-9.

Sharma, S., Sachdeva, P., and Virdi, J.S. (2003) Emerging water-borne pathogens.Appl Microbiol Biotechnol. 61, 424-428.

Shwimmer, A., Freed, M., Blum, S., Khatib, N., Weissblit, L., Friedman, S., and Elad, D. (2007) Mastitis Caused by Yersinia pseudotuberculosis in Israeli Dairy Cattle and Public Health Implications. Zoonoses and Public Health. 54, 353-357.

Skovgaard, N (2007) New trends in emerging pathogens. *International Journal of Food Microbiology.* 120, 217-224.

Selma, M.V., Allende, A., López-Gálvez, F., Elizaquível, P., Aznar, R., and Gil, M.I. (2007) *Journal of Applied Microbiology.* Potential microbial risk factors related to soil amendments and irrigation water of potato crops. 103, 2542-2549.

Shigematsu, M., Nagano, Y., Millar, B.C., Kenny, F., Lowery, C.J., Xiao, L., Rao, J.R., Nicholson, V., Watabe, M., Heaney, N., Sunnotel, O., McCorry, K., Rooney, P.J., Snelling, W., Dooley, J.S., Elborn, J.S., Matsuda, M., and Moore, J.E. (2007). Molecular detection and identification of Cryptosporidium species in lettuce employing nested small-subunit rRNA PCR and direct automated sequencing. *British Journal of Biomedical Science.* 64, 133-135.

Smith, H.V., Cacciò, S.M., Cook, N., Nichols, R.A., and Tait, A. (2007) Cryptosporidium and Giardia as foodborne zoonoses. Veterinary Parasitology. 149, 29-40.

Snelling, W.J., Moore, J.E., and Dooley, J.S.G. (2005). The colonization of broilers with Campylobacter. *World's Poultry Science Journal.* 61, 655-662.

Snelling, W.J., Matsuda, M., Moore, J.E., and Dooley JS (2005b) Campylobacter jejuni. Letters in Applied Microbiology. 41, 297-302.

Snelling WJ, McKenna JP, Lecky DM, Dooley JS. (2005b) Survival of Campylobacter jejuni in waterborne protozoa. Applied and Environmental Microbiology. 71, 5560-5571.

Snelling, W.J., Moore, J.E., McKenna, J.P., Lecky, D.M., and Dooley, J.S.G. (2006) Bacterial-protozoa interactions; an update on the role these phenomena play towards human illness. Microbes and Infection. 8, 578-587.

Solomon, J.M., and Isberg, R.R. (2000) Growth of Legionella pneumophila in Dictyostelium discoideum: a novel system for genetic analysis of host–pathogen interactions. Trends in Microbiology. 18, 478–480.

Steinert, M., Ott, M.M., Luck, P.C., Tannich, E., and Hacker, J. (1994) Studies on the uptake and intracellular replication of Legionella pneumophila in protozoa and in macrophage-like cells. FEMS Microbiology Ecology. 15, 299–307.

Steinert, M., Birkness, K., White, E., Fields, B., and Quinn, F. (1998) Mycobacterium avium bacilli grow saprozoically in coculture with Acanthamoeba polyphaga and survive within cyst walls. Applied and Environmental Microbiology. 64, 2256–2261.

Stelma, G.N. Jr., Lye, D.J., Smith, B.G., Messer, J.W., and Payment, P. (2004) Rare occurrence of heterotrophic bacteria with pathogenic potential in potable water. *International Journal of Food Microbiology.* 92, 249-254.

Stern, N.J., and Pierson, M.D. (1979) Yersinia enterocolitica: A review of the psychrotrophic water and foodborne pathogen. *Journal of Food Science.* 44, 1736-1742.

Stojek, N., and Dutkiewicz, J. (2006) Legionella and other gram-negative bacteria in potable water from various rural and urban sources. Annals of Agricultural and Environmental Medicine. 13, 323-335.

Stout, J.E., Yu, V.L., Muraca, P., Joly, J., Troup, N., and Tompkins, L.S. (1992) Potable water as a cause of sporadic cases of community-acquired Legionnaires' disease. The *New England Journal of Medicine.* 326, 151-155.

Straub, T.M., Dockendorff, B.P., Quiñonez-Díaz, M.D., Valdez, C.O., Shutthanandan, J.I., Tarasevich, B.J., Grate, J.W., and Bruckner-Lea, C.J. (2005) Automated methods for multiplexed pathogen detection. *Journal of Microbiological Methods.* 62, 303-316.

Sunnotel, O., Lowery, C.J., Moore, J.E., Dooley, J.S., Xiao, L., Millar, B.C., Rooney, P.J., and Snelling, W.J. (2006a) Cryptosporidium. Letters in Applied Microbiology. 43, 7-16.

Sunnotel, O., Snelling, W.J., Xiao, L., Moule, K., Moore, J.E., Millar, B.C., Dooley, J.S., and Lowery, C.J. (2006b). Rapid and sensitive detection of single Cryptosporidium oocysts from archived glass slides. *Journal of Clinical Microbiology.* 44, 3285-3291.

Sunnotel, O., Snelling, W.J., McDonough, N., Browne, L., Moore, J.E., Dooley, J.S., and Lowery, C.J. (2007) Effectiveness of standard UV depuration at inactivating Cryptosporidium parvum recovered from spiked Pacific oysters (Crassostrea gigas). Applied and Environmental Microbiology. 73, 5083-5087.

Tabbara, K.F. (2007) Tuberculosis. Current Opinions in Ophthalmology. 18, 493-501.

Thompson, H.P., Dooley, J.S., Kenny, J., McCoy, M., Lowery, C.J., Moore, J.E., and Xiao, L. (2007) Genotypes and subtypes of Cryptosporidium spp. in neonatal calves in Northern Ireland. Parasitology research. 100, 619-624.

Torvinen, E., Suomalainen, S., Lehtola, M.J., Miettinen, I.T., Zacheus, O., Paulin, L., Katila, M.L., Martikainen, P.J. (2004) Mycobacteria in water and loose deposits of drinking water distribution systems in Finland. Applied and Environmental Microbiology. 70, 1973-1981.

Van Poucke, S.O., and Nelis, H.J. (2000) Rapid detection of fluorescent and chemiluminescent total coliforms and Escherichia coli on membrane filters. *Journal of Microbiological Methods* 42, 233–244.

Wang, X.W., Zhang, L., Jin, L.Q., Jin, M., Shen, Z.Q., An, S., Chao, F.H., and Li, J.W. (2007) Development and application of an oligonucleotide microarray for the detection of food-borne bacterial pathogens. Applied Microbiology and Biotechnology. 76, 225-233.

Wassenaar, T.M. and Newell, D.G. (2000) Genotyping of Campylobacter spp. Applied and Environmental Microbiology. 66, 1–9.

Whan, L., Grant, I.R., and Rowe, M.T. (2006). Interaction between Mycobacterium avium subsp. paratuberculosis and environmental protozoa. BMC Microbiology. 13, 6:63.

World Health Organisation (WHO) (1998) Guidelines for Drinking-Water Quality, 2nd edn, Addendum to Vol. 2: Health criteria and other supporting information.Geneva: World Health Organization. http://www.who.int/water_sanitation_health/dwq/2edaddvol2a.pdf

World Health Organisation, (WHO), in: WHO (Eds.), Guidelines for drinking-water quality:Volume 1: Recommendations,WHO, Geneva, 1996.

http://www.who.int/water_sanitation_health/diseases/wsh0302/en/index7.html

Wu, L., Liu, X., Schadt, C.W., and Zhou, J. (2006) Microarray-based analysis of subnanogram quantities of microbial community DNAs by using whole-community genome amplification, Applied and Environmental Microbiology. 72, 4931-4941.

Xiao, L. and Ryan, U.M. (2004) Cryptosporidiosis: an update in molecular epidemiology. Current Opinion in Infectious Diseases 17, 483–490.

Xiao, L., Fayer, R., Ryan, U. and Upton, S.J. (2004a) Cryptosporidium taxonomy: recent advances and implications for public health. Clinical Microbiology Reviews. 17, 72–97.

Yáñez, M.A., Valor, C., and Catalán, V. (2006) A simple and cost-effective method for the quantification of total coliforms and Escherichia coli in potable water. J Microbiological Methods. 65, 608-611.

Yates, M.V., Gerba, C.P., and Kelley, L.M. (1985) Virus persistence in 629 groundwater. Applied Environmental Microbiology. 49, 778-781.

Zhang, T., and Fang, H.H.P. (2006) Applications of real-time polymerase chain reaction for quantification of microorganisms in environmental samples. Applied Microbiology and Biotechnology Applied Microbiology and Biotechnology. 70, 281–289.

In: Drinking Water: Contamination, Toxicity and Treatment ISBN: 978-1-60456-747-2
Editors: J. D. Romero and P. S. Molina

Chapter 4

THE RELIABILITY OF THE GRAB SAMPLE FOR THE DETERMINATION OF FLUORIDE IN POTABLE WATER SUPPLIES

J.A. Armstrong and S.A. Katz
Department of Chemistry, Rutgers University, Camden, NJ 08102-1411, USA

ABSTRACT

Water samples were collected three times each week for five weeks from the potable water supplies at domestic sites in five Southern New Jersey communities. The sampling regimen was designed to reflect periods of high and low water usage.

None of the water supplies was fluoridated, but fluoride is known to be a naturally-occurring water contaminant in some Southern New Jersey domestic water supplies.

The fluoride ion concentrations of the samples were determined by ion selective electrode potentiometry without prior distillation. The analytical error associated with the method was evaluated by repetitive measurements on reference samples, and the sampling errors were identified by statistical analysis of the results obtained from the samples.

The statistical analysis of the data showed the fluoride concentrations varied from site to site and from time to time. In some cases, the time-to-time variations exceeded the "two sigma" limits at a given site. This observation raises a serious question about the reliability of a grab sample for the determination of fluoride, and possibly other contaminants, in potable water supplies.

INTRODUCTION

There is no question that the grab sample reflects the composition of the system under investigation at only one point in the time – space continuum. The question of how well the grab sample represents the system is the focus of the research presented here.

Pursuant to the Safe Drinking Water Act (SDWA, 1974), the US EPA was authorized to establish drinking water quality standards. Subsequently, the states adopted these standards sometimes promulgating more rigorous regulations. The concentrations of over ninety contaminants are regulated. To meet these regulations, suppliers of potable water are required to monitor water quality. The US EPA and the NJ DEP have included fluoride among the potable water contaminants for which monitoring is required.

The monitoring requirement can be met with a surveillance protocol based on the collection of grab samples and their subsequent transport to the laboratory for measurement. The grab samples are usually collected from a relatively small number of locations within the distribution system at a predetermined frequency. Grab sampling may miss contamination events lasting only a few days, but grab sampling is more than adequate for detecting contaminants that have long term effects through chronic exposures. The fluoride ion is considered as such a contaminant.

In New Jersey, ground water supplies are sampled and analyzed at least once in three years. The maximum contaminant level for naturally occurring fluoride is 4.0 mg/L. The recommended upper limit for fluoride adjusted water supplies is 2.0 mg/L. The optimal concentration of fluoride in New Jersey drinking water is 0.9 mg/L in the southern part of the state, and 1.0 mg/L in the northern part of the state. The acceptable ranges are from 0.8 to 1.4 mg/L and from 0.9 to 1.5 mg/L, respectively (N.J.A.C. 70:10-7.2, 2000). The approved methods for the determination of fluoride in drinking water include ion chromatography, SPADNS colorimetry and ion selective electrode potentiometry (CFR 141.41.23). The former technique boasts multi-species identification and quantification capabilities while the latter has the advantages of speed, simplicity, sensitivity and selectivity.

More than thirty years ago, Katz and Scheiner (Katz and Scheiner, 1975) reported on the concentrations of six metal ions and of the fluoride ion in more than fifty southern New Jersey drinking water supplies. In the course of their investigation, the fluoride ion concentrations in different municipalities were found to range from less the 0.2 mg/L to over 2.5 mg/L. Monthly samples from a given site, however, showed fairly constant (± 20 %) fluoride ion concentrations. The fluoride ion concentrations of monthly samples collected from a single site in Woodbury, Gloucester County, New Jersey, during the first six months of 1974 ranged from 1.2 to 2.1 mg/L. The range of concentrations for samples collected similarly from a site in Woodstown, Salem County, New Jersey, was from 1.9 to 2.5 mg/L. More recently, Omena and her colleagues (Omena, et al., 2006) reported the mean fluoride ion concentrations of seven composite samples of the public water supply for Penedo, Alagoas, Brazil collected seven days apart ranged from 0.80 to 1.15 mg/L. The public water supply for the city of Penedo is artificially fluoridated, but the system lacks automatic dosage devices. This is not the case in Woodbury or in Woodstown. In these public water supplies, the fluoride ion is a natural contaminant. Sampling error, measurement error and actual variations in the fluoride concentration of the water are among the speculations on reasons for the observed values of from 1.2 to 2.1 mg/L at Woodbury from 1.9 to 2.5 mg /L at Woodstown.

Sample Collection

Domestic water supplies in Blackwood, Camden County, New Jersey (designated site C), Hammonton, Atlantic County, New Jersey (designated site B), Pitman, Gloucester County,

New Jersey (designated site E), Sicklerville, Camden County, New Jersey (designated site A) and Turnersville, Gloucester County, New Jersey (designated site F) were selected as sampling sites. Samples were collected in the mornings (between 06.00 and 10.00) and in the evenings (between 18.00 and 22.00) on Tuesdays, Thursdays and Saturdays for five weeks at each site. Water was run from the taps for approximately two minutes before filling 500 – mL polyethylene bottles to within 2 cm of the top. No preservatives were added to the samples. After entering the particulars in the sampling log, the samples were transported to and stored in the laboratory at ambient temperature.

Laboratory Measurments

Using polyethylene cylinders, 50.0 – mL aliquots of well – mixed samples were measured into polyethylene beakers and mixed with 50.0 mL of total ionic strength adjustment buffer (TISAB) containing cyclohexanedinitrilotetraacetate (CDTA) purchased from Red Bird Service. Simultaneously, working standards were prepared from dilutions of a 1000 – ppm Thermo-Orion fluoride ion reference standard solution in 50/50 v/v distilled water/TISAB. Calibration curves were prepared from the potentials measured for the working standards using an Orion Model 9409 fluoride ion selective indicator electrode – Orion Model 9001 s.j. Ag/AgCl reference electrode pair connected to an Orion Model 9609 Ionalyzer millivolt meter. A typical calibration curve for fluoride ion concentrations from 0.200 to 2.00 ppm is shown in Figure 1. The potentials of the sample solutions were measured, and their fluoride ion concentrations were determined by reference to the calibrations curves. Frequent measurements of the 1.00 ppm working standard were made to reassure the validity of the calibration curve. After removal of aliquots for the determination of fluoride in each individual sample, the water samples from each of the sampling sites were composited by pipetting 25.0 – mL aliquots into separate polyethylene beakers. After thoroughly mixing the contents of each beaker, 50.0 – mL aliquots were measured into polyethylene beakers and mixed with 50.0 mL TISAB for determinations of their fluoride ion concentrations. The fluoride ion concentrations in the composites were measured by ions selective electrode potentiometry as described above for the individual water samples. Measurements were made on nine aliquots from each composite. The details for these measurements are contained in SW-846 Method 9214 (US EPA, 1996).

Results

The results for the fluoride ion concentrations in individual water samples from the five sampling sites are shown in Figure 2. The mean values and the corresponding standard deviations for the individual results from each sampling site are presented in Table 1 as are the mean values and standard deviations (σ) for the composites.

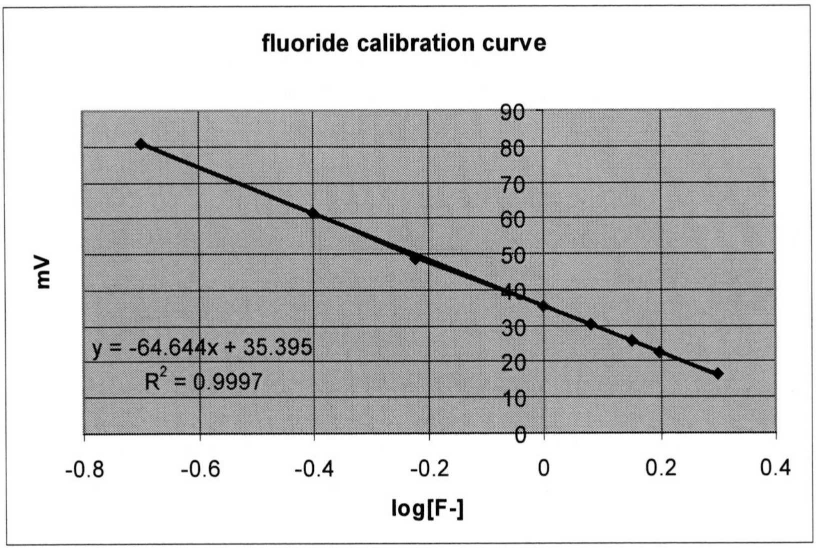

Figure 1. Typical calibration curve for fluoride ion concentrations from 0.200 to 2.00 ppm

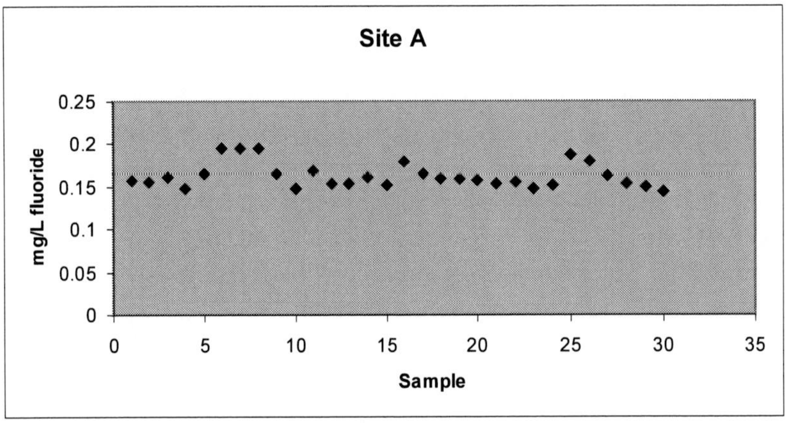

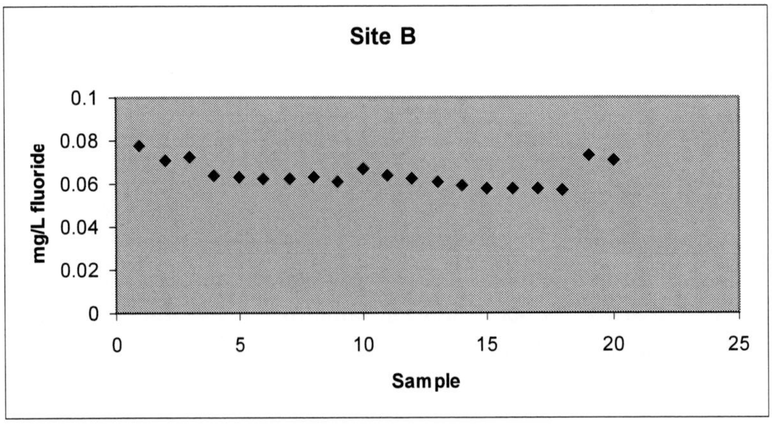

Figure 2. (Continued on next page)

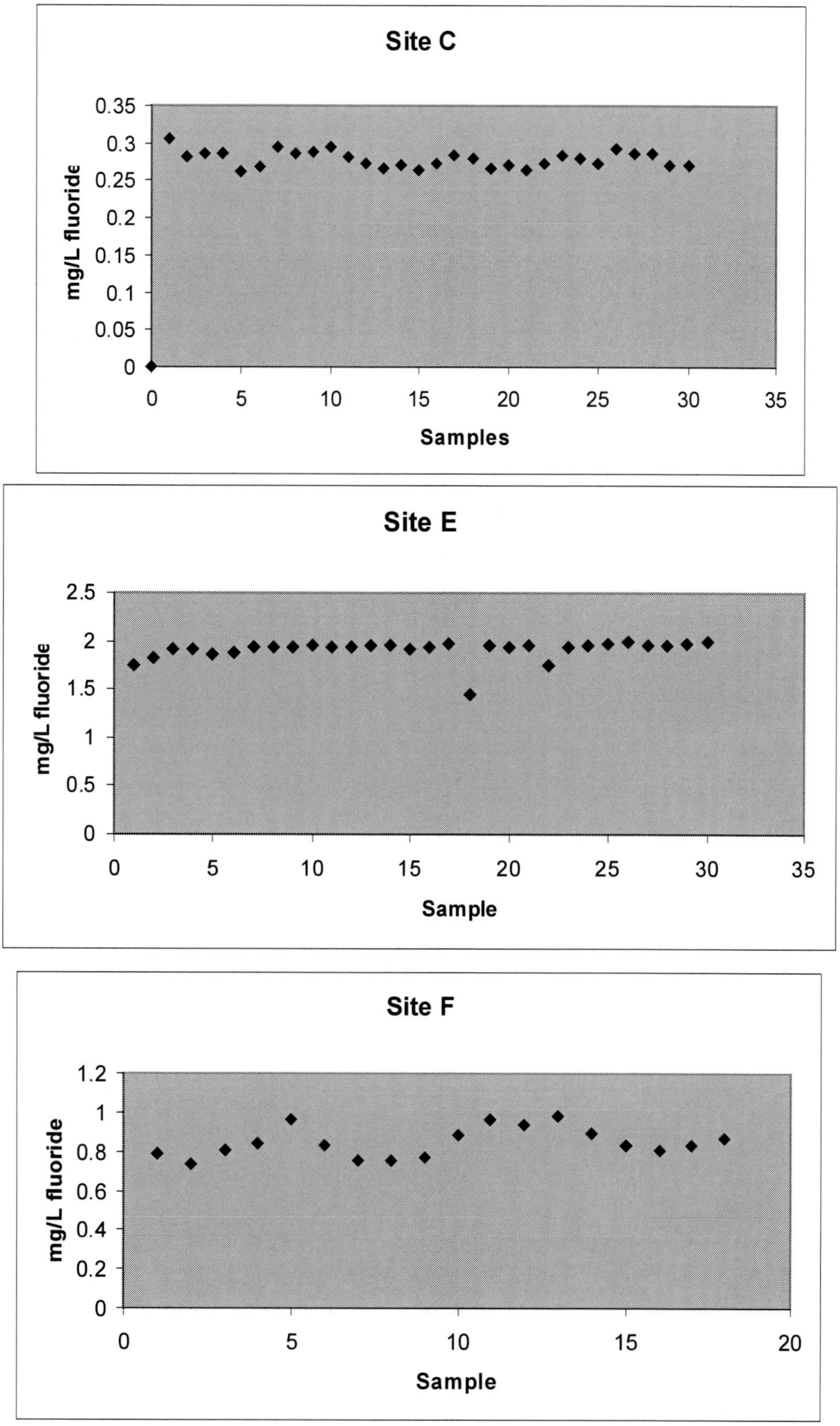

Figure 2. Fluoride ion concentrations in individual water samples from the five sampling sites

Table 1. Results for Fluoride Ion Concentrations in Water Supplies, mg/L

Sampling Site	Individual Samples Mean (σ)		Composited Samples Mean (σ)	
Sicklerville, Site A	0.163	0.0147	0.183	0.0027
Hammonton, Site B	0.0642	0.0059	0.0687	0.0010
Blackwood, Site C	0.279	0.0109	0.288	0.0057
Pitman, Site E	1.91	0.105	1.93	0.0300
Turnersville, Site F	0.850	0.0816	0.871	0.0178

DISCUSSION

Inspection of Figure 2 shows day-to-day variations in the fluoride ion concentrations at each of the five sampling sites. Some of these variations are extreme. For all of the samples from all of the sites, some 5 % of the individual results fell beyond the "two sigma" limits of their respective means. Approximately 10 % of the individual results from site A were found to lie beyond their "two sigma" limits while none of those from site F were found to be outliers. The precision of the mean results from the measurements made on the samples from these sites is comparable.

The data presented in Table 1 show, as predicted (Brumelle et al., 1984), "..., composite sampling does provide a smaller estimate of standard error than does grab sampling in the context of estimating population means." Standard deviations for the means of the results obtained with the composites were approximately five times smaller than those for the means of the results obtained with the individual samples.

Intuitively, the means of the results for the individual samples and the means of the results from repetitive measurements on their composites should coincide. All of the mean values for the composites appear to be greater than the corresponding mean values for the individual samples. Application of the t test to the results for the fluoride ion concentrations from the five sampling sites indicates the differences between the means of the individual samples and the composite samples from sites E and F are not statistically significant by conventional criteria. The respective two-tailed P values are 0.5790 and 0.4567. However, the corresponding differences from sites B and C are statistically significant with two-tailed P values of 0.0327 and 0.0233, respectively. In the case of site A, the two-tailed P value is 0.0003, and the difference is extremely statistically significant by conventional criteria.

Statistically significant differences occur between the means of the individual samples and the means of the composited samples at three of the five sampling sites; viz. site A, site B and site C. The means of the results for the individual samples collected at the sampling sites having the highest fluoride ion concentrations, site E and site F, do not differ statistically from the means of the results obtained with repetitive measurements on their respective composites.

Assuming the variances (σ^2) in the mean results for the samples from sites A, B, C, E and F are the sums of the variances in the sampling errors ($\sigma_S{}^2$) and the variances in the analytical errors ($\sigma_A{}^2$) allows estimation of the variances in the sampling errors. On this basis, $\sigma_S{}^2$ for sites A, B, C, E and F are an order of magnitude greater than the corresponding σ_A^2. This would indicate the sampling errors are greater than the analytical errors. This may be attributed to variations in the fluoride concentrations of the potable water.

Conclusion

The variations of the natural fluoride ion concentrations in some potable water supplies suggests the current monitoring requirement of at least once every three years may not provide adequate information about the concentration this contaminant.

References

Brumelle, S., Nemetz, F. and Casey, D., 1984, Estimating Means and Variances: The Comparative Efficiency of Composite and Grab Samples, Environmental Monitoring and Assessment, 4, 81-84.

CFR 141.41.23.

Katz, S.A. and Scheiner, D.M., 1975, Trace Element levels in Southern New Jersey Drinking Waters, in *Progress in Analytical Chemistry*, vol. 8, Selected Papers from the 1975 Eastern Analytical Symposium, I.L. Simmons and G.W. Ewing, eds., Plenum Press, New York, pp. 83-99.

N.J.A.C. 70:10-7.2, 2000

Omena, L.M.F., Silva, M.E.deA., Pinheiro, C.C., Cavalcante, J.C. and Sampiao, F.C., 2006, Fluoride Ion from Drinking Water and Dentifrice by Children Living in a Tropical Area of Brazil, J. Appl. Oral Sci., 14(5), 382-387.

SDWA, 1974, P.L. 93-523; U.S.C. 300, *et seq.*

US EPA, 1996, Method 9214, Potentiometric Determination of Fluoride in Aqueous Solutions with Ion – Selective Electrode, in *Test Methods for Evaluating Solid Waste* (SW-846), United States Environmental Protection Agency, Office of Solid Waste, Economic, Methods and Risk Analysis Division, Revision 1, December

In: Drinking Water: Contamination, Toxicity and Treatment ISBN: 978-1-60456-747-2
Editors: J. D. Romero and P. S. Molina

Chapter 5

RETENTION OF MICROBIAL PATHOGENS BY FILTRATION

Vitaly Gitis[1] and Elizabeth Arkhangelsky
Unit of Environmental Engineering, Ben-Gurion University of the Negev,
PO Box 653, Beer-Sheva 84105, Israel

ABSTRACT

Microbial contamination is the biggest concern for drinking water suppliers. A wide variety of microbial pathogens including enteric viruses, *E.coli* 0157:H7 and *Cryptosporidium parvum* are spread out through water sources even in developed countries such as United States. Hence, the primary goal of water treatment process is to ensure that the drinking water is free of pathogenic viruses, bacteria and protozoa. Because no single treatment process can be expected to remove all of the different types of pathogens, the international health authorities has promulgated a multiple barrier approach. The idea is to use at least one physical and one chemical process to ensure pathogen attenuation to a safe level.

One of the most employed physical water treatment processes is filtration. Filtration is a common name for a process of physical retention of pathogenic microorganisms by size rejection or by accumulation of the pathogens within its media. Water treatment facilities employ either granular media filters or membrane filters. Four of most known filtration processes are slow sand, rapid granular, microfiltration and ultrafiltration. Degree of microbial retention by a filter depends on pathogen's transport, accumulation and inactivation. Each process has many variables/characteristics and for the same application a retention level may differ by orders of magnitude. Some of the variables are macroscopic and others are microscopic, their interplay and relative influence on the degree of retention are discussed, and useful conclusions are drawn.

[1] Tel. +972-8-6479031, Fax +972-8-6472983, E-mail gitis@bgumail.bgu.ac.il.

1. Background

Microbiological contamination of drinking water is currently officially recognized as the most severe problem that jeopardizes safety and quality of tap water (AEESP, 1999). Microbial pathogens entering gastrointestinal tract through either food or water route may cause serious disorders potentially lethal for sensitive groups of population. In the United States alone intrusion of waterborne microbial pathogens caused at least 750,000 illnesses and 10,000 deaths (Payment et al., 1991, 1997). Hospitalization alone costs over $3 billion/year. Worldwide, diarrheal diseases are second only to respiratory diseases as a cause of adult death and are the leading cause of childhood death. The risk of infection increases with higher concentration of microbes in watershed, insufficient removal by natural environmental processes and treatment, and elevation of post-treatment levels in drinking water (Ratnayake and Jayatilake, 1999; Kirmeyer et al., 2001). Despite the fact that a link between presence of waterborne microbial pathogens and spread of infective diseases was establish 150 years ago, and despite the fact that for the last 100 years tap water has been intensively treated, outbreaks of waterborne disease often occur. When in the past a major infection event was a matter of decades, now several events are occurring yearly at different places around the world.

As an ultimate response to the threat water treatment authorities has promulgated a multiple barrier approach, demanding at least one chemical and one physical treatments to be implemented at each place where water treatment is required (Figure 1). It is assumed that a combined effect of chemical and physical treatments will provide a necessary protection against spread of pathogenic microorganisms.

When a chemical barrier in a majority of cases is disinfection by means of addition of chlorine, ozone or application of UV light (the latter is a physical disinfection process), a physical treatment method may vary between adsorption, ion exchange, oxidation, aeration, chemical precipitation, screening, coagulation, flocculation, sedimentation, filtration and membrane treatment (Figure 2). Some of the above processes are effective against pathogenic microorganisms, some are not effective at all or even have a harmful potential by means of accumulation and subsequent outbreak of a pathogen. The harmful effect of the latter is due to a minimum concentration of the pathogen that has to be reached inside human body. For example, the minimum of 10 oocysts of *Cryptosporidium parvum*, one of the most dangerous microbial pathogens, can lead to infection.

The most reliable, simple and inexpensive yet effective physical method to prevent penetration of disinfectant-resistant pathogenic microorganisms is filtration (U.S.EPA, 2002) Although some 30 years ago the name filtration was affiliated with granular filtration processes, nowadays that general term includes two equally important processes of granular and membrane filtration. Granular filtration can be further subdivided into slow and rapid applications, and membrane filtration encounters micro, ultra- and, to some point, nanofiltration. The application of the latter for water treatment applications however is generally limited to desalination processes.

Different filtration processes may achieve different credits in retention of microbial pathogens (Table 1).

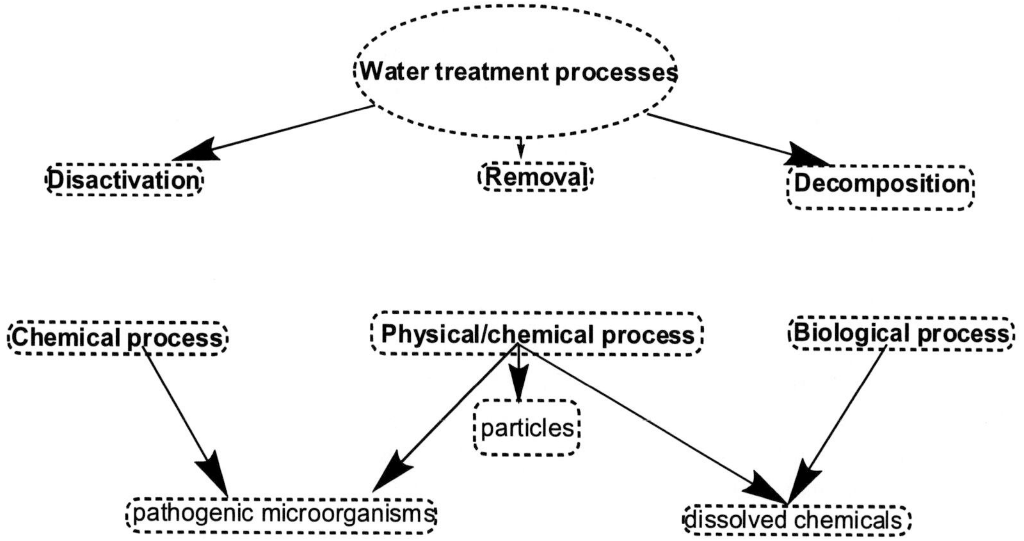

Figure 1. Classification of water treatment processes by type of operation.

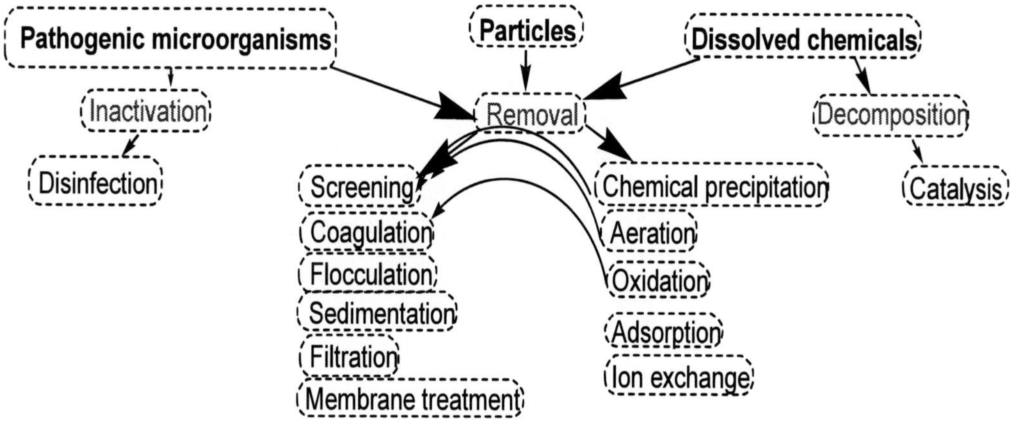

Figure 2. Classification of water treatment processes by type of impurity.

Table 1. Credits of filtration processes in retention of pathogenic microorganisms

Process	Bacteria retention	Viral retention	Protozoa retention	Filtrate turbidity (NTU)
Rapid granular filtration	Partial	Partial	99.9%	<0.5
Slow sand filtration	99.9%	<99%	99.9%	<0.1
Microfiltration	99.9-99.99%	<90%	99.9-99.99%	<0.2
Ultrafiltration	Unlimited	<99%	Unlimited	<0.2
Nanofiltration	Unlimited	Unlimited	Unlimited	<0.1
Reverse osmosis	Unlimited	Unlimited	Unlimited	<0.1

Water treatment authorities are not limited to a specific filtration process however they have to achieve a certain total level of attenuation in concentration of pathogenic microorganisms. For example, by the Surface Water Treatment Rule (SWTR) water treatment authorities in U.S. are required to provide 99% reduction of the initial concentration of pathogenic protozoa, retention of at least 99.9% of bacteria and 99.99% of viruses. As the required retention has to be obtained by a combination of chemical and physical treatment processes, water treatment companies are choosing a type of filtration process that can provide a necessary retention degree of 2-2.5 logs removal. Log removal is defined as a $\log_{10}\left(C_0/C\right)$ where C_0 and C are the concentrations of a pathogen in influent and effluent, respectively. The necessary total attenuation is achieved by a combination of retention provided by filtration and disinfection. The latter can add between log and two logs removal depends on application and specifications. It is therefore water treatment authorities will prefer a filtration process that can achieve the requested retention level at minimum investment, and on the balance of cost and efficiency the most reliable and highly implemented applications are rapid granular and ultra-filtration.

2. Pathogenic Microorganisms

Different types of pathogenic microorganisms, such as bacteria, virus, protozoa and fungi, can lead to infection. Each type contains many different pathogens having various human pathogenesis. Some of pathogenic microorganisms with a certain disease they might cause are referred in Table 2 (Stanier, 1986).

Table 2. Pathogenic microorganisms

Type of pathogen	Name of pathogen	Disease
Bacteria	Corynebacterium diphtheriae	Diphtheria
	Mycobacterium bovis	Tuberculosis
	Mycobacterium leprae	Leprosy
	Bacillus anthracis	Anthrax
	Clostridium botulinium	Botulism
	Clostridium tetani	Tetanus
	Streptococcus pneumoniae	Pneumonia
	Escherichia coli	Diarrhea
	Vibrio cholerae	Cholera
Viruses	HTLV-III	AIDS
	Varicella virus	Chickenpox
	Hepatitis A, B, C, D, E	Hepatitis
	Poliomyelitis virus	Poliomyelitis
Protozoa	Plasmodium malariae	Malaria
	Cryptosporidium parvum	Cryptosporidioisis
	Giardia lamblia	Giardiasis
Fungi	Blastomyces dermatitidis	Blastomycosis
	Microsporium canis	Tinea capitis

Retention of pathogenic microorganisms depends on filtration, water and pathogen characteristics. In general engineering applications family of pathogen or its exact genome are irrelevant, and pathogenic filterability can be assessed by four parameters – its size, morphology, hydrophobicity and chemical composition. Due to heterogeneity the latter is often substituted by more simple yet representative surface charge determination. Some of the filtration processes, such as simple screening or nanofiltration, are build on size exclusion principle where pathogens of bigger size are rejected by the filtration grid and pathogens of smaller size are squeezing through (Figure 3). For other filtration processes such as granular, micro and to some extend ultra-filtration, size is only one of the four parameters and not necessarily pathogens of bigger dimensions will be better retained by filter. There are two main reasons for such behavior, one will be explained later in review of filtration mechanisms and other is morphology of the pathogen.

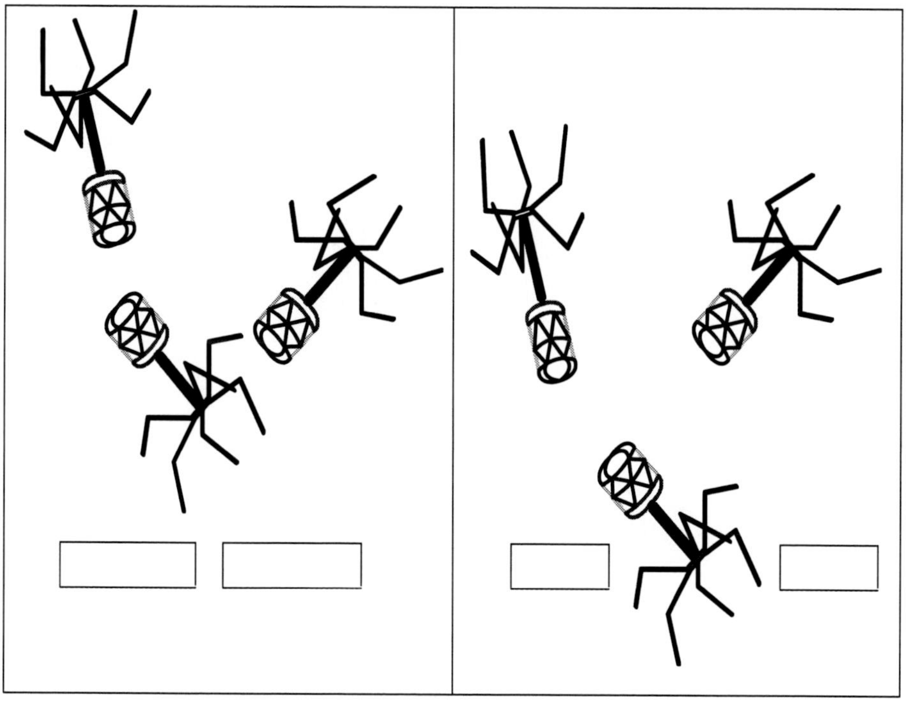

Figure 3. Size exclusion mechanism: a) pore to pathogen size ratio is small, b) pore size/virus size is large.

Pathogen form can vary between envelop, filament, rod, spindle-shaped, tailed, etc. types as depicted in Table 3 taking for example viruses. Although viruses as they are depicted in the Table, or as they are pictured by electron microscopy, in a real solution are surrounded by hydration layer that can be up to 30-40 nm tick, the whole envelop of virus + layer will resemble the general form of the virus itself. So, if viruses like PRD-1, MS2 and Ø6 can indeed be viewed as round particles, viruses like m13, T4, SIRV1 will unlikely obtain a sphere shape. Current filtration theory does not have a special expression of form, and all the pathogens are considered spherical.

To acquire a spherical shape the volume of non-spherical pathogens is equalized to that of sphere of certain diameter, and that diameter is considered the true dimension of the re-shaped pathogen. Thus, for example, a "new" size of m13 will be equal to that of phi X 174, and of course they behavior inside filter will be different. The shape, however, is a very important mechanism in retention of pathogens, especially when size exclusion can be considered as one of the main retention mechanisms. For example, membrane retention of linear microorganisms will be much lower than the rejection measured for globular pathogens. It is believed that linear particulates are able to squeeze through the membrane pores (Figure 4).

Table 3. Classification of viruses (Calendar, 2006)

Family name	Example	Morphology	Size, nm
Microviridae	X174	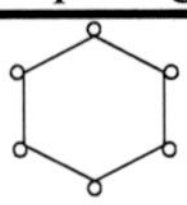	27
Inoviridae	m13		760-1950 x 7
Myoviridae	T4	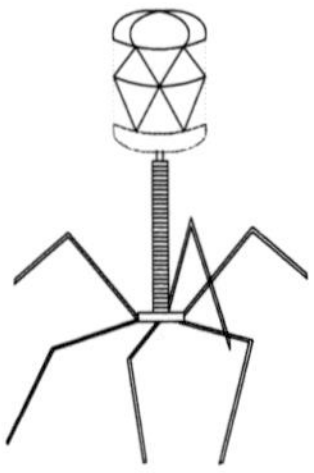	81 x 125
Corticoviridae	PM2	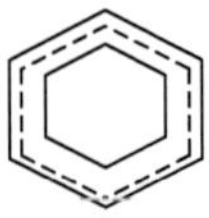	60
Tectiviridae	PRD1	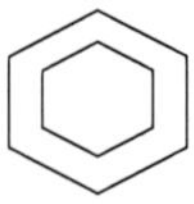	63
Fuselloviridae	SSV1	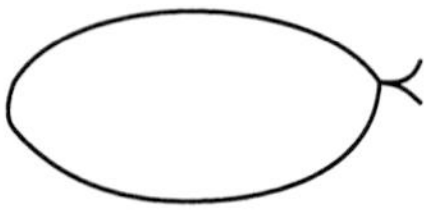	85 x 55
Plasmaviridae	L2	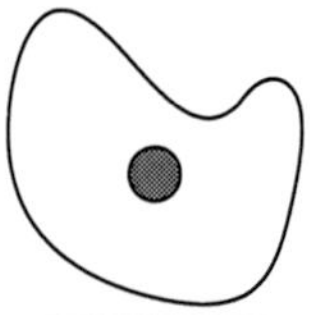	80
Lipothrixviridae	TTV1	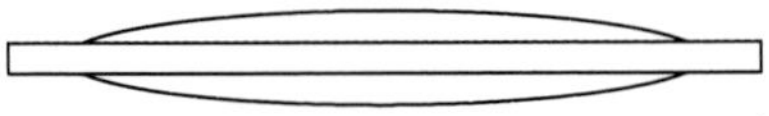	400-2400 x 20-40
Rudiviridae	SIRV1	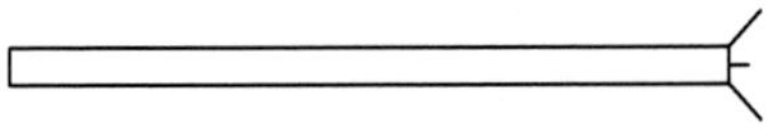	780-900 x 23
Leviviridae	MS2	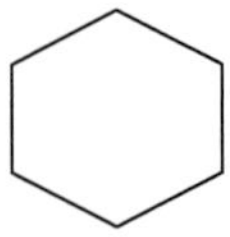	23
Cystoviridae	Ø6	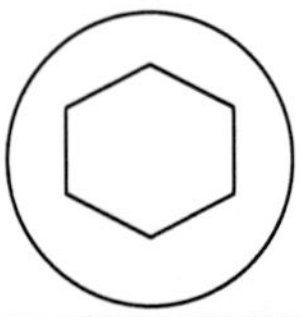	75-80

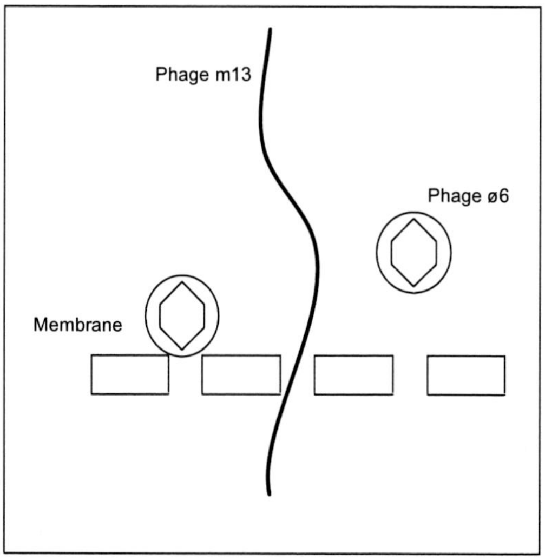

Figure 4. Influence of particulate shape on retention value.

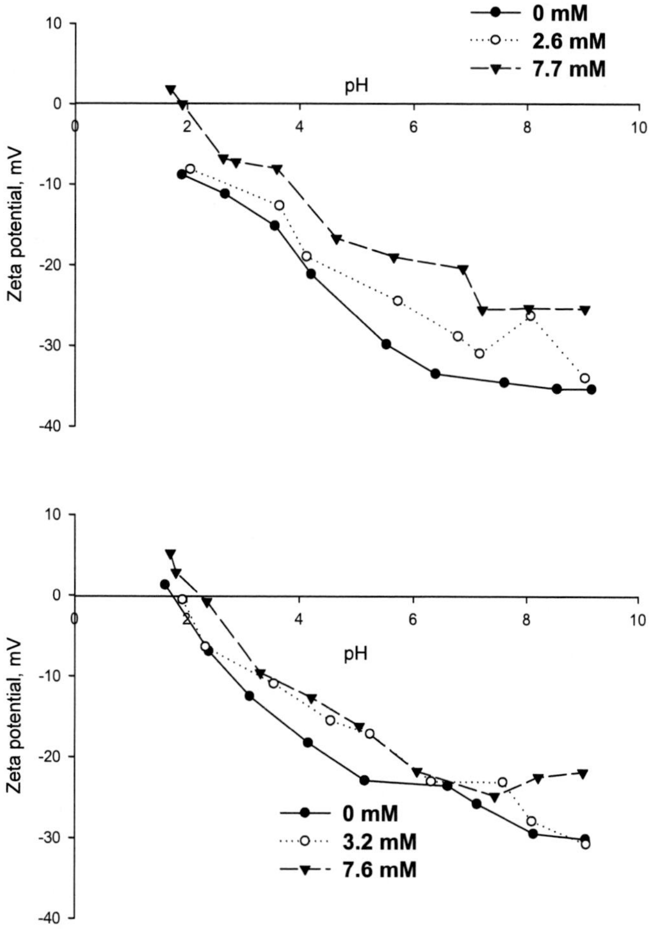

Figure 5. Zeta potential values of a) T4 and b) phi X 174.

Surface charge of a pathogen is very important and can give an indication of its fate inside filter. Significant charge can either promote or disapprove a contact with filter media, and weak or absence of charge will contribute to more interactions between pathogens and media, or between pathogens themselves. The surface charge on the pathogens cannot be measured due to presence of electric double layer, and pathogen's charge is usually expressed through zeta potential. Once again assuming a spherical shape of a pathogen and equivalent charge space distribution, measurements of zeta potential are giving a reasonable estimate of surface charge of different types of pathogens. Examples of zeta potential values for different viruses are given in Figure 5.

As easily detectable from the plots, zeta potential changes as a function of pH and for more alkaline conditions surface charge become more negative. Relative comparison of the measured zeta potential values combined with size distribution suggest absence of electrostatic repulsion for pH of isoelectric point and significant electrostatic repulsion for zeta potential values of -20 mV and higher. Majority of pathogenic microorganisms in water will carry a charge, and for typical surface water the pH will be high enough to consider electrostatic repulsion as one of significant retention mechanisms. Similar or opposite charge of filter media will result in considerable repulsion or attraction, respectively.

Hydrophobicity (fearing of water in *Attic Greek*) refers to a physical property of a pathogen (known as a hydrophobic) that will prefer not to be in the water if such a possibility exists. Water as a polar media tends to attract charged pathogens and to repel neutral and nonpolar ones. Majority of microbial pathogens are hydrophilic. Hydrophilicity or hydrophobicity of pathogen can be easy assessed by adsorption on hydrophobic or hydrophilic resins, and obtained log removal indicates how hydrophobic or hydrophilic the pathogen is (Shields and Farrah, 2002). There is no scale of higher or lower hydrophobicity, and that feature of the pathogens sometimes is used as a possible explanation of the pathogen behavior when other forecast has failed. Combination of the 4 factors – size, form, charge and hydrophobicity is minimal information relevant to waterborne pathogens to be removed by filtration.

3. FILTRATION

Filtration is a process of accumulation of pathogenic microorganisms on filter surface or inside its depth. That general definition can incorporate many variables, and it is mandatory to restrict further discussion to precise applications. In general, filtration can be classified by influent flow as downflow, upflow and cross-flow filtration (Figure 6). Filtration can be classified by driving force as pressure (including gravity), and vacuum filtration. Filtration can be performed through different materials such as granular materials (sand, anthracite coal, magnetite, garnet sand, and Coconut shells), diatomaceous earth, filter paper, sintered and perforated glass or metal screens, synthetic and ceramic membranes. Granular pressure filtration can be performed with constant flux or constant pressure, the latter will result in declining flux. Membrane filtration is usually performed with constant pressure. Granular constant flux filtration is subdivided by hydraulic load into slow, rapid and high rate filtration with typical approach velocities of 0.1-0.4, 5-20 and 25-40 m/h. Here approach velocity is defined as $V = Q/A$ where V = approaching velocity (m/h); Q = flux (m^3/h) and A = filter surface area (m^2). Filtration process includes both filtration and pretreatment, the latter might contain optional coagulation, flocculation and sedimentation following the scheme depicted on Figure 7.

Therefore implementation of a general term granular filtration will in fact mean granular (sand or anthracite) constant flux rapid down flow filtration performed under either direct or contact scheme. Membrane filtration will mean slow filtration through either synthetic or ceramic membrane in down flow (also called dead-end) or cross-flow direction performed under constant pressure contact scheme. Vast majority of full-scale filtration applications implement one of the two schemes generically named as rapid granular or micro-/ultra-filtration.

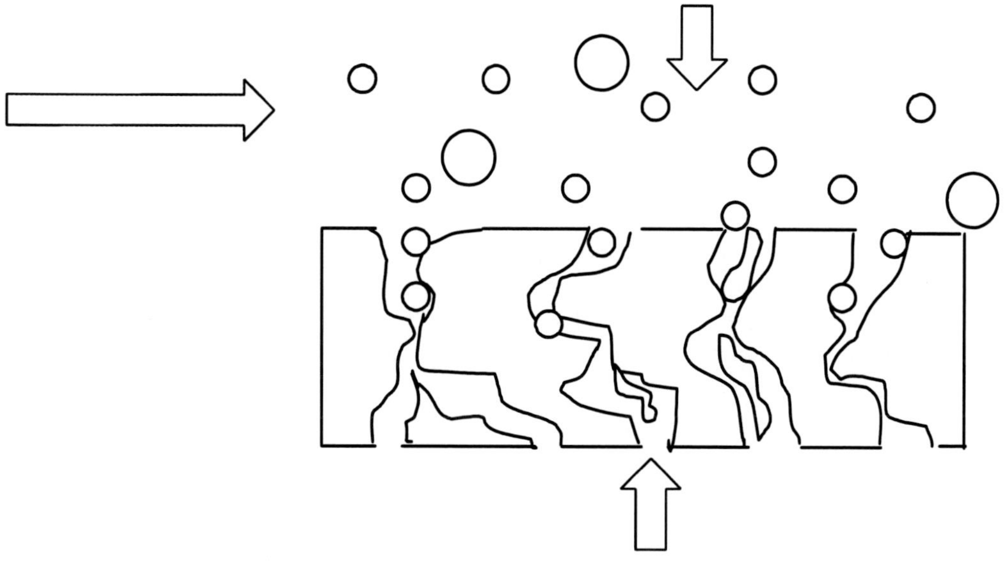

Figure 6. Filtration processes classified by flow direction.

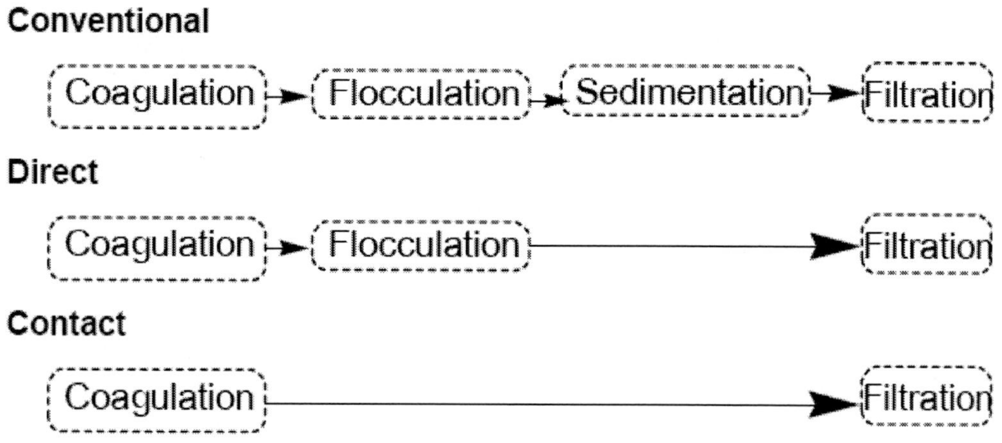

Figure 7. Filtration schemes.

Degree of microbial retention by filtration is defined by a combination of transport, accumulation and inactivation. Transport and inactivation mechanisms are different between membrane and granular filtration, and will be discussed later at separate chapters. The accumulation mechanism is controlled by the surface properties of filter media (O'Melia and Stumm 1967) and by chemical-physical interactions between a pathogen and the media surfaces (Elimelech and O'Melia 1990).*Electrical double layer interaction* (electrostatic interaction) is defined by the attraction or repulsion forces resulting from the interaction of ion clouds between pathogens in water creating repulsion if the filtration medium and particle zeta potential are alike and attraction if they have unlike potentials (Ives 1969). A more detailed explanation is given by O'Melia (1985), where the electrical double layer is defined as an accumulation of solute ions of opposite charge arranged in compact and diffuse layers near a pathogen surface so that each interfacial region is electrically neutral. The range of the double layer effect can be controlled by the presence of dissolved salts in the water (Chang and Vigneswaran 1990; Elimelech and O'Melia 1990; Huang et al. 1999; Nocito-Gobel and Tobiason 1996). Interaction of two similarly charged entities might result in attraction due to *van der Waals* attraction forces that act at close distance between charged bodies. High ionic strength, low pH or addition of multivalent ions (such as Al^{3+} or Fe^{3+}) may increase chances for attraction even in a case of two similarly charged pathogens.

Hydrogen bonding mechanism suggests that the attraction of pathogens to filter media could occur through hydrogen bonding of water molecules between surfaces (Elimelech and O'Melia 1990; Ives 1969; O'Melia and Stumm 1967).

Steric effect is another mechanism that can affect accumulation. When pathogen has different charge groups, filtration media may shield or interact with the particulate, in that way governing where and from what direction interaction will take place (Figure 8).

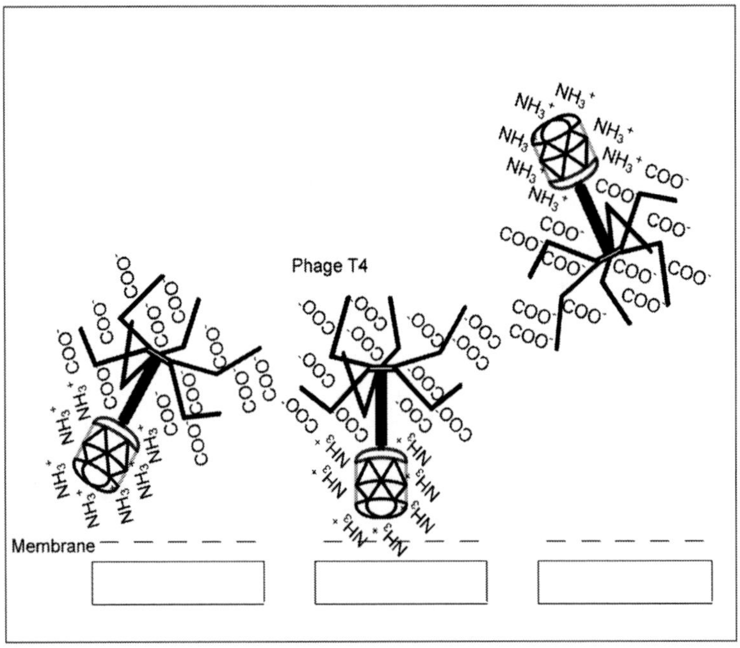

Figure 8. Steric interactions.

Hydrophobic interactions between pathogens and filter surfaces may also contribute significantly to adsorption (Huisman et al., 2000; Van Voorthuizen et al., 2001). Due to the increased electrostatic repulsion at higher pH levels, hydrophobic interaction could play the major role in maintaining pathogen-filter adsorption. Figure 9 shows hydrophobic interactions between membrane surface and a phage MS2. The virus is surrounded by water molecules, which orient spontaneously on boundary with non polar environment and form very tight monomolecular layer. After initial adsorption the structure that is formed around attached viruses keeps them close to membrane surface.

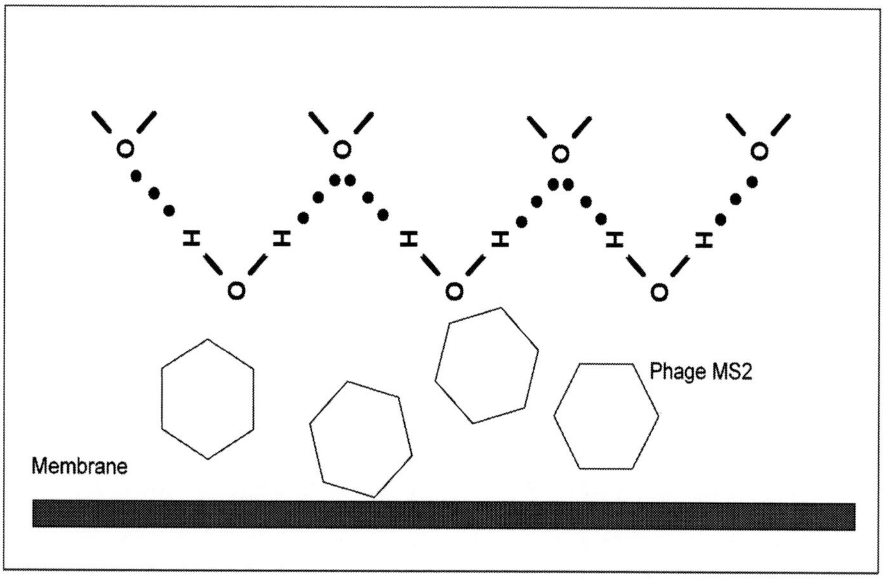

Figure 9. Hydrophobic interactions.

3.1. Granular Filtration

As was already stated above, degree of microbial retention by granular media is defined by a combination of transport, accumulation and inactivation mechanisms. Granular filtration transport process includes Brownian diffusion, interception, sedimentation, inertia and hydrodynamic action, as has been suggested by Yao et al. (1971).

Transport by Brownian diffusion (Figure 10a) is more effective for viruses with sizes much less than the one of minimal transport efficiency (1μm). The viruses are randomly bombarded by water molecules as they move through packed bed resulting in the pathogens moving out of the bulk streamline close to the media grain surface (Ives 1969; Yao 1968).

Removal by interception (Figure 10b) takes place when a pathogen of a finite diameter (d_p) travels on a flow stream line that comes within a distance from the media grain less or equal to $d_p/2$ thus contacting the collector surface and increasing the possibility of adhesion (Ives 1969; Yao 1968).

Transport by sedimentation (Figure 10c) will take place as gravity causes pathogens to move in a vertical direction with a corresponding settling velocity. This velocity will depend on pathogen's density and size as well as on water temperature (Amirtharajah 1988; Ives

1969; Yao 1968). Although in general that is one of three major transport mechanisms, for pathogens is become important for protozoa and fungi and less important for virus and bacteria. That is due to significantly lower pathogen density compared to that of minerals or metal particles. To become a significant transport mechanism the pathogens should be at least 40-50 μm size, and that is a rare feature.

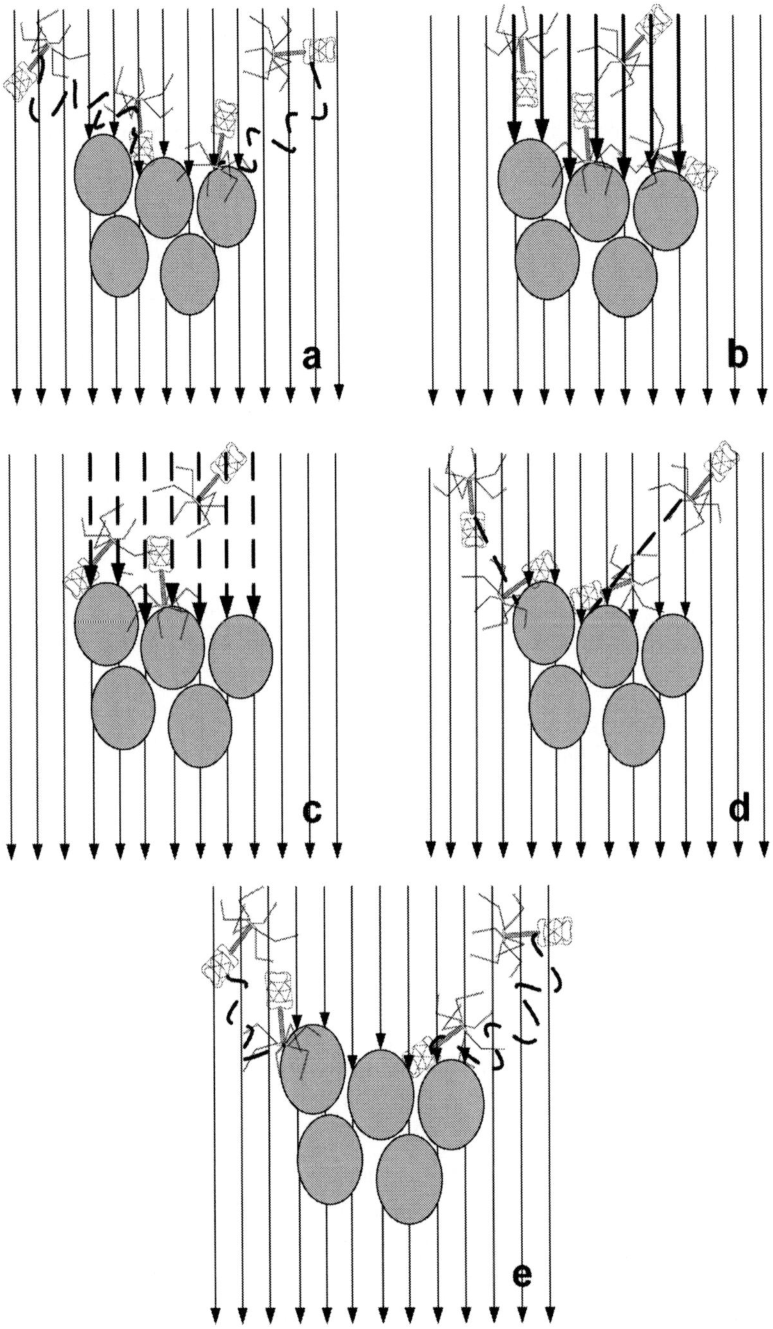

Figure 10. Transport mechanism: a) diffusion, b) interception, c) sedimentation, d) inertia, e) hydrodynamic action.

The inertial energy (Figure 10d) of the pathogens as they approach a collector will make them maintain their trajectory causing a collision with the collector. As has been observed for filtration of particles in air, transport by inertia should be considered as an important removal mechanism (Friedlander 1976). However, for pathogens suspended in water factors such as viscosity will prevent the pathogen from diverting out of its streamline as it passes around the collector making this removal mechanism unimportant under modern rapid filtration conditions (Ives 1969; Yao 1968).

Transport by hydrodynamic action (Figure 10e). It is known that pathogens in the presence of a shear force field will rotate and consequently experience a lateral force that can move them across streamlines (Ives 1969). This rotation is due to water shear force being lower on the side or the pathogen facing the collector than on the other side and will be enhanced if the pathogen is not spherical and the shear field is not uniform (O'Melia 1985). As the pathogen gets near the surface of the collector, the shear forces on the pathogen approach infinity as a result of the viscous fluid and the no-slip boundary condition at solid surfaces in fluid mechanics. This decreases the movement of the particle towards the grain. This phenomenon is known as hydrodynamic retardation (Amirtharajah 1988; Ives 1969; O'Melia 1985; Yao 1968). As we can infer, the effects of hydrodynamic retardation suggest the necessary existence of attachment forces in order to create a pathogen-grain contact.

From stated above it is easy to explain the factors that were reported to influence the interactions between the pathogens and filter media such as 1) characteristics of the suspension, including ionic strength and composition (Zhuang and Jin, 2003; Quignon et al., 1998; Lance and Gerba, 1984), pH (Loveland et al., 1996; Bales et al., 1989), and presence of dissolved organic matter (Chattopadhyay et al., 2002; Powelson et al., 1991); 2) characteristics of the pathogen, such as size (Woessner *et al.*, 2001; Dowd *et al.*, 1998), isoelectric point (Chattopadhyay and Puls, 2000; Dowd et al., 1998; Floyd and Sharp, 1979) and hydrophobicity (Shields and Farrah, 2002; Lukasik et al., 2000; Bales et al., 1993); **3)** medium features (Meschke and Sobsey, 2003; Zhuang and Jin, 2003).

The attenuation of pathogens as they pass through filter is governed not only by accumulation, but also by *inactivation*. There are four factors with the proven inactivation effect: 1) free water content - pathogens accumulate at or near the contact surface of water with air (Johnson and Gregory, 1993) and their position on air-water-solid interface promote their inactivation (Chu et al., 2003; Jin and Flury, 2002; Thompson and Yates, 1999); 2) temperature - the survival of pathogens is prolonged at lower temperatures while at elevated temperatures the inactivation is fairly rapid (Wait and Sobsey, 2001; Straub et al., 1993; Hurst et al., 1989), 3) sunlight - both visible and short wave ultraviolet (UV) components of sunlight are lethal to pathogens (Sinton et al., 2002; Wommack et al., 1996); and 4) pH – most pathogens are acid stable and high pH is considered important for their inactivation (Feng et al., 2003; Soule et al., 1998).

The beginning of deep-bed filtration theory dates back to more than half-century ago with works of Iwasaki (1937) and Minz (1951). Later phenomenological models of the filtration process can be easily classified by their further development of either one of the two basic concepts, as illustrated below. The provided brief analysis of pros and cons of the basic ideas does not constitute a comprehensive review that can be found elsewhere (Vigneswaran and Ben Aim, 1989; Tien, 1989). The mathematical description of filtration process as a change in concentration C of suspended particles with time t relies on a local mass conservation law combined with an accumulation kinetics equation. Assuming preferential transport in down flow direction z and neglecting spatial deposit difference as insignificant, a macroscopic volume-average mass balance equation reads

$$\frac{\partial \varepsilon\rho}{\partial t} + \Gamma = -V\frac{\partial}{\partial z}(\varepsilon\rho) + \frac{\partial}{\partial z}\left(D_{AB}\varepsilon\frac{\partial\rho}{\partial z}\right) + \frac{\partial}{\partial z}\left(D_{aim}^{E}\varepsilon\frac{\partial\rho}{\partial z}\right) \quad (1)$$

Here D_{AB} = diffusion coefficient, D_{aim}^{E} = dispersion coefficient, t = time, V = approach velocity, ε = porosity of the filter bed, ρ = mass density of pathogen and Γ is specific deposit = mass of pathogens adsorbed per unit volume of the filter.

The first two terms in Eq. (1) represent, respectively, the rate of change of the partial mass density of suspended and deposited pathogens. Terms on the r.h.s. of Eq. (1) represent the flux of particles through the filter due to advection, diffusion and hydrodynamic dispersion, respectively. Eq. (1) was derived under few assumptions such as particles are spherical, non-interacting and in low concentration. The flow is assumed to be laminar, and the porous media is homogeneous and isotropic.

The kinetic equation that expresses a specific pictorial view of the deposition process, generally reads $\Gamma = F(C,t)$. Controversy in a view on attachment – detachment kinetics resulted in two alternative basic expressions for $F(C, t)$ in attempt to model particle accumulation:

1) A suspended particle, once retained by the porous bed, is never entrained again by the flow (Iwasaki, 1937; Ives, 1969; O`Melia and Ali, 1978; Tien et al., 1979). The respective equations of attachment kinetics of this type have the general form $\Gamma = \lambda Cu$, with various models differing by the way of specifying λ.
2) Suspended particle is alternatingly deposited and ripped off by the flow (Minz, 1951; Adin and Rebhun, 1977). The respective equation of accumulation kinetics reads $\Gamma = K_a uC - K_d\sigma$, with the phenomenological coefficients K_a and K_d specified empirically. The second hypothesis might be considered as more comprehensive including the first one as a part of the general kinetic equation. Possible attachment – detachment filtration mechanism, corresponding to kinetic equation of second type, is depicted on Figure 11.

The controversial issue of whether or not detachment of previously removed pathogens takes place is still unresolved. According to the first kinetic hypothesis, experimentally evidenced breakthrough stage occurs when media pores become so narrow that under constant feed the interstitial velocity does not allow to suspended pathogens to adhere (Ives, 1969, 1973). Other investigators claim that since the accumulation process can be simplified into adhesion of destabilized flocs, tearing off discrete (original) pathogens by the shear forces is as obvious as in flocculation.

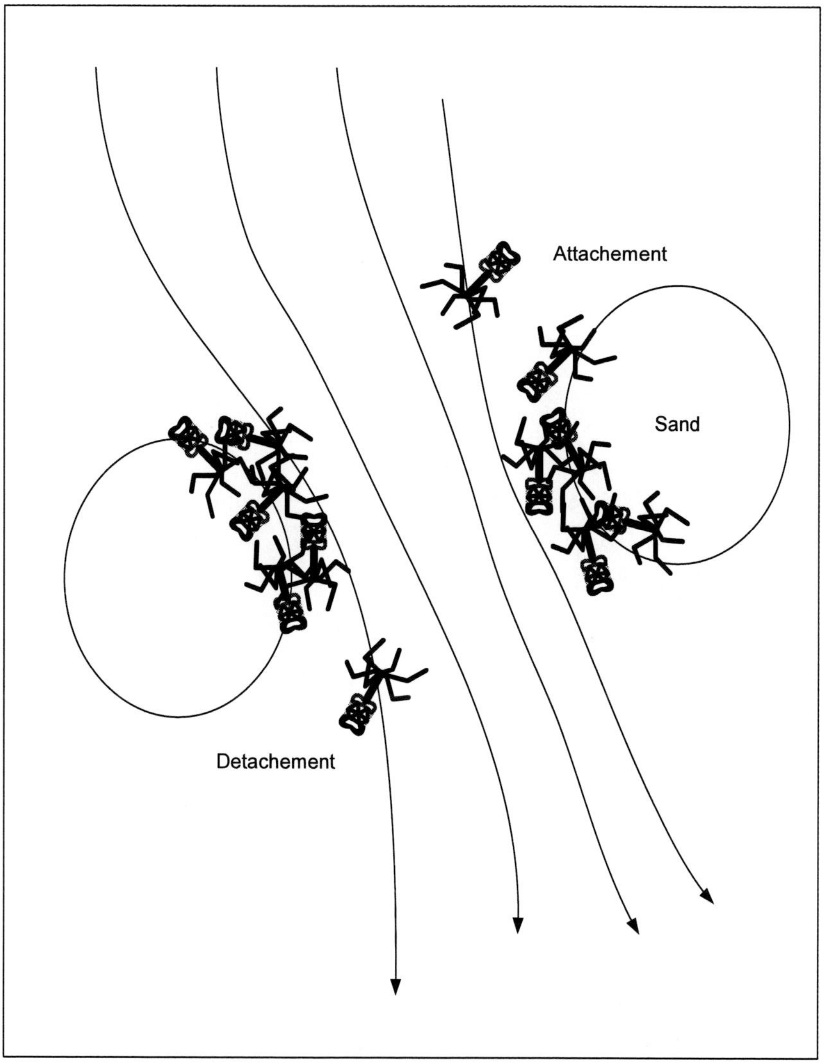

Figure 11. Attachment-detachment mechanism in granular filtration of pathogens.

3.2. Membrane Filtration

Degree of microbial retention by membrane filtration is defined by the combination of transport and accumulation. In general, a phenomenological theory of membrane filtration is still in its infancy and several new hypotheses will be postulated here. Despite a general determination as a slow filtration process the hydraulic retention time of pathogenic microorganisms in micro – and ultra- filtration (MF and UF) filters is relatively short. Therefore omitting potential biofilm formation that is more relevant to reverse osmosis applications, our hypothesis is that inactivation is not an important mechanism in membrane filtration. That is generally true for industrial applications performed in closed vessels with room temperature and pH in the range of surface water. None of the known potential inactivation factors – temperature, pH, presence of non dissolved gas, UV or other disinfection agent - will not be significant. Thus fate of pathogens will be influenced only by transport and accumulation mechanisms. Membrane filtration transport process will include Brownian diffusion, interception, and hydrodynamic action, with neglected sedimentation and inertia.

Transport by Brownian diffusion (Figure 12a) is an effective transport mechanism for small pathogens (viruses) that under influence of Brownian motion will deviate from water streamlines and will attain membrane surface. Efficiency of the mechanism can be expressed by the following relation (Equation 2)

$$\eta_D = 0.9\left(\frac{KT}{\mu d_p d_m u}\right)^{\frac{2}{3}} \tag{2}$$

where d_p = hydrodynamic diameter of the pathogen, d_m = membrane pore size, K= Boltzmann constant, T = absolute temperature, u = pore velocity, η_D is transport efficiency by Brownian motion and μ = absolute viscosity of water.

Transport by hydrodynamic action (Figure 12b) will occur through local changes of flow regime that will lead to lateral transport of pathogens across streamlines. The changes will cause a rotation of not spherical pathogens and result in transport of pathogens near membrane surface.

Transport by interception (Figure 12c) takes place when a pathogen travels on a flow stream line that comes within a close vicinity of the membrane surface thus increasing the possibility of adhesion. The relative efficiency of the mechanism can be expressed as

$$\eta_I = \frac{3}{2}\left(\frac{d_p}{d_m}\right)^2 \tag{3}$$

where d_p = hydrodynamic diameter of the pathogen, d_m = membrane pore size, and η_I is transport efficiency by interception.

Many of membrane features are a function of polymer from which they were obtained. Although other forms, including ceramic and metallic membranes, are available, almost all membranes manufactured for drinking water production are made of polymeric material. Basically, all polymers can be used but for processing and lifetime requirements of the membrane, only a limited number is used in practice.

MF and UF membranes may be synthesized from cellulose acetate (CA), polyvinylidene fluoride (PVDF), polyacrylonitrile (PAN), polypropylene (PP), polysulfone (PS), or polyethersulfone (PES). Each of these materials has different properties with respect to surface charge, degree of hydrophobicity, pH and oxidant tolerance, strength, and flexibility. Main features of the membrane materials are summarized in Table 4.

Now, after a brief introduction of background of filtration processes we would like to report and discuss application of rapid granular filtration for retention of oocysts of *Cryptosporidium parvum* (Chapter 4) and ultrafiltration for retention of enteric viruses (Chapter 5). The latter are represented by bacteriophages T4, MS2 and phi X 174.

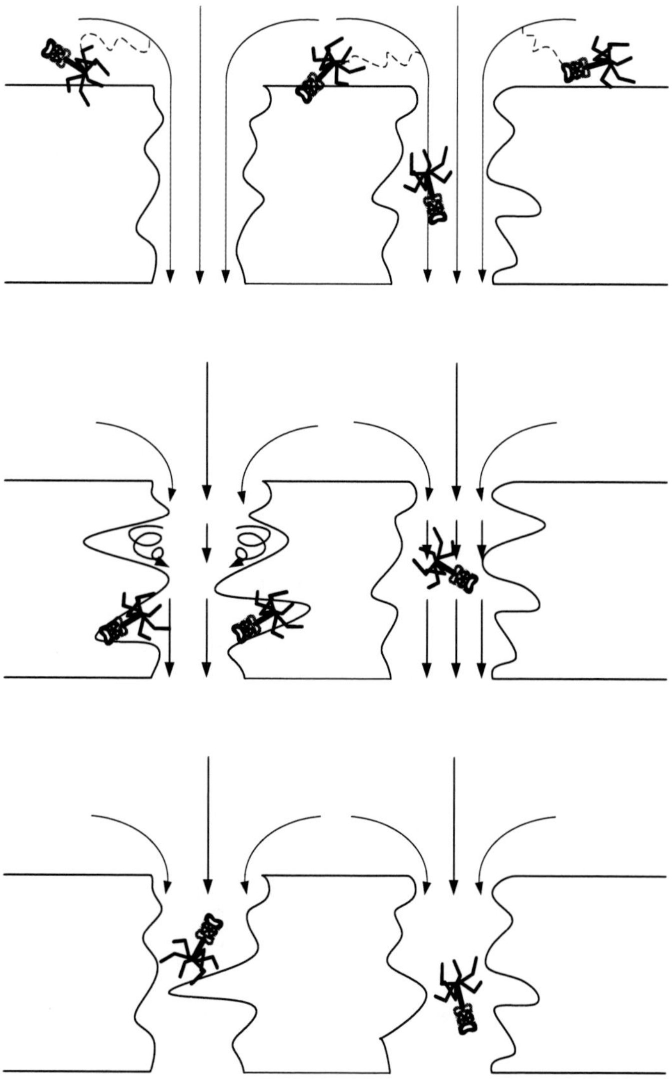

Figure 12. Transport mechanisms in membrane filtration: a) diffusion, b) hydrodynamic action, c) interception.

Table 4. Features of the most commonly used membrane materials: (+) tolerant, (-) non-tolerant

Membrane	Hydrophobicity	pH	Temperature	Biological stability	Chemical stability	Chlorine resistance
CA	hydrophilic	-	-	-	-	-
PS/PES	hydrophobic	+	+	+	+	+
PP	hydrophobic	+	+	+	+	-
PVDF	hydrophobic	+	+	+	+	+

4. Study on Rapid Sand Filtration of Oocysts of Crptosporidium Parvum

The study was conducted at the U.S. EPA Test and Evaluation (TandE) Facility in Cincinnati, Ohio, U.S.A., using specially designed and built filtration system depicted schematically on Figure 13. The slurry, consisting of Kaolin clay particles and dechlorinated city of Cincinnati tap water treated by Reverse Osmosis (R2D40, Ecodyne Water Treatment, Inc., Naperville, IL), was pumped into a 1000 L Cross-Linked Polyethylene (XLPE)-made feed water supply tank. The suspension in the tank was kept completely mixed by using a high-speed mixer (VL3501, Baldor Electric Co., Fort Smith, AR). A centrifugal pump (AC-3C-MD, March Manufacturing Inc., Glenview, IL) was used to lift the suspension into a 100 L head tank located 3.6 m above the filtration column. The water level in the feed tank was maintained at a constant height by an overflow line returning the suspension to the feed tank. To avoid remnant presence of oocysts in the feed tank, *C.parvum* was injected into the mainstream immediately before the static mixer (2850, Westfall Manufacturing Co., Bristol, RI). The major physical characteristics of suspension-forming Kaolin clay particles and *C.parvum* oocysts are presented in Table 5. The influent oocyst concentration of $8{\cdot}10^6$ oocysts/L was up to 7 logs higher than that found in surface water (Karanis *et al.,* 1998). *C.parvum* oocysts samples were obtained from six weeks-old immune-compromised female mice (Cicmanec and Reasoner, 1997) following a modified protocol reported by Yang et al., 1996. The samples were then purified using a cesium chloride solution resulting in 99% purity.

The final oocyst samples were suspended in a phosphate buffer saline solution (pH=7.4) containing antibiotics/antymicotic and stored at 4° C until used in the experiments. Enumeration of *C.parvum* oocysts was performed using Immuno Fluorescent Assay (IFA) technique, following the procedure described by Kao and Ungar (1994). The high feed concentration was needed in order to ensure detectable effluent oocyst concentration. Even at the feed concentrations used in that work, oocysts only ranged from 0.007 to 0.06 percent of total number of influent particles (greater than 0.5 μm). At these relative concentrations, oocysts were assumed to not significantly change expected accumulation dynamics.

Seventeen cm diameter, 2.6-m high acrylic transparent filter column was used for the study. The column consists of 9 sampling and 9 pressure ports, located in pairs on opposite sides of the column. Sampling points were inserted 0.05 m inside the column to avoid wall effects that could lead to misinterpretation of filtrate quality. The pressure transmitters (206, Setra Systems, Inc., Boxborough, MA) were constantly connected to the pressure ports. The column was packed with 1.6 m of uniform size sand having a geometric mean diameter of 1.05 mm and a uniformity coefficient of 1.55 (#5 Silica Torpedo Sand, Parry Company, Richmond Dale, OH). The media porosity, determined by volumetric measurements, was 0.44. The column was operated at a constant flow rate under contact (in line) filtration mode in a conventional downward direction. The filtration velocity was controlled and adjusted by a flowmeter (U-32470-02, Cole-Parmer Instrument Co.) connected to the column outlet. Immediately after each run, the filter was backwashed for 1 min by air at 200 kN/m^2 pressure, followed by 10 min upflow of Cincinnati tap water at 1.1 L/sec. Both caused 50% bed expansion. To avoid risk of cross-contamination, the filter was chlorinated and all removable parts of the system (pumps, baths, pipes) were sterilized after each run. *C.parvum* sampling jars were autoclaved.

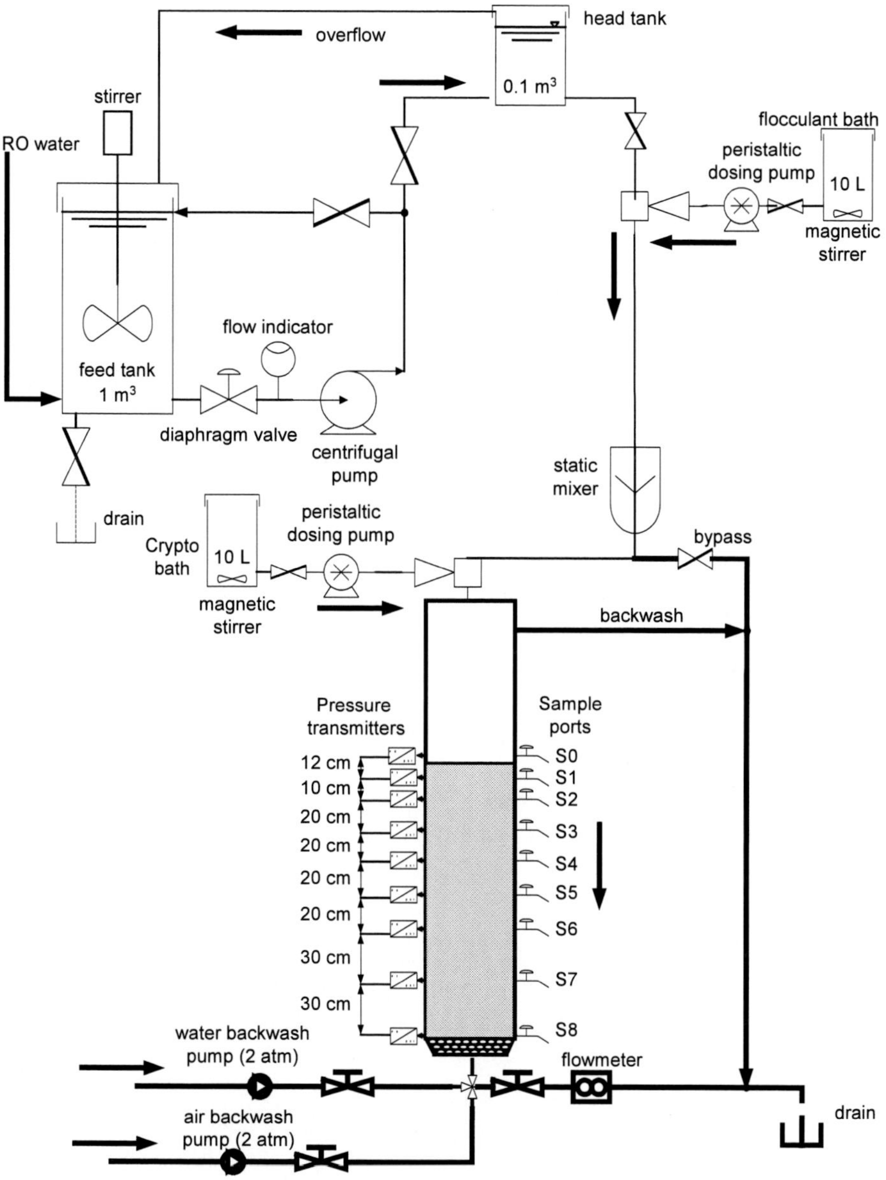

Figure 13. Schematic of apparatus for deep-bed filtration experiments

Table 5. Physical characteristics of suspension-forming entities

Particle	Size (μm)	Density (g/cm3)	Zeta-potential after RO, mV	Zeta-potential in Cincinnati tap after RO system with 20 mg/L Alum, mV
Kaolin	0.778 ±0.315	2.6	-10.33	-3.33
C.parvum	4 - 6	1.045	-16.0	-1.26

Flocculants Preparation. To imitate a filtration plant in the best possible manner, the suspension was destabilized using Alum $[Al_2(SO_4)_3 \cdot 18H_2O]$ (Sigma-Aldrich Inc.) solution, injected into the suspension using a peristaltic pump (Masterflex P-07553-70, Cole-Parmer Instrument Co., Vernon Hills, IL*)* through Tygon silicone pipes (Masterflex L/S16, Cole-Parmer Instrument Co.). The optimal dose of Alum to be added to slurry of Kaolin and *C.parvum* in water was established using a series of flocculation/sedimentation tests (Jar-tests). The following procedure was used in all tests: (i) Rapid mixing at 100 rpm for 1 min; (ii) Flocculation at 30 rpm for 20 min; and (iii) Quiescent settling for 30 min. The tests were performed at a room temperature (~ 20°C). Turbidity samples were obtained using a 20-mL pipette from a point located 0.04 m below the clarified suspension level. *C.parvum* samples were filled into 0.25-L polypropylene jars by careful decantation from the upper part of 1-L test beakers.

Quantitative analysis of the data was performed using filtration models. The ripening rate coefficient K_r was found by fitting theoretical curves to experimental data using the best-fit value method. Parameter variation in curves fitting was used to attain the closest possible approximation of experimental data by theoretical curves. In terms that are generally applicable to the ripening stage of the filtration process irreversible accumulation mode, the local mass balance equation in its idealized form (with no axial dispersion) yields

$$\frac{\partial}{\partial t}\left(\varepsilon_0 - {C_s}/{\rho_d}\right)C_l + K_r\left(1+\beta\frac{C_s}{\varepsilon_0}\right)uC_l + u\frac{\partial C_l}{\partial Z} = 0 \qquad (4)$$

Here C_l = oocyst concentration in water; C_S = concentration of oocysts attached to media grains; t = time; u = approach velocity; Z = position in the column measured as the distance from the entrance in the direction of the flow; β = geometrical constant; ε = porosity of filter bed; ε_0 = initial porosity of filter bed; K_r = attachment rate coefficient; ρ_d = physical deposit layer density.

The assumption that suspended microorganisms, once retained by the porous bed, are never entrained again by the flow, was widely shared during the last decade (Martin et al., 1992; Wan et al., 1995). Filtration removal efficiency was expressed in terms of the filter coefficient λ (Iwasaki, 1937; Ives 1969)

$$\lambda = \frac{\ln\left({C_{l0}}/{C_l} \right)}{L} \tag{5}$$

$$\lambda = \frac{3}{2}\frac{(1-\varepsilon)}{d_c}\eta\alpha \tag{6}$$

Here C_{l0} = initial value of oocyst concentration in the liquid phase; L = filter depth; α = the collision efficiency factor; η = single-collector transport efficiency.

The attachment efficiency α represents the fraction of pathogens collisions with collector surfaces resulting in attachment. The theory of interaction between two charged particles in solution, known as the DLVO (Derjagin, Landau, Vervey, Overbeek) theory, was approached in the current study using Debye length, thickness of the compensating ionic atmosphere

$$k^2 = \frac{F^2}{\mathrm{EE}_0 RT}\sum_j c_j z_j^2 \tag{7}$$

Here F = Faraday constant; R = Universal gas constant; T = absolute temperature; c = molar concentration of the electrolyte molecules which dissociate completely; z = valence of the molecules; E= relative static dielectric permittivity of water; E_0 = permittivity of vacuum.

When pathogens and collector surfaces have opposite charges, essentially all collisions result in attachment because of the strong attractive energy between pathogens and collectors. Under these conditions, deposition was assumed to be favorable, and a value of 1 was used for α. Thus, colloid transport efficiency η was expressed as

$$\eta^1 = \frac{2}{3}\frac{d_c}{(1-\varepsilon_0)}\lambda_0 = \frac{2}{3}\frac{d_c}{(1-\varepsilon_0)}\frac{\ln\left({C_0}/{C} \right)_0}{L} \tag{8}$$

Here d_c = media grain diameter; λ_0 = initial value of filter coefficient; L = filter depth. The experimentally observed values of η were compared to calculated values following theoretical expressions for the single-collector transport efficiency as suggested by Yao *et al.* (1971; Eq. 9) and Rajagopalan and Tien (1976; Eqs. 10-12).

$$\eta^2 = \alpha\left\{0.9\left[\frac{kT}{\mu d_p d_c u}\right] + \frac{3}{2}\left(\frac{d_p}{d_c}\right)^2 + \frac{(\rho_p - \rho)g d_p^{\,2}}{18\mu u}\right\} \tag{9}$$

Here k = Boltzman constant; μ = water viscosity; d_p = oocyst diameter; ρ_p = oocyst density; ρ = water density; g = gravitational constant.

$$\eta^3 = \frac{4\sqrt[3]{A_s}}{\left(\frac{3\pi d_p d_c u}{kT}\right)^{2/3}} + A_s\left(\frac{4A_H}{9\pi\mu d_p^2 u}\right)^{1/8}\left(\frac{d_p}{d_c}\right)^{15/8} + 0.00338A_s\left(\frac{g(\rho_p - \rho)d_p^2}{18\mu u}\right)^{1.2}\left(\frac{d_p}{d_c}\right)^{-0.4} \quad (10)$$

Here A_s = specific surface area of the solid phase available for adhesion

$$A_s = \frac{2(1-(1-\varepsilon)^{5/3})}{(2-3(1-\varepsilon)^{1/3}+3(1-\varepsilon)^{5/3}-2(1-\varepsilon)^2)} \quad (11)$$

And A_H = Hamaker constant ($1.26\bullet10^{-20}$ J), calculated using Berthelot principle

$$A_H = \sqrt{A_{qw}A_{wC}} \quad (12)$$

Here A_{qw} – Hamaker constant for interface water-quartz, $1.6\bullet10^{-20}$ (Lyklema, 1991) and A_{wC} – Hamaker constant for interface *C.parvum*-water, $1\bullet10^{-20}$.

Results and Discussion

Table 6 presents experimental conditions and values of filtration coefficients, both fitted and calculated. Calculations, performed based on the theoretical approximation by Yao *et al.* (1971; Eq. 9, η^2 in Table 6) and Rajagopalan and Tien (1976; Eq. 10, η^3 in Table 6), resulted in a much lower predicted removal efficiency compared to that derived experimentally (Eq. 8, η^1 in Table 6).

Table 6. Experimental conditions and values of calculated filtration parameters

Exper. #	Kaolin, mg/L	Humic acid, mg/L	u, m/h	Alum, mg/L*	C/C_0 x10^{-3}	K_r (m^{-1})	λ_0 (m^{-1})	η^1 x10^{-4}	η^2 x10^{-4}	η^3 x10^{-4}
SEP11-1	10	-	5	-	110	2.95	5.62	70.9	6.44	4.91
SEP15-1	10	-	10	-	240	4.15	3.53	44.5	3.39	3.97
SEP18-1	10	-	10	10	90	8.4	5.78	72.9	3.47	3.99
SEP22-1	10	-	5	20	30	24.9	8.62	109	6.12	4.82
SEP27-1	10	12	5	20	490	1.9	1.77	22.3	6.44	4.91
OCT2-1	20	-	10	20	2.9	25.4	14.58	184	3.39	3.97
OCT5-1	10	-	10	20	6.5	27	12.58	159	3.31	3.94
OCT13-1	10	-	17	20	2.8	26.2	14.68	185	2.13	3.52
OCT16-1	10	12	17	20	670	1.2	0.996	12.6	2.23	3.55

* Filtration parameters: media size 1.06 mm; oocyst initial concentration varies between $2\cdot10^8$ to $5\cdot10^9$ oocyst/L; temperature varies between 24 and 26 °C.

η^1 Experimental values using Eq. 8.

η^2 Predicted values using Eq. 9 under assumption of ideal destabilized system α=1.

η^3 Predicted values using Eqs. 10-12.

Predicted levels of transport efficiency could not be adjusted to those derived experimentally, even under the assumption of ideal attachment conditions. The observed discrepancy shows that a deviation exists between contemporary filtration theory and observed trends in *C. parvum* removal. However, basing speculations on existing theory, transport stage in removal of *C.parvum* has only a limited effect on *C.parvum* penetration probability. *C.parvum* single collector transport efficiency is high enough to not be considered as the deterministic factor in oocyst fate.

Thus, *proper chemical destabilization regime is important* to improve *C.parvum* filterability through enhanced number of fruitful collisions between oocysts and Kaolin clay particles resulting in floc formation. Comparisons, performed with the use of fitted filter coefficients λ_0 (Eq. 5) and K_r (Eq. 4), show significant improvement in filtrate quality with the addition of Alum. The ripening coefficient, K_r, was up to 7 times higher when proper chemical destabilization was implemented.

Figure 14 depicts removal ratio calculated on oocyst basis, as a function of filtrate volume through the bed, in the presence and absence of Alum. A concentration of 20 mg/L Alum was deemed as the optimal dose for filtration based on Jar-test experiments and previous recommendations (Adin *et al.*, 1979).

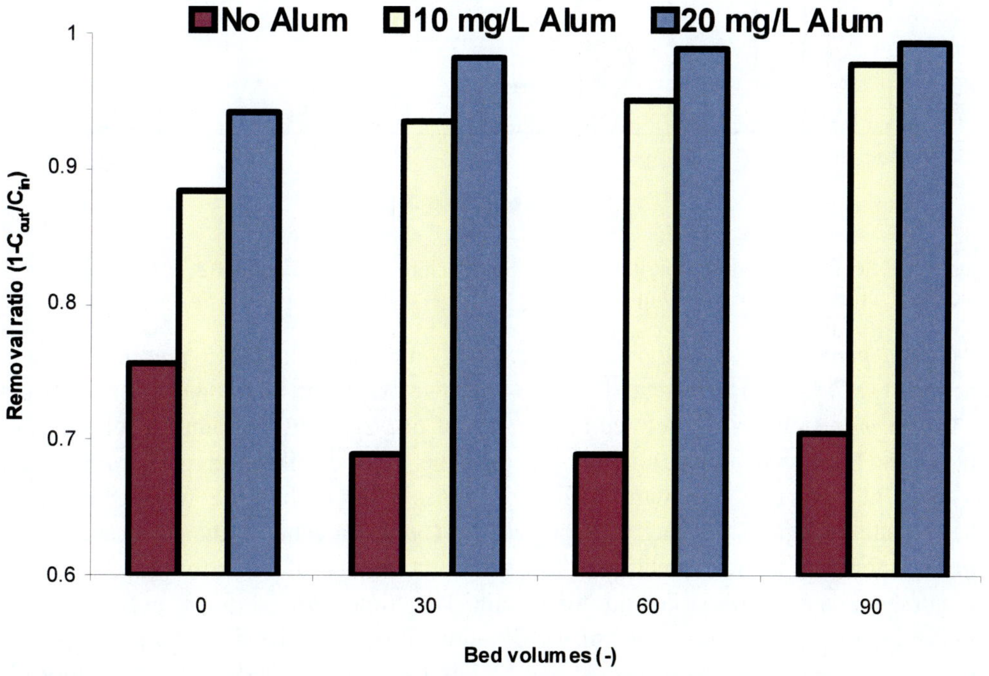

Figure 14. Removal ratio: effect of absence, under-optimal and optimal concentration of Alum on removal of C.parvum. Suspension consisted of 10 mg/L Kaolin + 109 oocysts, approached the filter at 10 m/h. Filtrate samples were collected at a depth of 0.4 m.

Filtration experiments performed with no chemical destabilization resulted in poor removal efficiency of *C.parvum.* Unstable filtrate quality, influenced by hydrodynamic conditions, was constantly more pronounced for small depths and became moderate as the filter became deeper. Addition of an optimal 20 mg/L dose of Alum improved the quality of filtrate obtained after implementation of a 10 mg/L Alum dose, and resulted in practical neutralization of surface charge. Performed calculations of Debye length suggested that at a distance of 3-4 nm from *C.parvum* surface, electric potential $\psi(r)$ dropped to approximately the zero level. At the same time, in system with a low ionic strength of 10^{-4} and 10^{-5} M, the cloud of compensating ions had a long tail up to 50 nm to the stream. Performed zeta-potential measurements (Table 5) confirmed theoretical calculations and showed values closed to electro-neutrality (-1.26 mV).

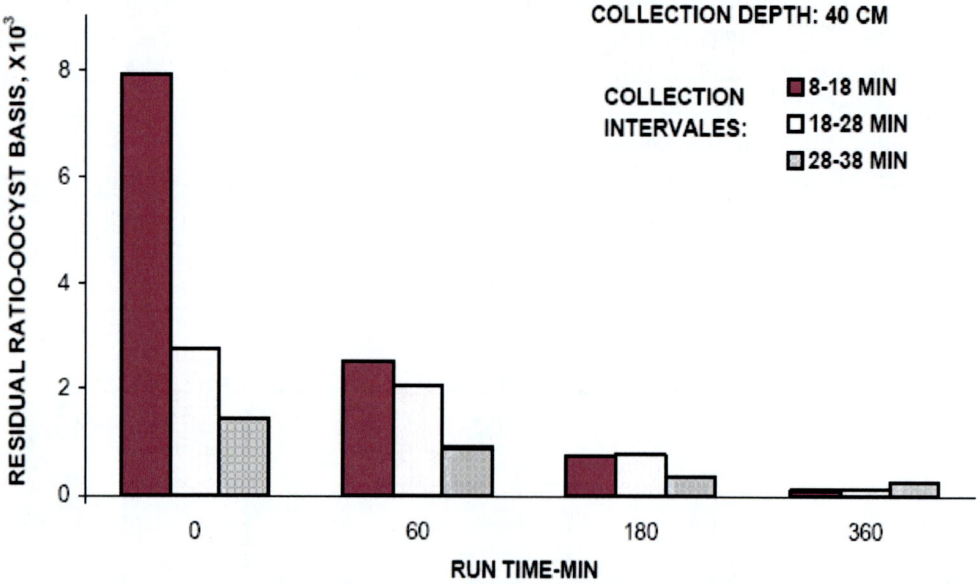

Figure 15. Retention time study for 40 cm depth. Suspension: 10 mg/L Kaolin + 2.3·108 oocysts/L; Flocculants: 20 mg/L Alum; u= 10 m/h.

Crypto penetration velocity was slower than expected. Figure 15 presents results of experiment OCT5-1, performed at 10 m/h approach velocity with addition of 20 mg/L Alum.

C.parvum residual ratio, defined as the ratio of oocyst counts in samples collected from ports S3 and S0, was plotted as a function of run time. Oocyst spikes were performed at 0, 60, 180 and 360 min pass the start. Effluent samples after each spike were consecutively collected into three jars 20 L each and analyzed for *C.parvum* concentration. The analysis was performed for each jar separately. To allow sufficient volume of water for detectable oocyst concentration, each jar was filled out for 10 min. The "time zero", 8 min after the spike, was calculated based on the dimensions of the filtration system. As the filtration progressed, the ratio between concentrations of oocysts in the jars has changed from distinct superiority in sample collected at 8-18 min interval at the beginning of the run, 0 min, to only slight concentration difference between samples collected at 8-18 and 18-28 min intervals at run times of 60 and 180 min toward bigger concentration in 28-38 min sample in samples collected 360 min after the start. Constant concentration of *C.parvum* above the media was observed.

The results of counting showed that the prediction based on plug-flow reactor (Lievenspiel, 1985) was true only for 10 cm depth. Here the concentration of *C.parvum* from sample collected at first time interval was constantly bigger or equal to the concentration collected at the second time interval. Both were sufficiently oocyst-concentrated than the sample collected at the third time interval. For 40 and 80 cm depths, the average retention time and its variance constantly increased as function of both filter depth and time from the beginning of the run. The discrepancy between actual and estimated oocyst penetration velocities resulted in 20 minutes delay from calculated arrival time. Calculated axial dispersion coefficient D showed values growing from $1.2 \cdot 10^{-4}$ to $7.6 \cdot 10^{-4}$ m^2/sec, questioning the usefulness of usual trend in calculations of particle retention time with soluble tracers (Deborde et al., 1999). Noteworthy that result of 80-cm samples is partially conclusive since absolute numbers of *C.parvum* in collected samples were low.

Growing dynamics of average retention time indicates shift from interactions of *C.parvum* particles with media grains to interaction with Kaolin particles coating the filter media. When the media grains are essentially bare of deposit, the *C.parvum* oocysts are either accumulate on media grains or pass the filter bed through. Kaolin particles flatten from dendrites (Rajagopalan and Tien, 1976) to over-media coat at the end of ripening period, present better accumulation opportunities from both electrostatic and hydrodynamic points of view. The number of interactions is growing and causes increase in average retention time, along with constant increase in removal efficiency.

Difference in Crypto and Kaolin removal dynamics has been emphasized. Figures 16 and 17 depict log removal, calculated on turbidity and oocyst basis, as function of run time. The log removal was calculated from the measured influent and effluent concentrations or detection limit instead effluent concentration, 0.05 oocyst/mL, when no oocysts were detected. Log of ratios S1/S0, S3/S0, and S5/S0 were plotted as 10, 40, and 80 cm, respectively. In all 3 depths turbidity removal efficiency (Figure 16) was bigger at the beginning of the run, 0.2, 0.7 and 1.2 log removal for 10, 40 and 80 cm depth, respectively. It is due to initial mix of remnant backwash water with down warding suspension. The removal efficiencies decreased one hour after initiation of the run, and reached a plateau an hour later. At the same time, removal of *C.parvum* (Figure 17) constantly improved for each port as filter run progressed.

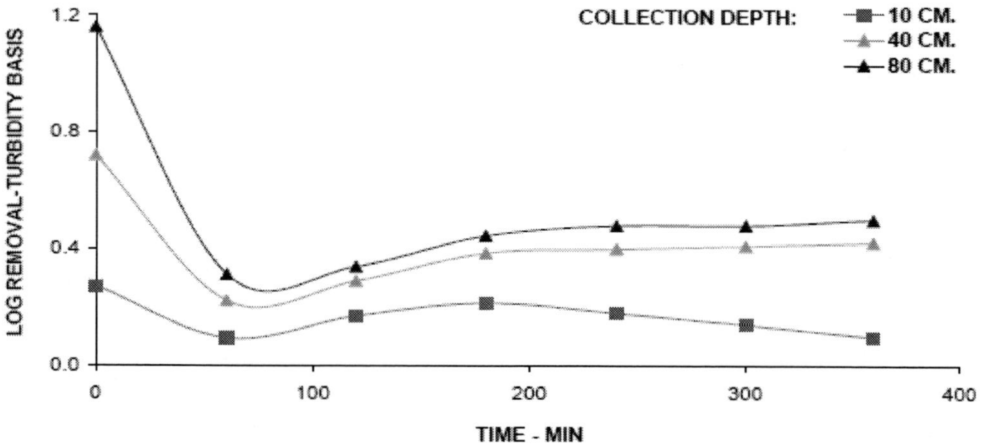

Figure 16. Removal efficiency: effect of filter depth on turbidity removal. (Parameter values as in Figure 15).

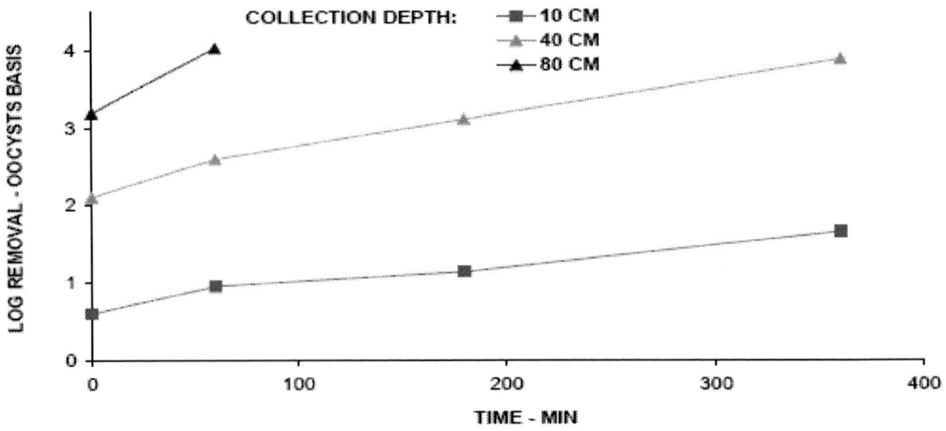

Figure 17. Removal efficiency: effect of filter depth on C.parvum removal. (Parameter values as in Figure 15).

For 10 cm depth, removal efficiency reached 0.5 log at the beginning of the run, and grew up to 1.6 log 6 hours after the start. In 40 cm depth, efficiency was found to rise from 2.1 to 3.9 logs of removal. On 80 cm, the removal enhanced from 3 to 4 logs during one hour. Continue of tracking for that depth became impossible due to limitation of used counting technique.

Differences in removal efficiency between ripening and working stages were noted throughout variety of experiments performed. Average level of 1.7 logs reduction was observed for 80 cm depth samples at the beginning of the run, t=0 min, compared to average 3 logs 60 min later.

Finding was supported by statistical analysis. More than 450 discrete filter efficiency data points were obtained with three flow velocities, three flocculant concentrations, five sampling depths, and ten sampling times. Two five-way Analyses of Variance (ANOVA) were conducted to investigate the effect of those independent variables on turbidity and oocyst removal ratios. The significance of the main effects alone was investigated: all interactions were found to be not statistically significant and were pooled with the error. The omnibus F-ratio was highly statistically significant for both turbidity and oocyst ratio, indicating that there are sources of variability in either model that cannot be attributed to chance alone (Winer, 1997). Indeed, as described previously, the sample depth was found to be statistically significant for both ratios (*p-value* < 0.0001). In addition, the flocculant concentration was found to affect significantly the oocyst ratio (*p-value* < 0.0001), but not the turbidity ratio. Flow velocity was not significant factor for both ratios. Yielded conclusions were supported by Student-Newman-Keuls (SNK) multiple comparison tests.

Run time, filter depth and flocculant addition are influencing Crypto removal. Figure 18 depicts effect of run time on minimum protection filter depth, calculated to adjust granted by LT1FBR 2-log reduction of initial oocyst concentration. The reduction was achieved only in conditions of optimal chemical regime, 20 mg/L Alum. Crypto removal remained below the required reduction level for experiments carried out without flocculants, as well as for the experiment performed with flocculant dosage below the optimal. Addition of optimal dose significantly improved total removal to average more than 3 logs (99.9%). Minimum protection depth decreased repeatedly as run progressed. The depth was found lower for 10 m/h (curve 4) than for 5 m/h (curve 1). Opposite to expected, addition of Kaolin did not decreased minimum protection depth (curve 3 vs. 4). Presence of organic material (curve 2) increased minimum protection depth compared to experiment performed without NOM (curve 1). In experiments without NOM removal efficiency was constantly higher for all depths and filtration velocities. This is due to increased colloidal stability caused by electrostatic stabilization. Dissolved humic substances, adsorbing on mineral surfaces (Petrovic et al., 1999), caused charge revolution of positively charged mineral edges (Brady et al., 1996) and increasing negative surface charge. Log reduction was constantly higher for experiments performed at 17 m/h than those performed at 5 m/h. Assuming equal accumulation, during 60 min 3.4 more Kaolin will be absorb at 17 m/h than at 5 m/h. Thus faster transition from ripening to working stage will increase Crypto absorption. The hypothesis was tested by simultaneous correlation analysis using sets of data for oocyst and turbidity removal. Pearson's R^2 correlation coefficient was 0.52 suggesting that under the conditions tested, enhanced *C.parvum* removal can be linked to amount of deposited Kaolin. However, absence of solid correlation indicates that low values of turbidity may cause misinterpretation of risk of microbiological contamination.

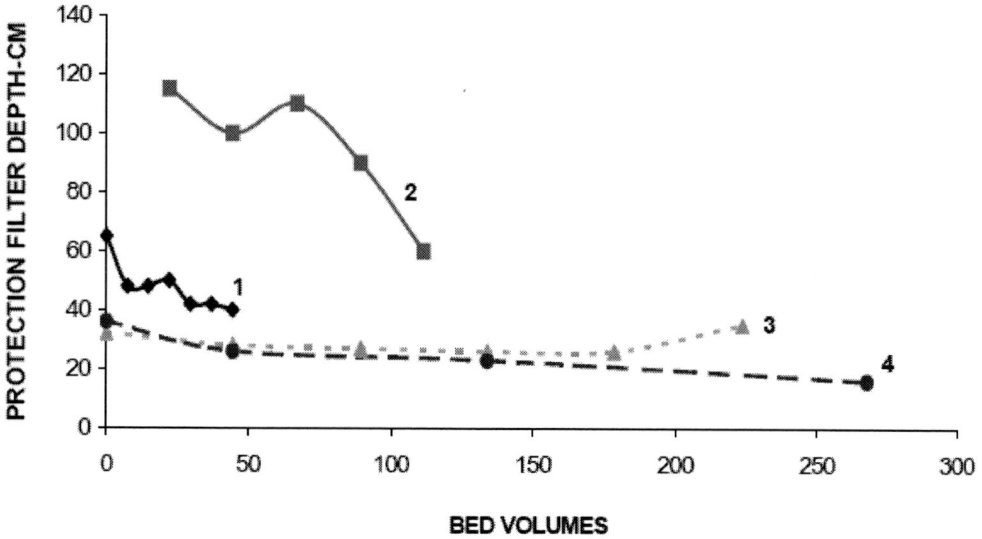

Figure 18. Effect of NOM (2), clay concentration (3) and velocity (1,4) on protection depth with time. C.parvum: 109 oocysts; Flocculants: 20 mg/L Alum; (1) Kaolin 10 mg/L; u=5 m/h; (2) Kaolin 10 mg/L + Humic acid 12 mg/L; u=5 m/h; (3) Kaolin 20 mg/L; u=10 m/h; (4) Kaolin 10 mg/L; u=10 m/h.

Several useful *conclusions* were drawn from the study:

- Removal of *C.parvum* using contact filtration scheme was affected by three factors – run time, coagulant/flocculant addition and filter depth. Best oocyst removal was found at operable stage using correct chemical regime on deep filters. However, in non-optimized conditions optimal values of two out of three factors will reduce risk to public health posed by *C.parvum*.

- Besides unquestionable importance of optimal destabilization regime, filter depth plays important role in removal of *C.parvum*. It importance becomes more critical in non-optimized chemical conditions or during ripening. Thus, in determination/optimization of filtration site parameters, reduced risk of contamination through increased filter depth should be considered as important factor along with filtrate goals, backwash strategy and intense. It is also important in terms of *C.parvum* velocity inside the media. In case of deep filters, constant sampling and analysis on presence of *C.parvum* may give necessary time to turn contaminated filtrate to reservoir instead supply it to customers.
- Performed study indicates that rapid contact granular filtration can provide satisfactory levels of reduction of influent *C.parvum* concentration. In highly turbid water average 3 logs *C.parvum* removal were found during steady state. Insufficient levels of 1.7 logs during ripening period were linked to reduced amount of accumulated material. Higher effluent concentrations of oocysts were observed during period of poor coagulation/flocculation regime.

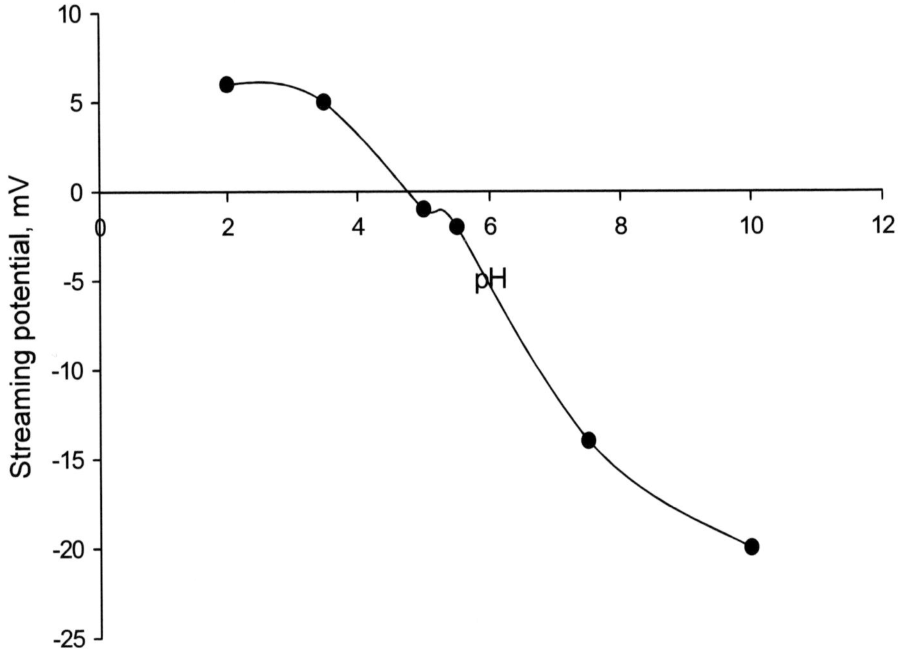

Figure 19. Streaming potential - pH dependence for PES membrane.

5. Membrane Filtration of Enteric Viruses

Prior to filtration experiments both polyethersulfone (PES) membrane with Molecular Weight Cut-Off (MWCO) of 20 kDa and pathogens were thoroughly characterized. Contact angle and streaming potential measurements were also used to characterize the membranes. The contact angle measurements gave a value 62±1.6°, indicating that the PES membrane is relatively hydrophobic. Streaming potential measurements were performed at pH values ranging from 2 to 9 (Figure 19). The obtained curve was divided into the regions of positive charge (from 6 to 5 mV), of electroneutrality (from 5 to -2.5 mV), and of negative charge (from -2.5 to -20). The point of zero charge was found at pH 4.8. Since PES has no dissociated functional groups, specific ion adsorption is the only possible process for the formation of surface charge, i.e., the initial charge can be attributed to the adsorption of hydroxyl ions on the PES membrane surface.

For investigation of rejection values of UF membranes we used T4, MS2, X174 bacteriophages and E. coli. It was found that while the T4 and phi X174 phages are considered to be hydrophilic, MS2 is hydrophobic (Shields and Farrah, 2002). Based on the hydrophobicity measurements, for similar sizes and zeta potential values, it is expected that the degree of MS2 retention will be higher than that for both T4 and for phi X174.

Zeta potential values for T4, MS2, phi X174 and E. coli in the pH range of 2 to 10 are shown in Figure 20.

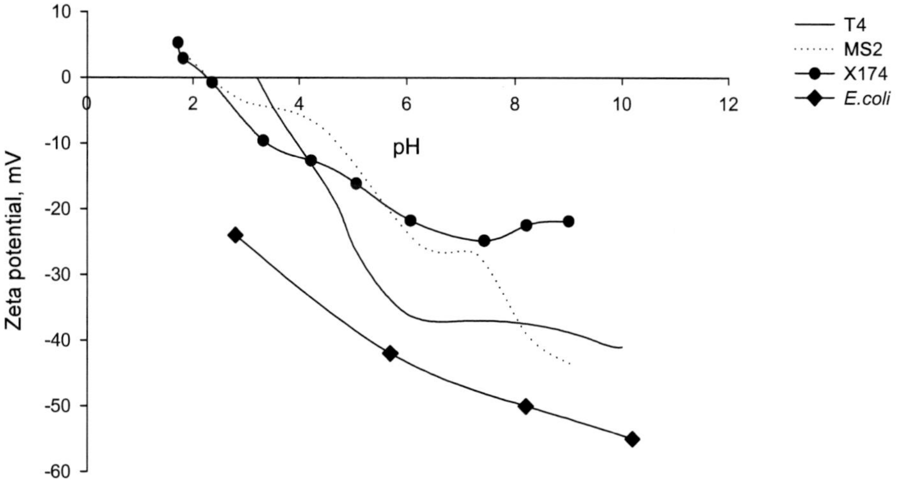

Figure 20. Zeta potential - pH dependence of T4, MS2, X 174 and *E.Coli*.

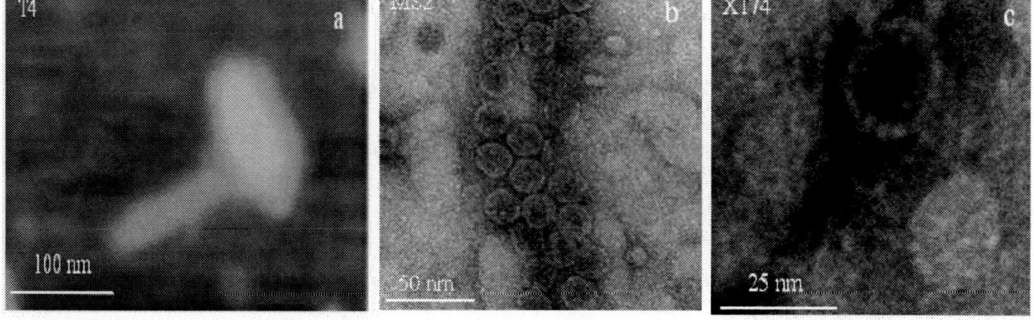

Figure 21. The microscope image of bacteriophages: a) AFM image of T4, b) TEM image of MS2 and c) phi X174.

Generally, the bacteriophages were positively charged at pH 2 and lower. As the pH increased, phage charges tended toward electroneutrality, with absolute zeta potential values in the range of 5 mV at pH 3. As the pH became even more alkaline, the phages usually took on a negative charge. The points of zero charge for T4, MS2, and phi X174 were found to be pH 3.2, 2.35, and 2.36, respectively. At pH 7, the zeta potential values of the phages were -38 mV, -30 mV, and -27 mV for T4, MS2, and phi X174, respectively. The zeta potential of *E. coli*, on the other hand, varied from -24 to -55 mV, which is lower (vs. -50 to -75 mV) than that previously observed (Madaeni, 1997). Based on the highly negative zeta potential of *E. coli* at pH 7, its point of zero charge is probably located in the hyperacidic environment, which the currently available instrumentation is incapable of measuring.

In the AFM images of the T4 bacteriophage, the typical icosahedral structures of the head (Brock et al., 1984; Stanier, 1986) and the tail were clearly visible (Figure 21a). The measured size of the phage was 200 nm, slightly smaller than the previously reported values of 220 and 250 nm (Stanier, 1986). The size difference was attributed to the flattening of the head during drying on the mica surface prior to AFM micrography. The flattening effect was more pronounced for the head than for the tail, probably due to the tail's increased compactness and flattening resistance (Ikai et al., 1993). The TEM images of the MS2 and phi X174 bacteriophages showed that these phages have spherical shapes with diameters of 30 and 26 nm, respectively (Figure 21b and c), values that are somewhat lower than those reported earlier (Dowd et al., 1998). In contrast to MS2, phi X174 exhibited spikes that were observed along the periphery of the virion.

Plots of the hydrodynamic radii of the T4, phi X174, and MS2 bacteriophages as functions of pH showed peaks around pH 3 to 4, where the hydrodynamic radii jumped to 140, 176, and 174 nm for T4, phi X174, and MS2, respectively (Figure 22).

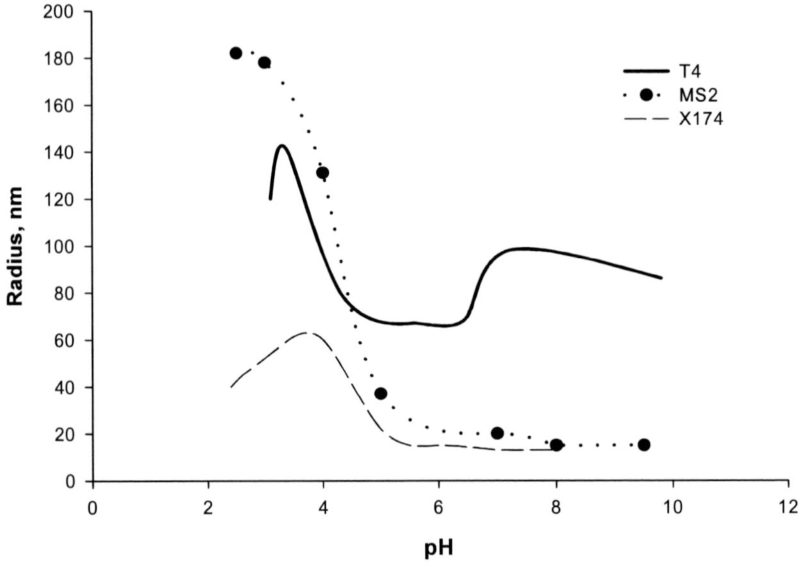

Figure 22. Size measurement of bacteriophages.

An examination of these sizes against the background of zeta potential measurements led us to conclude that these peaks were obtained at or near points of zero charge, indicating aggregation of the virus particles. At pH 7, the sizes of T4 and phi X174 were both 80 nm, and that of MS2 was 30 nm.

An estimate of *E. coli* dimensions, based on a digital picture from an epifluorescence microscope, gave a 1 μm width and 2-4 μm length. The hydrodynamic radius of *E. coli* at pH 7 was 776 nm, a measurement consistent with microscope observations. The observed values, however, are somewhat higher than those reported earlier (Miao et al., 2003). We attribute the observed differences to a stationary phase of bacterial growth in our measurements.

Our membrane analysis indicates that the UF membrane should form an efficient barrier against the penetration of bacteriophages and *E. coli*. The dimensions of the microorganisms used in the study were at least twice the average membrane pore size (10 nm) of the 20-kDa MWCO PES membrane. We should remember that our membrane retention measurements were performed with linear, water-soluble molecules (PEG and PEO), which are more easily able to pass through membrane pores than non-linear molecules.

Therefore, the membrane rejection levels for non-linear molecules of the same molecular weights as our linear polymers should, in fact, be much higher, i.e., the difference between the size of the microorganism and pore sizes is actually much larger. We may thus assume that the PES membrane constitutes an almost absolute barrier to the passage of the phages used in this study. The validity of this assumption gains additional strength from the zeta potential values. The values of -20 mV and higher for *E. coli* and bacteriophages in the typical tap water range of pH 6–8.5 *vs.* a streaming potential of –15 mV for the membrane would result in a strong repulsion. Consequently, the retention of *E. coli* and bacteriophages by a 20-kDa MWCO membrane is expected to be very efficient or even unlimited.

The first filtration test was performed with *E. coli* to test the overall setup and to verify the membrane's ability to retain microorganisms. UF of *E. coli* showed a 6 log removal of the bacteria, consistent with previous reports of a 7- to 8-log retention (Lovins, et al., 2002; EPA, 2001). The results indicate that size exclusion is the main mechanism for filtration of bacteria by UF membranes. Thereafter, the suitability of UF for virus retention was examined as a function of transmembrane pressure (TMP). LRV vs. TMP was plotted for the three phages at TMP values of 1 to 5 bars (Figure 23). As TMP was increased from 1 to 5 bars, the observed LRV for T4 fell from 3.83 to 3.35 (a half-log difference), for phi X174 from 3.15 to 2.35 (also a half-log difference), and for MS2 from 3.82 to 2.76 (>1 log removal difference). Despite the significant size differences between the viruses, similar LRVs were observed for a TMP of 1 bar. At higher TMPs, however, the LRVs of the smaller viruses dropped faster than those for T4. For example, at a TMP of 2 bars, T4 retention dropped from 3.8 to 3.7 while phi X174 retention fell from 3.1 to 2.4. At TMPs >2, the LRVs of MS2 and phi X174 remained almost constant, fluctuating only slightly. Despite the significant zeta potential values of the bacteriophages and the membrane, the LRVs obtained were smaller than those previously reported and therefore, electrostatic repulsion probably did not significantly influence virus removal. Hydrophobicity, however, does appear to be important, at least for low TMPs. The size and shape of MS2 and phi X174 are similar, but their corresponding virus retention levels differ significantly: MS2 is retained more efficiently than phi X174, an observation that may be attributed to the hydrophobicity of the MS2 virus.

The above notwithstanding, size can have an effect on the degree of retention: at all pressures, the degree of retention was found to behave according to the following order: T4>MS2>phi X174. Hence, it will be probably possible to assess the degree of retention of viruses by the smaller viruses available assuming that degree of retention of bigger viruses will be at least as big as the removal of the smaller representatives. At the same time MS2 probably is not the best choice due to high hydrophobicity and therefore enhanced removal at lower TMP.

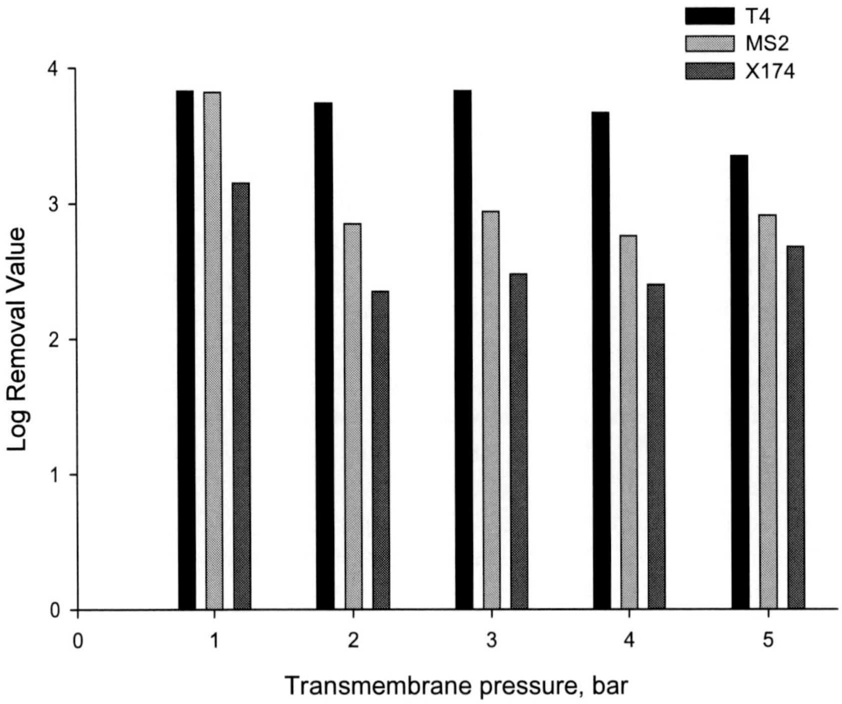

Figure 23. Retention of bacteriophages as a function of TMP.

It is likely that partial penetration of UF membranes by viruses is facilitated by enlargement of the membrane pores rather than by "shrinkage" of virus particles, since elastic changes in the viral forms have never been reported. Two hypotheses have been put forward to explain the phenomenon of selective solute removal by UF membranes. The hypothesis of Urase *et al.* (1994; 1996) states simply that the membranes have abnormally large pores that are not included in the main pore size distribution of UF membranes, but there is insufficient direct evidence to support this hypothesis. In addition, the hypothesis does not explain why the removal of bacteria and protozoa is incomplete despite pore sizes small enough to retain viruses.

The second hypothesis rests on the presence of physical defects and the lack of membrane integrity manifested as pinholes in the membrane (Kitis, et al., 2002). Although such problems may occur, their appearance under certain circumstances does not reflect a general phenomenon. Lastly, physical defects linked to the seal material are unlikely to significantly impair virus removal rates (Jacangelo et al., 1991; Otaki et al., 1998).

In an effort to find a more satisfying explanation, the current study focused on the retention of viruses as a function of TMP. Previous studies indicated that increases in TMP led to membrane compaction, ultimately resulting in the formation of a denser membrane with smaller pores, but this was not the case for the membranes employed in the current study. However, it is likely that the flux deterioration associated with increased TMP is linked to virus hydrophobicity rather than to shrinking pores. Our study did not show the expected increase in particle retention as TMP was increased. One possible explanation is a pressure-induced enlargement of membrane pores with time, which enables viruses to penetrate through the membrane matrix. The process is TMP-dependent, such that the greater the increase in pressure the more enlarged the pores become, appears to apply equally to viruses and to organic materials. Measurements indicated clearly that pore sizes in the membrane matrix increased with increasing TMP. The evidence thus indicates that viral penetration of the membrane is linked to the formation over time of abnormally large pores rather than to the initial presence of oversized pores. The total absence of viruses in the permeate of diffusion-driven membranes lends support to the pore-formation hypothesis. In terms of virus retention by UF membranes, pressure-driven changes in membrane pore sizes represent an additional virus retention mechanism that joins the well-known mechanisms of size-based retention, charge repulsion, and hydrophobic fouling.

Assessing virus retention rates via tests that utilize polymer materials is so far an unproven approach. The prediction of membrane viral retention capacity without filtration experiments with viruses may, therefore, be very difficult or even impossible. The results of the current research indicate the importance of developing new and better nanometric prediction tools for the safe use of UF in water treatment processes.

References

Adin A., Bauman E.R. and Cleasby J.L. (1979). The Application of Filtration Theory to Pilot-Plant Design. *J. Am. Water Works Assoc.* 71, 17-27.

Adin, A. and Rebhun, M. (1977). A Model to Predict Concentration and Head-Loss Profiles in Filtration. *Journal American Water Works Association* 69, 444-453.

Amirtharajah A. (1988). Some Theoretical and Conceptual Views of Filtration. *J. Am. Water Works Assoc.* 80, 36-45.

Association of Environmental Engineering and Science Professors (AEESP) Research Frontiers Conference. *Pennsylvania State University*, July 31 - August 3, 1999

Bales R.C., Gerba C.P., Grondin G.H., Jensen S.L. (1989). Bacteriophage transport in sandy soil and fractured tuff. *Applied and Environmental Microbiology* 55, 2061-2067.

Bales R.C., Li S.M., Maguire K.M., Yahya M.T., Gerba C.P. (1993). MS-2 and poliovirus transport in porous-media - hydrophobic effects and chemical perturbations. *Water Resources Research* 29, 957-963.

Brady P.V., Cygan R.T. and Nagy K.L. (1996). Molecular Controls on Kaolinite Surface Charge. *J. Colloid Interface Sci.* 183, 356-364.

Brock T. D., Smith D. W., Madigan M. T. (1984). Biology of microorganisms. Fourth edition. Prentice-Hall Inc.

Calendar R. and Abedon S. T. (2006). The Bacteriophages. Second edition. Oxford University Press.

Chang, J. S., and Vigneswaran S. (1990). Ionic Strength in Deep Bed Filtration. *Water Resources* 24, 1425-1430.

Chattopadhyay D., Chattopadhyay S., Lyon W.G., Wilson J.T. (2002). Effect of surfactants on the survival and sorption of viruses. *Environmental Science and Technology* 36, 4017-4024.

Chattopadhyay S., Puls R. W. (2000). Forces dictating colloidal interactions between viruses and soil. *Chemosphere* 41, 1279-1286.

Chu Y.J., Jin Y., Baumann T., Yates M.V. (2003). Effect of soil properties on saturated and unsaturated virus transport through columns. *Journal of Environmental Quality* 32, 2017-2025.

Cicmanec, J., and Reasoner, D. J. (1997). Enhanced Production of *Cryptosporidium parvum* oocyst in Immunosuppressed Mice. in Proc., *AWWA, International Symposium on Waterborne Cryptosporidium*.

Deborde D.C., Woessner W.W., Kiley Q.T., Ball P. (1999). Rapid transport of viruses in a flood plain aquifer *Water Research* 33, 2229-2238.

Dowd S.E., Gerba C.P., Pepper I.L. (1998). Confirmation of human pathogenic microsporidia. *Encterocytozoon bieneusi, Encephalitozoon intestinalis* and *Vittaforma cornea* in water. *Applied and Environmental Microbiology* 64, 3332-3335.

Dowd S.E., Pillai S.D., Wang S.Y., Corapcioglu M.Y. (1998). Delineating the specific influence of virus isoelectric point and size on virus adsorption and transport through sandy soils. *Applied and Environmental Microbiology* 64, 405-410.

Elimelech, M., and O'Melia, C. (1990). Kinetics of Deposition of Colloidal Particles in Porous Media. *Environmental Science and Technology* 24, 1528-1536.

Environmental Protection Agency (EPA) (2001) Low Pressure Membrane Filtration for Pathogen Removal: Application, Implementation, and Regulatory Issues, EPA 815-C-01-001. www.epa.gov/safewater/disinfection/lt2/pdfs/report_lt2_membranefiltration.pdf, accessed on June 18, 2007

Feng Y.Y., Ong S.L., Hu J.Y., Tan X.L., Ng W.J. (2003). Effects of pH and temperature on the survival of coliphages MS2 and Q beta. *Journal of Industrial Microbiology and Biotechnology* 30, 549-552.

Floyd R., Sharp D.G. (1979). Viral aggregation: buffer effects in the aggregation of poliovirus and reovirus at low and high pH. *Applied and Environmental Microbiology* 38, 395-401.

Friedlander, S. K. (1976). *Smoke, dust, and haze: fundamentals of aerosol behavior*. Wiley Interscience.

Huang, C., Rushing, P., and Huang, S. (1999). Collision Efficiencies of Algae and Kaolin in a Depth Filter: The Effect of Surface Properties of Particles. *Water Research* 33, 1278-1286.

Huisman I. H., Pradanos P., Hernandez A. (2000). The effect of protein- protein and protein-membrane interactions on membrane fouling in ultrafiltration, *Journal of membrane science* 179, 79-90.

Hurst C.J., Benton W.H., McClellan K.A. (1989). Thermal and water source effects upon the stability of enteroviruses in surface fresh-waters. *Canadian Journal of Microbiology* 35, 474-480.

Ives, K. (1969). Theory of Filtration. In Proc., *IWSA Congress*, Vienna.

Iwasaki, T. (1937). Some Notes on Sand Filtration. *Journal American Water Works Association* 29, 1591.

Jackangelo J. G., Laîné J.-M., Carns K. E., Cummings E. W., Mallevialle J. (1991). Low pressure membrane filtration for removing Giardia and microbial indicators, *AWWA* 83, 97-106.

Jin Y., Flury M. (2002). Fate and transport of viruses in porous media. *Advances Agronomy* 77, 39-51.

Johnson R.P.C., Gregory D.W. (1993). Viruses accumulate spontaneously near droplet surfaces - a method to concentrate viruses for electron-microscopy. *Journal of microscopy-Oxford* 171, 125-136.

Kao, T., and Ungar, B. (1994). Comparison of Sequential, Random and Hemacytometer Methods for Counting *Cryptosporidium* oocysts. *Journal of Parasitology* 80, 819-819.

Karanis P., Schoenen D. and Seitz H. (1998). Distribution and Removal of *Giardia* and *C.parvum* in Water Supplies in Germany. *Water Sci. Technol.* 37, 9-18.

Kirmeyer G.J., Friedman M., Martel K.D., Noran P.F. and Smith D. (2001) Practical guidelines for maintaining distribution system water quality. *Journal American Water Works Association* 93 (7), 62-76

Kitis, Mehmet, Lozier, James C., Kim, Jae-Hong, Mi, Baoxia, Marinas, Benito (2002). Evaluation of Biologic and Non-Biologic Methods for Assessing Virus Removal by and Integrity of High Pressure Membrane Systems, Proceedings of the 5th conference, Germany, 143-156.

Lance J.C., Gerba C.P. (1984). Effect of ionic composition of suspending solution on virus adsorption by a soil column. *Applied and Environmental Microbiology* 47, 484-488.

Levenspiel O. (1985). *Chemical Reactor Engineering,* Wiley: New York, 2nd Ed.

Loveland J.P., Ryan J.N., Amy G.L., Harvey R.W. (1996). The reversibility of virus attachment to mineral surfaces. *Colloids and Surfaces A-Physicochemical and Engineering Aspects* 107, 205-221.

Lovins, W.A. III, Taylor, J.S. and Hong, S.K., (2002). Microorganism rejection by membrane systems. *Environmental Engineering Science* 19, 453-465.

Lukasik J., Scott T.M., Andryshak D., Farrah S.R. (2000). Influence of salts on virus adsorption to microporous filters. *Applied and Environmental Microbiology* 66, 2914-2920.

Lyklema J. (1991) *Fundamentals of interface and colloid science.* London; San Diego: Academic Press.

Madaeni S. S. (1997). Mechanism of virus removal using membranes, *Filtration and Separation* 34, 61-65.

Martin R.E., Bouwer E.J., and Hanna L.M. (1992) Application of clean-bed filtration theory to bacterial deposition in porous media. *Env. Sci. Tech.*, 26 (5), 1053-1058.

Meschke J. S., Sobsey M. D. (2003). Comparative reduction of Norwalk virus, poliovirus type 1, F+ RNA coliphage MS2 and Escherichia coli in miniature soil columns. *Water Science and Technology* 47, 85-90.

Miao, J., Hodgson, K.O., Ishikawa, T., Larabell, C.A., LeGros, M.A. and Nishino, Y. (2003). Imaging whole Escherichia coli bacteria by using single-particle x-ray diffraction. Proceedings of the National Academy of Sciences PNAS 100, 110-112.

Minz D. M. (1951). Kinetics of filtration of low-concentration water suspensions in water purification filters. *Dokl. Akad. Nauk SSSR* 78, 315-318 (in Russian).

Nocito-Gobel, J., and Tobiason, J. (1996). Effects of ionic strength on colloid deposition and release. *Colloids and Surfaces A: Physicochemical and Chemical Aspects* 107, 223-231.

O'Melia, C. (1985). Particles, Pretreatment and Performance in Water Filtration. *Journal of Environmental Engineering* 111, 874-890.

O'Melia, C., and Ali, W. (1978). The Role of Retained Particles in Deep Bed Filtration. *Progress in Water Technology* 10, 167-182.

O'Melia, C., and Stumm, W. (1967). Theory of Water Filtration. *Journal American Water Works Association* 59, 1393-1412.

Otaki M.,. Yano K, Ohgaki S. (1998). Virus Removal in a Membrane Separation Process. *Water Science Technology* 37, 107-116.

Payment P., Richardson L., Siemiatycki J. *et al.* (1991) A randomized trial to evaluate the risk of gastrointestinal disease due to consumption of drinking water meeting currently accepted microbiological standards. *American Journal of Public Health* 81, 703-708

Payment P., Siemiatycki J. and Richardson L. (1997) A prospective epidemiological study of gastrointestinal health effects due to the consumption of drinking water. *International Journal of Environmental Health Research* 7, 5-31

Petrovic M., Kastelan-Macan M. and Horvat A.J.M. (1999). Interactive sorption of metal ions and humic acids onto mineral particles. *Water Air Soil Pollut,* 111, 41-56.

Powelson D. K., Simpson J. R., Gerba C. P. (1991). Effects of organic matter on virus transport in unsaturated flow. *Applied and Environmental Microbiology* 57, 2192-2196.

Quignon F., Thomas F., Gantzer C., Huyard A., Schwartzbrod L. (1998). Virus adsorption in a complex system: an experimentally designed study. *Water Research* 32, 1222-1230.

Rajagopalan, R., and Tien, C. (1976). Trajectory Analysis of Deep-Bed Filtration with the Sphere-in-cell Porous Media Model. *AIChE* 22, 523-533.

Ratnayake N. and Jayatilake I.N. (1999) Study of transport of contaminants in a pipe network using the model EPANET. *Water Science and Technology* 40 (2), 115-120

Shields P.A., Farrah S.R. (2002). Characterization of virus adsorption by using DEAE-sepharose and octyl-sepharose. *Applied and Environmental Microbiology* 68, 3965-3968.

Sinton L.W., Hall C.H., Lynch P.A., Davies-Colley R.J. (2002). Sunlight inactivation of fecal indicator bacteria and bacteriophages from waste stabilization pond effluent in fresh and saline waters. *Applied and Environmental Microbiology* 68, 1122-1131.

Soule H., Duc D.L., Mallaret M.R., Chanzy B., Charvier A., Gratacap-Cavallier B., Morand P., Seigneurin J.M. (1998). Virus survival in hospital environment: an overview of the virucide activity of disinfectants used in liquid form. *Annales de Biologie Clinique* 56, 693-703.

Stanier R., Ingraham J., Wheelis M., Painter P. (1986). The Microbial world. Fifth Edition. Prentice-Hall.

Straub T., Pepper Y., Gerba C. (1993). Virus survival in sewage sludge amended desert soil. *Water Science and Technology* 27, 421-424.

Thompson S.S., Yates M.V. (1999). Bacteriophage inactivation at the air-water-solid interface in dynamic batch systems. *Applied and Environmental Microbiology* 65, 1186-1190.

Tien C., Granular filtration of aerosols and hydrosols, Butterworth Publishers, Stoneham MA, USA, 1989

Tien C., Turian R.M., Pendse H. (1979). Simulation of the dynamic behavior of deep bed filters. *Journal of American Institute of Chemical Engineers* 25, 385-395.

U.S. EPA (2002) Treatability screening for CCL microbial contaminants. *EPA/822/R/01/002*, Washington, DC

Urase T., Yamamoto K., Ohgaki S. (1994). Effect of pore size distribution of UF membranes on virus rejection in crossflow conditions, *Wat. Sci. Tech.* 30, 199-208.

Urase T., Yamamoto K., Ohgaki S. (1996). Effect of Pore Structure of Membranes and Module Configuration on Virus Retention. *Journal of Membrane Science* 115, 21-29.

Van Voorthuizen E. M, Ashbolt N. J.,. Schafer A. I. (2001). Role of Hydrophobic and Electrostatic Interactions for Initial Enteric Virus Retention by MF Membranes. *Journal of Membrane Science* 194, 69-79.

Vigneswaran S., Ben Aim R. (1989) Water, wastewater and sludge filtration. CRC Press, Inc., Boca Raton, Fl, U.S.A.

Wait D.A., Sobsey M.D. (2001). Comparative survival of enteric viruses and bacteria in Atlantic Ocean seawater. *Water Science and Technology* 43, 139-142.

Wan J., Tokunaga T.K., and Tsang C.-F. (1995) Bacterial sedimentation through a porous medium. *Water Resources Research*, 31 (7), 1627-1636.

Winer B.J., Brown D.R. and Michels K.M. (1997). *Statistical Principles in Experimental Design*, 3rd Edition, McGraw-Hill, New York.

Woessner W.W., Ball P.N., DeBorde D.C., Troy T.L. (2001). Viral transport in a sand and gravel aquifer under field pumping conditions. *Ground Water* 39, 886-894.

Wommack K.E., Hill R.T., Muller T.A., Colwell R.R. (1996). Effects of sunlight on bacteriophage viability and structure. *Applied and Environmental Microbiology* 62, 1336-1341.

Yang, S., Healey, M., and Du, C. (1996). Infectivity of Preserved *Cryptosporidium parvum* oocyst for Immunosuppressed Adult Mice. *FEMS Immunos Medicine Microbiology* 13, 141-145.

Yao K. M. (1968). Influence of suspended particle size on the transport aspect of water filtration, Chapel Hill.

Yao K. M., Habibian M. T., O`Melia C. R. (1971). Water and wastewater filtration: concepts and applications. *Environmental Science and Technology* 5, 1105-1112.

Zhuang J., Jin Y. (2003). Virus retention and transport as influenced by different forms of soil organic matter. *Journal of Environmental Quality* 32, 816-823.

In: Drinking Water: Contamination, Toxicity and Treatment ISBN: 978-1-60456-747-2
Editors: J. D. Romero and P. S. Molina

Chapter 6

EFFECTIVE REMOVAL OF LOW CONCENTRATIONS OF ARSENIC AND LEAD AND THE MONITORING OF MOLECULAR REMOVAL MECHANISM AT THE SURFACE

Yasuo Izumi[1]
Department of Chemistry, Graduate School of Science, Chiba University
Yayoi 1-33, Inage-ku, Chiba 263-8522, Japan

ABSTRACT

New sorbents were investigated for the effective removal of low concentrations of arsenic and lead to adjust to modern worldwide environmental regulation of drinking water (10 ppb). Mesoporous Fe oxyhydroxide synthesized using dodecylsulfate was most effective for initial 200 ppb of As removal, especially for more hazardous arsenite for human's health. Hydrotalcite-like layered double hydroxide consisted of Fe and Mg was most effective for initial 55 ppb of Pb removal.

The molecular removal mechanism is critical for environmental problem and protection because valence state change upon removal of e.g., As on sorbent surface from environmental water may detoxify arsenite to less harmful arsenate. It is also because the evaluation of desorption rates is important to judge the efficiency of reuse of sorbents. To monitor the low concentrations of arsenic and lead on sorbent surface, selective X-ray absorption fine structure (XAFS) spectroscopy was applied for arsenic and lead species adsorbed, free from the interference of high concentrations of Fe sites contained in the sorbents and to selectively detect toxic As^{III} among the mixture of As^{III} and As^{V} species in sample.

Oxidative adsorption mechanism was demonstrated on Fe-montmorillonite and mesoporous Fe oxyhydroxide starting from As^{III} species in aqueous solution to As^{V} by making complex with unsaturated $FeO_x(OH)_y$ sites at sorbent surface. Coagulation mechanism was demonstrated on double hydroxide consisted of Fe and Mg from the initial 1 ppm of Pb^{2+} aqueous solution whereas the mechanism was simple ion exchange reaction when the initial Pb^{2+} concentrations were as low as 100 ppb.

---

[1] yizumi@faculty.chiba-u.jp.

Introduction

Recently, global environment induces even serious debate, *e.g.* the Novel Prize 2007 for Piece to "*An Inconvenient Truth*" by Al Gore [1]. The environmental problem is not only the global warming as the major claim in this movie/book by Gore. Contamination of water and soil is one of traditional, major environmental problems and directly affects the human health, *e.g.* carcinogenic risk via drinking water [2 – 4]. Cadmium contaminated in rice [5], copper contaminated in environmental water from mine, mercury contaminated in fish [6 – 8], and arsenic contaminated in powdered milk are most notorious environmental tragedies occurred in Japan between 1890 and 1960. These accidents are all related to contamination of water derived from human activity (industry) [2].

The health risk of poisonous elements in water has been studied and is becoming clear. Recent environmental regulation sets the minimum level of Mn, Cu, Cd, Pb, Cr, As, and Hg to 400, 125, 5, 10, 50, 10, and 0.5 ppb, respectively. Among these elements, lead is less focused and less intensively studied to adjust to the regulation. Arsenic can be contaminated in ground water not only anthropogenically but from the earth naturally [2 – 4]. Unfortunately, mines containing As mainly distribute in contact with the ground water in developing countries, *e.g.* Bangladesh, East Bengal, Argentina, Chile, and Vietnam [9]. Especially in Bangladesh, the shallow ground water with arsenic concentrations up to 1 ppm is often used pumped from tube wells without adequate treatment for drinking and thus leading to serious health problem. The arsenic release into ground water is related to microbial metabolism of organic matter [10]. The high capital and maintenance costs of piped water supply are still not acceptable in Bangladesh, especially most affluent villages compared to present individual tube wells [10]. Lead was used as gasoline additive and automobile tail pipes and may be included in drinking water from water supply pipes made of lead, soil contamination, and toys/tools made in developing countries [11, 12].

This chapter reviews recent development of sorbents for low concentrations of Pb and As to adjust to modern environmental regulation for drinking water and monitoring of the molecular removal process by selective spectroscopy in water [13]. The monitoring of surface uptake of Pb [14] and As includes local coordination structure and electronic structure changes in the inner sphere reaction.

Methods

This chapter focuses on the removal of Pb and As by sorption because sorption is economic process to be applicable in developing countries and the reuse is possible by desorption. The concentrations of Pb and As in test aqueous solutions were set between 55 ppb and 32 ppm in this chapter. Various sorbents were evaluated to maintain the concentrations of Pb or As less than 10 ppb or not.

Fe-montmorillonite was prepared by mixing 0.43 M ferric nitrate solution with Na-montmorillonite (Kunipia F; $Na_{1.5}Ca_{0.096}Al_{5.1}Mg_{1.0}Fe_{0.33}Si_{12}O_{27.6}(OH)_{6.4}$) [15]. A 0.75 M sodium hydroxide solution was added dropwise to the mixture until the molar ratio Fe^{3+} added and hydroxide reached 1:2. Iron cations and/or $FeO_x(OH)_y$ nanoparticles were inserted between negatively-charged montmorillonite clay layers. Recently, some chemical forms of

Fe^{III} species formed between montomorillonite layers were spectroscopically analyzed [16]. Monomeric and/or dimeric Fe^{III} species was active in oxidative dehydrogenation of propane. In contrast, polymeric $FeO_x(OH)_y$ nanoparticles were effective for arsenic sorption and unselective propane combustion.

$FeO_x(OH)_y$ porous material was prepared by mixing 0.10 M ferrous chloride with 0.070 M sodium dodecylsulfate followed by the addition of 0.25 M H_2O_2 [17]. Obtained $FeO_x(OH)_y$ material was mixed with 0.050 M sodium acetate in ethanol for anion exchange or with pure ethanol for washing. The micro/mesoporous $FeO_x(OH)_y$ material was characterized by X ray diffraction (XRD), specific surface area measurements and pore volume determination by N_2 adsorption/desorption, high-resolution transmission electron microscope (TEM), Fourier-transformed infrared absorption (FT-IR), inductively coupled plasma (ICP) combined with optical emission spectroscopy (OES), electron probe microanalysis (EPMA), thermo-gravimetric differential thermal analysis (TG-DTA), and Fe K-edge X-ray absorption fine structure (XAFS). Based on these analyses, detailed structural transformation was clarified for the sorbents as depicted in Figure 8 of Ref 17.

Hydrotalcite-like (pyroaurite) layered double hydroxide $Mg_6Fe_2(OH)_{16}(CO_3)\bullet 3H_2O$ was synthesized via the procedure described in Ref 18. The carbonate anions are sandwiched between positively-charged $[Mg_3Fe(OH)_8]^+$ layers.

To monitor low concentrations of Pb and As, XAFS spectroscopy is most appropriate technique. For several spectroscopic techniques of structural analysis in the application to nanotechnology, the advantage and drawback were summarized in Table 1 [19]. For non-crystalline or hybrid samples, EXAFS (extended X-ray absorption fine structure) gives direct structural information for X-ray absorbing local element sites. XANES (X-ray absorption near-edge structure) is a part of EXAFS spectrum near the X-ray absorption edge region ranging up to 100 eV and gives electronic and (indirectly) structural information [20]. Thus, XAFS spectroscopy (EXAFS, XANES) is essentially single technique for local structure analysis accompanied with valence and coordination symmetry information of nanoparticles and micro/mesoporous materials. The Pb and As adsorbed from low concentrations of aqueous solutions in this chapter are typical examples of nanoparticles and micro/mesoporous material samples.

Further, this chapter combines X-ray fluorescence (XRF) spectrometry with the XAFS spectroscopy [21 – 24]. Simply, XRF spectra support valence state information deduced from XAFS. Essentially, high-energy-resolution XRF spectrometry is able to discriminate valence state of Pb and As. In this chapter, the XRF signals originating from Pb^{II}, As^{III}, and As^{V} were monitored independently in the XAFS measurements to obtain each coordination structure of Pb^{II}, As^{III}, and As^{V} (*state-selective XAFS*). The experimental setup and measurement conditions for state-selective XAFS were depicted and described in Refs 21, 23, and 24. In brief, XRF spectra and state-selective XAFS spectra measurements were performed at Undulator beamline 10XU of SPring-8 (Sayo, Japan) by utilizing a homemade high-energy-resolution Rowland-type fluorescence spectrometer equipped with a Johansson-type Ge(555) crystal (Saint-Gobain) and NaI(Tl) scintillation counter (Oken). The monochromator of beamline used Si(111) double crystal and the X-ray beam intensity in front of sample was monitored using ion chamber (Oken) purged with N_2 gas.

Table 1. Various Analytical Methods for Nano Structure Classified Based on Directness of the Information and the Target to Be Analyzed

Method	Directness	Target
TEM (Transmission electron microscope)	Direct	Local
XRD (X-ray diffraction)	Direct	Local
EXAFS (Extended X-ray absorption fine structure)	Direct	Bulk
XANES (X-ray absorption near-edge structure)	Indirect	Bulk
Raman	Indirect	Bulk
Small angle scattering	Indirect	Bulk
NMR (Nuclear magnetic resonance)	Indirect	Local
Mössbauer	Indirect	Local
SPM (Scanning probe microscope)	Indirect	Local (surface)
Reflectivity	Indirect	Local (surface)

RESULTS AND DISCUSSION

Arsenic Problem. Adsorption isotherms of arsenite and arsenate at 290 K for 12 h on Fe-montmorillonite in batch setup are depicted in Figure 1. The Fe-montmorillonite was superior to α-FeO(OH) (göthite > 95%) both for arsenite and arsenate sorptions. Fe-montmorillonite consisted of 2-dimensionally distributed Fe^{3+} ions and $FeO_x(OH)_y$ nanoparticles between clay layers [16]. The saturated amount of As adsorbed was evaluated to 8.0 and 76 $mg_{As}\ g_{sorbent}^{-1}$ for arsenite and arsenate, respectively, on Fe-montmorillonite. The equilibrium adsorption constant was 1.4×10^6 ml g_{As}^{-1} for arsenite on Fe-montmorillonite.

Even better adsorption capacity was found on acetate-exchanged microporous $FeO_x(OH)_y$ as depicted in Figure 2 for arsenite. The saturated amount and the equilibrium adsorption constant of As adsorbed were evaluated to 21 $mg_{As}\ g_{sorbent}^{-1}$ and 1.0×10^7 ml g_{As}^{-1}, respectively. The high specific surface area of microporous $FeO_x(OH)_y$ (230 $m^2\ g^{-1}$) was advantageous compared to Fe-montmorillonite (100 $m^2\ g^{-1}$) with as much as 14 wt% of Fe. Utilizing template synthesis technique for microporous and mesoporous materials, lower coordination FeO_x sites were effectively exposed to surface to complex with $As^{III}(OH)_3$ [17, 26]. The acetate-exchanged and ethanol-washed $FeO_x(OH)_y$ consist of 3-dimensionally distributed wormholes exposed with coordinatively unsaturated FeO(OH) sites.

The surface uptake mechanism of most toxic arsenite was monitored by XRF and XAFS spectroscopy. Arsenic was adsorbed on Fe-montmorillonite from 200 ppb test aqueous solution of arsenite. The As Kα_1 emission spectrum was depicted in Figure 3.

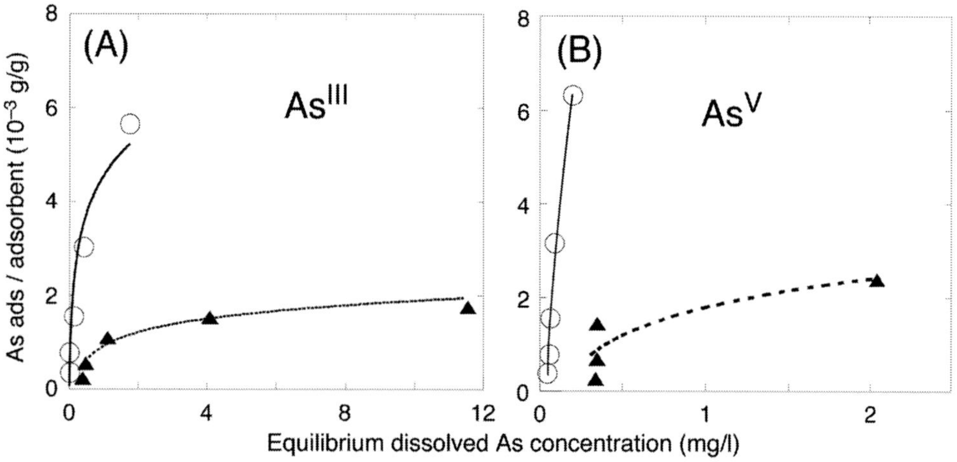

Figure 1. Adsorption isotherms of arsenite (A) and arsenate (B) at 290 K on Fe-montmorillonite (14.0 wt% Fe) (circles) and α-FeO(OH) (triangles). Batch tests for 12h. Observed data were plotted as points and the fits to first-order Langmuir equations were drawn as lines.

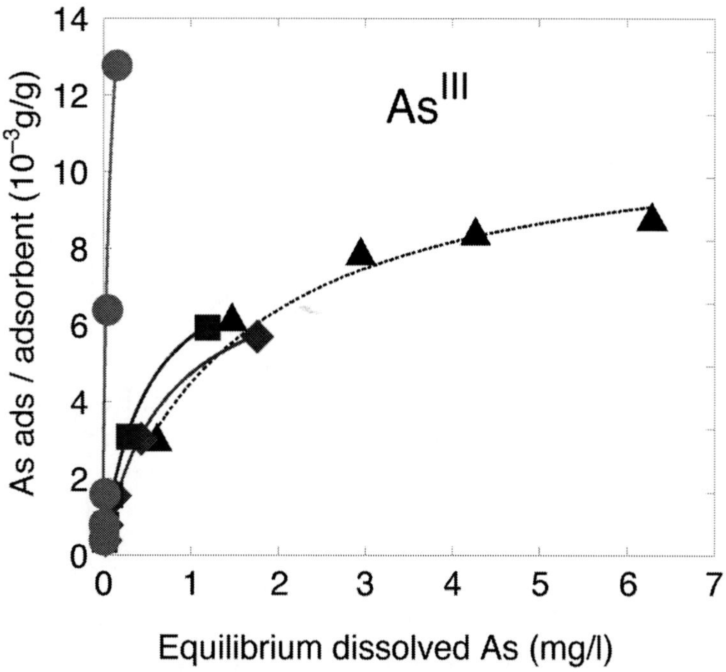

Figure 2. Adsorption isotherms of arsenite at 290 K on acetate-exchanged FeOx(OH)y previously heated at 423 K (circles), ethanol-washed FeOx(OH)y previously heated at 423 K (squares), Fe-montmorillonite (14.0 wt% Fe; diamonds), and α-FeO(OH) (triangles) [25]. Batch tests for 12h. Observed data were plotted as points and the fits to first-order Langmuir equations were drawn as lines.

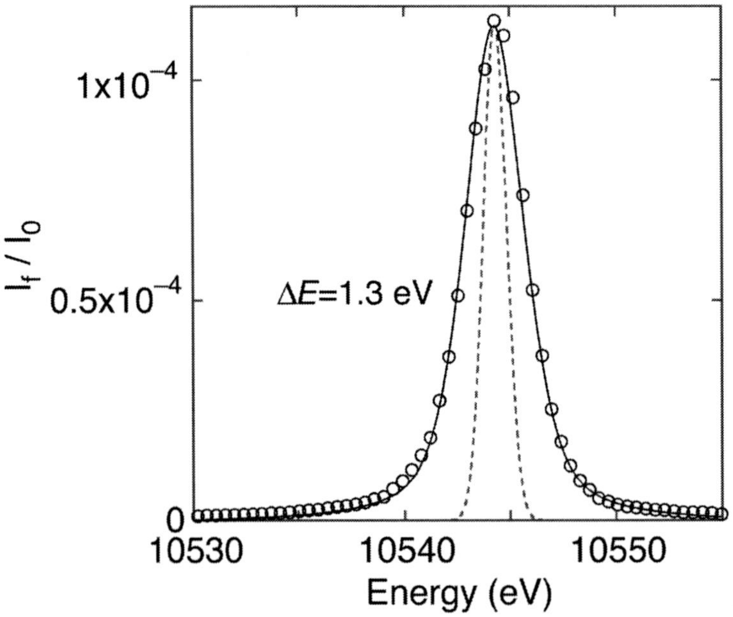

Figure 3. Arsenic Kα1 emission spectrum for As adsorbed on Fe-montmorillonite (14.0 wt% Fe) from 200 ppb test solution of arsenite (points). A fit to data with pseudo-Voigt function (solid line) and the energy resolution of fluorescence spectrometer (dotted line) were also drawn. The intensity ratio of the Lorentzian and Gaussian components was fixed to 1:1 for the pseudo-Voigt function.

The peak energy position suggested that the adsorbed As state changed to V, not remained at III, compared to the data for $KH_2As^VO_4$ and $As^{III}{}_2O_3$ [15].

As K-edge XANES spectra for standard inorganic compounds of As^0, As^{III}, and As^V consist of broad peak feature (Figure 4a – c) and it is complicating to evaluate each valence contribution to a spectrum for sample of mixed valence. In order to demonstrate directly the oxidative adsorption of arsenite suggested above, the author of this chapter observed the uptake of low concentrations of arsenite on Fe-montmorillonite by means of state-selective XAFS. Note that the energy resolution of fluorescence spectrometer (1.3 eV; Figure 3) was smaller than the core-hole lifetime width of As K level (2.14 eV) [27]. Based on the theory discussed in the Appendix section of Ref 24, the As $K\alpha_1$-selecting As K-edge XANES spectrum with the energy resolution of 1.3 eV would be shaper and more resolved.

The energy values of K absorption edge and first strong peak after the edge were essentially identical for As adsorbed on Fe-montmorillonite from 200 ppb – 16 ppm of As^{III} solutions (Figure 4e, f) and from 16 ppm of As^V solution (d). Thus, oxidative adsorption of 200 ppb – 16 ppm of arsenite on Fe-montmorillonite was confirmed based on As $K\alpha_1$-selecting XANES and As $K\alpha_1$ emission spectrum. Proposed molecular surface uptake mechanism was illustrated in Figure 5 over acetate-exchanged microporous $FeO_x(OH)_y$. The reaction formula was $As^{III}(OH)_3 + FeO(OH) \rightarrow (FeO)_2As^V(OH)_2 + H_2O$.

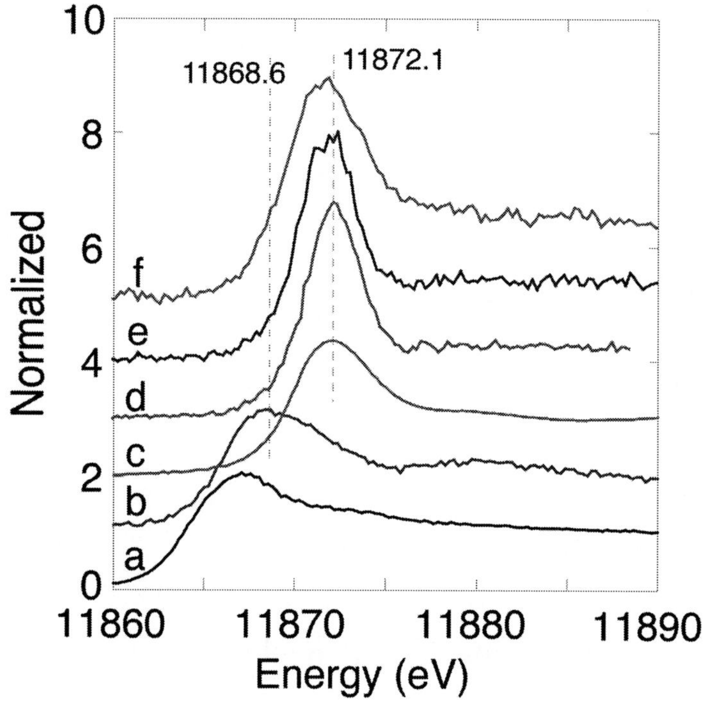

Figure 4. XANES spectra measured at 290 K in transmission mode for As metal (a), AsIII2O3 (b), and KH2AsVO4 (c). Arsenic Kα1-selecting As K-edge XANES spectra measured at 290 K for As adsorbed on Fe-montmorillonite (14.0 wt% Fe) (d – f) from aqueous test solutions of 16 ppm of KH2AsVO4 (d), 16 ppm of AsIII2O3 (e), and 200 ppb of AsIII2O3 (f). Tune energy of fluorescence spectrometer was 10544.3 eV for spectra d – f.

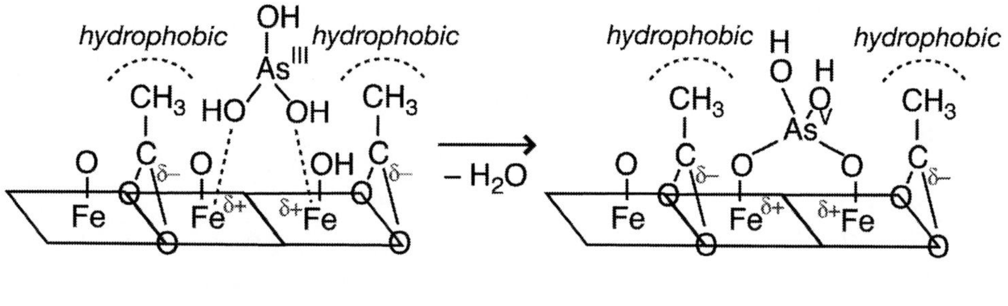

Figure 5. Proposed adsorption mechanism of low concentrations of arsenite on acetate-exchanged FeOx(OH)y previously heated at 423 K.

In summary, oxidative adsorption of low concentrations (200 ppb – 16 ppm) of arsenite was found on coordinatively unsaturated $FeO_x(OH)_y$ nanoparticles or micro/mesoporous $FeO_x(OH)_y$ partially covered with acetate anions. The oxidation to arsenate seems to be due to lower coordination of surface $FeO_x(OH)_y$ species. The lower coordination was also the reason to make the equilibrium sorption constant greater for acetate-exchanged $FeO_x(OH)_y$ and Fe-montmorillonite [15, 17].

Lead Problem. Sorption tests for low concentrations (55 ppb) of lead in flow setup were depicted in Figure 6 [18]. The superiority of $Mg_6Fe_2(OH)_{16}(CO_3)$•$3H_2O$ was clearly demonstrated to maintain the Pb^{2+} concentration less than modern environmental regulation (10 ppb) compared to commercially available activated carbon.

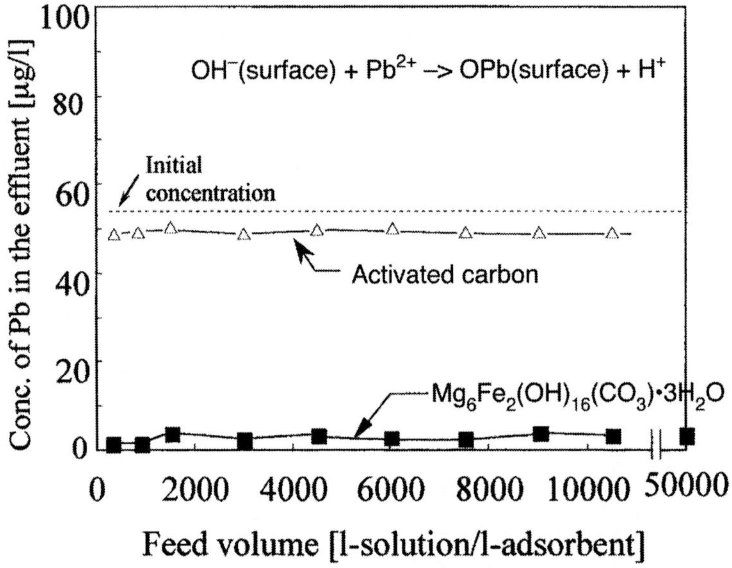

Figure 6. Results of sorption on Mg6Fe2(OH)16(CO3)•3H2O and on activated carbon from 55 ppb of Pb2+ aqueous test solution. The space velocity was 150 min–1.

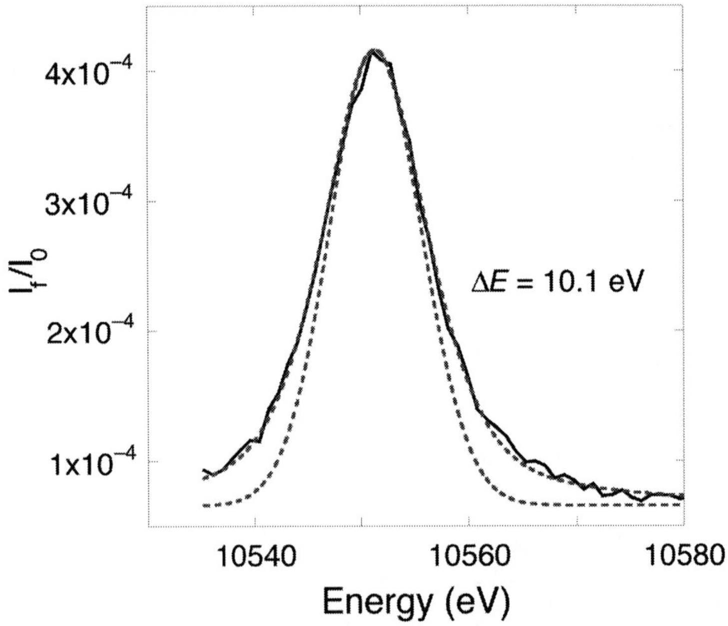

Figure 7. Lead Lα1 emission spectrum for 0.12 wt% of Pb adsorbed on Mg6Fe2(OH)16(CO3)•3H2O. The solid line is the experimental data, and (wider) dotted line is a fit with a pseudo-Voigt function. The intensity of the Lorentzian and Gaussian components was fixed to 1:1. The narrower dotted line is the energy resolution of fluorescence spectrometer (10.1 eV).

Lead $L\alpha_1$ emission spectrum for Pb adsorbed on $Mg_6Fe_2(OH)_{16}(CO_3)\bullet 3H_2O$ from 100 ppb Pb^{2+} test solution was depicted in Figure 7. The peak energy was identical to that for standard Pb^{II} compounds. The energy resolution of fluorescence spectrometer in this measurement condition was 5.0 – 10.1 eV dependent on the measurement conditions of fluorescence spectrometer (Figure 7) [24, 28]. Because the core-hole lifetime widths are 5.81 and 2.48 eV for Pb L_3 and M_5 levels, respectively [27], the width for $L\alpha_1$ is 8.29 eV comparable to the energy resolution of fluorescence spectrometer. Thus, sharper, more resolved spectral feature was expected in Pb $L\alpha_1$-selecting Pb L_3-edge XANES spectrum if the energy resolution of fluorescence spectrometer was smaller than 8.29 eV, similar to the case of As $K\alpha_1$-selecting XANES in previous section.

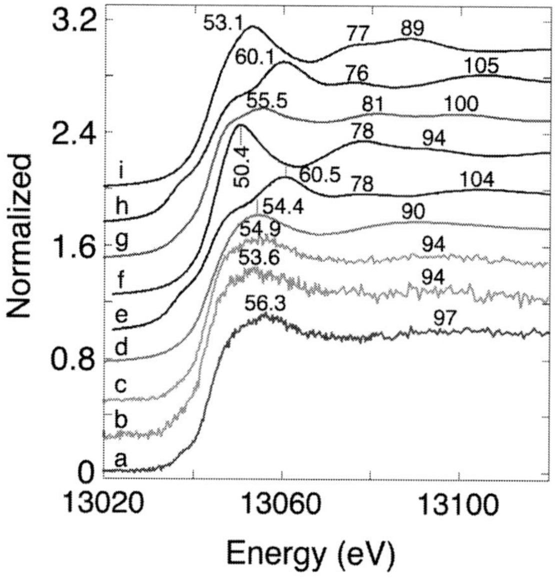

Figure 8. Lead Lα1-selecting Pb L3-edge XANES spectra measured at 290 K for Pb adsorbed on Mg6Fe2(OH)16(CO3)•3H2O (a – c). Tune energy of fluorescence spectrometer was 10551.5 eV. The Pb content was 1.0 wt% adsorbed from 1.0 ppm Pb2+ aqueous test solution (a) and 0.30 (b) and 0.12 wt% (c) from 100 ppb Pb2+ test solution. XANES spectra measured in transmission mode (d – i) for PbY zeolite (d), PbO (e), Pb(NO3)2 (f), 2PbCO3•Pb(OH)2 (g), Pb6O4(OH)4 (h), and PbCO3 (i).

Pb $L\alpha_1$-selecting XANES spectra are shown in Figure 8a – c. The spectral pattern of b and c resembled each other. The two samples contained 0.30 – 0.12 wt% of lead adsorbed from 100 ppb test aqueous Pb^{2+} solution. The Pb contents in samples were determined by ICP. Compared to XANES spectra for standard inorganic Pb compounds (Figure 8d – i) and supported Pb species on zeolite [29] or on Fe_2O_3 [30], the spectra b and c resembled most spectrum d measured for ion-exchanged PbY zeolite (d). Thus, surface uptake mechanism via ion exchange reaction was proposed on $Mg_6Fe_2(OH)_{16}(CO_3)\bullet 3H_2O$ from relatively low concentration of 100 ppb divalent lead solutions (Figure 9). The reaction formula is $[Mg_3Fe(OH)_8]^+ + Pb^{2+} \rightarrow [Mg_3Fe(OH)_7(OPb)]^{2+} + H^+$.

Pb $L\alpha_1$-selecting XANES spectrum for Pb adsorbed from 1.0 ppm test Pb^{2+} solution is depicted in Figure 8a. Compared to XANES spectra for standard inorganic Pb compounds (Figure 8d – i), the spectrum a resembled most spectrum g measured for $2PbCO_3\bullet Pb(OH)_2$.

Thus, coagulation uptake mechanism was proposed on $Mg_6Fe_2(OH)_{16}(CO_3)\bullet 3H_2O$ from relatively high concentration of 1 ppm Pb^{2+} test solution. Thus-identified reaction formula was $Pb^{2+} + nCO_3^{2-} + 2(1-n)OH^- \rightleftarrows nPbCO_3\bullet(1-n)Pb(OH)_2$.

With closer look of Figure 8a – c, a shoulder feature appeared at 13049 eV. Similar shoulder feature can be found in spectra for PbO, $2PbCO_3\bullet Pb(OH)_2$, and $Pb_6O_4(OH)_4$ (spectra e, g, and h, respectively). However, intense peak at 13060.1 – 13060.5 eV for PbO or $Pb_6O_4(OH)_4$ did not appear in spectra a – c. Thus, minor coagulation uptake mechanism was suggested from 100 ppb Pb^{2+} solutions in addition to major ion exchange process (Figure 9).

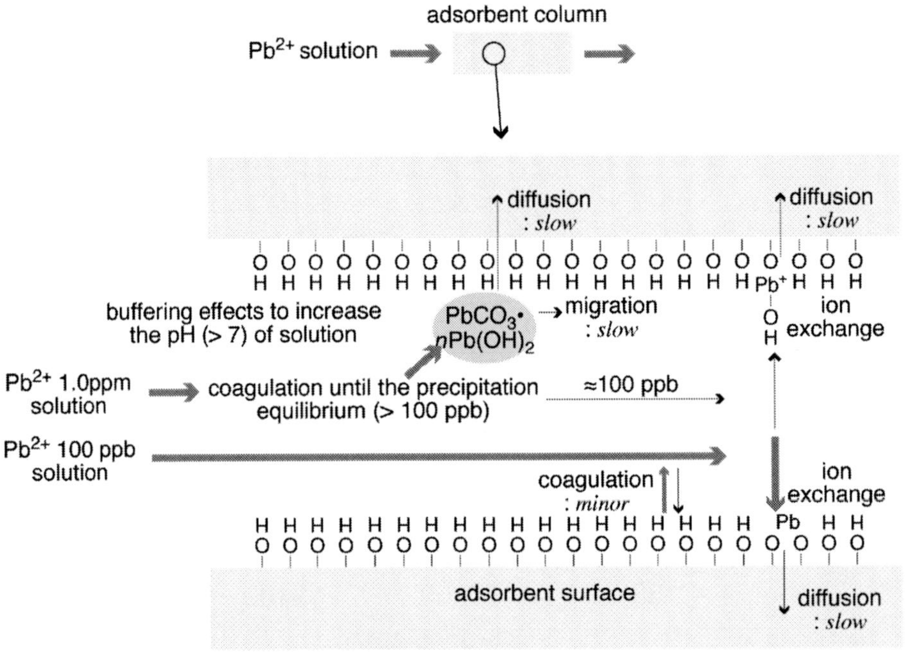

Figure 9. Pb2+ adsorption mechanism on Mg6Fe2(OH)16(CO3)•3H2O from Pb2+ 1.0 ppm and 100 ppb aqueous solutions.

In summary, Pb^{2+} uptake mechanism on $Mg_6Fe_2(OH)_{16}(CO_3)\bullet 3H_2O$ exhibited a switch-over from coagulation to (major) ion exchange reactions as the Pb^{2+} concentration decreased from 1.0 ppm to more environmentally plausible 100 ppb [28].

CONCLUDING REMARK AND FUTURE PROSPECTS

This chapter focused on the removal of low concentrations (55 – 200 ppb) of arsenite and lead by utilizing Fe-montmorillonite, micro/mesoporous $FeO_x(OH)_y$ effectively porous due to carboxyl-exchange method, and layered double hydroxide consisted of Fe and Mg (pyroaurite). It is still open to discuss the systematic survey of the removal process of other dilute toxic elements, *e.g.* Cr, Cu, Zn, Cd, or Hg. It is important to formulate the efficiency of surface uptake with respect to critical factors, *i.e.* pH, chemical species of toxic elements (naked cations, oxyanions, or oxyhydroxyl anions), initial concentrations (10 – 100 ppb) and space velocity of aqueous solutions to be processed, and chemical combination of surface *versus* chemical species (*e.g.* $Fe^{III}O(OH)$ *versus* $As(OH)_3$ and OH^-(clay surface)/CO_3^{2-} (between layers) *versus* Pb^{2+}).

In the analytical point of view to monitor the destiny of low concentrations of toxic elements, selective XAFS technique, with which the author of this chapter has also investigated surface catalytic sites of gold, platinum, tin, and vanadium [31 – 35], needs to be combined with other technique with nanoscale spacial resolution. Spectroscopy with nanoscale spacial resolution is under investigation, but not established to be applicable to nanotechnology (Table 2).

At present, spacial resolution of X-ray microscopy is 1 μm [36]. Several types of X-ray microscopy/imaging are under investigation, *e.g.* microbeam XAFS, photoemission electron microscope (PEEM), or phase contrast imaging [37, 38]. The spacial resolution of TEM is already smaller than 1 nm if the sample nanoscopic condition matches to the high-resolution measurement.

Table 2. Spectroscopy Needed to Be Developed to Give Direct Spacial Information of Surface Uptake Mechanism from Low Concentrations (10 – 100 ppb) of Toxic Metal Elements

Probe Method	Factor to be developed for this application	Refs
X-ray Microscope	Smaller X-ray beam (< 10 nm)	36 – 38
Electron Microscope	3-dimensional information	39
Tip Microscope	Scanning probe microscope to discriminate the kind of element	41, 42
X-ray Diffraction	Surface-sensitivity between nano-layers or in micro/mesospace	43
X-ray XAFS/XRF	On-site analysis in the field	44, 45

To monitor the sorption between 2-dimensinal layers (*e.g.* montmorillonite, hematite), in 2-dimensional mesopores (*e.g.* FSM-16, MCM-41), and in 3-dimensional micro/mesopores (*e.g.* ZSM-5, acetate-exchanged $FeO_x(OH)_y$ [17]), 3-dimensional TEM images would be very helpful by taking series of TEM snapshots from various angles to sample and organizing 3-dimensional image on computer [39].

Scanning probe microscope (SPM), especially scanning tunneling microscopy (STM) and atomic force microscope (AFM), has an advantage of atomic resolution for well-defined surface [40]. To utilize SPM technique to monitor the sorption from low concentrations of toxic elements, the combination with element specific spectroscopy, *e.g.* XPS, XAFS, is essential to describe the surface chemical mechanism. The author of this chapter is developing this combination (AFM and XPS) based on temporal electron trap phenomena in the metal nano-dots in the front of the AFM tip [41, 42].

Surface-sensitive XRD may be applicable to monitor the application of water purification [43]. The breakthrough is selectivity to internal surface of layered and micro/mesoporous materials used as sorbents. Portable XRF/XAFS apparatus that has been investigated for space science [44, 45] will be applicable to environmental on-site monitoring of the fate of toxic elements in the field.

ACKNOWLEDGMENTS

This article is part 26 in the series of state-sensitive XAFS. The X-ray experiments were performed under the approvals of the SPring-8 Program Review Committee and of the Photon Factory Proposal Review Committee. The works included in this chapter were financially supported by grants from the Grant-in-aid for Encouragement of Young Scientists (B14740401, A12740376) and the Grant-in-aid for Basic Scientific Research (B13555230, C17550073) from the Ministry of Education, Culture, Sports, Science, and Technology, Yamada Science Foundation (2000), Toray Science Foundation (98-3901), and Research Foundation for Opto-Science and Technology (2005 – 2006).

REFERENCES

[1] Holt, R. *Science* 2007, *317*, 198 – 199.
[2] Förstner, U. *Integrated pollution control (Umweltschutztechnik)*; Springer-Verlag: Berlin, 1998.
[3] Smith, A. H.; Lopipero, P. A.; Bates, M. N.; Steinmaus, C. M. *Science* 2002, *296*, 2145 – 2146.
[4] Nordstrom, D. K. *Science* 2002, *296*, 2143 – 2145.
[5] Schroeder, H. A.; Balassa, J. J. *Science* 1963, *140*, 819 – 820.
[6] Cox, C.; Davidson, P. W.; Myers, G. J.; Kawaguchi, T. *Science* 1998, *279*, 459.
[7] Stern, A. H.; Hudson, R. J. M.; Shade, C. W.; Ekino, S.; Ninomiya, T.; Susa, M.; Harris, H. H.; Pickering, I. J.; George, G. N. *Science* 2004, *303*, 763 – 766.
[8] Harris, H. H.; Pickering, I. J.; George, G. N. *Science* 2003, *301*, 1203.
[9] Smedley, P. L.; Kinniburgh, D. G. *Appl. Geochem.* 2002, *17*, 517 – 568.
[10] Ahmed, M. F.; Ahuja, S.; Alauddin, M.; Hug, S. J.; Lloyd, J. R.; Pfaff, A.; Pichler, T.; Saltikov, C.; Stute, M.; van Geen, A. *Science* 2006, *314*, 1687 – 1688.
[11] Gobeil, C.; Macdonald, R. W.; Smith, J. N.; Beaudin, L. *Science* 2001, *293*, 1301 – 1304.
[12] Nriagu, J. O. *Science* 1998, *281*, 1622 – 1623.
[13] Polvakov, E. V.; Egorov, Y. V. Russ. *Chem. Rev.* 2003, *72(11)*, 985 – 994.
[14] Huang, M. R.; Peng, O. Y.; Li, X. G. *Chem. Eur. J.* 2006, *12(16)*, 4341 – 4350.
[15] Izumi, Y.; Masih, D.; Aika, K.; Seida, Y. *J. Phys. Chem. B* 2005, *109*, 3227 – 3232.
[16] Grygar, T.; Hradil, D.; Bezdick, P.; Dousová, B.; Capek, L.; Schneeweiss, O. *Clays Clay Miner.* 2007, *55(2)*, 165 – 176.
[17] Izumi, Y.; Masih, D.; Aika, K.; Seida, Y. *Micropor. Mesopor. Mater.* 2006, *94*, 243 – 253.

[18] Seida, Y.; Nakano, Y.; Nakamura, Y. *Water Resear.* 2001, *35*, 2341 – 2346.
[19] Billinge, S. J. L.; Levin, I. *Science* 2007, *316*, 561 – 565.
[20] Koningsberger, D. C.; Prins, R., Eds. *X-ray Absorption – Principles, Applications, Techniques of EXAFS, SEXAFS, and XANES*; John Wiley and Sons: New York, 1988.
[21] Izumi, Y.; Oyanagi, H.; Nagamori, H. *Bull. Chem. Soc. Jpn.* 2000, *73(9)*, 2017 – 2023.
[22] Izumi, Y.; Nagamori, H. *Bull. Chem. Soc. Jpn.* 2000, *73(7)*, 1581 – 1587.
[23] Izumi, Y.; Kiyotaki, F.; Nagamori, H.; Minato, T. *J. Electro. Spectrsc. Relat. Phenom.* 2001, *119(2/3)*, 193 – 199.
[24] Izumi, Y.; Nagamori, H.; Kiyotaki, F.; Masih, D.; Minato, T.; Roisin, E.; Candy, J. P.; Tanida, H.; Uruga, T. *Anal. Chem.* 2005, *77(21)*, 6969 – 6975.
[25] Dixit, S.; Hering, J. G.; *Environ. Sci. Technol.* 2003, *37*, 4182 – 4189.
[26] Izumi, Y.; Masih, D.; Aika, K.; Seida, Y. *Micropor. Mesopor. Mater.* 2007, *99*, 355.
[27] Zschornack, G. *Handbook of X-Ray Data*; Springer: Berlin, 2007.
[28] Izumi, Y.; Kiyotaki, F.; Minato, T.; Seida, Y. *Anal. Chem.* 2002, *74(15)*, 3819 – 3823.
[29] Huang, F. T.; Jao, H. J.; Hung, W. H.; Chen, K.; Wang, C. M. *J. Phys. Chem. B* 2004, *108(52)*, 20458 – 20464.
[30] Pinakidou, F.; Katsikini, M.; Paloura, E. C.; Kalogirou, O.; Erko, A. *J. Non-Cryst. Solids* 2007, *353(28)*, 2717 – 2733.
[31] Izumi, Y.; Obaid, D. M.; Konishi, K.; Masih, D.; Takagaki, M.; Terada, Y.; Tanida, H.; Uruga, T. *Inorg. Chim. Acta*, in press (http://dx.doi.org/10.1016/j.ica.2007.09.027).
[32] Izumi, Y.; Masih, D.; Roisin, E.; Candy, J. P.; Tanida, H.; Uruga, T. *Mater. Lett.* 2007, *61(18)*, 3833 – 3836.
[33] Izumi, Y.; Masih, D.; Candy, J. P.; Yoshitake, H.; Terada, Y.; Tanida, H.; Uruga, T. "*X-Ray Absorption Fine Structure 13th International Conference*", Hedman, B.; Pianetta, P. Eds., AIP Conference Proceedings 2007, Vol. 882, 588 – 590.
[34] Izumi, Y.; Konishi, K.; Obaid, D. M.; Miyajima, T.; Yoshitake, H. *Anal. Chem.* 2007, *79(18)*, 6933 – 6940.
[35] Izumi, Y.; Kiyotaki, F.; Yagi, N.; Vlaicu, A. M.; Nisawa, A.; Fukushima, S.; Yoshitake, H.; Iwasawa, Y. *J. Phys. Chem. B*, 2005, *109(31)*, 14884 – 14891.
[36] Hokura, A.; Kitajima, N.; Terada, Y.; Nakai, I. *SPring-8 Research Frontiers* 2006, 120 – 121.
[37] Tuohimaa, T.; Otendal, M.; Hertz, H. M. *Appl. Phys. Lett.* 2007, *91(7)*, 074104.
[38] Koshikawa, T.; Guo, F. Z.; Yasue, T. *SPring-8 Research Frontiers* 2005, 52 – 53.
[39] Midgley, P. A.; Thomas, J. M.; Laffont, L.; Weyland, M.; Raja, R.; Johnson, B. F. G.; Khimyak, T. *J. Phys. Chem. B* 2004, *108(15)*, 4590 – 4592.
[40] Marti, O.; Möller, R. Eds. *Photons and Local Probes*; Kluwer Academic Publishers: Dordrecht, 1995.
[41] Klein, L. J.; Williams, C. C. *Appl. Phys. Lett.* 2001, *79(12)*, 1828 – 1830.
[42] Klein, L. J.; Williams, C. C.; Kim, J. *Appl. Phys. Lett.* 2000, *77(22)*, 3615 – 3617.
[43] Takahasi, M. *SPring-8 Research Frontiers* 2006, 56 – 57.
[44] Robinson, A. L. *Science* 1980, *208*, 163 – 164.
[45] Frierman, J. D.; Bowman, H. R.; Perlman, I.; York, C. M. *Science* 1969, *164*, 588.

In: Drinking Water: Contamination, Toxicity and Treatment ISBN: 978-1-60456-747-2
Editors: J. D. Romero and P. S. Molina

Chapter 7

DETOXIFICATION OF STEROIDAL HORMONES IN THE AQUATIC ENVIRONMENT

***Tomoaki Nishida*[1,*] *and Hideo Okamura*[2]**
[1]Faculty of Agriculture, Shizuoka University, Ohya 836, Surugaku, Shizuoka 422-8529, Japan
[2]Graduate School of Maritime Sciences, Kobe University, Fukueminami 5-1-1, Higashinadaku, Kobe 658-0022, Japan

1. INTRODUCTION

Lignin is one of the most abundant organic polymers, probably only second to cellulose among the renewable organic resources on Earth. In contrast to cellulose and hemicellulose in wood, lignin is resistant to degradation by microorganisms. But ligninolytic activity has been recognized by certain microorganisms. At present, the white rot fungi are the best known and the most lignin-degrading microorganisms.

There is a great interest in the white rot fungi and their ligninolytic enzymes, because their industrial potential for degrading and detoxifying recalcitrant environmental pollutions. Considerable concern has recently been expressed that steroidal hormones (estrogens) excreted into the environment by humans, domestic or farm animals, and other wildlife, in part via sewage treatment plants, may be disruptive to the endocrine systems. This article will review our recent research on detoxification of natural steroidal estrogens (17β-estradiol and estrone) (Suzuki et al., 2003; Tamagawa et al., 2006) and synthetic one (17α-ethynylestradiol) (Suzuki et al., 2003), focusing on the contamination and environmental toxicity of these estrogens and the detoxification of these estrogens by the treatment with ligninolytic enzymes from white rot fungi.

---

* Corresponding author: Tomoaki Nishida. Tel. and Fax: +81-54-238-4852, E-mail: aftnisi@agr.shizuoka.ac.jp

2. Contamination and Environmental Toxicity of Steroidal Estrogens

A multitude of chemicals have shown to be endocrine disrupters to date. Among these, natural and synthetic steroidal estrogens are effective at the lower ng l^{-1} level, while most other chemicals having estrogenic effects are biologically active at the μg l^{-1} level (Purdon et al., 1994; Routledge et al., 1998; Metcalfe et al., 2001). The occurrence of steroidal estrogens in the aquatic environment due to discharge of municipal wastewater has attracted considerable attention due to evidence that these compounds may cause disruption of endocrine function in some aquatic species in sewage-impacted receiving waters. Compounds of concern include the natural female hormone 17β-estradiol (E_2), its oxidation product estrone (E_1), and the synthetic contraceptive additive 17α-ethynylestradiol (EE_2) (Folmer et al., 1996; Harries, et al., 1996; Harries et al., 1997, Desbow et al., 1998). These compounds are extremely potent estrogen receptor modulators and therefore may cause deleterious effects on organisms at low concentrations. The levels of steroid estrogens are very low in most aquatic environments due to dilution, sorption, and degradation following wastewater discharge.

Alterations of the sexual function has been observed in fish living downstream of sewage treatment plants (STPs). Natural estrogens (E_2 and E_1) as well as the synthetic estrogen (EE_2) contribute to a large extent to the estrogenicity of STP effluents. These chemicals are active at trace levels; for instance, EE_2 has been reported to induce the production of vitellogenin, a female specific york protein precursor normally synthesized by female fish, in male rainbow trout at concetrations as low as 0.1 ng/L (Purdom et al., 1994). Effluents from STPs can discharge estrogenic contaminants into rivers at levels sufficient to induce vitellogenin biosynthesis in male fish (Jobling et al., 1998). Thus, considerable concern has been expressed that steroidal estrogens excreted into the environment by humans, domestic or farm animals, and other wildlife, in part via STPs, may be disruptive to the endocrine system. The residue levels of these estrogens in STP effluents are reported with sample numbers, nation detected, and analytical procedure (Table 1). Measured concentrations have been ranging from 0.2-88 ng/L for E_2, 1-220 ng/L for E_1, and 0.2-7.0 ng/L for EE_2 in STP effluent. The concentrations are decreasing orders in $E_1 > E_2 > EE_2$ and higher concentration of each estrogen are enough to contribute to cause endocrine disruption of some fish species.

It has been reported that E_2 is oxidized to E_1, which is further eliminated in aerobic batch experiments containing diluted slurries of activated sludge from an actual STP without any observed degradation products (Ternes et al., 1999). However, recent studies have claimed that the concentration of E_1 in final effluent at seven of the 25 STPs in Italy, Spain, and Canada is elevated above that in influent, and the removal efficiency of E_1 markedly lower than that of E_2 (Baronti et al., 2000; Carballa et al., 2004, Servos et al., 2005). At least part of the increase in E_1 concentration in effluent is thus explained by the accumulation of E_1 resulting from biological oxidation of E_2 at STPs. Furthermore, it has been demonstrated that the amount of E_1 discharged from STPs into receiving waters is more than ten times that of E_2 (Servos et al., 2005; D'Ascenzo et al., 2003). Both in vitro (Routledge and Sumpter, 1997) and in vivo (Routledge et al., 1998) experiments have shown that the estrogenic potency of E_1 is half that of E_2. These observations led to the conclusion that, among natural estrogens, E_1 is by far the most important endocrine disruptor in the aquatic environment (D'Ascenzo et al.,

2003). On the other hand, very little is known about the actual biodegradation of synthetic steroidal estrogen EE_2 (Ternes et al., 1999; Fujii et al., 2002), and no attempts have been made, other than with bacteria, to degrade EE_2 using microorganisms.

3. Removal of Estrogenic Activities of Steroidal Estrogens by Ligninolytic Enzymes from White Rot Fungi

Great interest is currently being expressed in lignin-degrading white rot fungi and their ligninolytic enzymes due to the recognized potential for degrading and detoxifying recalcitrant environmental pollutants such as dioxins (Bumpus et al., 1985), chlorophenols (Joshi and Gold, 1993), polycyclic aromatic hydrocarbons (Bezalel et al., 1996; Collins et al., 1996), and dyes (Ollikka et al., 1993; Nishida et al., 1999). Manganese peroxidase (MnP) and laccase produced extracellularly by white rot fungi have been shown to be involved in the degradation of lignin and these pollutants.

Table 1. Residue concentrations (ng/L) of steroidal estrogens in effluents from sewage treatment plants

E_2 17β-Estradiol	E_1 Estron	EE_2 17α-Etynylestradiol	n	Nation	Analytical method	Reference
<1-4.4	17-71	<1	7	France	GC/MS	Labadie and Budzinski (2005)
0.2-14.7	1-96	nd	18	Canada	GC/MS	Servos et al. (2005)
ldl-3.0	2.4-4.4	nd	5	Spain	GC/MS/MS	Carballa et al. (2004)
0.77-6.4	1.6-18	nd	2	USA	LC/MS	Ferguson et al. (2001)
0.35-3.3	2.5-82	ldl-1.7	30	Italy	LC/MS/MS	Baronti et al. (2000)
4-88	27-220	1.7-3.4	8	UK	GC/MS	Rodgers-Gray et al. (2000)
0.5-4.2	5.3-35.1	nd	5	Japan	LC/MS/MS	Tajima et al. (2000)
0.6-5.7	16-110	<1	6	Japan	LC/MS/MS	Ishii et al. (2000)
ldl-3.7	nd	ldl-0.76	8	USA	HPLC/fluorescence or RIA	Snyder et al. (1999)
2.7-48	14-76	0.2-7.0	21	UK	GC/MS	Desbrow et al. (1998)

nd: not determined
ldl: less than detection limit
RIA: radioimmunoassay

MnP is a heme peroxidase produced by white rot fungi and catalyzes the oxidation of Mn(II) to Mn(III) in the presence of H_2O_2. Malonate, oxalate, and α-hydroxy acids such as malate, lactate, and tartrate chelate the generated Mn(III) and release Mn(III) from the manganese-binding site of MnP. The released Mn(III)-organic acid complex in turn oxidizes various phenolic compounds, including lignin (Glen et al., 1986; Wariishi et al., 1989). Fungal laccase is a multicopper oxidase that catalyzes the single-electron oxidation of phenolic compounds by reducing molecular oxygen to water (Reinhammer, 1984). In the presence of a mediator, such as 1-hydroxybenzotriazole (HBT) or 2,2′-azinobis (3-ethylbenzothiazoline-6-sulfonate), laccase is capable of oxidizing nonphenolic compounds that it cannot oxidize alone (Bourbonnais et al., 1997).

Therefore, we applied the enzymatic treatment, MnP, laccase, or the laccase-HBT system, in order to remove the estrogenic activities of steroidal estrogens (Suzuki et al., 2003; Tamagawa et al., 2006). HPLC detection of residual E_2, E_1, or EE_2 on an ODS column (285 nm) during enzymatic treatment revealed that these estrogens substantially disappeared in the reaction mixture after 1 h of treatment. These indicate that the treatment with ligninolytic enzyme effectively decrease E_2, E_1, and EE_2. However, the greatest focus concerning the biodegradation of endocrine-disrupting (estrogenic) chemicals should be on the removal of estrogenic activity. We therefore attempted to assay the estrogenic activities of the reaction mixtures of E_2, E_1, and EE_2 during enzymatic treatment using the yeast two-hybrid estrogenic assay system. This system is newly developed and is based on the ligand-dependent interaction between the nuclear hormone receptor and its coactivator. The method is rapid and has been confirmed to be reliable for measuring estrogenic activity (Nishikawa et al., 1999).

The estrogenic activities of E_2, E_1, and EE_2, expressed as β-galactosidase activity, are compared with those of BPA and NP (Figure 1). The activities of E_2 ($\sim 5 \times 10^{-9}$ M), E_1 (10^{-8} M), and EE_2 (10^{-8} M) were almost the same as those of BPA (10^{-4} M) and NP (10^{-5} M), indicating that these steroidal estrogens have much higher estrogenic activities at much lower concentrations than do BPA and NP, which are widely used in a variety of industrial and residential applications. This finding is consistent with a previous report in which the relative estrogenicities of E_1, EE_2, BPA, and NP were found to be 0.6, 0.5, 5×10^{-5} and 4×10^{-4}, respectively, as compared with E_2 (Nakamuro et al., 2002).

The data depicted in Figure 2 demonstrate that MnP, laccase, and the laccase-HBT system reduced the estrogenic activity of E_1 by 99%, 97%, and 97% after a 1-h treatment, respectively, and completely removed the activity after 2 h of treatment. On the other hand, HPLC detection of the residual E_1 during enzymatic treatment revealed that E_1 completely disappeared in the reaction mixture after a 1-h treatment with either MnP, laccase, or the laccase-HBT system; the residual concentration of E_1 was below the HPLC detection limit (10^{-8} M). Figure 1 shows that E_1 exhibits estrogenic activity at very low concentrations. Thus, the fact that 1-3% of the estrogenic activity of E_1 remained after a 1-h treatment (Figure 2) may be due to residual traces of E_1. Furthermore, MnP and the laccase-HBT system substantially removed the estrogenic activities of E_2 (Figure 3) and EE_2 (Figure 4) after 4-h of treatment but the activities remained at 10-20% in the reaction mixture after 1 h of treatment, although HPLC detection of the residual E_2 and EE_2 during enzymatic treatment revealed that both estrogens almost completely disappeared in the reaction mixture after a 1-h treatment. These results strongly indicate that the estrogenic assay is necessary in studies of the bioremediation of endocrine-disrupting chemicals.

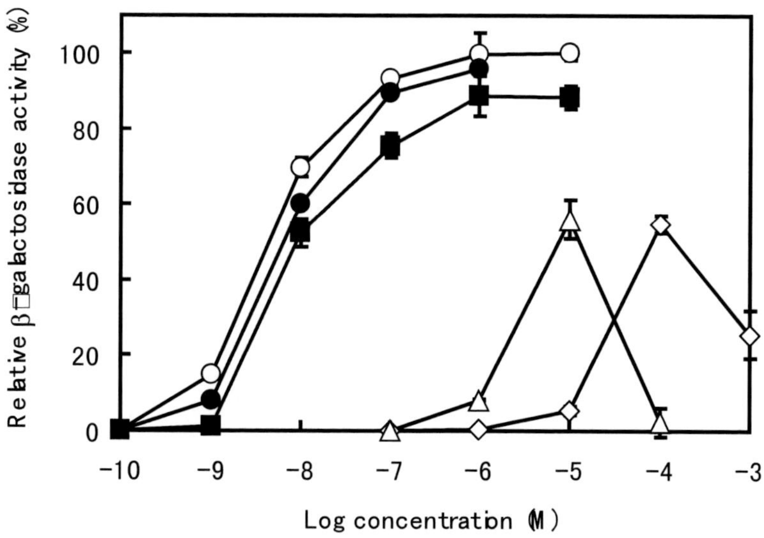

Figure 1. Dose response curve for estrogenic activity as measured by yeast two-hybrid assay. Activity of E2 at 10-5M is 100% standard of relative activity. Indicated for each point are the mean and standard deviation of five experiments for E2, E1, EE2, NP, and BPA. (○), E2; (●), EE2; (■), E1; (△), NP; (◇), BPA.

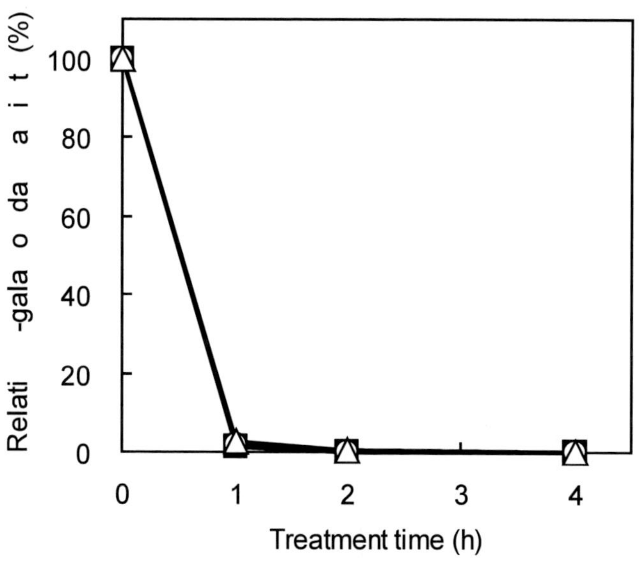

Figure 2. Removal of estrogenic activity of E1 by enzymatic treatment with MnP, laccase, or laaccase-HBT system. (■), MnP; (△), laccase ; (○), laccase-HBT system.

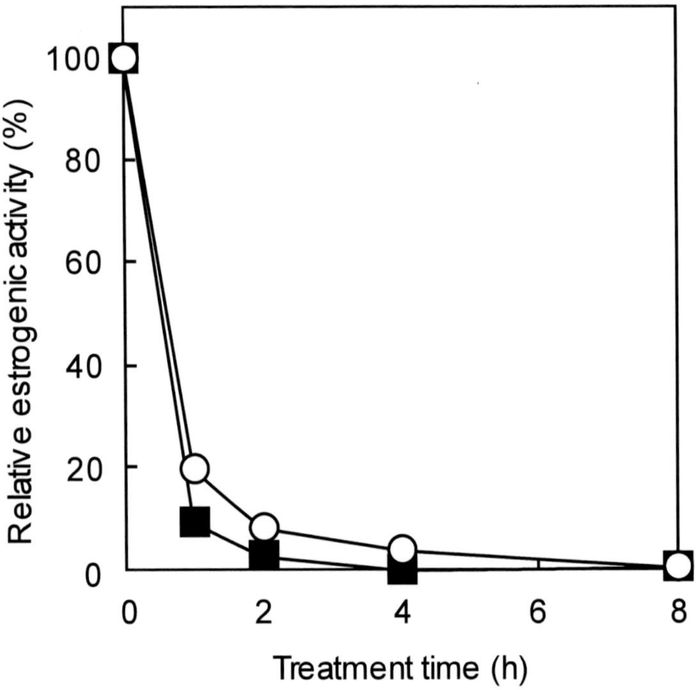

Figure 3. Removal of estrogenic activity of E2 by enzymatic treatment with MnP or laccase-HBT system. (■), MnP; (○), laccase-HBT system.

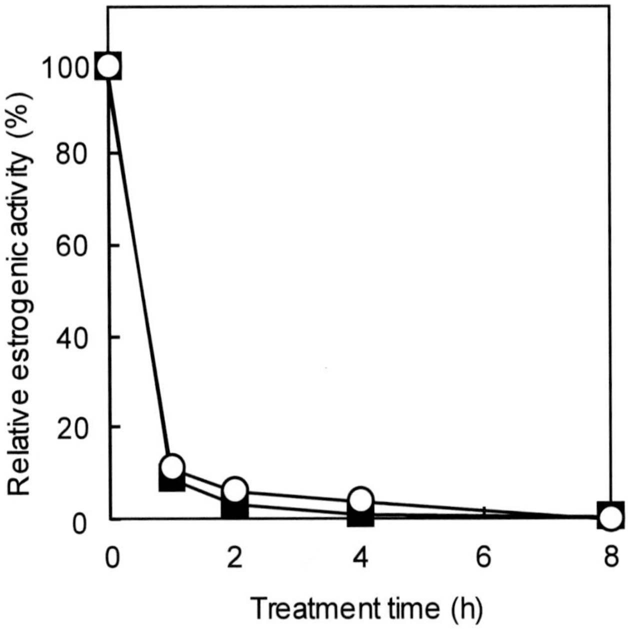

Figure 4. Removal of estrogenic activity of EE2 by enzymatic treatment with MnP or laccase-HBT system. (■), MnP; (○), laccase-HBT system.

In a previous report, we demonstrated that treatment of BPA and NP with MnP and laccase resulted in oligomeric reaction products through the formation of phenoxy radicals followed by radical coupling, and the reaction products must lose structural characteristics of acting as estrogens (Tsutsumi et al., 2001). Thus, the removal of estrogenic activities of E_2, E_1, and EE_2 may be due to polymerization brought about by enzymatic oxidation, because these estrogens have a similar *para*-substituted phenol structure to BPA and NP. Furthermore, the fact that 10-20% of the estrogenic activities of E_2 and EE_2 remained after 1 h of treatment with MnP and the laccase-HBT system (Figure 3 and 4) may be due to the reaction products of E_2 and EE_2 sharing a key structural feature for acting as estrogens.

In this article, we mentioned that the ligninolytic enzymes MnP and laccase from white rot fungi can effectively eliminate E_2, E_1, and EE_2 and remove their estrogenic activities. The past few years have been seen our laboratory applying the white rot fungi and their ligninolytic enzymes to the bioremediation treatment. We have also demonstrated that MnP and laccase are effective in removing the estrogenic activities of BPA (Tsutsumi et al., 2001), NP (Tsutsumi et al., 2001), 4-*tert*-octylphenol (Tamagawa et al., 2007), and phytoestrogen genistein (Tamagawa et al., 2005) and in degrading methoxychlor (Hirai et al., 2004) and the antifouling compound Irgarol 1051 (Ogawa et al., 2004). These findings indicate that white rot fungus and ligninolytic enzyme is one of the most attractive microorganisms and enzymes, respectively, for detoxification of recalcitrant environmental pollutants.

REFERENCES

Baronti, C., Curini, R., D'Ascenzo, G., Corcia, A. D., Gentili, A. and Samperi, R., 2000. Monitoring natural and synthetic estrogens at activated sludge sewage treatment plants and in a receiving river water. *Environ. Sci. Technol. 34,* 5059-5066.

Bezalel, L., Hadar, Y., Fu, P. P., Freeman, J. P. and Cerniglia, C., 1996. Initial oxidation products in the metabolism of pyrene, anthracene, fluorine, and dibenzothiophene by the white rot fungus *Pleurotus ostreatus*. *Appl. Environ. Microbiol. 62*, 2554-2559.

Bourbonnais, R., Paice, M., Freiermuth, B., Bodie, E. and Bornerman, S., 1997. Reactivities of various mediators and laccase with kraft pulp and lignin model compounds. *Appl. Environ. Microbiol. 63*, 4627-4632.

Bumpus, J. A., Tien, M., Wright, D. and Aust, S. D., 1985. Oxidation of persistent environmental pollutants by a white rot fungus. *Science 228,* 1434-1436.

Carballa, M., Omil, F., Lema, J. M., Llompart, M., García-Jares, C., Rodríguez, I., Gómez, M. and Ternes, T., 2004. Behavior of pharmaceuticals, conmetics and hormones in a sewage treatment plant. *Water Res. 38*, 2918-2926.

Collins, P. J., Kotterman, M. J., Field, J. A. and Dobson, A. W., 1996. Oxidation of anthracene and benzo(a)pyrene by laccases from *Trametes versicolor*. *Appl. Environ. Microbiol. 62*, 4563-4567.

D'Ascenzo, G., Di Corcia, D., Gentili, A., Mancini, R., Mastropasqua, R., Nazzari, M. and Samperi, R., 2003. Fate of natural estrogen conjugates in municipal sewage transport and treatment facilities. *Sci. Total Environ. 302*, 199-209.

Desbrow, C., Routledge, E. J., Brighty, G. C., Sumpter, J. P., Waldock, M., 1998. Identification of estrogenic chemicals in STW effluent. 1. Chemical fractionation and in vitro biological screening. *Environ. Sci. Technol. 32*, 1549-1558.

Ferguson, P. L., Iden, C. R., McElroy, A. E., Brownawell, B. J., 2001. Determination of steroidal estrogens in wastewater by immunoaffinity extraction coupled with HPLC-electrospray-MS. *Anal. Chem. 73,* 3890-3895.

Folmar, L. C., Denslow, N. D., Rao, V., Chow, M., Crain, D. A., Enbrom, J., Marcino, J. and Guillette, L. J., 1996. Vitellogenin induction and reduced serum testosterone concentrations in feral male carp (*Cyprinus carpio*) captured near major metropolitan sewage treatment plant. *Environ. Health Perspect. 104,* 1096-1101.

Fujii, K., Kikuchi, S., Satomi, M., Ushio-Sata, N., and Morita, N., 2002. Degradation of 17β-estradiol by a gram-negative bacterium isolated from activated sludge in a sewage treatment plant in Tokyo, Japan. *Appl. Environ. Microbiol. 68*, 2057-2060.

Glenn, J. K., Akileswaran, L. and Gold, M. H., 1986. Mn(II) oxidation is the principal function of the extracellular Mn-peroxidase from *Phanerochaete chrysosporium. Arch. Biochem. Biophys. 251*, 688-696.

Harries, J. E., Sheahan, D. A., Jobling, S., Matthiessen, P., Neall, P., Sumpter, J. P., Tylor, T. and Zaman, N., 1997. Estrogenic activity in five United Kingdom rivers detected by measurement of vitellogenesis in caged male trout. *Environ. Toxicol. Chem. 16*, 534-542.

Harries, J. E., Sheahan, D. A., Jobling, S., Matthiessen, P., Neall, P., Routledge, E. J., Rycroft, R., Sumpter, J. P. and Tylor, T., 1996. A survey of estrogenic activity in United Kingdom inland waters. *Environ. Toxicol. Chem. 15,* 1993-2002.

Hirai, H., Nakanishi, S. and Nishida, T., 2004. Oxidative dechlorination of methoxychlor by ligninolytic enzymes from white-rot fungi. *Chemosphere 55*, 641-645.

Ishii, Y., Okita, S., Torigai, M., Yun, S.-J., 2000. Determination of estrogens in environmental water samples by LC/MS/MS (in Japanese). *Bunseki Kagaku 49*, 753-758.

Jobling, S., Noran, M., Tyler, C. R., Brighty, G. and Sumpter, J. P., 1998. Widespread sexual disruption in wild fish. *Environ. Sci. Technol. 32*, 2498-2506.

Joshi, D. K. and Gold, M. H., 1993. Degradation of 2, 4, 5-trichlorophenol by the lignin-degrading basidiomycete *Phanerochaete chrysosporium. Appl. Environ. Microbiol. 59*, 1779-1785.

Labadie, P. and Budzinski, H., 2005. Determination of steroidal hormone profiles along the Jalle d'Eysines river (near Bordeaux, France) *Environ. Sci. Technol. 39*, 5113-5119.

Metcalfe, C. D., Metcalfe, T. L., Kiparissis, Y., Koenig, B. G., Khan, C., Hughes, R. J., Croley, T. R., March, R. E. and Potter, T., 2001. Estrogenic potency of chemicals detected in sewage treatment plant effluents as determined by in vivo assays with Japanese Medaka (*Oryzias latipes*). *Environ. Toxicol. Chem. 20,* 297-308.

Nakamuro, K., Ueno, H., Okuno, T., Sakazaki, H., Kawai, H., Kamei, T. and Ugawa, M., 2002. Contribution of endocrine-disrupting chemicals to estrogenicity of environmental water (in Japanese). *J. Jpn. Soc. Water Environ. 25*, 355-360.

Nishida, T., Tsutsumi, Y., Kemi, M., Haneda, T. and Okamura, H., 1999. Decolorization of anthraquinone dyes by white-rot fungi and its related enzymes (in Japanese). *J. Jpn. Soc. Water Environ. 22*, 465-471.

Nishikawa, J., Saito, K., Goto, J., Dakeyama, F., Matsuo, M. and Nishihara, T., 1999. New screening methods for chemicals with hormonal activities using interaction of nuclear hormone receptor with coactivator. *Toxicol. Appl. Pharmacol. 154*, 76-83.

Ogawa, N., Okamura, H., Hirai, H. and Nishida, T., 2004. Degradation of the antifouling compound Irgarol 1051 by manganese peroxidase from the white rot fungus *Phanerochaete chrysosporium*. *Chemosphere 55*, 487-491.

Ollikka, P., Alhonmäki, K., Leppänen, V-M., Glumoff, T., Raijola, T. and Suominen, I., 1993. Decolorization of azo, triphenyl methane, heterocyclic, and polymeric dyes by lignin peroxidase isoenzymes from *Phanerochaete chrysosporium*. *Appl. Environ. Microbiol. 59*, 4010-4016.

Purdom, C. E., Hardiman, P. A., Bye, V. J., Eno, N. C., Tyler, C. R., and Sumpter, J. P., 1994. Estrogenic effects of effluents from sewage treatment works. *Chem. Ecol. 8*, 275-285.

Reinhammer, B., 1984. Laccase. In: Lontie, R. (Ed.), Copper Proteins and Copper Enzymes, Vol. 3. CRC Press, Boca Raton, FL, pp. 1-35.

Rodgers-Gray, T. P., Jobling, S., Morris, S., Kelly, C., Kirby, S., Janbakhsh, A., Harries, J. E., Waldock, M.J., Sumpter, J.P., Tyler, C.R., 2000. Long-term temporal changes in the estrogenic composition of treated sewage effluent and its biological effects on fish. *Environ. Sci. Technol. 34*, 1521-1528.

Routledge, E. J., Sheahan, D., Desbrow, C., Sumpter, J. P. and Waldock, M., 1998. Identification of estrogenic chemicals in STW effluent. 2. in vivo responses in trout and roach. *Environ. Sci. Technol. 32*, 1559-1565.

Routledge, E. J. and Sumpter, J. P., 1997. Structural features of alkylphenolic chemicals associated with estrogenic activity. *J. Biol. Chem. 272*, 3280-3288.

Servos, M. R., Bennie, D. T., Burnison, B. K., Jurkovic, A., McInnis, R., Neheli, T., Schnell, A., Seto, P., Smyth, S. A. and Ternes, T. A., 2005. Distribution of estrogens, 17β-estradiol and estrone, in Canadian municipal wastewater treatment plants. *Sci. Total Environ. 336*, 155-170.

Snyder, S. A., Keith, T. L., Verbrugge, D. A., Snyder, E. M., Gross, T. S., Kannan, K., Giesy, J. P., 1999. Analytical methods for detection of selected estrogenic compounds in aqueous mixtures. *Environ. Sci. Technol. 33*, 2814-2820.

Suzuki, K., Hirai, H., Murata, H. and Nishida, T., 2003. Removal of estrogenic activities of 17β-estradiol and ethinylestradiol by ligninolytic enzymes from white rot fungi. *Water Res. 37*, 1972-1975.

Tajima, H., Tsujimura, K., Yamaguchi, M., 2000. Development of an analytical method of 17β-estradiol in river water by liquid chromatography/tandem mass spectrometry (in Japanese). *Bunseki Kagaku, 49*, 843-848.

Tamagawa, Y., Hirai, H., Kawai, S. and Nishida, T., 2007. Removal of estrogenic activity of 4-*tert*-octylphenol by ligninolytic enzymes from white rot fungi. *Environ. Toxicol. 22*, 281-286.

Tamagawa, Y., Yamaki, R., Hirai, H., Kawai, S. and Nishida, T., 2006. Removal of estrogenic activity of natural steroidal hormone estrone by ligninolytic enzymes from white rot fungi. *Chemosphere 65,* 97-101.

Tamagawa, Y., Hirai, H., Kawai, S. and Nishida, T., 2005. Removal of estrogenic activity of endocrine-disrupting genistein by ligninolytic enzymes from white rot fungi. *FEMS Microbiol. Lett. 244*, 93-98.

Ternes, T. A., Kreckel, P. and Mueller, J., 1999. Behaviour and occurrence of estrogens in municipal sewage treatment plants-II. Aerobic batch experiments with activated sludge. *Sci. Total Environ. 225*, 91-99.

Tsutsumi, Y., Haneda, T. and Nishida, T., 2001. Removal of estrogenic activities of bisphenol A and nonylphenol by oxidative enzymes from lignin-degrading basidiomycetes. *Chemosphere 42*, 271-276.

Wariishi, H., Dunford, H. B., MacDonald, I. D. and Gold, M. H., 1989. Manganese peroxidase from the lignin-degrading basidiomycete *Phanerochaete chrysosporium*. *J. Biol. Chem. 264*, 3336-3340.

In: Drinking Water: Contamination, Toxicity and Treatment ISBN: 978-1-60456-747-2
Editors: J. D. Romero and P. S. Molina

Chapter 8

INFLUENCE OF MINERALIZED/MINED AREAS IN THE QUALITY OF WATERS DESTINED FOR THE PRODUCTION OF DRINKING WATER

A. Ordóñez[1], R. Alvarez[1], E. De Miguel[2], J. Loredo[1] and F. Pendás[1]

[1]Dep. Explotación y Prospección de Minas, School of Mines
University of Oviedo, Spain
[2]Dep. Ingeniería Química y Combustibles, School of Mines
Polytechnic University of Madrid, Spain

ABSTRACT

Surface and groundwater quality is often altered in naturally mineralized areas. Mining activities can enhance these effects since they involve a number of processes that impact water quality. The type of water contamination is highly dependent on the mineralization, the reactants used in the ore processing activities and the type of mining: i) *Industrial minerals and rocks:* the absence of reactive mineral phases and chemical treatment implies that the typical affection to water bodies is caused by suspended solids; ii) *Metallic mining:* water contacting open-cut and underground mining of base metal deposits can show low pH values (*Acid Mine Drainage*) and high concentrations of many chemicals, leached from the ore and host rock (sulphate, Fe, Pb, Hg, As and many others), as well as substances used to extract metals from the ore (flotation reagents and leaching compounds); iii) *Fossil fuel mining:* In coal mining, low pH and high sulphate and metal content in water are very common, mainly due the oxidation of pyrite present in the coal bed. The most outstanding feature of waters related to oil and gas mining is the presence of brines (high salinity), CH_4 and other hydrocarbons, as well as H_2S. Abandoned pits and mine shafts are sometimes used for water supply after mine closure. Additionally, leachates from mining areas and/or spoil heaps, and affected groundwater are frequently received by water bodies which are used for drinking water production. In most cases, the input of water affected by mining and metallurgical works is negligible when compared with the total volume of input from the catchment and, therefore, an admissible risk is obtained as a result of a risk assessment in these sites. This is illustrated in several study cases presented in this paper, describing how different types of mining sites in Spain affect water bodies used for drinking purposes. That is the case of the "El

Aramo" mine on a Cu-Co-Ni deposit in Asturias: groundwater sampled at a nearby spring contained more than 4.9 mg l^{-1} Cu, well above the national standard for drinking water. Considering that there is a significant catchment point for drinking water downstream of the mining works, the presence of high As and Cu concentrations in some periods of the year can pose a potential health risk. The old "Santa Agueda" mine in León extracted an As ore; mining and smelting wastes remain in spoil heaps, exposed to weathering, and the leachates, containing up to 0.9 mg l^{-1} As, flow directly to a water reservoir constructed for production of hydraulic energy and drinking water. Mercury mining districts in NW Spain, including several mine sites, contribute mine drainage and leachates (containing up to 18 mg·l^{-1} As) to a catchment sometimes used for drinking water supply. Additionally, human and animal consume of mine-affected water in rural areas has been detected. These and other cases are analysed from a holistic point of view in this paper.

Introduction

Water is needed at mine sites for dust suppression, mineral processing, washing, and metallurgical extraction. For these applications, water is mined from surface water bodies and groundwater aquifers, or it is a by-product of mine dewatering processes. Open pit and underground mining operations commonly extend below the regional water table and require dewatering during mining. At some stage of the mining operation, unwanted or used water needs to be disposed of constantly. At modern mine sites, water is collected and discharged to settling ponds and tailings dams, whereas at historic mine sites, uncontrolled discharge of mine water commonly occurs from adits and shafts into the environment. Under these circumstances, water comes in contact with exposed minerals, may dissolve them partially or totally, and as a result it carries dissolved and particulate matter. When such laden waters reach receiving water bodies, such as lakes, streams or aquifers, they can cause undesirable turbidity and sedimentation, may alter their temperature and chemical composition, and may ultimately result in the occurrence of toxic effects on plants and animals (Lottermoser, 2007). In the United States, it has been estimated that 19,300 km of streams have been seriously damaged by mine effluents from abandoned coal and metal mines (Kleinmann, 1989).

Although in many cases, surface and groundwater quality becomes affected in natural, mineralized areas, mining activities can enhance these effects. Mining operations that can result in different impacts on water quality include drilling, spills, leaks, heap leaching of rock piles and open cuts, among others. "*Mine water*" originates as ground or meteoric water which undergoes compositional modifications due to mineral-water reactions at mine sites. This term includes any water which runs off or flows through a portion of a mine site and had contact with any of the mine workings, including surface water and subsurface groundwater (Morin and Hutt, 1997). Water of poor quality requires remediation as its uncontrolled discharge, flow, drainage or seepage from the mine site may be associated with the release of heat, suspended solids, bases, acids, and dissolved solids including process chemicals, metals, metalloids, radioactive substances or salts. Such a release could result in a pronounced negative impact on the environment surrounding the mine site (Lottermoser, 2007). The type of water contamination is highly dependent on the nature of the mineral deposit, the reactants used in the mineral processing activities, and the type of mining operation: i) Industrial minerals and rocks: the extraction and production of these materials, such as clay, sand, gravel, etc., generally do not produce any waste, as almost the entire mined material is extracted and processed; however, crushing, washing and sizing of rock aggregates generate minor amounts of unwanted fine-grained slimes and dust particles; therefore, considering the frequent absence of reactive mineral phases (with the exception of Ca and Mg carbonates and chloride and sulphate in salt mines), as well as chemical treatments, the most common attribute affecting water bodies quality are suspended solids, which are easily removable; occasionally, detectable levels of nitrogen compounds have been reported, mainly coming from the explosives used in the blasting; ii) Metallic mining: Only a very small valuable component is extracted from metalliferous ores during processing and metallurgical extraction; the great majority of the total mined material is gangue which is generally rejected as unwanted processing and metallurgical waste; water contacting open-cut and underground mining of base metal sulphide deposits, and related wastes, can exhibit very low pH values (*Acid Mine Drainage*, AMD) and high concentrations of many chemicals leached from the ore and host rock (sulphate, Fe, Pb, Hg, As, Ba, etc.), as well as substances used to extract metals from the ore (flotation reagents such as phenols and xantates and leaching compounds, like cyanide or thyocianate); iii) Fossil fuel mining: Coal mining commonly results in low pH and high sulphate and metal content in water, mainly due to oxidation of pyrite present in the coal bed. The most outstanding characteristics of waters associated with oil and gas mining are the presence of brines (high salinity), CH_4 and other hydrocarbons, as well as H_2S. In general, coal mining and processing generate the largest quantity of waste followed by non-ferrous and ferrous ores (Carretero and Pozo, 2007; Lottermoser, 2007).The worst example of poor mine water quality is AMD, which originates from the autocatalytic oxidation of sulphide minerals, whereby low pH mine water is formed. The term *acid rock drainage* (ARD) highlights the fact that there are naturally outcropping sulphide orebodies and sulphide-rich rocks, which actively weather, oxidize, and cause acidic springs and streams (Furniss et al., 1999; Posey et al., 2000; Munk et al., 2002). The acid streams draining such ores and rocks can contain high levels of metals and metalloids in excess of water quality standards and may result in toxic effects on aquatic biota. Although ARD occurs naturally and unrelated to mining activities, such natural situations are rare compared to those where mining has been directly responsible for the acidification of waters. AMD impacts more frequently on groundwater quality than on the surface drainage from a mine (Bennett and Ritchie, 1993). It is most severe after sulphide oxidation begins, but it can last from decades (Lambert et al., 2004; Demchak et al., 2004, Lottermoser, 2007) to thousands of years. That is the case of the Rio Tinto mining district, in SW Spain, where the Rio Tinto ores have been mined since the Copper Age, 5,000 years ago. These mining activities have left uncountable waste rock heaps, ore stockpiles, tailings dumps, slag deposits, and settling ponds, most of which do not support any vegetation. The exploitation of sulphidic ores has created a unique "mining landscape" and caused massive AMD flowing into the Tinto River, which drains the district and is one of the most polluted surface courses in the world. This river remains strongly acidic (pH <3) for its entire length (90 km) and carries excessive dissolved sulphate, metal and metalloid loads from the headwaters to its estuary (Nelson and Lamothe, 1993; Leblanc et al., 2000; Braungardt et al., 2003). However, according to Ariza (1998) and others, mining may not be entirely responsible for this impact on the Rio Tinto watershed, since weathering of outcropping sulphide ores could have generated ARD prior to the beginning of mining operations and the water conditions today could be a combination of natural ARD and mining induced AMD (Lottermoser, 2007). The water of the affected rivers exceed the drinking water standards for Cu, Pb, and Zn by several orders of magnitude and although it is not documented to be utilized as a drinking water source for humans, it serves as a water source for livestock that provide dairy products and meat for human consumption (Nelson and Lamothe, 1993).

The exploitation of mineral resources results in the production of large volumes of waste rocks as they have to be removed to access the resource, being usually placed in large heaps on the mining lease. Mining wastes either do not contain ore minerals, industrial minerals, metals, coal or mineral fuels, or their concentration is subeconomic, but every mine has different criteria for separating mining waste from ore. Thus, they are heterogeneous geological materials, with particle sizes ranging from clay size particles to boulder size fragments. Physical and chemical characteristics of mining wastes vary according to their mineralogy and geochemistry, type of mining equipment, particle size of the mined material, and moisture content (Hassinger, 1997; Lottermoser, 2007). *Processing wastes* include tailings, sludges and waste water from mineral processing, coal washing, and mineral fuel processing. They can be used for backfilling mine workings or for reclamation and rehabilitation of mined areas, but usually, they are disposed being dumped at the surface next to the mine workings. Extractive metallurgy results in the production of various waste products (*metallurgical wastes*) including atmospheric emissions, flue dust, slag, roasting products, waste water, and leached ore, which are deemed too poor to be further treated. At many historical metal mines, the ore or the concentrate were smelted or roasted in order to remove sulphur and to produce a purer marketable product. Consequently, roasted ore, slag, ash, and flue dust are frequently found at historical metal mine sites. More than 4,700 Mt of mining waste and 1,200 Mt of tailings are stored all over the European Union (BRGM, 2000). The global quantity of non-fuel mineral commodities removed from the Earth's crust each year by mining is of the order of 3,700 Mt and approximately 15,000 to 20,000 Mt of solid mine wastes are being produced annually around the world (Lottermoser, 2007). Much of the environmental impacts of mining are associated with the release of harmful elements from mine wastes. Mine wastes pose a problem not just because of their sheer volume and aerial extent, but because some of them may impact on local ecosystems. If uncontrolled disposal of mine wastes occurs, it can be associated with increased turbidity in receiving waters or with the release of significant quantities of potentially harmful elements, acidity or radioactivity. Anthropogenic inputs of metals and metalloids to atmospheric, terrestrial and aquatic ecosystems as a result of mining have been estimated to be at several million kilograms per year (Nriagu and Pacyna, 1988; Smith and Huyck, 1999; Lottermoser, 2007).

The long-term off-site release of contaminants is particularly possible from mining and related processing or metallurgical wastes. As a result, the operations of the mining industry have been criticized by the conservation lobby for some time. An understanding of the long-term release of contaminants requires a solid knowledge of the factors that control such discharge. The major factors that influence contaminant release from a specific mine site or waste repository, are the geology of the mined resource, climate and topography, and the applied mining and mineral processing activities (Lottermoser, 2007). Abandoned pits and mine shafts are sometimes used for water supply after mine closure. Additionally, leachates from mining areas and/or wastes disposed in spoil heaps, and affected groundwater are frequently received by water bodies which are used for drinking water production. Several study cases, describing in detail the factors why different Spanish mining sites affect water bodies used for drinking purposes, are presented in this paper.

CASE STUDY: ARSENIC MINE IN RIAÑO (LEÓN, SPAIN)

A water reservoir constructed for production of hydraulic energy, sport fishing activities and drinking water in the Riaño valley (León, N Spain) receives water from the upper catchments of the Esla, Yuso and Orza Rivers, but also leachates from polluted soils and spoil heaps from a Santa Águeda Mine site, located at a topographic level of 1,200 m, where small-scale As mining and smelting operations were developed in the first half of the 20th century. The first mining activities at the site consisted on a deep ditch to exploit the As-vein as an open pit. The underground works started from this ditch exploiting the vein by the chambers and pillars method. The most significant extraction period of the mine was between 1940 and 1960. A smelting furnace was constructed at the beginning of the 50s to treat the mineral on site. Currently, the legacy of these extractive activities remains as a network of old galleries and chambers (through which water flows) two spoil heaps with widely ranging mineralogical composition, polluted soils and remainders of the smelting plant. The main galleries of the mine have collapsed and the mine, except for the upper gallery, is flooded (Álvarez et al., 2006).

Concerning climate, it can be defined as continental in León. Annual average temperature is 11°C, ranging from 5 to 16°C, over the last 30 years with high thermal oscillations, from -1 to 22°C in winter and summer, respectively. The studied area, located at elevated height is one of the coldest and wettest of the region, with an average yearly rainfall of 1,500 mm. The annual relative average humidity is about 66%. From a geological-mineralogical point of view, the exploited epithermal As-mineralization occurs as veins or stocks in Westphalian silicified sediments, associated with a fault breccia. The ore is constituted by arsenopyrite, As-rich pyrite, marcasite, stibnite, bravoite, realgar, scorodite, Sb-ochres, and goethite, within a gangue of quartz and carbonate. The mineralization appears as a fine grained dissemination in a quartz-carbonated matrix constituted by jasperoid or millonitic to cataclastic fault breccias. The ore has a clear structural control in a strongly tectonized area and with a high level of secondary permeability. The mineralogical study of samples from the spoil heaps shows the abundant presence of arsenopyrite and pyrite in an early alteration state . In spite of the fact that the alteration grade of the ore is low, the geochemical processes developed as a consequence of sulphide minerals weathering in the spoil heaps cause the oxidation of As-bearing sulphides, forming a complex assemblage of secondary minerals, mainly composed of arsenates. The stability of these secondary As-bearing phases depends on parameters such as pH, cristalinity and molar Fe/As ratios (Krause and Ettel, 1989), but also on microbial activity (Pongratz 1998). In these spoil heaps, the weathering and oxidation of pyrite and arsenopyrite in low-grade ore and enclosing rocks have resulted in the formation of secondary minerals such as scorodite and goethite, and this is accompanied by the release of As to runoff waters. It is known that scorodite has a limited solubility under most pH/Eh conditions, and its formation constrains the generation of As-rich leachates in many case-study sites (Williams, 2001). The solubility of scorodite has been discussed by Dove and Rimstidt (1985), Robbins (1987), and Nordstrom and Parks (1987). According to Kavanagh et al.,(1997), from studies accomplished on mine wastes of the Tamar valley (SW England) the proportion of water extractable As in those wastes range from 0.02 to 1.2%. Arsenopyrite oxidises rapidly in the low-pH conditions that prevail in some spoil heap impoundments (Richardson and Vaugham, 1989) and the As released can remain in solution (as it occurs in the surface As-rich water downstream the mine) or be trapped by on secondary phases and colloids or by coprecipitation reactions.

Arsenic background concentrations in soils of the world range from 1 to 40 mg·kg^{-1}, with mean values often around 5 mg·kg^{-1} (Bowen 1979; Beyer and Cromartie, 1987). The local As geochemical background in soil at the mine site (150 mg·kg^{-1}) is higher than reference values for unpolluted soils in the world, but this naturally elevated As level in soils may be associated with geological substrata such as sulphide ores. Arsenic concentrations may exceed 27,000 mg·kg^{-1} for soils contaminated with mine or smelter wastes (U.S.EPA, 1982). In this case, a maximum value of 23,800 mg·kg^{-1} As has been found (mean concentration: 4,600 mg·kg^{-1} As). An important factor controlling the mobility and the concentration of As in spoil heap materials and soils is pH, because it affects all adsorption mechanisms and the complexing of metals in the soil solution. The sampled soils show a great variation of pH, ranging from acidic (4.1) to alkaline (8.0). These values are influenced by the dominant lithology (limestones), so the Ca concentration in soils appears to be highly correlated to the pH. At the studied site, soils show a high correlation between As and Fe ($r = 0.88$) and this is in agreement with the well known adsorption of As by iron hydroxides surfaces. Mobility of As in the environment is strongly affected by adsorption phenomena and there is a large body of literature addressing the mechanisms of As adsorption on mineral surfaces of Fe oxides and hydroxides (Gieré et al.,2003). The adsorption affinity is higher for arsenate under lower pH conditions and for arsenite under higher pH conditions (Stollenwerk, 2003). As the arsenate is less mobile that the arsenite, special attention should be put in those alkaline soils in the site, which are closer to the Riaño reservoir water.

Arsenic is outstanding among the most important water pollutants, and human activity has largely contributed, mainly through mining and metallurgy, to its dispersal in the environment (North et al., 1997). The toxic effects of As have long been known both to humans and other animals, and affect almost every major bodily function and metabolic pathway (Squibb and Fowler, 1983; IPCS, 2001). Important cases of As contamination in groundwater leading to human As poisoning are widely reported in the literature (Chakraborti et al., 1996). Average As content in groundwater is 1-2 µg·l^{-1}, excepting areas with volcanic rocks and sulphide mineral deposits. In some mining areas, As concentrations of up to 48 mg·l^{-1} have been reported (Page, 1981; Welch et al., 1988). Arsenic is widely distributed in surface freshwaters, and concentrations in rivers and lakes in the world are generally below 10 µg·l^{-1}, although individual samples may range up to 1 mg·l^{-1} (Page 1981; Smith et al.,1987; Welch et al.,1988). Water flowing through tailings impoundments, can reach As contents ranging from 300 to 100,000 µg·l^{-1} (Loredo et al., 2004). Based on information from EPA´s CERCLIS 3 database, As is the second most common contaminant of concern (COC) cited in Records of Decision (RODs) for sites on the Superfund National priorities List (NPL) (U.S.EPA, 2002), so much research has been recently carried out on As, as a response to the growing evidence of the element's detrimental health effects, and to reductions in regulatory limits for this element. In 2001, EPA published a revised Maximum Contaminant Level (MCL) for As in drinking water that requires public water suppliers to maintain As concentrations at or below 0.01 mg·l^{-1} by 2006 (the former MCL was 0.05 mg·l^{-1}). The current Spanish legislation on mining/industrial effluents limits the total As content to 0.5 mg·l^{-1}, and the As limit in waters destined for the production of drinking water has been reduced from 0.05 to 0.01 mg·l^{-1} (BOE 1985; RD 140/2003). This value is exceeded by spoil heaps and soils leachates in Santa Águeda mine site, which form sporadic courses flowing to the dam, and whose average As concentration is 0.9 $mg \cdot l^{-1}$ (Alvarez et al., 2006). Therefore, pollution migration in the main form of soluble As occurs from the mine site towards the Riaño reservoir, which is located at 300 m downstream of the mining works, and where fishing practices are currently developed in the reservoir. However, if surface water downstream of the mine works shows a significant pollution by oxidation of arsenopyrite, the absence of major deposits of this mineral or their oxidation products diminish the potential environmental risk.

The application of a computing based risk assessment software (RBCA Tool Kit for Chemical Releases) to the site, concluded that although As concentrations in soils and surface waters are higher than current regulatory levels, the absence of exposure (the site is not inhabited and difficult to access) results in a negligible risk for human health. In relation to water, the contribution of polluted water (spoil heap and soil leachates) to the dam does not pose a significant environmental risk because of its dilution with clean runoff water from the catchment. Quantitatively, the total contribution of runoff water to the reservoir in the catchment is 14,000 times higher than the contribution to the dam of the runoff water flowing through the site of the abandoned mine (mine catchment), since the area of the catchment affected by the old mining works is only 43 ha, in contrast to the 60,400 ha of the reservoir catchment. The average surface flow from the mined area catchment to the dam is $241 \cdot 10^6$ litres/year, which represents 0.07 % of the total water flow to the Riaño reservoir. This fact, together with the moderate As concentrations, in general, of leachates and mine drainage does not involve a real significant contribution of the polluted water entering in the reservoir. This was proven when the As content in the reservoir water was found to be below the detection limit of the analytical technique employed, and the quality of the water in the reservoir has been classified as 'excellent'. The influence of other metals analysed are not considered here, as they were not found in high concentrations in the water and only two of the soil samples (located where the ore used to be stored) showed elevated Hg contents.

Case Study: Copper Mine in Riosa (Asturias, Spain)

The Cu-Co-Ni Texeo mine is located in Riosa (Asturias, NW Spain), in a steep scarcely populated area on the northern slope of the Aramo Mountains. It has been the most important source of Cu in NW Spain since Roman times, with intermittent periods of mining activity, and maximum annual productions of 370 tons of Cu metal between 1947 and 1956. The main mining operations developed in the area consisted of three pit heads located at different heights from 715 to 1,207 m above sea level, and transversals and galleries with a total length of about 1,750 m. The mining site extends about 300,000 m^2. Four uncovered spoil heaps of different age and size are located on the slope of the mountain, in the immediate vicinity of the mine works. The materials in these spoil heaps range from coarse-grained waste rocks and low grade ore to fine-grained particles from processing and metallurgy, with a total volume estimated at 40,000 m^3 and average concentrations of 9,263 $mg \cdot kg^{-1}$ Cu, 1,100 $mg \cdot kg^{-1}$ As, 549 $mg \cdot kg^{-1}$ Co, and 840 $mg \cdot kg^{-1}$ Ni. The release of toxic elements from the abandoned works, facilities and waste piles has been posing a threat to the environment since the cessation of the mining activities (Loredo et al., 2007).

In contrast to other regions of Spain, Asturias has a humid and temperate climate characterized by abundant rainfall during a large part of the year. Annual average temperature is 13ºC, ranging from 8 to 17ºC, over the last twenty years. The annual relative average humidity is about 78%. Average yearly rainfall is 966 mm and evapotranspiration calculated by Thorntwaite expression is 691 mm/year. In consequence, the average annual effective precipitation is about of 275 mm/year. The "one hundred-year 1-hour rainfall" is 40 mmh^{-1}. The combined factors of temperature and rainfall, largely defined in terms of a precipitation/evaporation ratio, affect the amount of water percolating into the soils and mining wastes, where vertical infiltration can be expected to be dominant.

The local geology corresponds to the Aramo Unit, which comprises a sequence of Devonian and Carboniferous sediments that include shales, sandstones, limestones, and dolostones. The mineralization is linked to the intersection of the Aramo Fault with the Aramo Thrust front (Julivert, 1971). An important dolomitisation and minor silicification was developed around the Aramo Fault, concerning the ore body. The host rocks of the mineralization are limestones of Namurian age. The ore deposit consists of mineralized veins with an average thickness of 25 cm and argillaceous infilled zones within the karstic cavities. The mineralization genetic model is the epithermal type, related to the circulation of ore brines at temperatures of about 90-130ºC along the Late-Hercynian faults (Paniagua et al., 1987). The paragenetic study of the mineralization reveals a great number of ore minerals (up to 29 different mineral phases) in a dolomite-quartz and later calcite gangue (Paniagua et al., 1987). The mineral assemblages are present as three successive primary stages, with a well developed supergene stage. The early association is formed by pyrite and bravoite, with Co-Ni arsenides and sulphoarsenides and later marcasite. During the intermediate primary stage, major tennantite and sphalerite were deposited as idiomorphic crystals in dolomite and/or quartz matrix; this assemblage was replaced by chalcopyrite and talnakite. The last primary association is formed by Cu-Fe sulphides (chalcopyrite, talnakite, bornite). An extensive supergene alteration of the primary ore deposit gave rise to an oxidation sequence, including numerous secondary minerals such as native copper, cuprite, tenorite, azurite, malachite, and erythrite. First mining works took place on the oxidised zone of the ore, which is well developed in the site of the mine. Subsequently, a well developed cementation zone has been exploited until the closure of the mine at the end of the 1950´s. The average grade of mined ore was about 12% Cu, 2-3% Co, and 2-3% Ni (Gómez-Landeta and Solans, 1981; Gutiérrez-Claverol and Luque, 1993).

The mining area is crossed by tributaries of the Llamo River, which drains a total catchment of 9.7 km^2, being only 0.75 km^2 directly affected by the mining works. The rainwater that does not evaporate nor flows superficially is infiltrated entering the mine by natural downward movement, and then, by lateral movement through fissures, faults, and worked voids. The enclosing rocks of the mineralization and the geological substrate at the site are permeable carboniferous limestones which constitute an important aquifer at a regional level. High "secondary" porosity associated to the underground mine works and surface spoil heaps favours the infiltration of water. According to the mentioned climate data in the affected area, and considering an infiltration coefficient of 0.7 for a spoil heap without vegetal cover, the infiltrated water entering in contact with the wastes stocked in a total area of 4,000 m^2 is about 770 $m^3 \cdot year^{-1}$. Copper and other metals tend to enter streams via leaching and erosion of mine wastes and polluted soils (Loredo et al., 2007).

Metal concentrations in the soils are highly above the local background, reaching up to 9,921 mg·kg^{-1} Cu, 1,373 mg·kg^{-1} As, 685 mg·kg^{-1} Co and 1,040 mg·kg^{-1} Ni, among others. Copper concentrations in stream sediments downstream of the mine works reach values up to 240 mg·kg^{-1}. Concentrations up to 0.16 mg·l^{-1} As and 1.87 mg·l^{-1} Cu were found in surface waters downstream of the mining and metallurgical area in the rainy period (spring), when abundant waste leaching and runoff over the mineralized and mined areas, occurs, although leachate contribution is reduced by dilution. During dry periods, in contrast, these leachates are not produced, so the water quality of the receiving bodies improves, having lower metal content and electrical conductivity, as well as slightly higher pH. All the sampled water is alkaline, especially groundwater, as a result of the carbonate buffering. Groundwater sampled in the wet season at a spring in Llamo village contained more than 4.9 mg·l^{-1} Cu. These values are above the National standard for drinking water (2 mg·l^{-1} Cu and 0.01 mg·l^{-1} As). It can be stated that, since there is an important catchment point for drinking water supply in a spring downstream of the mining works, the presence of high As and Cu concentrations in some periods of the year can pose a potential health risk (Loredo et al., 2007).

Computer based risk assessment for the site gives a carcinogenic risk associated to the presence of As in surface waters and soils, and a health risk for long exposures, so trigger levels of Cu, Co, Pb and As are high enough to warrant further investigation. In this context, the most hazardous pollution source is that constituted by the abandoned spoil heaps that still contain large amounts of potentially leachable toxic metals and metalloids.

CASE STUDY: HG MINING DISTRICT IN CAUDAL RIVER CATCHMENT (ASTURIAS, SPAIN)

Asturias has a long mining tradition and Hg mining was an important industry -Asturias was an important Hg producer on a world scale-, until the beginning of the 1970s (SADEI, 1993), when mercury prices began to decline due to falling demand as a consequence of emerging toxicological problems associated with the use and production of Hg (Crounse et al., 1983; Callister and Winfrey, 1986; Francis, 1994). Falling Hg prices led ultimately to the total interruption of the extractive activity in many up till then important Hg mining districts in the world (Ferrara, 1999). In Asturias, this crisis brought about to not only the cessation of some important mining projects, but the successive closure between 1973 and 1974 of all the active mines ((Loredo et al., 2004). These mines are distributed all over the region, but the most important ones occur within two districts (Mieres and Pola de Lena), which are included in the Caudal River catchment. Mercury mining represents a major environment threat because of the high toxicity, persistence, biogeochemical cycling, accumulation and concentration of this metal in the food chain (Ditri and Ditri, 1977; Ferrara et al., 1991; Palinkas et al., 1995; Biester et al., 1996; Miklavcic, 1999; Turner and Southworth, 1999). The potential environmental risk of these mine sites is enhanced by the presence of As in the mined ore, which is also abundant in mine wastes. These derelict Hg mines have been abandoned without remediation, and the legacy of these mine works remain currently in the form of significant quantities of mine and metallurgical wastes stocked at surface in spoil heaps of different age and size, and old galleries and transversals, which are potential pollution pathways for the spreading of As to the environment (Loredo, 2000; Loredo et al.,

2001; Loredo et al., 2002; Loredo et al., 2003). Mine drainage and leachates from spoil heaps are incorporated to streams tributaries of Caudal River, –a river with salmonids, which are fished and eaten by the local population (Loredo et al., 2004).

These Asturian Hg deposits are located in Precambrian to Carboniferous formations, mostly hosted by Carboniferous sediments, but also in conglomeratic horizons or siliceous breccias in sandstones. There are also mineralizations impregnating fractured lutitic carbonaceous sequences. These mineralizations usually show clear lithologic and tectonic controls; they are epithermal, epigenetic deposits and there is a clear spatial association with late-hercynian fractures. Although Hg is present in the form of cinnabar, metacinnabar and native Hg are also occasionally found. Arsenic is quite abundant in these ores in the form of sulphides (As-rich pyrite), sulphoarsenides (arsenopyrite, realgar and orpiment), and supergenic minerals (scorodite). Electronic microprobe analyses of pyrites confirm their high content of As (up to 10%). Pyrites are generally in an advanced state of oxidation and framboidal pyrite is quite frequent; it is composed by fine particles with ample reactive surface -the more reactive type of pyrite-, with high acid generating potential. Other primary metallic minerals, which are present in the paragenesis of the ore deposits, are pyrite, melnikovite, sphalerite, marcasite, chalcopyrite, arsenopyrite, galena, and stibnite. Smithsonite, hemimorphite, cerusite, goethite, malachite, jarosite, melanterite and gypsum can be present as secondary minerals. Electronic microprobe analyses of iron oxides (goethite) show As contents ranging from 0.6% to 9.4%. The gangue constituents are quartz, carbonates (calcite, dolomite and ankerite) and argillaceous minerals, such as kaolinite and dickite (Luque et al., 1989; Luque, 1992; Loredo et al., 1998).

From a hydrogeological point of view, the substrate of the mined areas consists of an alternation of limestones, sandstones and shales. Whereas the shales can be considered impermeable, limestones and sandstones are permeable formations, and they constitute small aquifers. The springs in the area are scarce and most of them are associated with the limestone levels. Additionally, flooded underground mining works and spoil heaps may be considered as a pseudokarstic aquifer, where flow is essentially characterized by multiple flow paths, extreme ranges of hydraulic conductivity, and a high degree of unpredictability (Loredo et al., 2004).

At the most important mine sites ("La Peña - El Terronal" and "La Soterraña"), smelting plants were in operation for years and significant volumes of metallurgical wastes remain on site in spoil heaps, together with rejected material from the underground mining operations, i.e. enclosing rocks containing low metal grades. These spoil heaps have variable As concentrations ranging from 800 $mg \cdot kg^{-1}$ to 6.2% (Loredo et al., 2004). Excepting one of the biggest spoil heaps in the area, containing 300,000 tons, which has been recently isolated in an on-site safety deposit, waste impoundments abandoned 30 years ago have been subjected to the erosion and the washing effects of rainfall for years, intense chemical and mechanical dispersions are occurring, and as a consequence the potential mobility of As is high. Airborne dispersion of the finest particles of wastes is also taking place. The potential of these abandoned mining wastes to pollute the environment is enhanced by the abrupt topography of the site and their proximity to streams. In some cases the spoil heaps are located very close to urban areas, increasing the risk of exposure to water and airborne contaminants (Loredo et al., 2002; Loredo et al., 2003).

As a consequence of the weathering of As-rich minerals, As contents are high in mine effluents and spoil heap leachates, which finally end in Caudal River. In the spoil heaps, waste materials are subjected to the action of a humid environment, and in consequence the instability of sulphides and sulphoarsenides leads to the production of acid and metal rich leachates which are incorporated into surface water or infiltrate through the land reaching groundwater. Mine effluents and spoil heap leachates show As concentrations that can range from 1.4 to 9.2 $mg \cdot l^{-1}$ and high concentrations of As have been detected also in stream sediments, reaching up to 3.2%. The circulation of water through the underground mining works and spoil heaps promotes in some cases the formation of acid mine drainage.

Baseline concentrations of As in world river waters can range from 0.8 (Ferguson and Gavis, 1972) to 2 $\mu g\ l^{-1}$ (Smedley and Kinniburgh, 2002), but As concentrations in waters affected by contamination related to mining activity can be 5 to 8,000 times higher than those typical concentrations, whereas mine drainage can reach up to 850,000 $\mu g \cdot l^{-1}$ (Azcue and Nriagu, 1995; Williams et al., 1996; Plumlee et al., 1999). These concentrations can vary depending on the composition of the surface recharge and the nature of the bedrock lithology. Although the local hydrochemical background of As is naturally high in the Caudal River catchment (due to the presence of As-rich iron sulphides associated to the local rocks), an As enrichment in surface and groundwater is exacerbated by the presence of Hg mines in state of abandonment. In some particular tributaries of Caudal River, significant increments of As concentrations occur as a result of pollution from leachates of mine/metallurgical waste spoil heaps and mine water discharges. Acid mine drainage, with pH ranging from 1.6 to 2.1, 9 to 18 $mg \cdot l^{-1}$ As and high concentrations in metals and sulphates is generated at one of the mining sites, and it is incorporated to San Juan River (tributary of Caudal River) where the As contents are diluted to 5 - 8 $\mu g \cdot l^{-1}$. Surface water reach As concentrations up to 13 $mg \cdot l^{-1}$ downstream "La Soterraña" Hg site. In the San Tirso River, a direct tributary of the Caudal River, which flows through the historical mining site of "La Peña-El Terronal", As concentrations increase from values lower than 0.005 $mg \cdot l^{-1}$ in samples upstream the mine works, to values between 0.9 to 14 $mg \cdot l^{-1}$ in samples collected downstream. It is interesting to note that the As concentration is higher during the low flows of dry periods in summer (4 - 14 $mg \cdot l^{-1}$) than during the high flows in winter (0.9 - 1.1 $mg \cdot l^{-1}$), due to dilution effects. Average As concentrations in the Caudal River after its confluence with San Tirso River is 1.6 $mg \cdot l^{-1}$. The baseline concentrations of As in surface water of Caudal River catchment are about 9 $\mu g \cdot l^{-1}$ (Loredo et al., 2004). Hg is not usually found in the sampled water, due to its low solubility.

It is known that As(V) is generally the dominant species in river waters, although concentrations and relative proportions of As(V) and As(III) vary according to changes in input sources, redox conditions and biological activity (Seyler and Martin, 1990; Pettine et al., 1992; Smedley and Kinniburgh, 2002). Under the aerobic and acidic to near-neutral conditions, typical of many natural environments, As is very strongly adsorbed by oxide minerals. When pH increases, especially above 8.5 units, As desorbs from the oxide surfaces, thereby increasing the concentration of As in solution. The As background level in sediments of streams of the Caudal River catchment (deduced from analysis of samples collected in different areas upstream of mining works) is 186 $mg \cdot kg^{-1}$. In contrast, the concentrations of As in sediments downstream from different mining sites range from 190 to 41,400 $mg \cdot kg^{-1}$. Very high concentrations of Hg (320 $mg \cdot kg^{-1}$), Fe (12%), and S (4%) have been also found. Arsenic contents in sediments collected downstream from a site can be reduced by 92% at a distance of 300 m. The decrease of As content with distance from the source can be related to its predominant dispersion in solution and its sorption by iron hydroxides –mainly goethite-, derived from the sulphides weathering. The coprecipitation of As and Fe and the adsorption of the As onto the Fe oxides, is enhanced in oxidising environments, within neutral to slightly acidic conditions, as in this case.

A risk assessment at the mining sites revealed that the target carcinogenic risk was exceeded for As. However, the As input from the streams draining the Hg mining sites to the Caudal River is disguised by the increase of water flow and also the retention in sediments, although this element can be easily desorbed from them if geochemical conditions change. There are dams used for drinking water supply in this river, but they are located upstream of the mining works. However, there are catchment points that collect water from the alluvium aquifer of the river, downstream of the mining district, that are used for drinking purposes. Arsenic contents in groundwater downstream of mine works in the Caudal River catchment range from <0.001 to 6 $mg \cdot l^{-1}$ (in direct contact with mining wastes). Groundwater sampled in wells used for irrigation and as drinking supply for animals, show As contents ranging from 1.2 to 15 $\mu g \cdot l^{-1}$.

Case Study: Gold Mine in Salas (Asturias, Spain)

Carlés underground and open pit gold mine, located in Salas (Asturias), was exploited between 2002 an 2006, although exploration and preparation works had been started 14 years before. The ore, rich in Au, Cu and As, is related to a skarn type mineralization, which was generated by contact metamorphism over the Rañeces Formation materials (limestones, dolomites and shales). This is the result of the intrusion of Carlés granodiorite into Silurian-Devonian materials, as part of the incursion of small intrusive rocks stocks, occurred during the Hercynian orogeny, in this area of the Cantabrian Zone (Martin-Izard, 2000; Arquer et al., 2007).

From a hydrogeological point of view, granodiorites and skarn materials can be considered as originally impermeable, although secondary permeability can be developed by fracturing and limited karstification in marbled limestones and also in unmetamorphosed carbonates. Additionally, interbedded detrital (argillaceous) beds have low hydraulic conductivity. The aquifers in the area, giving rise to small springs with high flow only in wet periods, correspond to carbonate levels of reduced thickness and generally isolated, although they can be connected by fractures. Mining galleries and exploration drills intersecting fractures in these materials facilitate their drainage. This fact is more evident near the igneous-metamorphic contact, where the fracture network is denser and where the -fishing-local Narcea River flows. Water sampled from the carbonate levels has in some cases high electrical conductivity and appreciable As contents, ranging from 0.4 to 1 $mg \cdot l^{-1}$, probably related to superficial weathered skarn materials. However, these concentrations gradually decrease inside the mine and downward to the river, possibly favored by the infiltration of river water into the fracturing network, as most mining voids are below the river water level (Arquer et al., 2007).

In this case, against what could have been expected, mining activity is not responsible for the presence of As in waters with potential drinking water use. On the contrary, water circulating through preferential paths (fractures and fissures) in igneous and metamorphic materials and host rocks in the area is naturally enriched in As, especially after leaching the alteration zone, and this phenomenon occurred before the mine started its activity.

FINAL COMMENTS

Environmental contamination as a result of mining is not new to the industrialized world, as mining has been closely linked to human devepoment for thousands of years. In historic times, mine wastes were released into the environment, causing in some cases contamination on a local or regional scale (Agrícola, 1556). Until then, environmental impacts of mining and mineral processing were poorly understood, not regulated, or viewed as secondary in importance to resource extraction and profit maximization. The introduction of environmental regulations in the 20th century made the mining industry more accountable and enforced environmental protection. In many industrialized countries, the majority of mining environmental problems are legacies from the past, and improper disposal practices of wastes only continue to occur in nations where unregulated mining and uncontrolled release of mine wastes cause environmental harm. Technologies can be used to prevent pollution and to ensure that the environmental performance of mining companies is adequate, operating in a sustainable manner. At present the environmental protection aspects are essential in determining the viability of a modern mining operation (Maxwell and Govindarajalu, 1999; Lottermoser, 2007).

Monitoring and progressive rehabilitation of mine sites is an integral part of mine planning, development and final closure. These include modelling future water quality in underground workings and aquifers. Similar to open pits, the closure of an underground mine leads to mine flooding. Deep saline ground waters may rise to shallow levels or contaminants may be leached from the workings into shallow aquifers. Therefore, an assessment of the potential contamination of shallow groundwater is needed (Lottermoser, 2007). In the rehabilitation context, open pits may be used as water storage facilities, wetland/wildlife habitats, aquaculture ponds for fish and crustaceans, recreational lakes, heritage sites, engineered solar ponds to capture heat for electricity generation, heating, or desalinisation and distillation of water, or as repositories for mining, industrial or domestic waste. Additionally, mine waters can be converted into drinking water (Schwartz and Ploethner 2000; Smit 2000; Varnell et al.,2004); and also be used for heating or cooling purposes (Banks et al.,2004; Watzlaf and Ackman 2006). For example, in the mining valleys of the Asturian Central Coal Basin (NW Spain), about 40 Hm^3 of groundwater are pumped out every year from active and inactive shafts; considering the current prices of conventional fuels, the thermal energy of this mine water (average temperature of 20°C) can be competitively used by means of the heat pump for heating and refrigerating close buildings in the area, reducing comparatively CO_2 emissions (Jardón et al., 2007); additionally, considering the physico-chemical characteristics of the mine water and the proximity of the mines to populated areas in the region, its "exploitation" as water resource for public supply, constitutes a appealing alternative, particularly when the mining voids can be used as a "regulator underground dam", where water can be stored in surplus periods and withdrawn in times of shortage.

Conclusion

Mining operations are responsible for water pollution in some cases, particularly in those old sites which were exploited when the environmental protection aspects were not considered as essential. The release of toxic elements from the abandoned works, facilities and waste piles has been posing a threat to the environment since the cessation of the mining activities in these sites. Mining galleries and shafts constitute artificial aquifers, with unpredictable flows and variable hydraulic conductivity, whose quality may be altered by contact with exposed minerals. This affected groundwater, as well as leachates from polluted soils in mining areas, and especially wastes disposed in uncovered spoil heaps, are frequently received by water bodies which are used for drinking water production. In this paper, several examples of how abandoned mines can compromise potential uses of affected water have been discussed. On the other hand, there are cases where water becomes "naturally contaminated", by flowing through mineralized areas without the influence of mining activity or being enhanced by it. In all cases, preventive/corrective measures to protect water quality should be undertaken. Nowadays, mining activity is regulated so absolute protection to surface and groundwater is theoretically guaranteed. The quality of mine water has not necessarily to be poor, so it can be recycled for many uses, including drinking supply, becoming then a resource instead of a waste.

References

Agricola, G. (1556). De re metallica. Translated by Hoover HC, Hoover LH (1950). Dover Publications, New York.

Álvarez, R., Ordóñez, A., Loredo, J. (2006). Geochemical assessment of an arsenic mine adjacent to a water reservoir (León, Spain). Environ. Geol., 50, 873–884.

Ariza, L.M. (1998). River of vitriol. *Sci. Amer.*, 279(3),15–18.

Arquer, F., Meléndez, M., Nuño, C., Rodríguez, M.L. (2007). Aspectos hidrogeológicos e hidroquímicos de las aguas subterráneas de la mina de oro de Carlés (Asturias). Consejo Superior de Ingenieros de Minas (ed.), Proceedings of the 12th International Congress on Energy and Mineral Resources. Oviedo.

Azcue, J.M., Nriagu, J.O. (1995) Impact of abandoned mine tailings on the arsenic concentrations in Moira Lake, Ontario, *J. Geochem. Explor.*, 52, 81-89; Amsterdam, Elsevier.

Banks, D., Skarphagen, H., Wiltshire, R., Jessop, C. (2004) .Heat pumps as a tool for energy recovery from mining wastes. In: Gieré R, Stille P (eds) Energy, waste, and the Environment: a geochemical perspective. Geological Society London Special Publications, 236, 499–513.

Bennett, J.W., Ritchie, A.I.M. (1993). Bio-oxidation of pyrite in mine wastes. Mechanisms which govern pollutant generation. In: Biomine'93, International conference and workshop applications of biotechnology to the minerals industry. Australian Mineral Foundation, Glenside, pp 12.1–12.9.

Beyer, W.N., Cromartie, E.J. (1987). A survey of Pb, Cu, Zn, Cd, Cr, As, and Se in earthworms and soil from diverse sites. *Environ. Monit. Assess.,* 8(1), 27-36.

Biester, H., Hess, A., Müller, G. (1996). Mercury phases in soils and sediments in the Idrija mining area. In Report of the Meeting Idrija as a Natural and Anthropogenic Laboratory; Dirija, Idrija Mercury Mine, Idrija, Slovenia.

BOE (Boletín Oficial del Estado) (1985). Ley de aguas 29/1985 [Water Law 29/1985]. BOE nº189, Madrid.

Bowen, H.J.M. (1979). Environmental chemistry of the elements. Academic Press, New York.

Braungardt, C.B., Achterberg, E.P., Elbaz-Poulichet, F., Morley, N.H. (2003). Metal geochemistry in a mine-polluted estuarine system in Spain. *Appl. Geochem.*, 18, 1757–1771.

BRGM (2000). Management of mining, quarrying and ore-processing waste in the European Union. European Commission, Directorate-General Environment, ENV.E.3 – Waste Management (BRGM/RP-50319-FR, http://europa. eu.int/comm/ environment /waste/ mining_report.pdf).

Callister, S.M., Winfrey, M.R. (1986). Microbial methylation of mercury in Upper Wisconsin river sediment, *Water Air Soil Pollut.*, 29, 1873-1887.

Carretero, M.I., Pozo, M. (2007). Mineralogía aplicada; salud y medioambiente. Thomson, Madrid, Spain, 406 p.

Crounse, R.G., Pories, W.J., Bray, J.T., Mauger. R.L. (1983). Geochemistry and man: health and disease, 1. Essential elements. In: Applied Environmental Geochemistry (ed. I. Thornton), 267-308; London, Acad. Press.

Das, D., Samanta, G., Mandal, B.K., Chowdhury, T.R., Chanda, C.R., Chowdhury, P.P., Basu, G.K., Chakraborti, D. (1996). Arsenic in groundwater in six districts of west Bangal, India, Environ. *Geochem. Health*, 18, 5-15.

Demchak, J., Skousen, J., McDonald, L.M. (2004). Longevity of acid discharges from underground mines located above regional water table. *J. Environ. Qual.*, 33,656–668.

Ditri, P.A., Ditri, F.M. (1977). Mercury contamination, New York, John Willey and Sons.

Dove, P.M., Rimstidt, J.D. (1985). The solubility and stability of scorodite, $FeAsO_42H_2O$. American Mineralogist, 70, 838-844.

Ferguson, J.F., Gavis, J. (1972). A review of the arsenic cycle in natural waters, *Water Res.,* 6, 1259-1274.

Ferrara, R. (1999). Mercury mines in Europe: Assessment of emissions and environmental contamination. In: Mercury contaminated sites (ed. R. Ebinghaus et al.), 51-72; Springer-Verlag.

Ferrara, R., Maserti, B.E., Edner, H., Ragnarson, P., Svanberg, S., Wallinder, E. (1991). Mercury in abiotic and biotic compartiments of an area affected by a geological anomaly (Mt. Amiata, Italy). *Water Air Soil Pollut.*, 56, 219-233.

Francis, B.M. (1994). Toxic substances in the environment, 360; Wiley and Sons Inc., New York.

Furniss, G., Hinman, N.W., Doyle, G.A., Runnells, D.D. (1999). Radiocarbon-dated ferricrete provides a record of natural acid rock drainage and paleoclimatic changes. *Environ. Geol* ., 37,102–106.

Gieré, R., Sidenko, N.V., Lazareva, E.V. (2003) .The role of secondary minerals in controlling the migration of arsenic and metals from high-sulfide wastes (Berikul gold mine, Siberia). *Appl. Geochem.,* 18, 1347-1359.

Gómez-Landeta F., Solans, J. (1981). Procesos supergénicos en la mina de cobre del Aramo, Asturias. Bol. *Geol. Min.*, 92, 429-436.

Gutiérrez-Claverol, M., Luque, C. (1993). Recursos del subsuelo de Asturias. Universidad de Oviedo, 392p.

Hassinger, B.W. (1997). Erosion. In: Marcus JJ (ed) Mining environmental handbook: effects of mining on the environment and American environmental controls on mining. Imperial College Press, *London, pp* 136–140.

IPCS (2001). Arsenic and Arsenic Compounds, Environmental Health Criteria, 224, Finland, World Health Organization.

Jardón, S., Pendás, F., Ordóñez, A., Cordero, C., Álvarez. C., Garzón, B. (2007). Aprovechamiento de las aguas de mina en la Cuenca Central Asturiana como recurso hídrico y energético. Consejo Superior de Ingenieros de Minas (ed.), Proceedings of the 12th International Congress on Energy and Mineral Resources. Oviedo.

Julivert, M. (1971).Decollement tectonics in the hercynian Cordillera of Northwest Spain. Am. *J. Sci.*, 270, 1-29.

Kavanagh, P.J., Farago, M.E., Thornton, I., Braman, R.S. (1997). Bioavailability of arsenic in soil and mine wastes of the Tamar valley, SW England. Chem. *Speciation Bioavailability* 9(3), 77-81.

Kleinmann, R.L.P. (1989). Acid mine drainage in the United States: controlling the impact on streams and rivers. In: 4th world congress on the conservation of built and natural environments. University of Toronto, pp 1–10.

Krause, E., Ettel, V.A. (1989). Solubilities and stabilities of ferric arsenate compounds. *Hydrometallurgy,* 22, 311-337.

Lambert, D.C., McDonough, K.M., Dzombak, D.A. (2004). Long-term changes in quality of discharge water from abandoned coal mines in Uniontown Syncline, Fayette County, PA, USA. *Water Res.,* 38, 277–288.

Leblanc, M., Morales, J.A., Borrego, J., Elbaz-Poulichet, F. (2000). 4500-year-old mining pollution in southwestern Spain: long-term implications for modern mining pollution. *Econ. Geol.*, 95, 655–662.

Loredo, J. (2000). Historic unreclaimed mercury mines in Asturias (Northwestern Spain): Environmental approaches. In: Assessing and Managing Mercury form Historic and Current Mining Activities, 175-180; San Francisco. U.S. Environmental Protection Agency. USA.

Loredo, J., Álvarez, R., Ordóñez, A., Bros, T. (2007). Mineralogy and geochemistry of the Texeo Cu-Co mine site (NW Spain): screening tools for environmental assessment. *Environ. Geol.* Published on line. (DOI: 10.1007/s00254-007-1078-y), 12 p.

Loredo, J., Luque, C., García Iglesias, J. (1998). Conditions of formation of mercury deposits from the Cantabrian Zone (Spain). *Bull. Minéral.,* 111, 393-400.

Loredo, J., Ordóñez, A, Pendás, F. (2002). Hydrogeological and geochemical interactions of adjoining mercury and coal mine spoil heaps in the Morgao catchment (Mieres, North-Western Spain). In: Mine water hydrogeology and geochemistry (ed. P. Younger and N.S. Robins), Geol. Soc. *London Special Publications,* 198, 327-337.

Loredo, J., Ordóñez, A., Álvarez, A., García Iglesias, J. (2004). The potential for arsenic mobilisation in the Caudal River catchment, North-West Spain. *Appl. Earth Sci.* (Transactions of the Institution of Mining and Metallurgy, Section B), 113 (1), 65-75.

Loredo, J., Ordóñez, A., Baldo, C., García Iglesias, J. (2003). Arsenic mobilization from waste piles of the El Terronal mine, Asturias, Spain. Geochem. Explor. *Environ. Anal.*, 3, 1-9.

Loredo, J., Ordóñez, A., Gallego, J., Baldo, C., García Iglesias, J. (1999). Geochemical characterisation of mercury mining spoil heaps in the area of Mieres (Asturias, northern Spain). J. Geochem. Explor., 67, 377-390; Ámsterdam, Elsevier.

Loredo, J., Ordóñez, A., Pendás, F. (2001). La escombrera de Morgao como acuífero: Estudio hidrogeoquímico. In: I International Workshop on Investigation, Management and Remediation of Contaminated Aquifers; Alicante. IGME. España.

Loredo, J., Ordóñez, A., Pendás, F. (2002). Soil pollution related to historic mercury mining in northern Spain and treatment technologies, Environ. Sci. Poll. Res, *Spec. Issue,* 3, 79.

Lottermoser, B.G. (2007). Mine Wastes. Characterization, Treatment, Environmental Impacts. 2nd Edition. Springer-Verlag, 304 pp.

Luque, C. (1992). El mercurio en la Cordillera Cantábrica. In: Recursos minerales de España (ed. García Guinea and Martínez Frías), Textos Universitarios nº15, 803-826; Madrid. C.S.I.C.

Luque, C., García Iglesias, J., García Coque, P. (1989). Características geoquímicas de los minerales de mercurio de la Cordillera Cantábrica (NW de España). *Trab. Geol.*, 18, 3-11. Univ. Oviedo.

Martin-Izard, A., Paniagua, A., García-Iglesias, J., Fuertes, M., Boixet, Ll., Maldonado, C., Varela, A. (2000). The Carlés copper–gold–molybdenum skarn (Asturias, Spain): geometry, mineral associations and metasomatic evolution. *J. Geochem. Explor.*, 71, 153–175.

Maxwell, P., Govindarajalu, S. (1999). Do Australian mining companies pay too much? Reflections on the burden of meeting environmental standards in the late twentieth century. In: Azcue, J.M. (ed.), Environmental impacts of mining activities. Springer, Berlin Heidelberg New York, pp 7–17.

Miklavcic, V. (1999). Mercury in the town of Idrija (Slovenia) after 500 years of mining and smelting. In: Mercury contaminated sites (ed. R. Ebinghaus et al.), 259-269; Springer-Verlag.

Morin, K.A., Hutt, N.M. (1997). Environmental geochemistry of minesite drainage. MDAG Publication, Vancouver.

Munk, L.A., Faure, G., Pride, D.E., Bigham, J.M. (2002). Sorption of trace metals to an aluminum precipitate in a stream receiving acid rock-drainage: Snake River, Summit County, Colorado. *Appl. Geochem.*, 17, 421–430.

Nelson, C.H., Lamothe, P.J. (1993). Heavy metal anomalies in the Tinto and Odiel river and estuary system, *Spain. Estuaries,* 16, 496–511.

Nordstrom, D.K., Parks, G.A. (1987). The solubility and stability of scorodite, $FeAsO_42H_2O$: Discussion. *American Mineralogist,* 72, 849-851.

North, G., Gibb, H.J., Abernathy, C.O. (1997). Arsenic: past, present and future considerations. In: Arsenic exposure and health effects (ed. Abernathy, Calderon and Chappell), 406-423; London, Chapman and Hall.

Nriagu, J.O., Pacyna, J.M. (1988). Quantitative assessment of world-wide contamination of air, water and soils by trace metals. *Nature*, 333, 134–139.

Oriol, R. (1893) .Criaderos de cobalto del Aramo. Revista Minera, XI, 390p.

Page, G.W. (1981). Comparison of groundwater and surface water for patterns and levels of contamination by toxic substances, Environ. *Sci. Technol.*, 15 (12), 1475-1481.

Palinkas, L.A., Pirc, S., Miko, S.F., Durn, G., Namjesnik, K., Kapelj, S. (1995). The Idrija mercury mine, Slovenia, a semi-millennium of continous operation: an ecological impact, Environ. Toxicol. Assess., 317-341.

Paniagua, A., Loredo, J., García-Iglesias, J. (1987). Epithermal mineralization in the Aramo mine (Cantabrian Mountains, NW Spain): Correlation between paragenesis and fluid inclusions data. Bull. Minéral., *Paris* 111, 383-391.

Pettine, M., Camusso, M., Martinotti, W. (1992). Dissolved and particulate transport of arsenic and chromium in the Po River, Italy, Sci. *Total Environ.*, 119, 253-280.

Plumlee, G.S., Smith, K.S., Montour, M.R., Ficklin, W.H., Mosier, E.L. (1999). Geologic controls on the composition of natural waters and mine waters draining diverse mineral deposit types. In: Filipek, L.H., Plumlee, G.S. (eds.), Environmental Geochemistry of Mineral Deposits, Part B- Case Studies and Research Topics, Chapter 19, 373-432.

Pongratz, R. (1998). Arsenic speciation in environmental samples of contaminated soil. Sci. Tot. *Environ.*, 224(1-3), 133-141.

Posey, H.H., Renkin, M.L., Woodling, J. (2000). Natural acid drainage in the upper Alamosa River of Colorado. In: Proceedings from the 5th international conference on acid rock drainage, vol 1. Society for Mining, Metallurgy, and Exploration, *Littleton, pp* 485–498.

REAL DECRETO 140/2003, de 7 de febrero, por el que se establecen los criterios sanitarios de la calidad del agua de consumo humano. BOE núm. 45, de 21 de Febrero de 2003.

Richardson, S., Vaugham, D.J. (1989). Arsenopyrite a spectroscopic investigation of altered surfaces. *Min. Mag.*, 53, 213-222.

Robbins, R.G. (1987). The solubility and stability of scorodite, $FeAsO_42H_2O$: Discussion. Am. Mineral., 72, 842-844.

SADEI (1993). Datos y cifras de la economía asturiana; 1968-1991, Oviedo, Sociedad Asturiana de Estudios Económicos e Industriales.

Schwartz, M.O., Ploethner, D. (2000). Removal of heavy metals from mine water by carbonate precipitation in the Grootfontein-Omatako Canal, Namibia. *Environ. Geol.*, 39, 1117–1126.

Seyler, P., Martin, J.M. (1990). Distribution of arsenite and total dissolved arsenic in major French stuaries: dependence on biogeochemical processes and anthropogenic inputs. Mar. Chem., 29, 277-294.

Smedley, P.L., Kinniburgh, D.G. (2002). A review of the source, behaviour and distribution of arsenic in natural waters, Appl. Geochem., 17, 517-568.

Smit, J.P. (2000). Potable water from sulphate polluted mine sources. *Min. Environ. Man.*, 8, 7–9.

Smith, K.S., Huyck, H.L.O. (1999). An overview of the abundance, relative mobility, bioavailability, and human toxicity of metals. In: Plumlee, G.S., Logsdon, M.S. (eds), The environmental geochemistry of mineral deposits. Part A: Processes, techniques and health issues. Society of Economic Geologists, Littleton (Reviews in economic geology, vol 6A, pp 29–70).

Smith, R.A., Alexander, R.B., Wolman, M.G. (1987). Water-quality trends in the Nations rivers. *Science,* 235, 1607-1615 .

Squibb, K.S., Fowler, B.A. (1983). The toxicity of arsenic and its compounds. In: Biological and environmental effects of arsenic (ed. B.A. Fowler), 233-269; New York, Elsevier.

Stollenwerk, K.G. (2003). Geochemical processes controlling transport of arsenic in groundwater a review of adsorption. In: Welch, A.H. and Stollenwerk, K.G. (eds), Arsenic in Ground Water Geochemistry and occurrence. Kluwer Academic Publishers, Boston, MA, USA, pp 351-379.

Turner, R.R., Southworth, G.R. (1999). Mercury-contaminated industrial and mining sites in North America: an overview with selected case studies. In: Ebinghaus et al. (eds.), Mercury contaminated sites, 89-112; Springer-Verlag, Germany.

U.S.EPA (1982). An exposure and risk assessment for arsenic. EPA report 440/4-85-005. US Environmental Protection Agency, Washington, DC, USA .

U.S.EPA (2002). Proven alternatives for aboveground treatment of arsenic in groundwater, EPA-542-S-02-002.

Varnell, C.J., Brahana van, J., Steele, K. (2004). The influence of coal quality variation on utilization of water from abandoned coal mines as a municipal water source. *Mine Water Environ.*, 23, 204–208.

Watzlaf, G.R., Ackman, T.E. (2006). Underground mine water for heating and cooling using geothermal heat pump systems. *Mine Water Environ.*, 25, 1–14.

Welch, A.H., Lico, M.S., Hughes, J.L. (1988). Arsenic in groundwater of the western United States, Ground Water, 26 (3), 333-347.

Williams, M. (2001). Arsenic in mine waters: an international study. *Environ. Geol.*, 40(3), 267-278.

Williams, M., Fordyce, F., Paijitprapapon, A., Charoenchaisri, P. (1996). Arsenic contamination in surface drainage and groundwater in part of the southeast Asian tin belt. Nakhon Si Thammarat Province, southern Thailand, *Environ. Geol.*, 27, 16-33.

In: Drinking Water: Contamination, Toxicity and Treatment ISBN: 978-1-60456-747-2
Editors: J. D. Romero and P. S. Molina

Chapter 9

ELECTROCHEMICAL STRIPPING ANALYSIS IN WATER QUALITY CONTROL

Jaroslava Švarc-Gajić[1]
Bulevar cara Lazara 1, Faculty of Technology
Department for Applied and Engineering Chemistry
21 000 Novi Sad, Serbia

ABSTRACT

Water quality is defined by the presence and quantity of the contaminants, nutrients and by the physical and chemical properties, such as pH and conductivity. Man significantly influences all these factors. Toxicants enter the natural waters as a result of anthropogenic and natural processes, retain in the ecosystem and enter the marine food chain. Among them, heavy metals, recognised as high toxic water contaminants which are not biodegradable, attract the attention of the analysts, because they further proliferate their negative effect to other environmental compartments, finally reaching men. Accurate determination of metal traces in water is very important because even smallest concentration can have profound influence on human health. Heavy metal traces are nowadays most often determined with the application of atomic absorption spectrometry (AAS), inductively coupled plasma spectrometry (ICP) and electrochemical stripping techniques (ESA). The application of certain technique depends not only on general analytical requirements (sensitivity, reproducibility, accuracy, selectivity), but also on specific ones, such as the price of the instrumentation and analysis cost, simplicity and safety, required sample size, sample preparation procedure, etc. Taking into consideration these requirements, flameless atomic absorption spectrometry and electrochemical stripping techniques fulfil most of the criteria for the technique selection.

In past few decades electrochemical stripping analysis becomes one of the most frequently exploited techniques for trace metal determination in various samples. It is most often used for the determination of toxic metals in environmental samples (water, soil and air), biological samples, food and pharmaceutical product. Its application for the determination of different organic and pharmacologically active compounds, mostly with the application of chemically-modified electrodes has risen as well. The reason for ESA

[1] Tel: +00 381 21 485 3662, Fax: +00381 21 450 413, E-mail: jaroslava@tehnol.ns.ac.yu.

so strongly competing with the most commonly used atomic absorption spectrometry, regarding metal determination, are multiple. High sensitivity, the possibility of simultaneous determination of more then one analyte and features of on-line analysis for continual monitoring are just some of the attributes. Analysis itself is non-destructive and can be performed in some cases directly (simple liquid sample) with the modest price. Selectivity, regarding the physico-chemical and thermodynamic properties, important for element speciation and for understanding their reactivity, toxicity and transport, is one of the distinguishing parameters among the other instrumental techniques.

The intention of this chapter is to present the basic principles of different electrochemical stripping techniques and its significance in water analysis. Interferences, difficulties, advantages and disadvantages will be discussed in order to reveal a place of this attractive technique in analytical practice, with emphasis on water quality control. In water analysis, more than in any other type of the sample, the advantages of the technique are expressed in its full light and will be given through the review of the developed and applied methods for the determination of different elements in different types of the water.

Part of the research work confirming the attributes of chronopotentiometric stripping technique will be presented in order to promote its wider application and to break misapprehensions. Namely, voltammetric stripping techniques have a leading position and a stabile place among other stripping techniques, due to their highest sensitivity. Chronopotentiometric stripping analysis have been ignored for a long time, but with the development of the digital techniques, when time measurement became more accurate, they reached the sensitivity of the competitive voltammetry. Nevertheless, its application remained dissimulated. The work describing the development of various systems for the chronopotentiometric stripping determination of problematic analytes such as mercury and arsenic is elaborated in this chapter. Different flow-through electrochemical cells for mercury determination are described as a result of the research work focused on the development of the methods which could be used for continual monitoring.

INTRODUCTION

Rapid development of high capacity industry significantly influences the quality of water resources and is a background for the implementation of sustainable development in many state regulations. Drinking water is produced mainly from various surface water and groundwater. Due to the many anthropogenic processes, source water is usually contaminated in lower or higher extent. Industrial chemicals are highly toxic water contaminants with long-lasting and diverse influence, not only to human health, but also to biodiversity. Persistent organic pollutants such as polychlorinated biphenyls, polychlorinated aromatic hydrocarbons, dioxins, pesticides, chlorinated short chain paraffin's, organotin compounds etc. are major class of microcontaminants concerning anthropogenic sources. Another serious group of source water contaminants are inorganic ones, among which heavy metals carry serious risk and impact to human health because they are not biodegradable and are highly accumulated in aquatic organisms, entering food chain. Metal ions can also dissolve in groundwater and surface water as a result of the contact with rocks or soil containing the metals, usually in the form of metal salts. Discharges from sewage treatment and industrial plants also reflect the total metal content in source water. The toxicity of metals is dependent on their chemical form and solubility and this in turn, depends heavily on pH and on the presence of other ions. For example, calcium and magnesium are non-toxic and are normally absorbed by living organisms more readily than other metals. Therefore, if the water is hard, the toxicity of a given concentration of a toxic metal is reduced. On contrary, in soft, acidic water the same concentrations of metals may be more toxic.

Chemical form also determines metal toxicity. It is well known that lead, mercury, cadmium and arsenic are water pollutants of high concern. Their toxicity, transport and absorption behaviour depends on the oxidation state, thus making the speciation analysis very important. For example, As(III) is far more toxic (1000 times) than As(V), Tl(III) is less (10 times) toxic than Tl(I) and V(IV) species are more toxic (5 times) than V(III). Consequently, besides the determination of total metal content, speciation analysis is mandatory for defining the toxicity of the metals in water. This task can be resolved by the application of the ESA. Besides, techniques of ESA offer the possibility of simultaneous determination of more than one analyte and the possibility of the repetition of the analysis of the same solution. Flameless AAS, frequently exploited for metal determination in water, requires sample preparation. This technique has the sensitivity close to the sensitivity of the most sensitive techniques of ESA, but the analysis cost is far beyond the price of ESA exploitation.

In the last decade the application of electrochemical stripping techniques has risen significantly in all fields of analytical practice. Techniques are widely applied in water and food analysis as well as in analysis of pharmaceutical products. Practically all types of water can be analysed with success, such as sea, river, rain, sewage water, and snow. Low price of the instrumentation, easy handling, mobility of the instrumentation are only some of the reasons destining the ESA.

ESA is one of the few techniques that enable direct water analysis. Even simplest preparation procedure can cause the change of the analyte physico-chemical properties, its loss or contamination of the sample. On-line analysis is performed for continual monitoring of water quality in order to prevent further contamination and to react beforehead. Due to the small size of the instrumentation for ESA, easy handling and its mobility, it is possible to perform *in situ* analysis and to construct the flow systems for continual monitoring of metal content in water.

PREPARATION OF WATER SAMPLES

Water is heterogeneous, both spatially and temporally, thus making extremely difficult to obtain the reproductive samples. Great number of samples should be analysed in order to make reliable conclusions. Variation with respect to time can occur because of seasonal change, heavy precipitation or anthropogenic influence.

In stripping analysis of more complex samples or in the case when the interfering substance is present, the adequate separation technique should be applied. Separation methods can follow the decomposition step, or can be applied directly on a sample prior analysis. Target of the applied separation technique can be either trace component to be analysed or some interfering component. For particular application systematic research of the combination of the stripping technique and suitable separation technique is required. Precaution should be taken, because separation and preconcentration procedures are accompanied with the risk of analyte loss or contamination. Besides modern separation methods, the use of complexion agents also facilitates the broader application of stripping techniques in water analysis. Complexion agents provide improved selectivity of the stripping techniques in many ways, but the effects of the complexion agent on the electrode reaction and electrode reaction of the complexion agent itself should be well known and defined. In the presence of complexion agent the deposition of the interfering element can be avoided, because of the deposition potential shift.

Knowing the excellent sensitivity of the electrochemical stripping techniques, reagents used for the sample preparation should be of highest purity and should be used, when ever possible, in smallest possible amount in order to minimise the concentration of interfering substance and the possible contamination by the reagent. However, the concentration should be sufficient, since the stripping signal size depends on an ionic strength of the solution. Precaution should be taken in order to avoid the introduction of the impurities from the atmosphere or from the vessels. Vessels for stripping analysis should be cleaned with diluted acids in order to remove the metal traces from the vessel walls. The samples are recommended to be kept in the vessels made of inert material, such as polypropylene, to suppress the adsorption of the analyte on a vessel walls. Adsorption on the container wall can be minimised by conditioning the container with solution that contain adequate concentration of inorganic salts (sulphates, chlorides). By acidifying the samples to pH 1.5 rights after the sampling, adsorptive losses are minimised. Change of natural pH can cause the change of the physico-chemical properties of the analyte and can disturb the speciation process. Freezing of the sample is showed to be very efficient way of the preservation of the water samples for a long term [1].

Prior analysis solid particles are removed with membrane (0.45 μm) filtration. These filters often contain substantial amounts of contaminating metals and, therefore, should be precleaned with the acids. Residue after filtration contains organic metal compounds, stable organic salts and the amount of the metal adsorbed on inorganic, organic and mixed colloidal particles. After the digestion of the residue with strong mineral acids and dilution, the techniques of ESA can be applied. In this way the amount of the metal retained on a particular matter and organic metal species can be quantified.

Prior analysis the analyte can be concentrated and released from the interfering surrounding by the extraction procedure with the acids or organic solvent. A method of direct electrochemical stripping determination of metal traces in non-aqueous phase was reported [2]. The lack of information on the electrochemistry of metal chelates in non-aqueous solvents makes this approach less attractive and rarely applied. If the analysis is performed in organic phase, efficient removal of the oxygen is necessary.

Not many analytical techniques can offer convenience of direct analysis like ESA. When analysing waters, other than drinking, precaution should be taken, since present organic substances can disturb electrode processes. Adsorbed molecules influence the kinetics of the electrode reactions and cause the distortion of the potentiograms or voltammograms and are often a reason of nonlinear dependence of the analytical signal with the electrolysis time. All organic substances affect the response and sensitivity. It is recommended to perform the destruction of the organic matter prior analysis. Various procedures can be applied, such as ozonisation, γ-irradiation, UV-irradiation or mineralization. Analyst should consider possible reduction of the electropositive elements with UV rays. One of the critical elements that are reduced by UV rays is mercury. The irradiation step can be performed directly in the electrochemical cell, so additional manipulation steps are avoided. UV irradiation ensures the release of the metal from colloidal particles as well as from other metal bonds. To improve the decomposition process with UV rays, additional portions of extra pure hydrogen-peroxide can be added [1]. Removal of the organic substances from the water prior analysis lowers the detection limit and improves the accuracy. Complex forming organic substances and surfactants can also be eliminated by the electrochemical oxidation. The interval of the potentials 1.4 V – 1.5 V (3.5 Ag/AgCl) and chloride containing medium showed to be the best reducing conditions. Chloride concentration should not exceed 1 mol/l due to the chlorine generation during the preelectrolysis step, which can influence the analyte form and the electrode reaction. The positive outcome of the electrochemical oxidation procedure is observed if the slope of the calibration curve increases after the procedure.

Electrochemical Stripping Analysis

The history of electrochemical stripping analysis (ESA) started in 1931. when Zbinden [3] deposited copper at platinum working electrode and subsequently determined it by measuring the current during the electrochemical dissolution. Development of the operational amplifiers in 60s contributed to the further development of the technique. Sensitivity improvement was achieved due to the better accuracy and reproducibility of the potential measurement. Current and time could have been measured with better accuracy and reproducibility provoking further sensitivity improvement and leading to the construction of commercial, reliable and relatively cheap instrumentation.

Sensitivity limits were further shifted for 2-3 orders of magnitude with the development of the digital technology which initiated the development of the pulsed techniques in stripping voltammetry. Novel stripping technique – potentiometric stripping analysis (PSA), was presented in 80s by Jagner and Graneli [1]. Microprocessor development introduced the number of the more sensitive modified techniques especially with the respect to the analytical step.

Electrochemical stripping analysis is performed in three consequent steps and can be applied for the determination of all substances that can be subject of the electrode processes directly or indirectly. First step represents the deposition of the analyte in the electrode medium in conditions of convective mass transfer. After the deposition step, solution is left quiescent during the 15 – 30 s in order to enable the diffusive mass transfer in the next step. Consequent dissolution of the deposit is performed either electrochemically or chemically. Stripping techniques differ in this analytical step, which can be voltammetric, potentiometric, chronopotentiometric and rarely chronoamperometric. The sensitivity of ESA techniques is 3-5 orders of magnitude better in comparison with the applied diffusive technique in the analytical step because analytical signal depends directly on a concentration in the electrode medium and, indirectly, on a solution concentration. Certain amount of the analyte is preconcentrated from relatively large volume of the analised solution in significantly smaller volume of the electrode. It is very important to optimise the preconcentration step, taking into account the nature of the electrode processes, hydrodynamic conditions, electrode material, and electrochemical properties of the analyte as well as physico-chemical interferences that can occur. The preconcentration step is either performed electrochemically or by some nonelectrochemicall processes. Electrochemical preconcentration of the analyte is being conducted either in galvanostatic or potentiostatic conditions, where later is more frequently applied because of the easier choice of adequate electrolysis potential and the possibility of selective electrolysis performance [4]. In this case current changes during the time according to:

$$i_t = i_0 \cdot e^{-kt} \quad (1)$$

where t is – time; i_t – current in the time t; i_0 – initial current value and k – constant of the electrolysis rate.

This is also valid for the concentration:

$$C_t = C_0 \cdot e^{-kt} \quad (2)$$

where C_t is – analyte concentration in the electrode medium in time t and C_0 – initial analyte concentration in the electrode medium.

It can be seen from the equation (2) that higher k values contributes to more deposit in the electrode.

The amount of the deposited analyte in time t, depends on the constant of the electrolysis rate. This constant has greater value for the thinner diffusion layer, so preelectrolysis step is performed in the conditions of convective mass transffer, in order to minimise it:

$$k = AD/V\delta \quad (3)$$

where A is – surface of the working electrode; V – volume of the analised solution; D – diffusion coefficient of the analyte and δ - diffusion layer thickness.

During the electrolysis step convective mass transfer is enabled by the stirring of the solution (mechanical or magnetic stirrer), rotation of the working electrode or by the solution flow in flow systems. Mass transfer and hydrodynamic conditions have great impact on the sensitivity and the reproducibility of the determination. The amount of the formed deposit on/in the electrode depends on the efficacy of convective mass transfer. After the electrolysis step, prior deposit dissolution, the solution is left quiescence in order to enable the condition of diffusive mass transfer in the analytical step. In the case of the mercury working electrode this step is also important in order to obtain the uniform and homogeneous amalgam. During the electrolysis the metal distribution in mercury drop or film electrode is approaching to parabolic distribution with higher concentrations in the layers closer to the solution. Rest period enables the homogenisation and concentration uniformity, which is especially important for mercury drop electrodes, where the concentration gradient is more profound.

Deposit formation depends on the type and the physico-chemical properties of the analyte which, in return, determines the choice of the electrode material and the preconcentration procedure. In the case of the mercury working electrodes metals, which have the redox potential more negative than mercury are cathodically deposited in the form of the amalgam. Amalgamation depends on the metal solubility in the mercury and can be improved by the application of greater electrolysis currents or by adapting the experimental conditions. Elements which do not form the amalgams and those with the positive redox potentials are deposited at the solid inert electrodes in the form of the film. Electrodes made of noble metals and various carbon modifications can be used. Accuracy and reproducibility of the determination in these cases depends mostly on the quality of the electrode surface. Anions determination, determination of organic substances and some cation can be performed by its deposition in the form of the low soluble compound which is formed at the working electrodes. The deposition is performed in anodic range of the electrode potentials. Anions, such as Cl(-I), J(-I) and oxions are determined indirectly after the dissolution of the low-soluble compound that is formed with the metal ion (Hg(II), As(I)) originating from anodically dissolved electrode material. In the analytical step formed low-soluble compound is dissolved by cathodic potential sweep, reduction current or some chemical reductant. Some cations, such as Mn (II), Pb(II), Cu(II), Fe(II) can be indirectly determined by dissolving low soluble compounds, such as oxides and hydroxides, formed during the anodic deposition step with the supporting electrolyte components. Applicability of the mercury electrodes for this purpose is limited due to the anodic oxidation of mercury, so mostly graphite and glassy carbon electrodes are implied for the preconcentration in the form of low soluble compound. The deposition capacity increases with the surface of the working electrode for this type of analyte preconcentration, since the deposit is formed "on" the electrode, on contrary to amalgam formation, where electrode surface have less influence. Exception is with respect to mercury film electrodes which are only few atom layers thick, so surface/volume ratio determines high determination sensitivity with their application.

Adsorption or the chemical reaction on chemically-modified electrodes is used for the accumulation of the elements which can not be electrochemically deposited or for the purposes of sensitivity improvement. Organic substances or complexes are deposited at the electrode surface by non-specific adsorption in the conditions of convective mass transfer. After the rest period, deposit is desorbed electrochemically most frequently applying the voltammetric stripping techniques. Adsorption intensity of the analyte depends on many factors such as the supporting electrolyte, electrode material, pH and the temperature, mass transfer and time. Interferences in this type of the accumulation are caused by the presence of other surface-active components. Interfering substance can radically decrease the analyte accumulation and the intensity of the interference depends on the concentration ratio of the analyte and the interfering species, as well as on the type of the adsorption bonds. For the analytes with close desorption potentials, overlapping of the stripping peaks can occur.

Working electrodes made of mercury, carbon paste or glassy carbon can be modified with various ion-exchange resins, complexion agents etc. in order to provoke specific chemical reaction between the analyte and the electrode. Specific chemical agent can be covalently bond to the electrode surface, adsorbed or adhered.

Analyte, deposited in the electrode medium, is dissolved during the stripping step by potential sweep of the working electrode, chemical agents (Hg(II), oxygen, MnO_4^-, sodium, hydroquinone), constant current and rarely by the constant potential (stripping chronoamperommetry). In voltammetric stripping techniques the dissolution of the analyte is performed by potential sweep in cathodic (cathodic stripping voltammetry – CSV) or anodic (anodic stripping voltammetry – ASV) range; in potentiometric stripping analysis (PSA) by chemical oxidising or reducing agent while in chronopotentiometric stripping technique (CSA) by the constant current. In stripping chronoamperommetry the dissolution is performed on a constant potential.

STRIPPING VOLTAMMETRY

In voltammetric stripping techniques the dissolution of the preconcentrated analyte occurs electrochemically by changing the electrode potential. Potential sweep can be linear or with variously shaped impulses superposed on a linear potential component (Figure 1). Deposit dissolution initiates Gaussian shaped currents in relation to the potential of the working electrode (Figure 2). Dissolution potential, corresponding to peak maximum is qualitative characteristic of the analyte and the determination is performed on the basis of the peak current. Peak current changes linearly with the concentration and for the mercury drop electrode is equal:

$$i_p = K_1 \cdot t_{el} \cdot n^{3/2} \cdot r_0 \cdot D^{1/2} \cdot C_0 - K_2 \cdot t_{el} \cdot n \cdot D^{1/2} \cdot C_0 \quad (4)$$

where t_{el} – is electrolysis time; n – charge; r_0 –radius of the mercury drop; D – diffusion coefficient of the metal in mercury; C_0 – initial metal concentration in mercury; K_1 *and* K_2 –electrode reaction constants.

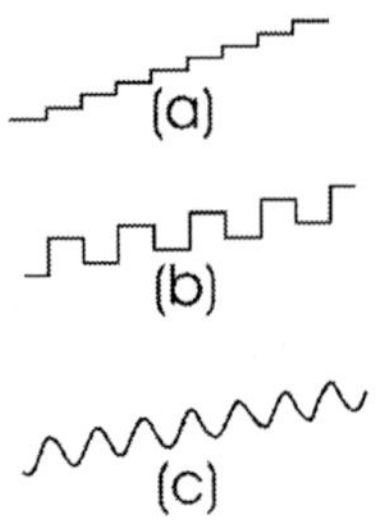

Figure 1. Potential-time signal in (a) staircase; (b) square-wave and (c) ac stripping voltammetry.

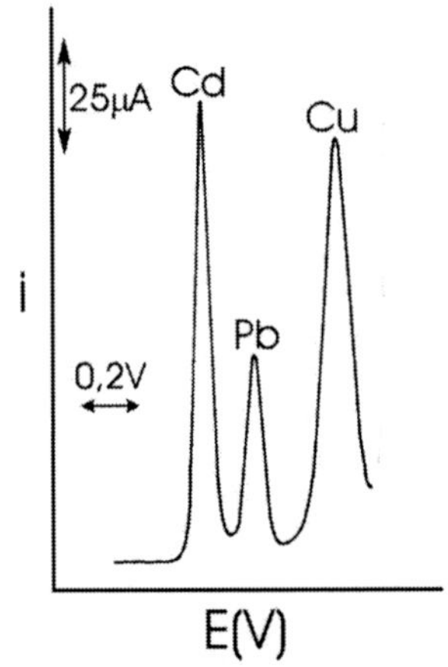

Figure 2. Typical voltammogram obtained with mercury drop electrode.

In the case of mercury film electrodes for the film thickness less than 10 μm and for small potential sweep rates (W<16.7 mV/s) as well as diffusive mass transfer in the analytical step:

$$i_p = n_2 \cdot F^2 \cdot A \cdot l \cdot C_r \cdot W / e \qquad (5)$$

where l is – mercury film thickness; C_r – concentration of the reduced metal in mercury;e – logarithm base; W –potential sweep rate and F - Faraday constant.

Total registrated current (i) is composed of a peak current (i_p) and beckgroung current (i_b):

$$i = i_p + i_b \qquad (6)$$

Beckground current negativly influences the analyte quantitation, decreasing the peak current, peak sharpness and the resolution. It consists of charging current (i_c) and Faradeic component(i_f):

$$i_b = i_c + i_f \qquad (7)$$

By changing the electrode potential, charging current charges the electrical double layer formed on the electrode/solution surface. It is significant component of the beckgroung current and depends on a differential capacity of the double layer for certain potential, C_d:

$$i_c = A \cdot C_d \cdot W \qquad (8)$$

Electrode reaction of the impurities present in the solution, electrolyte reactions or the reactions related directly to the electrode material are responsible for the Faradeic component of the beckgroung current .

In differential pulse voltammetry (DPV), the most sensitive voltammetric technique, the character of potential change during time is shown in Figure 3.

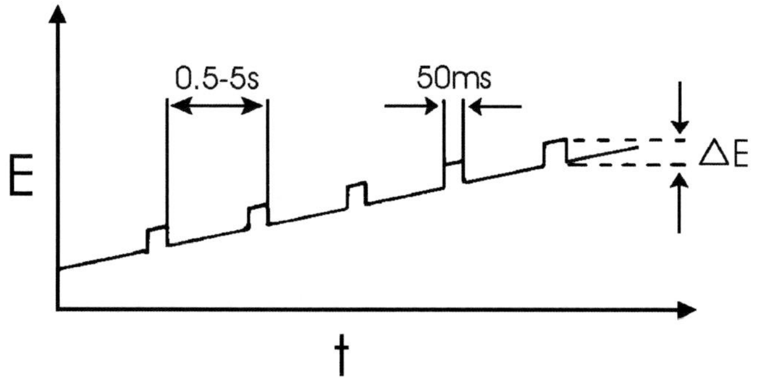

Figure 3. Potential-time sequence in DPV.

Amplitude of the potential impulse influences directly on a peak current. Disadvantages of DPV are relatively long analytical step (2-3 minutes) and significant influence of the Faradic component of the background current caused by the impurities deposition at the electrode surface. Square-wave stripping voltammetry (Figure 1) is another approach to minimise the charging currents. By sampling the current just before the square wave changes the polarity, the charging current is corrected. In alternating current (ac) stripping voltammetry (Figure 1) small amplitude of sinusoidal potential is added to a slow linear potential ramp. Discrimination of the charging currents is achieved using a phase-sensitive detector, which is capable of detecting only the current component of a specific phase. By immersing two identical working electrodes in a sample solution and imposing the same potential to both of the electrodes, but for different deposition times in electrolysis step, it is possible to measure the background current and to subtract it. This is used in subtractive striping voltammetry. During the stripping step the difference between the currents of the two electrodes is recorded. Anodic stripping voltammetry with collection utilizes dual working electrode system, usually ring-disc electrodes. Two electroactive surfaces are separated by insulator and are connected to separate polarising and measuring circuits. Electrolysis and stripping is performed on a disk surface. After a stripping step, a fraction of the analyte is redepositet at the ring electrode, where it is detected.

POTENTIOMETRIC STRIPPING ANALYSIS

Potentiometric stripping analysis (PSA) is the youngest among the all stripping techniques. The preconcentration step can be performed in same manner as in any stripping technique. After the electrolysis, the potentiostatic control is stopped and the potential of the working electrode is recorded. The potential change is due to the chemical dissolution of collected deposit. Dissolution time (τ) enables the quantitation of the analyte and the mean value of the dissolution potential is the qualitative characteristic of the analyte (Figure 4).

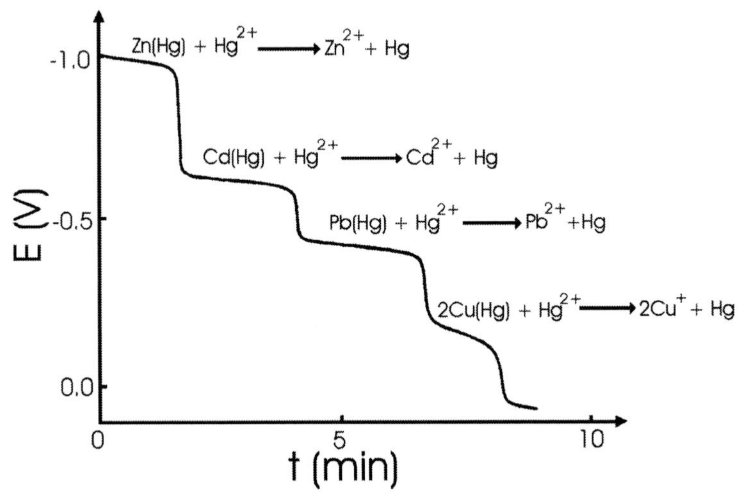

Figure 4. Classical potentiogram in PSA with Hg(II) as an oxidant. Mercury drop electrode.

In PSA most often used chemical oxidants are Hg(II) for the determination of the elements with more negative redox potentials than mercury, or MnO_4^- and $Cr_2O_7^-$ when mercury and more noble metals are to be determined. Concentration of the oxidant should be sufficient enough to oxidise noblest element in the solution. Too high concentrations of the oxidant, besides introducing the contamination risk, speedens the oxidation process decreasing the sensitivity. It is considered that the optimal concentration of the oxidant is 80 times higher comparing to the concentration of the analyte which is reversibly deposited at the working electrode. For some elements dissolved oxygen can be used in PSA as an oxidant. The advantages of this PSA modification are avoidance of the time consuming deaeration step, the minimisation of the contamination risk and avoidance of electrode surface blocking from the gas bubbles. From the same reasons this modification of the technique is convenient for the continual analysis. Relatively high oxygen concentration (8-9 mg/l) causes poorer sensitivity (faster oxidation) in comparison with the modification where Hg(II) is used as an oxidant. Oxygen concentration in the analised solution can be up to 1000 times higher then the analyte concentration. In those cases the oxidation process is the reaction of the first order, i.e. the oxidation time of the analyte doesn't depend on the oxygen concentration. For the analyte concentration above 50 μg/l the temperature and electrolyte influence can be significant. With temperature increase oxygen concentration decreases for 150 μg/l/°C in the range of 20 – 30 °C. Every g/l of the Cl^- at t = 25 °C causes the 80 μg/l concentration change of the oxygen in the range of 5 -20 g/l. For these reasons PSA in nondeaerated solutions should be performed with respect to temperature change less than ±0.5 °C if the calibration curve method is implied. Standard addition method eliminates both of the influences. With pH approaching the neutral, during the electrolysis the concentration of hydroxyl ions can be increased in the electrode area due to the reduction of the oxygen and the formation of metal sediment can occur. Also, calomel formation at mercury working electrodes in the modification of the PSA with an oxygen as an oxidant is more probable and favourable.

Sensitivity can be improved by the application of subtractive, differential or multiple-scanning PSA. In subtractive PSA mode, the stripping curve of the analyte is firstly recorded and then the curve obtained after the 1 s of deposition which corresponds to the background curve. Two curves are automatically subtracted. Analytical information can be presented in the form of peaks in differential PSA after the microcomputer has been transformed potential in dE/dt form. Multiple stripping and re-reduction can also be applied for signal enhancement. In this technique amount of newly oxidised analyte is redepositet and reoxidised in subsequent stripping step, so detection limit close to those attributed to most advanced voltammetric techniques can be achieved.

Chronopotentiometric Stripping Analysis

In chronopotentiometric stripping analysis (CSA) the dissolution of formed deposit is performed by applying the constant current, during which the potential of the working electrode versus time is being recorded. Oxidation time is analogous quantitative characteristic of the analyte as in potentiometric stripping techniques. Unlike the chronopotentiometry, the linear dependence of the dissolution time on a concentration has been confirmed in CSA, and this fact contributed to its wider application. In analytical practice it is less frequently applied comparing to voltammetric and potentiometric stripping techniques and it is used for the determination of the elements which can not be comfortably determined with other stripping techniques (Hg, Ag, Se, S, Ni, Co, As, Th, Sc) or when the sample matrix is complex so the dissolution processes can be assisted with proportionally greater currents. In comparison with voltammetric stripping techniques, CSA enables almost the same sensitivity of the determination due to less influence of the charging currents and computerised analytical step, namely the automatic and accurate measurement of the dissolution time. Selectivity of the chronopotentiometric stripping techniques is close to the selectivity of PSA. It should be pointed out that the application of CSA is related to the use of the solid working electrodes, such as gold, platinum, glassy carbon, so the reproducibility of the technique most closely depends on the reproducibility of the electrode surface.

In potentiometric and chronopotentiometric stripping techniques the quantitative characteristic of the analyte is dissolution time, which can be measured with high accuracy. In this sense the advantage comparing with the voltammetric stripping technique is unambiguous. In voltammetric stripping techniques as a quantitative characteristic of the analyte the dissolution current is measured, with lower accuracy and sensitivity comparing to time measurement. Consequently, similar sensitivity is attributed to chronopotentiometric and voltammetric stripping techniques, but much better selectivity for the CSA due to the possibility of the application of small dissolution currents and reduced influence of charging currents. Another analytical convenience of the chronopotentiometric stripping technique is that the linear dependence of the dissolution time on the charge required for the analyte deposition, i.e. on the electrolysis time, is observed for wider range comparing to analogous dependence of the dissolution current in voltammetric stripping techniques.

It was also empirically demonstrated that in the case of CSA at planar disc electrode, dissolution potential of the analyte practically does not depend on the analyte concentration [2]. For the concentration change of 10^2-10^3 mol/l in most cases the dissolution potential does not change more than 50 mV. On contrary, in stripping voltammetry, concentration shift is accompanied with obvious differences in the dissolution potential of the analyte. Such CSA characteristics are especially important for multielement analysis.

Glassy Carbon as an Electrode Material

For electrochemical anodic dissolution of some electropositive elements (Ag, Au) mercury electrodes are avoided because of their anodic oxidation and only solid electrodes can be applied. They are also used for the determination of the elements which do not form amalgams, i.e., are poorly soluble in mercury (As) or for mercury determination itself. Solid electrodes can be used as an inert support for the formation of the metal film electrodes (Hg, Ag, Au). Most commonly used materials for the solid electrodes are gold, platinum and various carbon modifications. The active surface of all solid electrodes is much bigger in comparison with the macroscopic one. For platinum electrode the ratio is approximately three, while for the porous graphite it is even higher. At such non-uniform surface, the adsorption is very profound and adsorbed molecules are difficult to remove. Hydrogen layer can be formed in the range of the positive potentials, while in the rage of the negative potentials, the oxygen layer adsorbs at the electrode surface. Oxygen adsorption in much stronger comparing to hydrogen, and it disturbs the electrode processes. Almost all electrode materials, even the noblest ones, are oxidised at positive potentials, forming the oxide layer. Potentials of the oxide layer formation depend on pH and the intensity of the oxygen and hydrogen evolution. Glassy carbon electrodes are much less prone to oxidation processes. This carbon modification is produced by the carbonisation of the organic polymers such as phenol resins, cellulose, PVC etc. Starting polymer has to have three-dimensional structure and lowest possible content of the heteroatoms. The polymerisation take place in the solid phase inducing the formation of the sp^2 hybridised carbon orbitals and the formation of the hexagonal planar structure. Unlike the graphite in which all carbon hexagonal layers are parallelly oriented, in glassy carbon these layers are unarranged (Figure 5).

Electrodes made of the glassy carbon are nowadays produced in many different dimensions and forms. They are used for the determination of the large number of the elements which are difficult to determine with the application of other electrode materials. Glassy carbon itself can be modified with various agents and specific sensors, such as sensors for proteins, DNA or antibiotics can be produced. Modified or in its native form, the main problem with glassy carbon is the difficulty to obtain the uniform structure and the composition due to the complexity of the technological procedure for its production. Renewal and obtaining the reproductive surface is its main disadvantage. Each mechanical procedure applied on a glassy carbon surface reveals the new layer with different mechanical and electrochemical properties. Glassy carbon electrodes, due to their high inertness, can be used in wide potential and temperature range (-30 °C – 1010 °C) in various electrolytes. Even though the glassy carbon according to its electrochemical properties approaches the "ideal" electrode material, some authors claim that the certain functional groups are present at the surface and that they disturb the electrode processes. Nevertheless, good electrical conductivity, inertness and absence of the intermetallic species formation, makes glassy carbon the advantageous electrode material.

For reproductive electrochemical processes at glassy carbon it is essential to regularly perform the electrochemical cleaning of the electrode. Cycles (20 to 60) of potential sweep from -0.7 V to +0.7 V with high oxidation current are sufficient to refresh the glassy carbon properties.

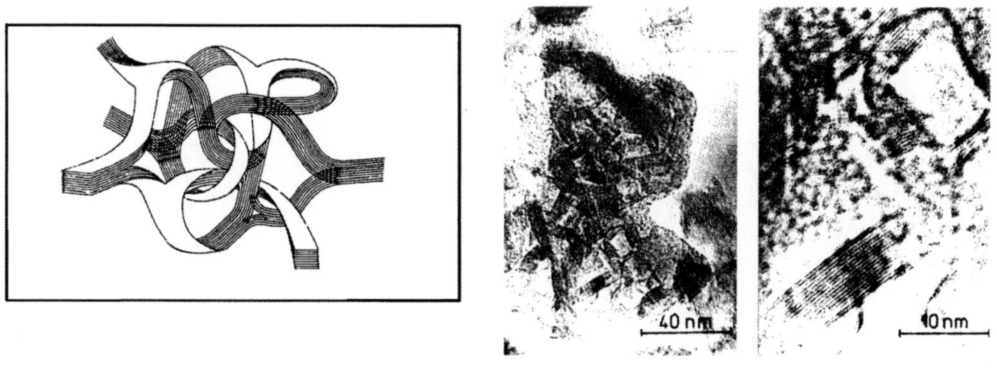

Figure 5. The structure of the glassy carbon.

In some cases for the activation of the glassy carbon electrochemical procedure is not sufficient, so cleaning in the ultrasonic water bath at 50 °C during 1 hour in water:ethanol (1:1) mixture is required. The activation of the glassy carbon is necessary due to the saturation of its surface and the change of the functional groups properties at the surface. In these cases regular mechanical cleaning with the acetone is not sufficient. It is assumed that the surface is somehow saturated due to the chemical bonding of the substances or adsorption that occurs in the deeper layers of the glassy carbon. Polishing procedures with diamande paste or Al_2O_3 suspension of decreasing granulation, followed by the ultrasonic cleaning, are applied if the loss of the electrode activity caused by its saturation occurs.

WATER ANALYSIS

Determination of the traces of toxic elements and compounds is one of the main objectives in water quality control. Monitoring the toxic substances content is mandatory not only in drinking water, but also in all natural waters, such as rivers, lakes, sea etc. in order to react beforehead in cases of unexpected situations. Among the organic and microbiological contaminants, attention should be focused on inorganic contaminants as well, among which, toxic metals and metalloids, are of highest interest (Table 1). Concerning heavy metals, special attention should be paid on the determination of metals such as thallium, lead, cadmium and mercury, i.e. the metals that bioacumulate in food chain.

In waters metals can be bound to the inorganic ligands, such as water, sulphates and carbonates or to the organic ligands, such as aminoacids, sugars, carbohydrates, humic acid etc. The direct stripping procedure enables the determination of hydrated free metal ion and labile metal complexes (Table 2), as well as metal ions adsorbed on both organic and inorganic colloidal particles

Table 1. Allowable metal contents in water

Element	Background level for surface water[5] (μg/l)	Allowed levels for drinking water[6] (μg/l)
Aluminium	<5	200
Iron	<30	200
Manganese	<10	50
Copper	0.6	2000
Zinc	5	300
Silver		10
Arsenic		10
Cadmium	0.04	5
Chromium	0.2	50
Mercury	0.002	1
Nickel	0.5	20
Lead	0.5	10
Antimony		5
Selenium		10

Table 2. Inorganic metal species in natural fresh waters

Metal	Main Species	Proportion of Free Ions
Ag(I)	Ag+, AgCl	0.6
Zn(II)	Zn2+, ZnCO3	0.4
Cd(II)	Cd2+, CdCO3	0.5
Co(II)	Co2+, CoCO3	0.5
Ni(II)	Ni2+, NiCO3	0.4
Pb(II)	PbCO3	0.05
Cu(II)	Cu(OH)2, CuCO3	0.01
Hg(II)	Hg(OH)2	1·10-10
Al(III)	Al(OH)3(s), Al(OH)2+, Al(OH)4-	1·10-9
Fe(III)	Fe(OH)3(s), Fe(OH)2+, Fe(OH)4-	2·10-11
Mn(IV)	MnO2(s)	
As(V)	HAsO42-	
Cr(VI)	CrO42-	
Se(VI)	SeO42-	

Metal complexes easily dissociate with the acidification of the sample and by negative enough potential employed in the electrolysis step of stripping analysis. Since free metal is often assumed to be most bioavalaible fraction of the metal, it is often correlated with its toxic fraction. However, this is not always the case. Typical example is mercury, which can be converted to methyl-mercury in water under certain conditions by the biochemical action of the microorganisms present in the water. Methyl-mercury is considered most toxic form of the mercury for human. It is concentrated in fish tissue and in this way it enters the man food chain. Once entered in human body, it covalently bonds to the human brain and causes irreversibly toxic effects and paralysis. In this sense, it would be more toxicologically justified to stress the methyl-mercury content, however tight liaisons with concentration of ionic mercury exist and the two contents are dependent.

PSA techniques are less susceptible to the drawback of the presence of organic matter in comparison with the voltammetric stripping techniques. Dissolution step performed in same conditions of the mass transfer as in the electrolysis step (convective mass transfer) partially compensates the influence of the organic matter. By stirring the solution, inflow of the oxidant to the electrode surface is intensified and the dissolution process is assisted even when adsorption of the surfactants and organic matter occurs.

High content of inorganic salts in some samples like sea water represent significant interference in most of the microanalytical techniques, while in ESA, due to the conductivity properties, it is considered as analytical convenience.

MULTIELEMENT ANALYSIS

The possibility of multielement analysis is one of the advantages of ESA and was widely exploited in water analysis with the application of voltammetric, potentiometric and chronopotentiometric stripping techniques. Elements that were most often simultaneously determined are Zn, Cd, Pb and Cu.

ZINK is naturally present in water. The average zinc concentration in seawater is 0.6-5 μg/l and in estuarine 10-40 μg/l. Rivers generally contain between 5 μg/l and 10 μg/l of zinc. Zinc salts cause a milky turbidity in water in higher concentrations. At concentrations of about 2 mg/l zinc may add an unwanted flavour to the water. This element appears to accumulate in some organisms. The human body contains approximately 2-3 g of zinc, and it has a dietary value as an essential element. Its functions involve mainly enzymatic processes and DNA replication. The human hormone insulin contains zinc, and it plays an important role in sexual development. Minimum daily intake that prevents deficiencies is 2-3 g. The human body absorbs only 20-40% of zinc from food. Many people intentionally drink mineral water rich in zinc. Symptoms of zinc deficiencies are tastelessness and loss of appetite. Children's immune and enzyme systems may be affected. Higher zinc intake appears to protect people from cadmium poisoning and it also decreases lead absorption. Symptoms of high zinc intake include nausea, vomiting, dizziness, colic's, fevers and diarrhoea and mostly occur after intake of 4-8 g of zinc.

CADMIUM. About half of the cadmium is released into rivers through weathering of rocks and some cadmium is released into air through forest fires and volcanoes. The rest of the cadmium is released through human activities, such as industry or motor vehicles. In fresh water, cadmium toxicity is influenced by the hardness of the water, the softer the water the greater the toxicity. It has high short-term and long-term toxicity to aquatic life. People expossed to cadmium in a short periods of time express symptoms of nausea, vomiting, diarrhea, muscle cramps, salivation, sensory disturbances, liver injury, convulsions, shock and renal failure. Cadmium is first transported to the liver through the blood. There, it is bond to the proteins and these complexes are transported to the kidneys. Cadmium accumulates in kidneys, where it damages filtering mechanisms. This causes the excretion of essential proteins and sugars from the body and further kidney damage. It takes a very long time before cadmium that has accumulated in kidneys is excreted from a human body. Long-term exposure to cadmium potentially cause the effects of kidney, liver, bone and blood damage.

LEAD is one of the most important of all heavy metals because it is both very toxic and very common. It eneters into drinking water from the corrosion of plumbing materials, where lead is commonly used. This is more likely to happen when the water is slightly acidic. This is a reason for the requirements for public water treatment systems for pH-adjustments in water that will serve drinking purposes. The lead dissolution from the water pipes depend on chloride and oxygen concentration, pH, temperature and water retention. Children are more susceptible to lead then the adults. Iminished learning abilities, behavioural disruptions, such as aggression, impulsive behavior and hyperactivity can occur in cases of lead poissoning. About 50% of lead is easily absorbed from the gastrointestinal tract in children, while 10-15% in adults. In surface and esturian water average lead content is 0.01 mg/l. Lead can causes number of unwanted effects, such as disruption of the biosynthesis of haemoglobin and anaemia, a rise in blood pressure, kidney damage and disruption of nervous systems. Lead can enter a foetus through the placenta of the mother. Because of this it can cause serious damage to the nervous system and the brains of unborn children.

In water analysis voltammetric stripping techniques with linear or pulsed potential sweep are most often applied. The determination is most frequently performed with the application of mercury drop or mercury film electrode, gold or gold film electrodes and electrodes of different carbon modifications. Platinum, silver and silver film electrodes are less commonly used. By applying the different procedures for analyte concentration in electrode medium, the technique was applied for Pb, Cd, Zn, Cu, Se, As, Hg, Co, Ni, Mn, Fe, Bi, Sb, Sn, Tl, Ag and Cr determination. Elements like Au, Pt, U, Ga, Al, Mo, V, Ti, Pd can also be determined.

By applying the anodic stripping voltammetry and the mercury film electrode, deposited on a glassy carbon, simultaneous determination of zinc, cadmium, lead, copper and bismuth in sea water have been reported [7]. All listed elements and antimony were determined in one analysis by direct anodic stripping voltammetry with both linear and pulsed potential sweep [8]. The effect of recollection was used for sensitivity enhancement in the case of stripping voltammetry with linear potential sweep. The analyte was recollected on a ring-disc electrode made of glassy carbon. By combining the DPASV and DPCSV multielement analysis in different water samples were accomplished [9]. Cathodic scan was applied for selenium determination and anodic for the determination of Pb, Cu and As. The combination of anodic and cathodic stripping voltammetry with linear potential sweep was also used for the analysis of waste and surface water regarding tin and antimony (ASV) as well as arsenic (CSV) determination [10]. Absolute method for the calculation of lead and copper content in spring and rain water was performed assuming 100% electrolysis efficacy for the microelectrodes [11]. Up to seven elements (Cd, Zn, Pb, Cu, Mn, Ni, Co) were determined in one analysis of river water applying the DPCSV and mercury drop electrode [12]. By coupling flow-through ASV in pulsed mode and inductively coupled mass spectrometry, Cr(VI) and V(V) were discriminated [13]. Most frequently, the content of Cr(III) and Cr(VI) was determined after the adsorptive preconcentration on a mercury working electrode. As a complexion agent's pyrocatechol-violet [14, 15], cupherron [16], triethylenetetramine-*N,N,N',N'',N''',N'''*-hexaacetic acid (TTHA) [17], hydroxyethyl ethylenediamine triacetic acid (HEDTA) [18] and biethylentriaminopentacetic acid [19] were used.

In potentiometric stripping analysis the dissolution of the deposited analyte is performed most often with chemical oxidatants, and rarely by chemical reduction of the analyte (reductive stripping analysis). For most metals dissolved oxygen has sufficient oxidising power to provoke the analyte dissolution. In some cases in order to increase the sensitivity, the deaeration of the solution is performed and the dissolution occurs due to the added oxidising agent (Hg(II), Cr(VI), MnO_4^{2-}...). The deaeration step is performed prior electrolysis by inert gas flow (helium, nitrogen). In sea and spring water oxidation of the cadmium, lead and copper was performed with added Hg(II) [20]. Same oxidant was used for the determination of Zn, Cd, Pb and Cu in potable water [21]. It was necessary to add Ga(III) prior zinc determination. Namely, copper and zink form an intermetallic compound ($CuZn_x$) which is dissolved at the potential close to the oxidation potential of copper. The addition of the "third" element (usually gallium(III)) which can bind the copper is necessary in order to enable the undisturbed deposition and subsequent dissolution of zinc. Copper was determined first under adequate experimental conditions, while in the next step the group of three remaining elements (Zn, Cd, Pb) were determined applying more negative electrolysis potential. The technique of PSA with the dissolved oxygen as an oxidant can be modified in order to improve the sensitivity of the determination. By performing the short deaeration (15 s) prior analysis, the concentration of the oxidant in the solution decreases, what reflects the longer dissolution time and, thus, better sensitivity. Constant small negative current set during the dissolution step, can also provoke the sensitivity improvement. The technique was applied for the determination of cadmium and lead in potable water with the mercury film working electrode and classical electrochemical cell [22, 23]. In flow systems, after stopping the solution flow and setting the reduction current, which partially reduces the oxidant and partially re-reduces dissolved analyte, significant sensitivity improvement is observed [23].

For the improvement of AAS, coupling with flow PSA was reported [24]. The hyphenated technique was applied for lead and cadmium determination in waters with PSA microcell and porous working electrode.

In the chronopotentiometric stripping analysis the generated deposit is dissolved with the constant current. For the oxidation of the analyte, positive currents are applied, and for the reduction, analogously, negative currents. The feature that differentiates this stripping technique from the others is the flexibility due to the possibility of the application of wide currents window, reducing the matrix influence and reaching the great sensitivity. In the analytical practice, this technique has a shadowy position and should be imposed and promoted, because it merges the advantages of the voltammetric techniques and PSA.

The preconcentration of Zn, Cd, Pb and Cu was performed in flow system on a mercury film, deposited on a glassy carbon powder. The deposit was dissolved applying the constant current. The content of the elements in different water samples, in the samples of milk, wine, and plants was calculated applying the absolute method [25]. The absolute method was also applied by Sahlim and Jagner for the determination of zinc, cadmium, lead, and copper in potable water applying the mercury film electrode [26]. The calculation based on a Faraday low enabled resolving the major problem of zinc determination in the presence of copper as mentioned before.

Speciation Analysis

In natural waters metals are present in different forms. The form of the metal influences its toxicity, bioavailability and transport. In speciation analysis, the whole analytical concept must be oriented toward the contamination minimisation and the avoidance of the change of any parameter that could cause the misbalance between the present metal species. For a speciation purposes the voltammetric stripping techniques are most suitable, due to their selectivity as well as sensitivity.

It is considered that the smallest part of the metal content is in the form of inorganic soluble ions, which pass the pores of the membrane filter 0.45 μm [27].

However, the distribution of metal species in water depends on many factors. Free metal ions are most toxic form of the metal present in the water. Hydrophobic complexes of some metals such as cadmium and copper easily pass the cell membrane and are considered as toxic as its free ions. Molecular both organic and inorganic compounds of the metal and its complexes make significant part of the total element content in water. Solid and colloidal particles suspended in the water can adsorb part of the metal and microorganisms can incorporate the microelement. Metals bound to colloidal particles and metal complexes are considered less toxic.

It is expected that with the increase of water density complex formation and bonding to inorganic ligands will be favourable and more probable. Consequently, in sea and river water only minor amount of the metal is present in the free form. Fulvic and humic substances can entrappe the metal in less toxic complexes. Interaction with the fulvic acid depends of the metal. Cadmium practically doesn't interact, but the interaction with zinc is significant. For the copper it is specific that it interacts depending on the structure of the fulvic acid. Most of the mercury (70%) and copper is bound in strong organic complexes, while lead is mostly (20-70%) bound in weak labile forms. Cadmium, zinc and copper form mostly moderately labile complexes [28]. Adsorption of free metal can occur on inorganic (MnO_2, Fe_2O_3), organic (humic, fulvic) or on a mixed (Fe_2O_3 particles coated with humic acid) colloidal particles. With fulvic acid, amino acids, sugars, hydrocarbons etc. the metal can form also a stable organic salt (Cu-fulvate). Many complexes are not electroactive and, therefore free metal should be released from it in order to determine the total amount. Metal ions adsorbed on inorganic or organic particles take apart in the electrode reaction and are determined by direct stripping analysis. The amount of the metals bound in complexes changes during the time and it should be taken into consideration if the element speciation is to be performed. The analysis should be performed as soonest as possible. Complex formation can vary depending on the depth and biological efficiency of the region. It also changes with the pH during the time.

Among the microanalytical techniques, stripping techniques are one of the most powerful for the purposes of element speciation. Different procedures are possible to apply for the speciation, depending on the analyte, but the simplest of them is selection of adequate solution composition and electrochemical parameters. One of the simplest procedures for the determination of different metal forms is by combining the filtration procedure, usually using a membrane filter with 0.45 μm pores, separation on an ion-exchange column, UV irradiation and the technique. Discrimination is possible between free metal ions, labile organic and inorganic complexes, labile forms adsorbed on organic and inorganic colloidal particles and stabile organic compounds. One of the major limitations of the speciation analysis with the application of ESA is the impossibility to distinguish labile forms which coexist in the river water. For example Cd^{2+}, $CdCl^{+}$ and $CdCl_3^{-}$ will produce one signal.

The advantage of the ESA concerning the speciation possibility is probably most widely exploited in water analysis. The examples of the Se(IV) determination in the presence of Se(VI) using mercury film electrode [29], As(III) in the presence of electroninactive and less toxic As(V) using the gold electrode [30], Cr(VI) in the presence of Cr(III) with the application of ion exchange in the preconcentration step have been reported [31]. Similarly, applying the mercury film electrode the determination of Sb(III) in the presence of Sb(V) can be performed [32].

Electrochemical activity of different metal ions varies, depending on acidity. The content of Sb(III) can be determined in 0.2 mol/l HCl, because the Sb(V) is practically inert due to high level hydrolysis. Total concentration of antimony can be determined in much more acidic solution [33]. One of the metal forms such as As(V), Cr(III) or Sb(V) can act inert under experimental conditions. Electroactive form can be determined applying the technique in a common way. After the appropriate chemical treatment of the sample, where inert form is transformed in the electroactive one, the results of the analysis will correspond to the total analyte content.

Selenium Determination

Selenium is naturally occuring element, found in natural deposits and ores and it is an essential nutrient at low levels. In some naturally rich regions concentration in water can reach 6000 μg/l. At such high concentrations and low pH, selenium species can be converted in more water-soluble forms. In potable water concentration of selenium is usually below 10 μg/l. In China, where selenium rich ores are present in high amount, the concentration in drinking water is between 50 μg/l and 160 μg/l. In natural estuarine or surface waters selenium concentration is usually between 0.06 and 400 μg/l. Selenium compounds are widely used as electronic components in glass, rubber and textile industry often being antropogenic factor influencing increased selenium content in water. One of the selenium's essential role is being a constituent of selenium-containing enzyme, glutathione-peroxidase, that catalyses the degradation of reactive organic hydrocarbons and protects the cell from oxidative damage. Many positive effects of selenium are also well known, such as influence on immune system, fertility etc. Increased selenium intake for relatively short periods of time causes hair and fingernail changes, damage to the peripheral nervous system, fatigue and irritability. For lifetime exposure to levels above maximum level set by EPA, hair and fingernail loss; damage to kidney and liver tissue, as well as the nervous and circulatory systems are observed. For selenium determination in waters cathodic stripping voltammetry was most frequently applied, more specifically differential pulse techniques, due to the required high sensitivity. Both adsorptive and electrochemical accumulation procedures enable similar sensitivity of the determination, however, when applying the adsorptive collection, false results can occur due to the nonlinearity of the calibration curve. Therefore for selenium contents below 0.8 ng/l adsorptive stripping voltammetry is recommended, but for higher concentration range, voltammetry with electrochemical deposition should be applied. DPCSV was applied for the speciation of Se(IV), Se(VI) and Se(-II) in snow and natural waters applying the mercury drop electrode [34]. The speciation of selenium species was also performed in sea water with the application of different chromatographic techniques [35, 36]. Since only tetravalent selenium is electroactive, after the sample is passed through adsorbent resin XAD 2, which retained the organic compounds, such as humic and fulvic acid above 90% and 70%, respectively, the content of Se(IV) could be determined by direct analysis. Organic selenium species, such as seleno-methionine and other organic forms of selenium (Se(-II)) are not retained by the ion-exchange resin IRA 400, which bonds only selenite and selenate. Percolated Se(-II) species were transformed into the electroactive Se(IV) form by thermal or photochemical procedures. UV irradiation is convenient since, besides oxidising Se(-II), it also destroys the organic matter and so eliminates the interferences due to natural organic surfactants. Transformation procedure is highly dependant on pH. By performing UV irradiation at pH 1, Se(-II) species are converted into the electroactive tetravalent form and could be determined by the analysis. The reduction of Se(VI) into Se(IV) was performed after the elution from IRA 400 by UV irradiation in conditions if adequate pH and medium composition. Irradiation at pH 10 enabled the reduction of Se(VI). Direct analysis performed after UV irradiation of the sample at pH 9 leaded to the total selenium concentration, i.e. the organic forms of selenium (Se(-II)) were released from the strong bonds and oxidised, while hexavalent selenium was reduced to tetravalent. When organic resins are employed in the analysis the risk of introduction of surface-active compounds into the analyte solution is encountered. Discrimination between organic and inorganic selenium species was also reported by other authors [37]. For sequential determination of different selenium species in water, flow-injection cathodic stripping analysis was performed [38]. Even though for selenium determination in water analysis mercury electrodes are most commonly applied, copper microelectrode [39] and gold fibre electrode [40] were reported as well. Stripping voltammetry with adsorptive concentration was also frequently applied for selenium determination in natural waters [41-43]. The selenium complex which was preconcentrated on a working electrode was most often formed with 2,3-diaminonaphtalene (DAN). Formed piazselenol complex can be adsorbed on a working electrode, for example graphite paste electrode, and subsequently dissolved with potential shift [43]. Regardless the advantageous feature of the PSA, that after the deposition step, the potentiostatic control is stopped, so the analysis is not prone to the interfering electroactive species as the voltammetric stripping techniques are, PSA has rarely been applied for selenium determination. Reductive PSA enables the selenium determination with poorer sensitivity in comparisson with voltammetric techniques, but with much better selectivity.

Copper Determination

In natural waters copper is present in average concentration of few μg/l, but depending on water properties, such as hardness, pH, anions and oxygen concentration, temperature and technical detailes of water pipes, it can reach even mg/l level in drinking water. Copper is an essential element because it is a constituent of many body enzimes, but at very high levels it is toxic and can cause vomiting, diarrhea, loss of strength or, for serious exposure, cirrhosis of the liver. Water can become blue-green in colour as the copper dissolves from the pipes and precipitates in the water in the form of its salts. Many authors were dealing with the determination of copper in different types of the water [44-47]. Specific biosensor based on copper-sensitive bacteria was used in order to determine the bioavailability of the copper in river waters. Cathodic stripping voltammetry was the applied technique [46]. For copper preconcentration from contaminated waters and sludge, carbon paste can be modified with alpha-benzoin oxime [48]. In order to determine copper from its labile complexes present in river water, the modification was performed with the nafion [49].

Mercury Determination

Mercury is naturally found in combination with other elements in various ores and as a trace component in many minerals. Fossil fuels and lignite contain mercury often at levels of 100 parts per billion and it is a fact that shouldn't be neglected due to the wide use of these fuels as energy resources. As a heavy metal pollutant, mercury generates a great problem, because of its high toxicity and the mobilisation of methylated forms by anaerobic bacteria. Electrical products such as dry-cell batteries, fluorescent light bulbs, switches, and other control equipment accounts for 50% of used mercury. Organic mercury compounds are widely used as a pesticides, particularly fungicides. Mercury concentration in natural waters is usually below 0.5 μg/l, but certain mineral layers can locally increase the concentration. Sewage effluents sometimes contain up to 10 times more mercury then typical natural waters. The Maximum Contaminant Level Goals (MCLG) concerning mercury in drinking water and set by EPA is 2 μg/l. Some of the toxicological effects of mercury include neurological damage, paralysis, blindness, insanity, birth defects and chromosome damage. Milder symptoms of mercury poisoning include depression and irritability and are of psychopathological character. In neutral and alkaline water the formation of volatile dymethylmercury ($(CH_3)_2Hg$) is favoured, so mercury can move into another parts of the ecosystem. Soluble methyl mercury ion (CH_3Hg^+) is produced by anaerobic bacteria from sediments with methylcobalamin (methylated vitamin B_{12} analogue). It is assumed that bacteria which produce methane also produce a methylcobalamine as an intermediate product. In fish fat tissue concentration factor for methyl-mercury can exceed 1000.

Mercury(II) was determined in natural waters with the gold film electrode and ASV with the detection limit of 0.4 μg/l [50]. Lower detection limit ($5 \cdot 10^{-14}$ mol/l) was achieved applying the pulsed technique and thyocianate as an electrolyte. The deposition of the mercury was performed on a glassy carbon [51]. For continual monitoring of mercury content in waste water, wall-jet flow cell was used with the application of chronopotentiometric stripping analysis and gold film working electrode [52]. In the conditions of stechiometric electrolysis, the absolute method, based on a Faradays law could be applied. Application of flow-through wall-jet cells was most often reported in combination with gold [53] or gold film working electrode [52], but mercury was also deposited from water on a platinum working electrode by non-specific adsorption without setting the electrode potential during the preconcentration step [54]. Applying the stripping voltammetry with linear potential sweep, carbon working electrode and flow-through cell for mercury determination in water, the detection limit of 0.05 μg/l was achieved [55]. In sea water anodic subtractive stripping voltammetry enabled the detection limit of 1 ng/l [53]. Anodic stripping voltammetry was also applied for mercury determination in sea waters with the use of graphite working electrode [56]. After 60 minutes of electrolysis dissolution of generated mercury deposit was performed in 0.005 mol/l $HClO_4$ (medium exchange technique) so detection limit of 5 ng/l was achieved. In order to avoid the exchange of the medium, thyocianate can be used as an electrolyte [55, 57, 58]. In latest time the real revolution occurred in the development and application of chemically-modified electrodes. They are characterised by high sensitivity and selectivity, fast response and low price. Mercury can be concentrated at such electrodes by selective complexation, ion-exchange, adhesion or adsorption. Great number of various chemically-modified electrodes has been developed for mercury determination. One of such electrodes was used for mercury determination in water and was produced by modifying the screen-printed electrode with Sumixelat Q R10 resin, which contained dithyocarbamate groups, responsible for complex formation with mercury [59]. In this way better sensitivity was achieved in comparison with the non-modified screen-printed electrodes. Glassy carbon was modified with poly(4-vinylpyridine) and Kryptofix-222 in order to preconcentrate Hg(II) [60]. For the determination of methyl-mercury carbon paste was modified with thyol resin [61].

ARSENIC DETERMINATION

Arsenic is known as a poison since ancient times. It occurs in the Earth crust at an average level of 2-5 mg/kg and it is most significant water metalloid pollutant. Estuarine waters can contain high concentration of arsenic, originating from geological sources. Typical example is arsenic contamination in Bangladesh. Concentrations of arsenic in open ocean seawater are typically 1-2 μg/l. Arsenic is also widely distributed in surface freshwaters, and concentrations in rivers and lakes are generally below 10 μg/l, although individual samples may range up to 5 mg/l near anthropogenic sources. Volcanic activities release arsenic in the atmosphere influencing its distribution. Usage of pesticides and various wood preservatives also increases arsenic contents in the environment. Combustion of fossil fuels, particularly coal, introduces large amounts of arsenic in the environment. Like mercury, arsenic can be converted in more toxic methyl derivatives by bacteria. Same anaerobic bacteria, which produce methylcobalamine as an intermediate, are involved in the transformation step. Soluble inorganic arsenic is acutely toxic and ingestion of large doses leads to disturbances in cardiovascular, gastrointestinal and nervous system functions, and eventually to death [62, 63]. Long-term exposure to arsenic in drinking water is related to increased risk of skin, lungs, bladder and kidney cancer [64]. Chronic arsenic exposure has been shown to cause "blackfoot" disease [65], a severe form of peripheral vascular illness which leads to gangrenous changes. Distribution of arsenic species in the environment is presented in Figure 6.

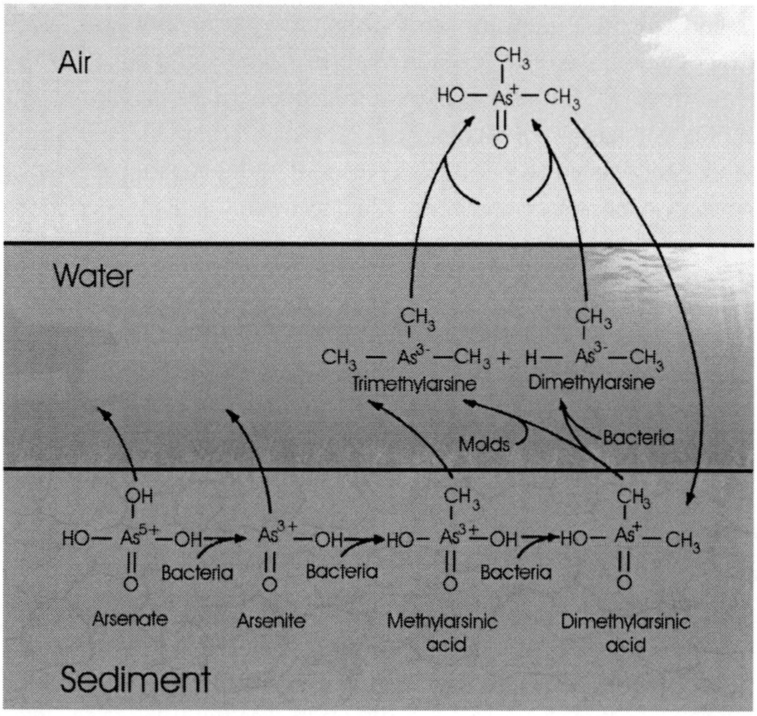

Figure 6. Distribution of arsenic species in environment.

When applying electrochemical stripping analysis for arsenic determination, one should note that only trivalent arsenic is electroactive. In order to determine the total arsenic content, the reduction procedure should be applied. Pentavalent arsenic can be reduced at room temperature with the mixture of potassium-iodide and ascorbic acid. The reduction can also be performed with sulphur-dioxide [66]. The reduction efficiency with the hydrazine hydrochloride can be improved by microwave heating [67]. The method was applied prior chronopotentiometric stripping determination of arsenic in potable, surface and waste water applying the flow system and gold film formed on a graphite powder. For concentration calculation, the absolute method was used. For arsenic determination cathodic mode in stripping voltammetry is most frequently applied [68-73]. Deposited arsenic is being dissolved by its reduction to As(-III) form.

In CSV arsenic was preconcentrated on a mercury electrode in the form of intermetallic compound with copper or selenium that was added to the analysed solution. Copper addition is necessary because of arsenic low solubility in mercury. Differential pulse anodic stripping voltammetry (DPASV) has also been used for arsenic determination in polluted waters [74, 75]. Anodic voltammetry is used in combination with platinum, gold or gold film electrodes [76-78]. DPASV offers higher sensitivity than linear scan anodic stripping voltammetry. A way of sensitivity improvement of anodic stripping techniques is the use of a heated gold wire as a working electrode [79] or use of rotating electrodes [80]. By applying the DPASV and the gold disc electrode, the detection limit of As(III) of 0.15 μg/l in natural waters can be reached [67]. Some authors have achieved extraordinary detection limit (5 ng/l) by applying the square-wave cathodic stripping voltammetry and mercury drop electrode with the addition of copper(II) chloride [81]. Square-wave stripping voltammetry was also reported for selective As(III) determination (0.7 ng/l) in synthetic and natural waters with *in situ* plated bismuth-film electrode [75]. For arsenic determination in contaminated waters constant current in stripping step has also been reported [82, 83]. Arsenic was preconcentrated in flow-through cell with a gold plated porous carbon electrode and then stripped anodically, although application of negative currents and mercury film electrode was also reported [84].

Thallium Determination

Thallium is a metal found in natural ores. The greatest use of thallium is in specialized electronic research equipment. It does not persist for a long time if released to water, but does have a strong tendency to accumulate in aquatic life. If released to land, it binds to alkaline soils. Main syptoms of a short-term exposure to tallium level below Maximal Contaminant Level, set by EPA(2 μg/l), include gastrointestinal irritation and nerve damage. Thallium has the potential to cause the severe effects from a lifetime exposure at levels above the Maximum Contaminant Level. This include changes in blood chemistry, damage to liver, kidney, intestinal and testicular tissues and hair loss.

By applying flow-injection analysis with tubular working electrode made of graphite impregnated with the epoxy resin and mercury film, thallium has been determined in the samples of tap and river water [85]. In DPASV mode low detection limits were obtained (2 ng/l) due to high flow rate and electrolysis of 100 cm^3 of the sample during 90 s. The method of potentiometric stripping analysis with Hg(II) as an oxidant was also defined for the determination of thallium traces (LOD = 0.01 μg/l) in tap and rain water as well as river sediments [86]. Mercury film electrode was formed by *in situ* electrolysis and, since the Hg(II) was used as an oxidant in the analytical step, it was necessary to deaerate the solution prior analysis. Applying the same technique Tl(I) was determined in Rayne river [87]. Mercury film electrode modified with poly-(4-vinylpyridine) enabled the preconcentration of Tl(III) in the anionic form ($TlCl_4^-$) due to the ion-exchange effect [88].

BISMUTH DETERMINATION

Bismuth is not considered highly toxic. Generally, Its compounds have very low solubility in water but should be handled with care, as there is only a limited information on their effects and fate in the environment. Bismuth and its salts can cause kidney damage, although the degree of such damage is usually mild. Large doses can be fatal. Industrially, it is considered one of the heavy metals with the mildest toxic effect. It is shown that it may affect the function of the liver and the kidneys and may cause anemia, black line on gums and ulcerative stomatitis. No chronic health effects were recorded. Direct Bi(III) determination was performed in sea water with the application of PSA and rotating gold disk electrode formed on a glassy carbon in 0.1 mol/l hydrochloric acid as an electrolyte [89]. In the upper layer of the sea water determined content of Bi(III) was 1-2.5 ng/l. Bismuth (III) can also be complexed with alizarin red [90] or cupherron [91], adsorbed on a mercury electrode and subsequently dissolved with potential shift. Bismuth(III) from sea water was deposited on a mercury film modified with poly-(4-vinylpyridine) and dissolved in square-wave voltammetric mode [92]. A highly selective and sensitive anodic stripping differential pulse voltammetry has been developed for the determination of trace amounts of bismuth in various samples of water after solid phase extraction of its 2-(5-bromo-2-pyridylazo)-5-diethylaminophenol complex on amberlite XAD-2 resin in the pH range of 2.0–3.0. The analyte retained on the resin was recovered with 2 mol/l hydrochloric acid so bismuth could be determined with the detection limit of 0.14 μg/l [93]. Adsorption of bismuth complex with bromopyrogallol red was also performed on a carbon paste electrode and consequently, anodic dissolution was peformed. For the accumulation time of 3 minutes, determined detction limit was $5 \cdot 10^{-10}$ mol/l [94]. Due to the strong cation-exchange ability and adsorptive characteristics of sodium montmorillonite, modified carbon paste electrode enabled the enhancement of the sensitivity of Bi(III) determination [95].

MANGANESE DETERMINATION

Manganese is an essential trace element and it plays an important role as a cofactor of many enzyme systems. A long term exposure to the concentrations higher than maximum allawable for drinking water set by EPA (50 μg/l) causes the symptoms symilar to those of Parkinson`s disease. It has also been suggested that Parkinsonism in patients with liver disease may be due to the manganese accumulation in basal ganglia, because liver can not metabolise the metal. Manganese exist in eight different valence states. It is assumed that cationic forms are more toxic than anionic and than Mn(II) is more toxic than Mn(III). In natural waters it occurs in two forms (divalent and trivalent), and it often accompanies iron in ground waters. Manganese is rarely found in natural surface waters in concentrations above 1.0 mg/l. However, in ground waters due to reducing conditions, manganese can be leached from the soil and can occur in high concentrations. The chlorides, nitrates, and sulphates of manganese are soluble in water, while the oxides, carbonates and hydroxides are only sparingly soluble. Present in water at high concentrations, it causes an unpleasant taste, deposits on food during cooking, stains on sanitary ware, discolouration of laundry and deposits on cooking utensils.

For the determination of manganese content in tap, river and lake water by chronopotentiometric stripping analysis the absolute method was applied [96]. Applying the reduction current in the stripping step, the deposit of Mn(II) that was formed by anodic electrolysis (E = 1.2 V (Ag/AgCl)) was dissolved. Determination was performed in flow-through cell applying the working electrode made of powdered glassy carbon.

Silver Determination

Silver is not a dietary nutrient for humans. The body of an adult contains approximately 2 mg of silver. Average daily intake of silver is 20-80 μg, of which approximately 10% is absorbed. In larger amounts, some silver compounds may be toxic, because silver ions have a high affinity for sulphur hydryl and amino groups, and therefore complexion with amino acids, nucleic acids and other essential compounds occurs in the body. The toxicity is relatively small at oral uptake due to the low absorption in the body. Silver is generally deposited in connective tissue, skin and eyes and causes a grey to black colouring. Person can accumulate approximately 9 mg of silver during long period of time.

Silver is a bactericide, and may therefore be applied in water disinfection. River water generally contains approximately 0.3-1 μg/l of silver. Seawater contains approximately 2-100 μg/l of silver and the concentration in surface layer may be even higher. In phytoplankton it accumulates with the concentration factor of 10^4-10^5. In water silver mainly occurs as Ag(I), and in seawater as $AgCl_2^-$ ion. Numbers of silver compounds, such as silver sulphide, nitrate and sulphur compounds are water soluble. For example, silver chloride has a water solubility of 0.1 mg/l, while silver nitrate has water solubility of 2450 g/l at 25 °C. Silver fluoride is much more water soluble comparing to other silver halogenides.

Ionic silver may be removed from water by ion exchange. Some silver compounds may be precipitated by coagulation. Two other efficient methods for silver removal include active carbon and sand filtration. For swimming pool water disinfection silver is applied in relatively small concentrations.

For silver determination in river and lake water voltammetric stripping techniques with the carbon paste working electrode [97] or carbon fibre [98] were applied. 8-Mercaptoquinoline was electropolymerised by a cathodic process at a glassy carbon electrode. The resulting modified electrode was used for selective preconcentration of Ag(I). Anodic stripping analysis with medium exchange provided a highly selective and sensitive method for Ag(I) determination. Procedures for both batch and flow analysis were developed and optimised [99]. For simultaneous and selective determination of copper and silver at sub-μg/l level differential pulse anodic stripping voltammetry (DPASV) with the working electrode of a gold alloy containing 4% bismuth was used [100].

COBALT AND NICKEL DETERMINATION

Cobalt is an element that occurs naturally in the environment in water, soil, air, rocks, plants and animals. It can be introduced into the air and water by men`s activity. Soil and rock containing cobalt can be washout by rainwater, so element can enter the surface water. Cobalt is beneficial for humans because it is a part of vitamine B_{12} which is essential for human health. It is used to treat anaemia in pregnant women, because it stimulates the production of red blood cells. The total daily intake of cobalt is variable and may be as much as 1 mg, but almost all amount pass through the body unabsorbed, except that in vitamine $B_{12.}$ However, too high concentrations of cobalt may influence negatively human health. Health effects that are result of its uptake in high concentrations include vomiting, nausea, vision and heart problems as well as tyroid demage.

Nickel has no biological function and it is of medium toxicity to humans. It occurs in the environment naturally, but it is also discharged in the environment by man's activity. The organic forms are much more toxic than inorganic. It is toxic to cardiovascular system, as well as being carcinogenic.

For the determination of these elements adsorption of their complexes was used in order to concentrate them on a working electrode, usually mercury drop electrode. As a complexion agent dimethylglyoxim [102], bypiridine [103], nioxim (cyclohexane-1,2-dion dioxime) [104] and 2-qiunolinethiol [105] were used. Nickel and cobalt can be determined in sea water in flow cell with the working electrodes made of carbon fibres, or the fibres can be modified with nafion. The application of CSA was successfully applied in such flow systems [106].

IRON DETERMINATION

Iron is an essential element, incorporated in the oxygen-carrying protein of the red blood cells, haemoglobin. In natural water it is either completly dissolved, suspended or in colloidal form. Organic iron is usually bound to tannins, lignins and other organic matter giving water colored appearance. Organic iron originates from decayed vegetation. Such water is usually found in shallow wells or surface water. The Maximum Contaminant Level for iron in drinking water, set by EPA is 0.3 mg/I. When the level of iron in water exceeds the 0.3 mg/l, red, brown, or yellow staining of laundry, glassware, dishes and household fixtures such as bathtubs and sinks can be expected. The water may also have a metallic taste and an offensive odor. Water system piping and fixtures can also become restricted or clogged. In drinking water iron can be found as dissolved or undissolved. Insoluble ferric oxide is observable as a perticipitate in water. Dissolved or ferrous iron is usually found only in water that has not been exposed to oxygen. Water containing ferrous ion is clear and colorless. Upon oxidation with atmospheric oxygen, red ferric oxide is formed. Dissolved iron from water can be used by some water bacteria in their metabolism. They oxidize iron into insoluble ferric state and deposit it in gelatinous material. Unpleasant taste and odor can develop in water containing iron bacteria, as well as piping clogging. Furthermore, multiplication of these bacteria can develop anaerobic conditions and cause the growth of sulphate-reducing bacteria.

In natural waters iron can be determined by voltammetric stripping of its complexes with 1-nitroso-2-naphtol [107, 108], catechol [109] and 2-(2-thiazolyazo)-p-cresol [110] from mercury drop electrode. Simultaneous determination of iron (III) and titanium (IV) in tap and rain water was reported. Chronopotentiometric stripping techniques, solochrome-violet as a complexion agent and mercury film electrode were used for the determination [111].

Uranium Determination

The natural weathering of rocks dissolves the natural uranium, which then enters the groundwater. Once in the water, uranium does not transfer into the air. Consequently, uranium can be find in water supplies. The tolerable daily intake of uranium is 6 μg/kg of body weight and for average water consumption of 2 litres per day, 20 μg/l is the safety limit set by WHO. The chemical toxicity of uranium is similar for all isotopes, but radiological toxicity differs among the isotopes and depends on the half-life. It is speculated that long-term exposure to uranium can cause kidney damage.

For uranium determination in natural water voltammetric stripping techniques and mercury drop electrode were most frequently applied. Uranium was preconcentrated on a working electrode in the form of its complex with cupherron [16, 112], catechol [113], arsenazo (III) [114], 4-(2-hydroxyethyl)-1-piperazineethanesulfonic acid (HEPES) [115] and potassium hydrogen phthalate [116]. Chronopotentiometric stripping determination of uranium in surface water was also reported with the application of screen-printed working electrode [117]. Such working electrodes have low price and can be used as disposable sensors for sensitive on-site control of the nuclear contamination in waters.

Aluminium Determination

Aluminium is permitted food additive and it is used in colourings, emulsifiers, stabilisers and anti-caking agents. Drinking water contributes to less than 2% of total aluminium intake. Aluminium from food and water is poorly absorbed through the gastrointestinal tract. Less than 1% of aluminium from water is absorbed and the rest is being excreted. Absorbed aluminium is mostly excreted via the kidneys and only a very small amount accumulates in bone, liver and brain tissues. In general population, the major concern is related to the association between the intake of aluminium and the neurodegenerative diseases such as Parkinson's disease. Higher than average levels of aluminium have been found in the brains of Alzheimer patients.

In sea water aluminium was determined after the complexion with dyhydroxyantraxinone on a mercury drop electrode [118]. CSA and mercury film electrode were applied for aluminium determination in tap water after the complexion with solochrome violet [119].

Adsorptive stripping voltammetry is also widely exploited for the determination of other elements in water. Antimony was complexed with catechol [120] and chloranilic acid [121], vanadium with 2,3-dihydroxynaphtalene [122], molybdenum with nitrates [123] and palladium with dimethylglyoxim [124] on a mercury drop electrode.

It is important to mention that not only metals, metalloids and ions are subject of stripping analysis, but also many other substances. In tap water the concentration of complexion agents such as EDTA, EGTA or DCTA were determined after the reaction with Hg(II) obtained by anodic dissolution of the working electrode [125]. Folic acid [126], ammonium hydroxide [127] and glutathione [128] were also successfully determined in sea water applying the cathodic stripping voltammetry and mercury working electrodes. Determination of some sulphur-containing pesticides in water and soil was reported with the application of stripping techniques, usually with mercury working electrodes [129].

DETERMINATION OF ANIONS

The ions that were most often determined in natural waters with the application of electrochemical stripping techniques are chlorides, iodides, bromides, sulphides and nitrates. Chloride is one of the major anions present in water. Since majority of chloride salts are highly soluble in water, the chloride content in water ranges from 10 mg/l to 100 mg/l. Sea water contains over 30 000 mg/l as NaCl. High chloride concentration is associated with the corrosion of the piping. For example, magnesium chloride can generate hydrochloric acid when heated. Corrosion rates and the amount of dissolved iron from piping increases with the chloride content. For these and other reasons the suggested Maximum Contaminant Level (EPA) for chloride in drinking water is 250 mg/l. Higher contents can cause salty taste in drinking water.

Iodine is vital for good thyroid function, which in turn is essential for health. It deficiency during pregnancy and early infancy can result in cretinism, i.e. the irreversible mental retardation and severe motor impairments. In adults low iodine intake or very high intakes can cause hypothyroidism. Hypothyroidism can manifest as low energy levels, dry, scaly or yellowish skin, tingling and numbness in extremities, weight gain, forgetfulness, personality changes, depression and anaemia. Hypothyroidism can lead to significant increases in cholesterol levels. For iodide, a relatively narrow range of intakes reliably supports good thyroid function (100 to 300 micrograms per day). Excess of iodide has a complex disruptive effect on the thyroid gland and may cause either hypothyroidism or hyperthyroidism in susceptible individuals, as well as increased risk of thyroid cancer. Hyperthyroidism may manifest as an enlarged thyroid, heart rate irregularities, tremor, sweating, palpitations, nervousness, increased activity and eye abnormalities.

For iodide determination in mineral water as well as in the table salt, DPCSV and working electrode of carbon paste was applied [130]. Chlorides can be determined in tap water with a silver film electrode deposited on a glassy carbon applying the chronopotentiometric stripping analysis [131]. Sulphides can be deposited on a mercury electrode in the form of mercury(II) sulphides and subsequently dissolved by a negative potential shift either in linear [132] or square-wave mode [133]. Excessive levels of nitrate in drinking water can cause serious illness and sometimes death, due to the oxidation of ferrous iron from heme into the trivalent form, which can not bind the oxygen. The infants are very susceptible to nitrates because they have a lack of enzimes which reduce methemoglobin beck to hemoglobin. This can be an acute condition in which health deteriorates rapidly over a period of days. Symptoms include shortness of breath and blueness of the skin (methemoglobinemia).

Long-term exposure to nitrates and nitrites causes diuresis, increased starchy deposits and hemorrhaging of the spleen. Ascorbic acid diminishes influence of nitrites on methemoglobinemia apperence by preventing its oxidation to nitrates in gastrointestinal tract. Nitrates are usually determined by voltammetric stripping techniques and silver working electrode. Planar disk silver electrode or composite electrodes made of graphite powder, silver and metacrilic resin were applied [134].

Bromide is found in seawater at a level of about 65 mg/l. Bromine has been used in swimming pools and cooling towers for disinfection. However, its use in drinking water is not recommended. Bromate is not present in natural waters but can be formed from bromide in ozonisation process. It can be found in the concentrations of 60-90 μg/l in waters after the ozonisation. More than 0.05 mg/l of bromine in fresh water may indicate the presence of industrial wastes, possibly from the use of pesticides or biocides containing bromine.

Sodium-sulphate and magnesium-sulphate levels above 250 mg/l in drinking water may produce laxative effects. Excess sodium may affest persons with hypertension and pregnant women.

CSA of Arsenic and Mercury in Water

Many social tragedies, such as methyl-mercury poisoning in Minamata Bay and most severe population exposure to arsenic in Bangladesh are written down in the annals of toxicological history. Mercury and arsenic, among the most dangerous water contaminants, should continuously be in the focus of the toxicological, scientific and analytical attention. These elements can be a subject of bioconversion by the microorganisms. Organic species, formed in these biochemical processes tend to accumulate in the tissues of aquatic species. As an analytes, they are often considered very problematic due to their high volatility and diversity of their chemical forms. Scientists, when meeting the problem of their determination, should consider requirements for high sensitivity of the analysis, since μg/l concentration levels in water already have serious impact on human health and can be easily further biomagnified.

As a part of performed research work related to the application of different electrochemical stripping techniques in food analysis, environmental analysis and the analysis of the pharmaceutical product, great work was consecrated to the development of the methods which could enable rapid, simple and reliable control of mercury and arsenic content in water samples, considering their toxicological and environmental importance. The aim of the research work described in this chapter was to develop a simple method for mercury and arsenic determination, avoiding tedious preparation of the working electrode as well as time-consuming sample preparation step, without neglecting the sensitivity criteria. Such methods could be used for continual and semicontinual water monitoring. Reliable methods of CSA of mercury and arsenic, characterised with its high sensitivity, selectivity and accuracy were defined for the screening of water contamination.

Chronopotentiometric Stripping Determination of Mercury Using a Glassy Carbon

Flow-through stripping systems are suitable for *in situ* analysis of natural waters. Continual on-line monitoring of metallic toxicants in the environment is important in the cases of sudden contamination. Basic advantages of flow systems are simplicity, low price and analysis duration. Big sample input is possible in a short time. In systems for on-line electrochemical analysis the stability and durability of the working electrode is often a limiting factor. The problem can be resolved by applying the stabile and inert electrodes, such as glassy carbon electrodes. Chronopotentiometric stripping techniques have been less frequently applied for water analysis regarding mercury determination in comparison with the voltammetric stripping techniques. When applied, gold working electrodes were most often used [135]. Sensitivity improvement up to eight times can be achieved if rotating gold disc electrode is used instead of the stationary one [135]. Besides the commonly used inert supports of the gold film electrodes, made of graphite or glassy carbon, specific support made of the graphite impregnated with the epoxy resin and polyethyleneamine were also used [136]. The data concerning the application of the working electrodes other than gold in chronopotentiometric stripping determination of mercury are not well elaborated in the literature. Chronopotentiometric stripping determination of mercury in waters in flow systems has been reported, but not in combination with the glassy carbon working electrode [136-138]. Within this research which glassy carbon was used as an electrode material, due to its stability, inertness and duration. Comparing to gold electrodes, maintenance of more inert glassy carbon is easier and the reproducibility of its surface is better.

The aim of this research work was not only to define the CSA of mercury on the glassy carbon, but to improve the sensitivity of mercury determination by chronopotentiometric stripping analysis. To reach this goal working electrode of a large active surface (process vessel working electrode) was used [139]. Preliminary investigation was directed toward the comparison with common electrochemical cells. Systems for continual and semicontinual water monitoring were investigated, developed and defined. Optimal experimental parameters were defined for various electrochemical systems: classical electrochemical cell with stationary and rotating disc electrode, system with a process vessel as a working electrode, thin-layer and tubular flow-through cells. The developed method was applied for direct determination of mercury(II) in different types of water.

From the solutions of low concentration, mercury is poorly deposited on a glassy carbon. Copper deposition is likewise better. A first amount of the deposited copper forms the film at which the mercury can be deposited easier. Mercury can, therefore, be deposited on a glassy carbon from solutions of low concentration only in codeposition process with some other element that easily deposit on a glassy carbon, like copper or gold.

In the preconcentration step convective mass transfer is assured by the stirring of the analised solution, by the rotation of the working electrode or by the solution flow. Intensity of convective mass transfer, i.e. the stirring rate in classical electrochemical cell during the preconcentration step influences directly on the amount of the formed deposit. With the increase of the stirring rate mercury analytical signal increases until the 5000 r.p.m. for classical electrochemical cell. Above 5000 r.p.m. at the surface of the working electrode air bubbles are attained so the active surface of the electrode is diminuated and this reflects the poorer sensitivity.

The electrolysis potential is very important experimental factor for the methods selectivity as well for the height and sharpness of the analytical signal. More negative electrolysis potentials are contributing to the higher amounts of mercury deposited at the glassy carbon surface. Reproducibility of mercury determination is significantly lower at the potentials more negative than -1.1 V due to the hydrogen evolution.

Applying the longer electrolysis during the preconcentration, the splitting of the mercury analytical signal can be observed (Figure 7).

One of the explication is that these two plateaus correspond to mercury dissolution from more and less active cites of the glassy carbon. During the mercury deposition on the glassy carbon, the active cites of higher energy/activity are occupied firstly. For the mercury bonding on a less active cites, higher energy, i.e. longer electrolysis time is required. Mercury deposition on a glassy carbon starts in the form of small droplets. Further deposition on already formed mercury droplets is more probable and faster then on a pure glassy carbon, so longer electrolysis time causes the enlargement of the mercury droplets and their cohesion. With the prolongation of the electrolysis, new small mercury droplets are formed on a less active cites in a second layer. During the mercury dissolution, mercury is firstly dissolved from less active cites (more negative plateau) and then from more active cites (more positive plateau) (Figure 7). Signal splitting is actually attributed to the formation of multiple mercury layers at the electrode surface.

Dissolution current is one of the most important experimental factors in chronopotentiometric stripping analysis. Adequate dissolution current enables the sharp and reproducible analytical signal.

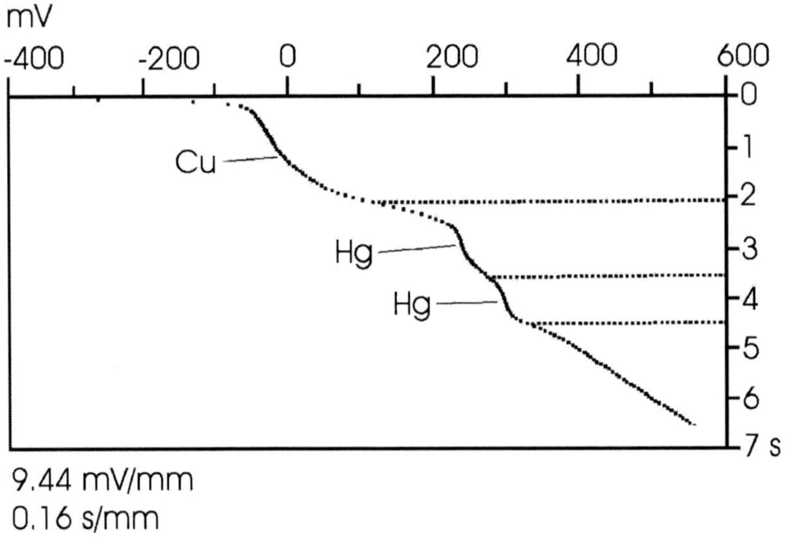

Figure 7. Splitting of the mercury analytical signal. Classical electrochemical cell. Stationary glassy carbon electrode. Cm(Cu2+) = 100 μg/l; Cm(Hg2+) = 5 μg/l; E = -0.9 V; t = 600 s; i = 4.4 μA.

Mercury dissolution time decreases exponentionally with the higher dissolution currents irrespective to the mercury concentration. The reproducibility of mercury determination is better if higher dissolution currents are applied. However, for lower mercury contents smaller dissolution currents are recommended. Dissolution potential of mercury in chronopotentiometric stripping analysis at a glassy carbon is not significantly affected by the dissolution current.

Often polynomial dependence of the mercury analytical signal on the concentration (Figure 8) can be observed in the cases when the determination is performed from the solutions with higher mercury concentrations. Even though for narrow concentration range the linearity is rather good, standard addition method should not be a method of choice, since significant intercepts at “Y” axis is affecting the accuracy of the determination. Standard addition method which, in most of the cases compensates the matrix influence and the interferences, can only be applied if the linear dependence and small “Y” intercepts are observed. Otherwise false results can be calculated by the analyser. In some cases the problem of the isignificant intercept at “Y” axis can be resolved if the calculated result is corrected.

Small “Y” intercepts and good linearity regardless the concentration ranges were observed for the rotating disc electrode. These are the important advantages of the rotating disc electrode comparing to the stationary one.

Copper addition is necessary for mercury deposition on a glassy carbon. Another elements, such as bismuth can also be added for that purpose [140]. Addition of the second element also improves the accuracy of the dissolution time measurement due to the existance of the previous analytical signal (plateau). To define the optimal concentration of the „helper“ element, the copper concentration was gradually increased for constant mercury content.

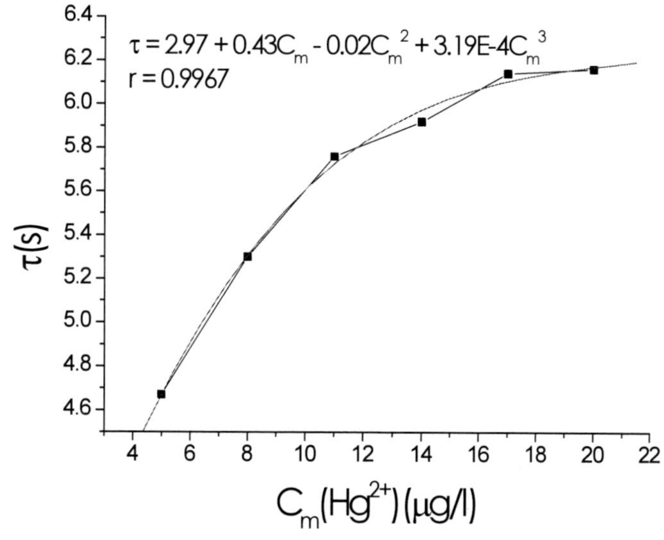

Figure 8. Dependence of mercury analytical signal on the content. Classical electrochemical cell. Stationary glassy carbon elevtrode. C (HCl) = 0.006 mol/l; Cm(Cu2+) = 100 μg/l; t = 120 s; i = 50 μA.

High copper content, in relation to mercury content, caused the loss of mercury analytical signal, possibly because copper deposition on a glassy carbon was favorable. Since copper is present in most of the waters, it is necessary to determine its content prior its addition because too large amounts of copper can leed to the loss of the mercury analytical signal or to the sensitivity decrease when mercury is present in low concentrations. In the analytical procedure, the range of the copper conentration in which mercury analytical signal remains approximatelly uncheanged should be used. For wider mercury concentration window adequat copper conentration can not be defined, but should be adjusted to the exemined system. For to high copper contents mercury analytical signal can be lost, and too low copper concentration will not sufficienly initiate the mercury deposition. Splitting of mercury analytical signal can also be observed in the cases of high copper concentrations. It can be assumed that for higher copper concentrations "copper" electrode is formed. Dissolution of mercury from such surface and from the rest of the glassy carbon is appearing at close potentials. From these experiments it can be concluded that for wider range of merury content, optimal copper content can not be defined. For the mercury contents in the range 1-5 μg/l that can be expected in some real samples, it is recommended to add ~85 μg/l in order ti initiate mercury deposition.

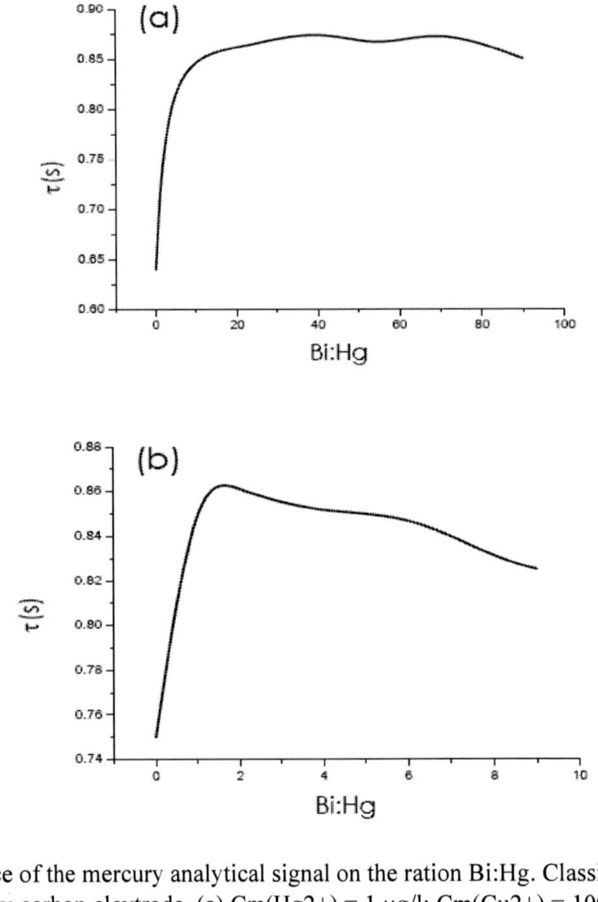

Figure 9. Dependence of the mercury analytical signal on the ration Bi:Hg. Classical electrochemical cell. Stationary glassy carbon elevtrode. (a) Cm(Hg2+) = 1 μg/l; Cm(Cu2+) = 100 μg/l (b) Cm(Hg2+) = 10 μg/l; Cm(Cu2+) = 100 μg/l.

Bismuth is an element very often present in natural water. Therefore it is important to define its influence on mercury analytical signal. Added bismuth can codeposit with the mercury and improve the relative sensitivity of the mercury determination. For adequate amount of added bismuth, mercury analytical signal is approximately constant and higher comparing to the signal obtained without bismuth addition (Figure 9).

Mercury analytical signal dissaperes when critical concentration of bismuth (3 mg/l) for the experimental conditions is added due to the formation of bismuth electrode. It can be assumed that the addition of small amounts of bismuth can possibly improve the relative sensitivity of mercury determination on a glassy carbon.

Silver deposits very well at a glassy carbon. Silver present in the mercury solution codeposits with the mercury at a glassy carbon and interferes at every concentration due to the analytical signal overlapping. The dissolution potential of a common analytical signal shifts to more negative values with the higher silver concentrations. For certain silver concentration, depending on a mercury concentration, splitting of the common plateau is observed (Figure 10). In splitted signal more negative one corresponds to the mercury dissolution and does not change with the concentration of the silver. When silver is present in proportionally higher concentration than mercury it completely covers the mercury analytical signal so one plateau at ~300 mv (3.5 mol/l Ag/AgCl), corresponding mainly to the silver dissolution is observed. Mercury determination at glassy carbon can not be performed in the presence of silver due to the formation of a common analytical signal. In those cases measured dissolution time is false.

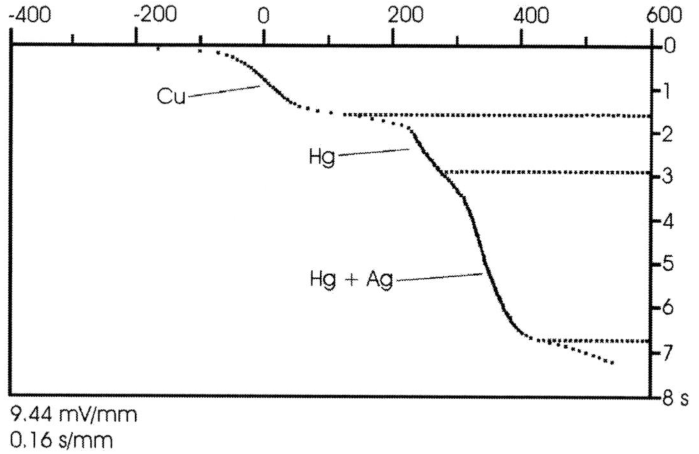

Figure 10. Splitting of the common signal of mercury and silver. Classical electrochemical cell. Glassy carbon disc electrode. Cm(Cu2+) = 100 μg/l; Cm(Hg2+) = 1 μg/l; E = -0.9 V; t = 600 s; i = 9.3 μA.

Manganese is almost always present in all types of waters. No significant influence on chronopotentiometric stripping determination of mercury on a glassy carbon was observed. However, presence of manganese is undesireble when the glassy carbon is used as a working electrode due to its influence on the reproducibility of the electrode surface. After analysis of solutions containing lower manganese concentrations ultrasonic cleaning in 50% ethanol is sufficient for obtaining the surface of good electrochemical properties. For solutions containing higher manganese concentrations this procedure is unefficient and it is necessary to polish the surface of the electrode with the suspension of Al_2O_3.

PROCESS VESSEL AS A WORKING ELECTRODE

Analyte deposition on a solid electrode is a very complex process. For higher concentrations of the deposited metal, the shift of the dissolution potential can occur towards more positive values. Brainina et al resolved the equations defining the dependence of the dissolution time and current on the concentration for reversal and ireversal processes in voltammetric stripping techniques, considering the amount of the formed deposit and the rate of potential sweep [140]. They came to the conclusion that the dissolution potential is independent on the amount of the formed deposit for reversal processes in the cases of deposit thickness less than a monolayer. For thicker films of the analyte, the dissolution potential is proportional to the logarithm of the amount of the deposit. When the analyte is preconcentrated in the form of the metal film, precaution and the analysts experience is required. Accuracy and the reproducibility of the determination in those cases depend on a reproducibility of the electrode surface. Compact and reproducible analyte layers can only be obtained on a homogenous surface of the working electrode. Reduction potential at solid electrodes can differ from the Nernst value and should be experimentally defined. More negative potential can be expected in the cases of thicker layers, while more positive potentials should be applied in the cases when the electrode surface is only partially covered with the deposit. Thicker layers are characterized with nonlinearity of the analytical signal for higher analyte concentrations, when the surface is completely covered with the analyte, and by the occurrence of the multiple analytical signals caused by the different strength of the bonds between the monolayer atoms and the electrode, and between the atoms from the other layers of the deposit and the electrode. Character of the analytical signal in these cases depends on a type and metal concentration, deposition time and surface characteristics of the electrode material.

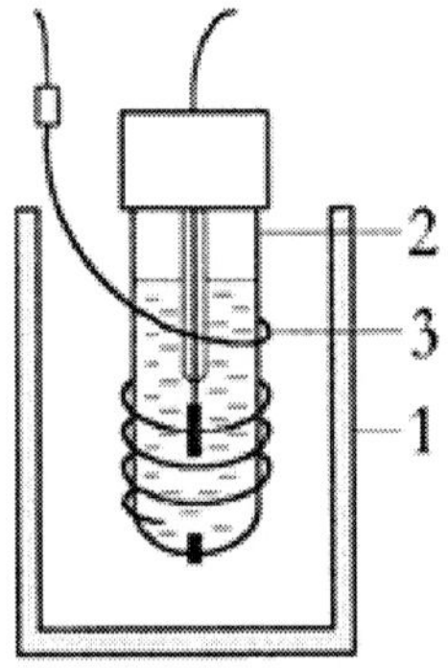

Figure 11. Electrochemical cell with the process vessel as a working electrode. 1. Process vessel of glassy carbon; 2. Reference electrode, Ag/AgCl (3.5 mol/l; KCl); 3. Auxiliary electrode – platinum wire.

In this study the role of the vessel made of the glassy carbon was dual, to serve as a process vessel and as a working electrode (Figure 11) [139].

This idea is in correlation with the basic principles of ESA concerning the film formation i.e. the bigger surface is assessable for the analyte to be abstracted. In this specific case, the surface of the working electrode depended on the volume of the analised solution. When working electrodes of big surface are used, sufficient current density must be applied.

The electrolysis of the analised solution in the process vessel can be performed in the conditions of diffusive mass transfer as well as in the conditions of convective mass transfer if the rotation of the process vessel is performed. Rotation of the electrode contributes significantly to the sharpness and height of the analytical signal, eliminating the problems associated with the determination of the inflection point at the chronopotentiogram and the oxidation time measurement. Ratio (R) of mercury signals obtained in conditions of convective mass transfer and in steady conditions for different mercury contents is given in Table 3.

From Table 3 it can be seen that the positive influence of convective mass transfer was less prone in solutions with lower mercury content and pointed at the conclusion that electrolysis efficiency depends on the concentration under applied experimental conditions. This was also confirmed by examining the dependence of mercury analytical signal on electrolysis time.

Electrolysis efficiency depended on the concentration. At lower mercury content (20 ng/l) the curving of the dependence of mercury analytical signal on the electrolysis time appeared approximately at the same electrolysis times for both steady and rotating vessel. At higher content (3 μg/l) and the conditions of convective mass transfer, the curving was observed at much shorter electrolysis time. In the conditions of convective mass transfer less active sites of the glassy carbon were also covered with the mercury and caused the splitting of the mercury signal.

Table 3. Ratio of mercury signals obtained in conditions of convective mass transfer and in steady conditions of the process vessel working electrode

	Mercury content (μg/l)			
	1	5	10	15
R	1.5	1.9	3.1	4.8

Applying the working electrode of such big surface, achieved sensitivity was significantly better comparing to disc and rotating disc electrodes made from the glassy carbon (Figure 12). As can be seen, mercury analytical signal is much higher when the process vessel is used as a working electrode, even though electrolysis time is shorter and the volume of the analised solution is much smaller comparing to two other systems

Comparing different electrochemical systems, the best sensitivity of mercury determination at glassy carbon was observed in the system where the process vessel is used as a working electrode (Table 4). This system enables semicontinual water monitoring at ng/l mercury levels. It should be noted that this system requires very small volume of the analised sample (5 – 9 ml) which is advantageous fact when sample size is limited. However, the reproducibility of such large surface can represent a major problem in analytical practice and requires great experience of the analyst. Maintaince of the active electrode surface is much simpler in the cases of the disc electrodes (stationary and rotating). Rotating disc electrodes enables approximately 25 times more sensitive mercury determination in comparison with the stationary disc electrode of the same active surface area. Best reproducibility of the determination was observed in the classical system with the stationary working electrode. Linearity, as one of the most important analytical requirements, is best for the rotating disc electrode, since it is observed in wide concentration range. No curving of the dependence was observed for this system in examined concentration range. The system with rotating disc electrode enabled the application of the standard addition method, which significantly compensates the matrix influence, since the intercept at "Y" axis is not significant. The dependence of the analytical signal on the electrolysis time was linear in wide range only for the system with the rotating disc electrode.

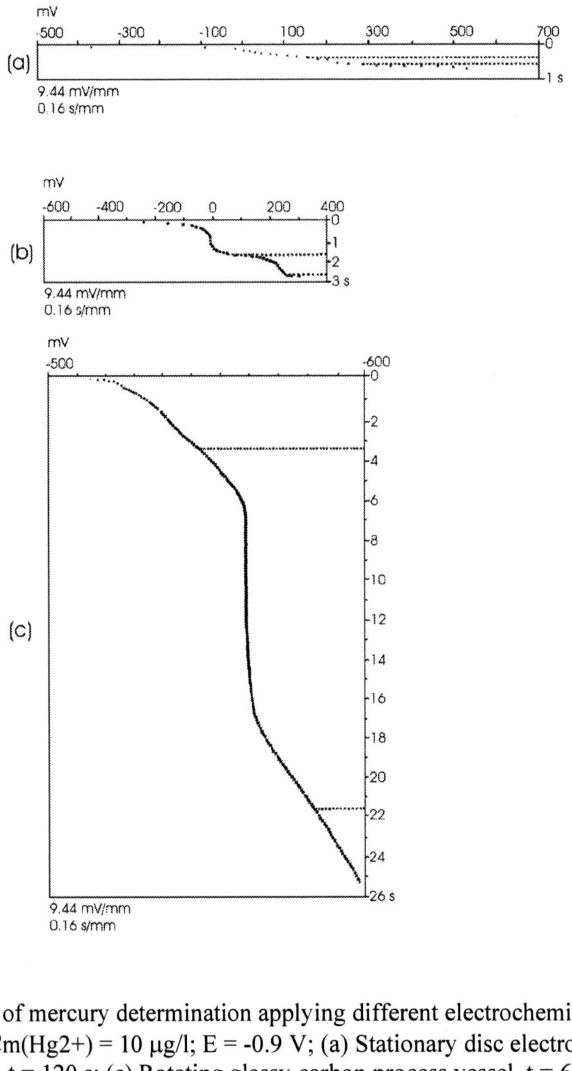

Figure 12. Comparison of mercury determination applying different electrochemical systems. Cm(Cu^{2+}) = 70 µg/l; Cm(Hg^{2+}) = 10 µg/l; E = -0.9 V; (a) Stationary disc electrode, t = 240 s; (b) Rotating disc electrode, t = 120 s; (c) Rotating glassy carbon process vessel, t = 60 s.

Table 4. Comparison of different electrochemical systems for mercury determination

Electrochemical cell	Working electrode	LOD	Reproducibility at the detection limit Coefficient of variation (%)	Experimental conditions for LOD	Linearity
Classical	Disc electrode d = 3 mm	1.1 μg/l	7	t_{el} = 600 s i = 9.3 μA	20 μg/l – 40 μg/l
Classical	Rotating disc electrode d = 3 mm	27 ng/l	8.1	t_{el} = 600 s i = 2.2 μA	10 ng/l – 10 μg/l
Process vessel	Process vessel H = 3.5 cm, d_{inner} = 1.9 cm	0.1 ng/l	9	t_{el} = 600 s i = 48 μA	Rotating vessel 5 ng/l – 30 ng/l
Flow system	Tubular l = 1 cm; d = 0.7 mm	0.3 μg/l	7.9	Electrolysis in stationary conditions t_{el} = 600 s i = 48 μA	5 μg/l – 25 μg/l
Flow system	Thin layer cell, d = 3 mm; l^* = 0.31 mm	0.2 μg/l	8.3	Flow rate 0.4 ml/min t_{el} = 600 s i = 2.2 μA	5 μg/l – 20 μg/l

*Thickness of the cell

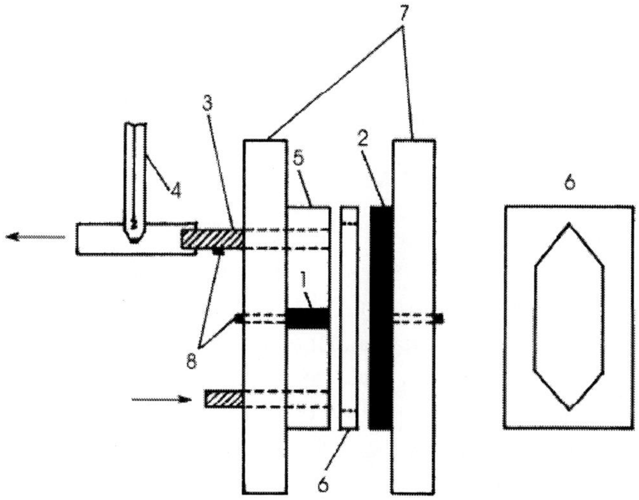

Figure 13. Thin-layer flow-through cell. 1. Working electrode - glassy carbon disc electrode, A = 7.06 mm2; 2. Working electrode 2 – glassy carbon, A = 280 mm2; 3. Auxiliary electrode, stainless steel, ϕ = 1.9 mm, l = 50 mm; 4. Reference electrode, Ag/AgCl (3.5 mol/l KCl); 5. Teflon body of the working electrode; 6. Teflon foil (0.2 – 0.8 mm); 7. Plexiglas plates; 8. Electrical junction.

Flow systems (Figures 13, 14) can be very useful for continual monitoring of the analyte traces in water. They are applicable only in the cases when the sample preparation procedure is simple or when the technique can be applied directly on a sample. This is the main advantage of the electrochemical stripping techniques in comparison with other instrumental techniques, since it enables direct water analysis due to the high sensitivity and selectivity and can be implied in flow systems.

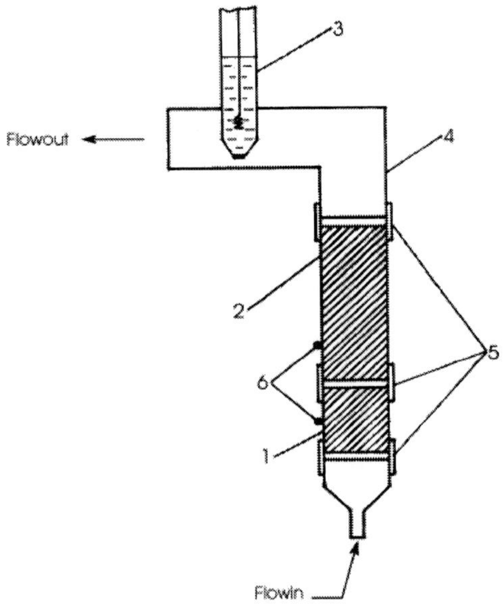

Figure 14. Tubular flow-through cell. 1. Working electrode (glassy carbon); 2. Auxiliary electrode (glassy carbon); 3. Reference electrode; 4. Reference electrode carrier; 5. Silicone tube; 6. Electrical junction.

In flow systems detection limits below 1 μg/l were determined with good linearity in the range of expected mercury contents in natural waters. Even though the durability of the glassy carbon is high, both flow systems require regular dissemble and the polishing of the working electrodes.

The content of the inorganic mercury and labile mercury complexes can be determined by the direct analysis of water samples by developed chronopotentiometric stripping technique. Different water samples were analised applying different electrochemical systems: Atlantic Ocean (Madeira cost), Adriatic sea (Montenegro), River water (Danube, Tisa, Drina, Sava), well water, rain water, tap water and waste water from the industry in order to verify the applicability of the method for different water types. The water samples were preserved in 0.01 HCl. Obtained results are presented in the Table 5. In most of the natural water samples copper is present in sufficient concentration to promote mercury deposition, so additional amounts were not added. It is recommended to define the calibration curve in water sample with similar matrix where mercury is not detected. In water samples copper content was previously determined (PSA) and the calculated content was added in the matrix for calibration curve definition. Due to the influence of the organic matter caused by its adsorption at the electrode surface, stripping step was performed in conditions of convective mass transfer.

By applying this method only inorganic mercury and mercury from its labile complexes was determined. In aerobic conditions in natural waters the greatest amount of the present mercury is in the form of inorganic compounds so this data can be very useful for detecting water contamination [141].

Table 5. The content of Hg(II)

Sample	Rotating disc electrode (μg/dm3) KK1 SA2	Rotating process vessel (μg/dm3) KK
Tap water 1	Nd3 Nd	Nd
Tap water 2	Nd Nd	Nd
Danube 1	0.25±0.044 0.40±0.04	0.40±0.05
Danube 2	0.17±0.03 0.30±0.02	0.40±0.02
Danube 3	Nd Nd	Nd
Tisa 1	Nd Nd	Nd
Tisa 2	0.60±0.03 0.90±0.03	0.90±0.04
Drina	Nd Nd	Nd
Sava 1	0.30±0.06 0.45±0.05	0.50±0.02
Sava 2	Nd Nd	Nd
Atlantic Ocean	Nd Nd	Nd
Adriatic sea	Nd Nd	Nd
Waste water	600±50 735±70	886±37

[1] The content determined by calibration curve method;
[2] The content determined by the standard addition method; [3] Not detected; [4]2SD

Chronopotentiometric Stripping Determination of Arsenic Using Gold Film Electrode

For arsenic determination most often gold working electrodes are used. At such electrodes arsenic is most probably deposited as an elemental. The more negative electrolysis potential is, the higher amount of total As(III) is reduced to As^0 at the electrode. Electrolysis potentials more negative than -0.35 V causes the evolution of the hydrogen at the electrode surface due to the electrolyte reduction. Hydrogen bubbles damage the gold film and, consequently, poor reproducibility is observed. Application of the electrolysis potentials more positive than -0.15 V usually don't provide sufficient sensitivity of the determination.

Enabling the convective mass transfer in the electrolysis step, high stirring rates are not recommended if the gold film electrodes are used. Turbulence causes the accumulation of air bubbles at the electrode surface, which interfere the arsenic deposition. Higher stirring rates cause the damage of the gold film and so diminishes the reproducibility as well as sensitivity. Small slopes and significant "τ" intercepts are observed for arsenic determination on a gold film electrode, so inadequate enlargement of the analytical signal is obtained after the addition of the standard causing the enlarged results calculated by the analyser. In such cases calibration curve method should be applied for content calculation.

Arsenic analytical signal exponentionally decreases with higher dissolution currents. In the cases when too small currents are applied the extension of the chronopotentiogram and difficulties in the determination of inflection point can occur. As a rule, higher currents can be used for higher concentrations leading to sensitivity decrease but better sharpness of the analytical signal. Generally, lower dissolution currents should be used for lower arsenic contents.

In most of the real samples containing arsenic, copper is almost always present as well. Copper and arsenic form an intermetallic compound. In stripping voltammetry arsenic accumulation in mercury electrodes is performed mostly in the form of the intermetallic compound and copper is therefore intentionally added to the analysed solution [71-74]. Copper influence on arsenic deposition at gold electrodes is not clearly described in the literature. By varying copper concentration in solutions containing As(III), it was concluded that copper addition contributed to the significant improvement of the sensitivity of arsenic determination (Figure 15).

In order to define behaviour of As:Cu system, by varying arsenic content in solutions containing constant copper content, the obtained dependence was analysed (Figure 16).

Dependence presented in Figure 16 with higher slope in the first part of the curve (~15 $\mu g/dm^3$ of As^{3+}) and much lower slope in the second part of the curve show that the content of added copper was sufficient for sensitivity improvement only for lower arsenic contents. By maintaining the As:Cu ratio constant and equal to one, assumption that for higher arsenic contents higher amounts of copper are required was confirmed. The dependence of arsenic analytical signal on content was linear with good correlation coefficient. Despite the fact that copper caused the increase of arsenic analytical signal, copper addition did not influence the detection limit of arsenic. Copper addition can have significant practical use in chronopotentiometric stripping determination of arsenic at gold electrodes due to shortening of the needed electrolysis time, though it should be considered that better reproducibility of arsenic determination is achieved without copper addition.

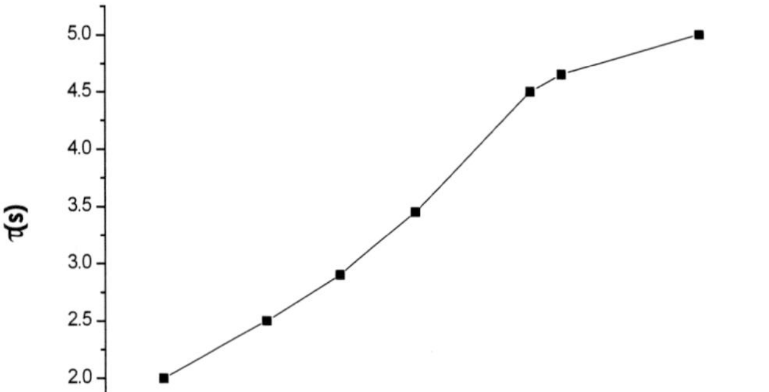

Figure 15. Copper influence on arsenic analytical signal (Cm(As3+) = 60 μg/dm3). Gold film electrode. 60 s deposition time, 4.8 μA stripping current, 2000 r.p.m. stirring rate.

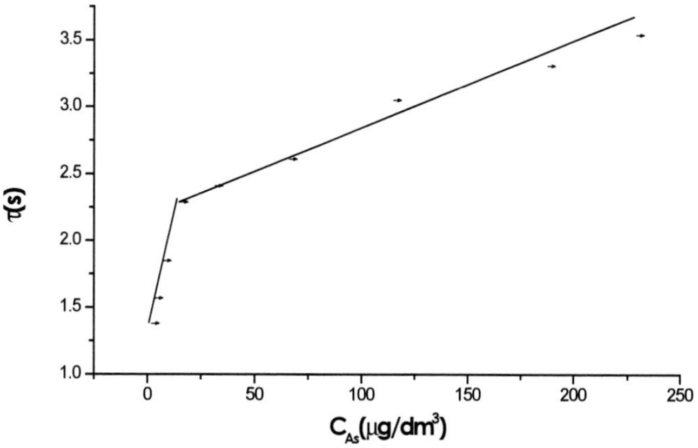

Figure 16. Copper influence on CSA of arsenic ($C_m(Cu^{2+})$ = 15 μg/dm^3). Gold film electrode. 240 s deposition time, 4.8 μA stripping current, 2000 r.p.m. stirring rate.

The chronopotentiometric stripping method can be used for the determination of As(III) and As(V) in various water samples. Since only trivalent arsenic is electrochemically active, in order to determine As(V) it is necessary to perform the reduction. The content of As(V) can then be calculated on the basis of the difference between total and As(III) content. Efficiency of various reducing agents such as hydroxylamine-hydrochloride, hydrazine-sulphate + HCl, mixture of ascorbic acid and potassium-iodide, Na_2SO_3, 0.1 mol/dm^3 HCl, L-cysteine and ascorbic acid was investigated. Best results were obtained with Na_2SO_3. Reduction can be assisted by microwave heating of the water samples in polyethylene bottles with 0.3% Na_2SO_3 during 4 minutes at 350 W. Prior analysis solution should be cooled. Excess Na_2SO_3 after the addition of the supporting electrolyte produces SO_2 which cause significant interferences at the electrode. By prolongating the deaeration step, excess SO_2 can be completely removed from the solution and interferences can be avoided.

By applying the developed technique, the arsenic content with respect to As (III) and As (V) was determined in tap water samples, samples of well water from the depth of 100 m and 300 m, sample of rain water and the samples of river water (Tisa, Danube). Water samples were analysed after acidifying with sulphuric acid (3 mol/dm^3). Prior analysis solutions were deaerated by purging the nitrogen. During the deaeration step it was necessary to hold the gold film electrode out of the process vessel in triply distilled water in order to avoid the damage of the electrode surface by nitrogen bubbles. After the deaeration step electrode was returned into the solution. For concentration calculation the calibration curve was used. It was necessary to add the copper to the standard solutions in the same concentration as in real samples. The copper content that should have been added was previously determined by potentiometric stripping analysis by means of standard addition method using mercury film electrode and dissolved oxygen as an oxidising agent. Contents of As(III) and As(V) in different water samples are shown in Table 6 . Values in parenthesis are the results of recovery assays.

Table 6. Contents of As(III) and As(V) in waters determined by CSA at gold film electrode

	As(III) (μg/dm3)	As(V) (μg/dm3)
Tap water 1	6.6±0.3a (103.0)	0.6±0.3 (85.0)
Tap water 2	8.0±0.2 (101.0)	3.2±0.3 (96.8)
Tap water 3	42.0±0.1 (101.5)	60.0±0.1 (89.0)
Well water 1 (100 m)	16.6±0.2 (88.9)	4.3±0.3 (88.0)
Well water 2 (100 m)	12.3±0.2 (97.0)	6.1±0.2 (97.5)
Well water 3 (300 m)	nd	nd
Well water 4 (300 m)	nd	5.3±0.3 (89.7)
Rain water	nd	13.3±0.2 (99.3)
River water 1 (Tisa)	24.2±0.1 (97.8)	10.7±0.3 (98.0)
River water 2 (Tisa)	32.0±0.1 (98.8)	10.2±0.2 (101.0)
River water 3 (Tisa)	19.1±0.2 (96.7)	8.6±0.4 (98.0)
River water 4 (Danube)	nd	10.5±0.2 (97.8)
River water 5 (Danube)	5.3±0.3 (88.0)	nd
River water 6 (Danube)	nd	nd

aresults of 2SD (2 standard deviations)

Arsenic content varies depending on the water type, region and depth. Special attention should be paid to the monitoring of arsenic content in natural waters in the regions where frequent natural fluctuations of arsenic content are observed (Tap water 3).

CONCLUSION

Preparation of quality drinking water is a very complex and expensive task, which requires multidisciplinary activity of professionals and scientists. In last two decades serious research work has been done in this field due to increasing requirements for quality, as well as quantity of the drinking water from the one side, and increasing contamination of water resources from another side, due to rapid industry development and increasing capacities of new technologies. Seeking for the solution for producing enough quantity of quality water, many new technologies are being introduced. Water preparation processes are becoming more and more complex due to poor quality of the starting natural water. In last decade continuously increased water contamination is observed. Significant part of the efforts is focused on the removal of inorganic water impurities (manganese, iron, ammonia etc.), pathogens (bacteria's and viruses), heavy metals, organic pollutants as well as compounds which negatively influence the sensory properties of the water (odour, colour, taste, turbidity etc.). For drinking water production, rationalisation of the production procedures is mandatory. First point of this task is to continuously monitor the quality of estuarine and surface waters which will be used for the water production. Science, low and political action should be integrated to develop more efficient system for water management.

Among various organic micropollutants which draw attention of the scientific population due to their high toxicity, persistency and complexity of the analytical procedures for their detection, heavy metals are also a contaminants of high concern and constantly in the focus of attention. Unlike all other contaminants they are not biodegradable and they tend to accumulate in aqueous flora and fauna, finally reaching man and highly affecting the biodiversity. Heavy metals enter marine and estuarine ecosystems through the discharge of industrial waste, treated sewage, mining operations and other antropogenic sources. They persist in the environment and so tend to accumulate in soils, sediments and living organisms eventually reaching men. For these reasons reliable, sensitive and accurate metal determination represents not only the rationalisation of production process, but is obligatory from toxicological point of view. Through the given review of the application of different electrochemical stripping techniques in water analysis, their relevance for the purpose has been confirmed. Wide array of different water types has been analised, thus demonstrating their universality and wide window of application. The possibility of direct water analysis, small and simple instrumentation, but high sensitivity and selectivity establishe electrochemical stripping techniques as unique in the domain of continual water monitoring and offres a wide application in on-line analysis and flow systems. However, there is no ideal analytical technique. Each of them suffres from different interferences, disadvantages and problems, but taking them into consideration and first of all, knowing them, analyst can make reliable conclusions and proceed them forward in order to reach the final goal.

References

[1] Wang, J. Analytical Electrochemistry, Second Edition. New York: A John WileyandSons, Inc. Publications; 2000.

[2] Vydra, F; Štulik, K; Julakova, E. Electrochemical stripping analysis, Ellis Horwood Limited; 1976.

[3] Wang, J. Stripping analysis. USA: VCH Publishers, Deerfield Beach; 1985.

[4] Suturović, ZJ. Elektrohemijska striping analiza, Monograph, Faculty of Technology, 1998.

[5] Andersen, JH, Conley, DJ, Hedal, S. SFT Classification of Environmental quality in fresh water, 1997.

[6] Regulation for drinking water in EU area, Council directive 98/83/EF, 1998.

[7] Jagner, D; Kryger, L. Computerized electroanalysis part III multiple scanning anodic stripping and its application to sea water. *Analytica Chimica Acta*, 1975, 80, 255-266.

[8] Gillain, G; Duyckaerts, G. Direct and simultaneous determination of Zn, Cd, Pb, Cu, Sb and Bi dissolved in sea water by differential pulse anodic stripping voltammetry with a hanging mercury drop electrode. *Analytica Chimica Acta*, 1979, 106, 23-37.

[9] Locatelli, C. Anodic and cathodic stripping voltammetry in the simultaneous determination of toxic metals in environmental samples. *Electroanalysis*, 1997, 9, 1014-1017.

[10] Bubnik, J. Voltammetric determination of small amounts of As, Sb and Sn in waters, leaches and materials with a complex matrix. *Chemické listy*, 1997, 91, 200-207.

[11] Daniele, S; Bragato, C; Baldo, MA. An aproach to the calibrationless determination of copper and lead by anodic stripping voltammetry at thin mercury film microelectrodes. Application to well water and rain. *Analytica Chimica Acta,* 1997, 346, 145-156.

[12] Komy, ZR. Determination of trace metals in Nile River and ground water by differential pulse stripping voltammetry. *Microchimica Acta*, 1993, 111, 239-249.

[13] Pretty, JR; Blubaugh, EA; Caruso, JA; Davidson, TM. Determination of Chromium(VI) and Vanadium(V) Using an Online Anodic Stripping Voltammetry Flow Cell with Detection by Inductively Coupled Plasma Mass Spectrometry. *Analytical Chemistry,* 1994, 66, 1540 – 1547.

[14] Domínguez, O; Arcos, MJ. Speciation of Chromium by Adsorptive Stripping Voltammetry Using Pyrocatechol Violet. *Electroanalysis*, 2000, 12, 449-458.

[15] Vukomanovic, DV; Vanloon, GW; Nakatsu, K; Zoutman, DE. Determination of Chromium (VI) and (III) by Adsorptive Stripping Voltammetry with Pyrocatechol Violet. *Microchemical Journal*, 1997, 57, 86-95.

[16] Wang, J; Lu, J; Luo, D; Wang, J; Tian, B. Simultaneous adsorptive stripping voltammetric measurements of trace chromium, uranium, and iron in the presence of cupferron. *Electroanalysis*, 1997, 9, 1247-1251.

[17] Paneli, M; Voulgaropoulos, AV.; Kalcher, K. The catalytic adsorptive stripping voltammetric determination of chromium with TTHA and nitrate. *Microchimica Acta*, 1993, 110, 205-215.

[18] Domínguez, O; Sanllorente, S; Alonso, MA; Arcos, MJ. Application of an Optimization Procedure for the Determination of Chromium in Various Water Types by Catalytic-Adsorptive Stripping Voltammetry. *Electroanalysis*, 2001, 13, 1505-1512.

[19] Dobney, AM; Greenway, GM. On-line determination of chromium by adsorptive cathodic stripping voltammetry. *Analyst*, 1994, 119, 293 – 297.

[20] Riso, RD.; Le Corre, P; Chaumery, CJ. Rapid and simultaneous analysis of trace metals (Cu, Pb and Cd) in seawater by potentiometric stripping analysis. *Analytica Chimica Acta,* 1997, 351, 83-89.

[21] Adeloju, SB; Sahara, E; Jagner, D. Anodic stripping potentiometric determination of Cu, Pb, Cd and Zn on a novel combined electrode system. *Analytical Letters*, 1996, 29, 283-302.

[22] Suturović, ZJ. Ispitivanje uslova predelektrolize kao prve faze elektrohemijskom striping analizom. *Magistarski rad,* Tehnološki fakultet, Novi Sad, 1985.

[23] Suturović, ZJ; Marjanović, NJ. Određivanje teških metala elektrohemijskom striping analizom. *Voda i sanitarna tehnika*, 1993, 3-4, 20.

[24] Bulska, E; Wałcerz, M.; Jędral, W; Hulanicki, A. On-line preconcentration of lead and cadmium for flame atomic absorption spectrometry using a flow-through electrochemical microcell. *Analytica Chimica Acta,* 1997, 357,133-140.

[25] Beinrohr, E.; Čakrt, M.; Džurov, J.; Jurica, L.; Broekaert, JAC. Simultaneous Calibrationless Determination of Zinc, Cadmium, Lead, and Copper by Flow-Through Stripping Chronopotentiometry. *Electroanalysis*, 1999, 11, 1137-1144.

[26] Sahlin, E; Jagner, D. Calibration-free determination of copper, zinc, cadmium and lead in tap water using coulometric stripping potentiometry. *Electroanalysis,* 1998, 10, 532–535.

[27] Stumm, W; Morgan, JJ. Aquatic chemistry. Chemical Equillibria and Rates in Natural Waters. Wiley Interscience, Third Edition, 1996.

[28] Salbu, B; Steinnes, E. *Trace elements in natural waters.* CRS Press; 1995.

[29] Matisson, G.; Nyholm, L.; Olin, A.; Ornemark, U. Determination of selenium in freshwaters by cathodic stripping voltammetry after UV irradiation. *Talanta,* 1995, 42, 817-825.

[30] Subramanian, JC; Leung, PC; Meranger, JC. Determination of Arsenic (III, V, Total) in Polluted Waters by Graphite Furnace Atomic Absorption Spectrometry and Anodic Stripping Voltammetry. *International Journal of Environmental Analytical Chemistry*, 1982, 11, 121-130

[31] Cox, JA.; Kulesza, PJ. Stripping voltammetry of chromium (VI) at a poly(4-vinylpyridine)-coated platinum electrode. *Analytica Chimica Acta*, 1983, 154, 71-78.

[32] Batley, GE; Florence , TM. An evaluation and comparasion of some techniques of anodic stripping voltammetry. *Electroanalytical Chemistry and Interfacial Electrochemistry*, 1974, 55, 23-43.

[33] Florence, TM. Electrochemical approaches to trace element speciation in waters: A review. *Analyst*, 1986, 111, 489-505.

[34] Papoff, P; Bocci, F; Lanza, F. Speciation of selenium in natural waters and snow by dpcsv at the hanging mercury drop electrode, *Microchemical Journal*, 1998, 59, 50-76.

[35] Quentel, F; Elleouet, C. Speciation analysis of selenium in seawater by cathodic stripping voltammetry. *Electroanalysis,* 1999, 11, 47–51.

[36] Elleouet, C; Quentel, F; Madec, C. Determination of inorganic and organic selenium species in natural waters by cathodic stripping voltammetry. *Water Research,* 1996, 30, 909-914.

[37] Rurikova, D; Kunakova, I. Determination of selenium in soil by cathodic stripping voltammetry. *Journal of Trace and Microprobe Techniques*, 2000, 18, 2, 193-199.

[38] Bryce, DW; Izquierdo, A; Luque de Castro, MD. Sequential speciation of selenium by flow injection cathodic stripping voltammetry. *Fresenius Journal of Analytical Chemistry*, 1995, 351, 433-437.

[39] Sužnjević, D; Blagojević, S; Vidić, J; Erceg, M; Vučelić, D. Determination of Selenium(IV) by Cathodic Stripping Voltammetry Using a Copper Microelectrode. *Microchemical Journal,* 1997, 57, 255-260.

[40] McLaughlin, K; Boyd, D; Hua, C; Smyth, MR. Anodic stripping voltammetry of selenium (IV) at a gold fiber working electrode. *Electroanalysis*, 1992, 4, 689-693.

[41] Lange, B; Van den Berg, CMG. Determination of selenium by catalytic cathodic stripping voltammetry. *Analytica Chimica Acta*, 2000, 418, 33-42.

[42] Van den Berg, CMG.; Khan, SH. Determination of selenium in sea water by adsorptive cathodic stripping voltammetry. *Analytica Chimica Acta*, 1990, 231, 221-229.

[43] Ferri, T; Guidi, F; Morabito, R. Carbon paste electrode and medium-exchange procedure in adsorptive cathodic stripping voltammetry of selenium(IV). *Electroanalysis* 1994, 6, 1087-1093.

[44] Muñoz, E; Palmero, S; García-García, MA. Fast Linear Scan Stripping Voltammetry of Copper Using a Glassy Carbon Rotating Disk Electrode. *Electroanalysis*, 2000, 12, 774-777.

[45] Prakash, R; Srivastava, RC; Seth, PK. Estimation of copper in natural water and blood using anodic stripping differential pulse voltammetry over a rotating side disk elektrode. *Electroanalysis*, 2002, 14, 303-308.

[46] Davies, CM, Apte, SC; Johnstone, AL. A Bacterial bioassay for the assessment of copper bioavailability in freshwaters. Environmental. *Toxicology and Water Quality*, 1998, 13, 263-271.

[47] Van Staden JF; Matoetoe, M. Determination of copper by anodic stripping voltammetry on a glassy carbon electrode using a continuous flow system. *Fresenius Journal of Analytical Chemistry*, 1997, 357, 624-628.

[48] Peng, T; Sheh, L; Wang, G. Linear scan stripping voltammetry of copper(II) at the chemically modified carbon paste electrode. *Microchimica Acta*, 1996, 122, 125-132.

[49] Labuda, J; Vanńičková, M; Uhlemann, E; Mickler, W. Applicability of chemically modified electrodes for determination of copper species in natural waters. *Analytica Chimica Acta*, 1994, 284, 517-523.

[50] Viltchinskaia, EA; Zeigman, LL; Garcia, DM; Santos, PF. Simultaneous determination of mercury and arsenic by anodic stripping voltammetry. *Electroanalysis*, 1997, 9, 633-640.

[51] Majer, S; Scholz, F; Trittler, R. Determination of inorganic ionic mercury down to 5×10^{-14} mol l^{-1} by differential-pulse anodic stripping voltammetry. *Analytical and Bioanalytical Chemistry*, 1996, 356, 247-252.

[52] Beinrohr, E; Čakrt, M; Dzurov, J; Kottaš, P; Kozakova E. Calibrationless determination of mercury by flow-trough stripping coulometry. *Fresenius Journal of Analytical Chemistry*, 1996, 356, 253-258.

[53] Sipos, L; Nurnberg, HW; Valenta, P; Branica, M. The reliable determination of mercury traces in sea water by subtractive differential pulse voltammetry at the twin gold electrode. *Analytica Chimica Acta*, 1980,115, 25-42.

[54] Kolpakova, NA; Borisova, NV; Nevostruev, VA. Nature of positive anodic peak in the stripping voltammogram of binary platinum metal systems. *Journal of Analytical Chemistry*, 2001, 8, 744-747.

[55] Bilewicz, R; Kublik, Z. The influence of various metals on the anodic oxidation of traces of mercury from carbon electrodes in thiocyanate media. *Analytica Chimica Acta*, 1983,152, 203-214.

[56] Fikai, R; Huynh-Negoc, L. Direct determination of mercury in sea water by anodic stripping voltammetry with a graphite electrode. *Analytica Chimica Acta*, 1976, 83, 375-379.

[57] Faller, C; Stojko, NyuN; Henze, G; Brainina, Kh Z. Stripping voltammetric determination of mercury at modified solid electrodes. Determination of mercury traces using PDC/Au(III)-modified electrodes. *Analytica Chimica Acta*, 1999, 396,195-202.

[58] Luong, L; Vydra, F. Voltammetry with disc electrodes and its analytical application. The selectivity of the stripping determination of mercury and the determination of some metals in the presence of mercury at glassy carbon electrodes. *Electroanalytical Chemistry and Interfacial Electrochemistry*, 1974, 50, 379-386.

[59] Ugo, P; Morreto, LM; Bertoncello, P; Wang, J. Determination of trace mercury in saltwaters at screen printed electrodes modified with sumichelate Q10R. *Electroanalysis*, 1998, 10, 1017-1021.

[60] Turyan, I; Erichsen, T; Schuhmann, W; Mandler, D. On-Line Analysis of Mercury by Sequential Injection Stripping Analysis (SISA) Using a Chemically Modified Electrode, Voltammetric quantification and speciation of mercury compounds. *Electroanalysis*, 2001, 13, 79-82.

[61] Agraz, R; Sevilla, MT; Hernández, L. Pulsed voltammetric stripping at the thin-film electrode. *Journal of Electroanalytical Chemistry*, 1995, 390, 47.

[62] Engel RR; Smyth AH, Arsenic in drinking water and mortality from vascular disease: an ecologic analysis in 30 counties in the United States. *Archives of Environmental Health,* 1994, 49, 418-427.
[63] [63] between ischemic heart disease mortality and long-term arsenic exposure. *Artherosclerosis, Thrombosos and Vascular Biology,* 1996,16, 504-510.
[64] Navas-Acien, A; Sharrett, AR; Silbergeld, EK; Schwartz, BS; Nachman, KE; Burke, TA, Guallar, E. Arsenic exposure and cardiovascular disease: a systematic review of the epidemiological evidence. *American Journal of Epidemiology*, 2005, 162, 1037-1049.
[65] Chi, IC; Blacwell, RQ. A controlled retrospective study of blackfoot disease, an endemic peripheral gangrene disease in Taiwan. *American Journal of Epidemiology,* 1968, 88, 7-24.
[66] Sun, YC; Mierzwa, J; Yang, MH. New method of gold-film electrode preparation for anodic stripping voltammetric determination of arsenic (III and V) in seawater. *Talanta*, 1997, 44, 1379-1387.
[67] Jurica, L; Manova, A; Dzurov, J; Beinrohr, E; Broekaer JAC. Calibrationless flow-trough stripping coulometric determination of arsenic(III) and total arsenic in contaminated water samples after microwave assisted reduction of arsenic(V). *Fresenius Journal of Analytical* Chemistry, 2000, 366, 260-266.
[68] Henze, G; Wagner, W; Sander, S. Speciation of arsenic(V) and arsenic(III) by cathodic stripping voltammetry. *Fresenius Journal of Analytical Chemistry*, 1997, 358, 741-744.
[69] Sadana, RS. Determination of arsenic in the presence of copper by differential pulse cathodic stripping voltammetry at a hanging mercury drop electrode. *Analytical Chemistry*,1983, 55, 304-307.
[70] Bubnik, J. Voltammetric determination of small amounts of As, Sb, and Sn in waters, leaches and materials with a complex matrix. *Chemichesky Listy* 1997, 91, 200-207.
[71] Henze G; Wagner, W; Sander, S. Speciation of Arsenic (V) and Arsenic (III) by cathodic stripping voltammetry in fresh water samples. *Fresenius Journal of Analytical Chemistry*, 1997, 358 741-744.
[72] He, Y; Zheng, Y; Ramanaraine, M; Locke, DC. Differential pulse cathodic stripping voltammetric speciation analysis of trace level inorganic arsenic in natural waters. *Analytica Chimica Acta*, 2004, 511, 55-61.
[73] Barra, CM; dos Santos Correia, MM. Speciation of inorganic arsenic in natural waters by square-wave cathodic stripping voltammetry. *Electroanalysis,*2001, 13, 1098-116.
[74] Kopanica, M; Novotny, L. Determination of traces of arsenic(III) by anodic stripping voltammetry in solutions, natural waters and biological material. *Analytica Chimica Acta*, 1998, 368, 211-218.
[75] Long, J; Yukio, N. Cathodic stripping voltammetric determination of As(III) with *in situ* plated bismuth-film electrode using the catalytic hydrogen wave. *Analytica Chimica Acta*, 2007, 593, 1-6.
[76] Leung, PC; Subramanian, S; Meranger, JC. Determination of arsenic in polluted waters by differential pulse anodic stripping voltammetry. *Talanta*, 1982, 29, 515-518.
[77] Chadim, P; Švancara, I; Pihlar, B; Vytras; K. Gold plated carbon paste electrodes for anodic stripping determination of arsenic. *Collected Czech Chemical Communications*, 2000, 65, 1035-1045.

[78] Viltchiskaya, EA; Zeigman, LL; Garcia, DM; Santos PF. Simultaneous determination of mercury and arsenic by anodic stripping voltammetry. *Electroanalysis,* 1997, 9, 633-640.
[79] Flechsid, GU; Korbut, O; Grundler, P. Investigation of deposition and stripping phenomenon at the heated gold wire electrode in comparison to the rotating disc electrode: copper (II), mercury(II) and arsenic (III). *Electroanalysis,* 2001, 13, 786-793.
[80] Rajiv Prakash , R; Srivastava, RC; Seth, PK. Direct Estimation of Total Arsenic Using A Novel Metal Side Disk Rotating Electrode. *Electroanalysis*, 2003, 15, 1410-1414.
[81] Li, H; Smart, RB. Determination of sub-nanomolar concentration of arsenic(III) in natural waters by square wave cathodic stripping voltammetry. *Analytica Chimica Acta*, 1996, 325, 25-32.
[82] Huang, H; Jagner, D; Renman, l. Flow potentiometric and constant current stripping analysis for arsenic(V) without prior chemical reduction to arsenic(III). *Analytica Chimica Acta*, 1988, 207, 37-46.
[83] Adeloju SB; Young, TM. Assessment of constant current anodic stripping potentiometry for determination of As in fish and water samples. *Analytical Letters*, 1997, 30, 1, 147-161.
[84] Adeloju, SB, Young, TM, Jagner, D; Batley, GE. Constant current cathodic stripping potentiometric determination of arsenic on a mercury film electrode in the presence of copper ions. *Analytica Chimica Acta*, 1999, 381, 207-213.
[85] Lukaszweski, Zl; Zembrzuski, W; Piela A. Direct determination of ultratraces of thallium in water by flow-injection--differential-pulse anodic stripping voltammetry. *Analytica Chimica Acta, 1996*, 318, 159-165.
[86] Švancara, I; Ostapczuk, P; Arunachalam, J; Emons, H.; Vytřas K. Determination of thallium in environmental samples using potentiometric stripping analysis. Method development. *Electroanalysis,* 1997, 9, 26-31.
[87] Cleven, R; Fokkert, L. Potentiometric stripping analysis of thallium in natural waters. *Analytica Chimica Acta*, 1994, 289, 215-221.
[88] Zen, JM; Wang, WM; Kumar, AS. Potentiometric Stripping Analysis of Traces of Thallium(III) at a Poly(4-Vinylpyridine) / Mercury Film Electrode. *Electroanalysis*, 2001, 13, 321-324.
[89] Eskilsson H; Jagner, D. Potentiometric stripping analysis for bismuth (III) in seawater. *Analytica Chimica Acta*, 1982, 138, 27-33.
[90] Shams, E. Determination of Trace Amount of Bismuth(III) by Adsorptive Stripping Voltammetry by Alizarine Red S. *Electroanalysis*, 2001, 13, 1140-1142.
[91] Wang, J; Lu, J.; Setiadji, R. Adsorptive stripping measurements of bismuth. *Electroanalysis*, 1993, 5, 319-324.
[92] Jyh-Myng, Z; Mu-Jye, C. Square-wave voltammetric stripping analysis of bismuth(III) at a poly(4-vinylpyridine) / mercury film electrode. *Analytica Chimica Acta*, 1996, 320, 43-51.
[93] Ali Taher, M; Ebrahim Rezaeipour, E; Afzali, D. Anodic stripping voltammetric determination of bismuth after solid-phase extraction using amberlite XAD-2 resin modified with 2-(5-bromo-2-pyridylazo)-5-diethylaminophenol. *Talanta*, 2004, 63, 3, 797-801.

[94] Huishi Guo, H; Li, Y; Xiao, P; He, N. Determination of trace amount of bismuth(III) by adsorptive anodic stripping voltammetry at carbon paste electrode. *Analytica Chimica Acta*, 2005, 534, 1, 143-147.
[95] Wensheng, H. Voltammetric Determination of Bismuth in Water and Nickel Metal Samples with a Sodium Montmorillonite (SWy-2) Modified Carbon Paste Electrode. *Microchimica Acta*, 2004, 144, 1-3, 125-129.
[96] Beinrohr, E; Csemi, P; Rojas, FJ; Hofbauerova, H. Determination of Manganese in Water Samples by Galvanostatic Stripping Chronopotentiometry in a Flow-trough Cell. *Analyst*, 1994, 119, 1355-1359.
[97] Schildkraut, DE; Dao, PT; Twist, JP; Davis, AT; Robillard, KA. Determination of silver ions at sub microgram-per-liter levels using anodic square-wave stripping voltammetry. *Environmental and Toxicological Chemistry* 1998, 17, 642–649.
[98] Wang, J; Lu, J; Percio, A; Farias, M. Remote electrochemical monitoring of trace silver. *Analytica Chimica Acta*, 1996, 318, 151-157.
[99] Guo, SX; Khoo, SB. Highly Selective and Sensitive Determination of Silver(I) at a Poly(8-mercaptoquinoline) Film Modified Glassy Carbon Electrode. *Electroanalysis*, 1999, 11, 1, 891-898.
[100] Mikkelson, O; Skogvold, SM; Schroder, KH. Determination of Silver and Copper at Low ppb and sub-ppb Concentration Ranges in Mixture in Natural Water by Using Anodic Stripping Voltammetry at a Gold-bismuth Alloy Electrode. *On-line journal of SMCBS Workshop*, 2003.
[101] Qian, J; Xue, HB.; Sigg, L; Albrecht, A. Complexation of Cobalt by Natural Ligands in Freshwater. *Environmental Science and Technology*, 1998, 32, 2043-2050.
[102] Gil, EP; Ostapczuk, P. Nickel and cobalt determination by constant current potentiometry. *Fresenius' Journal of Analytical Chemistry*, 1993, 346, 952-956.
[103] Sawamoto, H. Adsorption preconcentration for the trace analysis of nickel. *Journal of Electroanalytical Chemistry*, 1983, 147, 279-288.
[104] Vega, M; Van den Berg, CMG. Determination of Cobalt in Seawater by Catalytic Adsorptive Cathodic Stripping Voltammetry. *Analytical Chemistry*, 1997, 69 , 874 - 881.
[105] Paneli, MG; Voulgaropoulos, AN. Simultaneous voltammetric determination of cobalt, nickel and labile zinc, using 2-quinolinethiol in the presence of surfactants without prior digestion. *Fresenius' Journal of Analytical Chemistry*, 1994, 348, 837-839.
[106] Huiliang, H; Hua, C; Jagner, D; Renman; L. Computerized flow constant-current stripping analysis for nickel(II) and cobalt(II) with carbon fibre and modified carbon fibre electrodes. *Analytica chimica Acta*, 1988, 208, 237-246.
[107] Aldrich, AP; Van den Berg, CMG. Determination of Iron and Its Redox Speciation in Seawater Using Catalytic Cathodic Stripping Voltammetry. *Electroanalysis*, 1998, 10, 369-373.
[108] Gelado-Caballero, MD; Hernández-Brito, JJ; Herrera-Melián, JA; Collado-Sanchez, C; Pérez-Peña, J. Fast adsorptive stripping voltammetry of iron in oxygenated seawater. *Electroanalysis*, 1996, 8, 1065-1071.
[109] Gawryś, M; Golimowski J. Application of Adsorptive Stripping Voltammetry for Iron Determination with Catechol in Quartz and Silica Glass Samples. *Electroanalysis*, 1999, 11, 1318-1320.

[110] Croot, PL; Johansson, M. Determination of Iron Speciation by Cathodic Stripping Voltammetry in Seawater Using the Competing Ligand 2-(2-Thiazolylazo)-p-cresol (TAC). *Electroanalysis*, 2000, 12, 565-576.

[111] Jagner, D; Renman, L; Stefansdottir, SH. Determination of iron(III) and titanium(IV) as their solochrome violet rs complexes by constant-current stripping potentiometry. Partial least-squares regression calibration procedure for iron(III) and titanium(IV). *Analytica Chimica Acta*, 1993, 28, 315-321.

[112] Economou, A; Fielden, PR; Packham, AJ.Batch and flow determination of uranium (VI) by adsorptive stripping voltammetry on mercury-film electrodes. *Analyst*, 1994, 119, 279 – 285.

[113] Van Den Berg, CMG.; Huang, ZQ. Determination of uranium(VI) in sea water by cathodic stripping voltammetry of complexes with catechol. *Analytica Chimica Acta*, 1984, 164, 209-222.

[114] Komersová, A; Barto, M; Kalcher, K; Vytřas; K. Indirect Determination of Uranium Traces by Adsorptive Stripping Voltammetry with Arsenazo III. *Electroanalysis*, 1998, 10, 442-445.

[115] El-Maali, NA; El-Hady, DA. Square-Wave Stripping Voltammetry of Uranium(VI) at the Glassy Carbon Electrode. Application to Some Industrial Samples. *Electroanalysis*, 1999, 11, 201-206.

[116] Farghaly, OA; Ghandour, MA. Cathodic adsorptive stripping voltammetric determination of uranium with potassium hydrogen phthalate. *Talanta*, 1999, 49, 31-40.

[117] Wang, J; Baomin, T; Setiadji, R. Disposable electrodes for field screening of trace uranium. *Electroanalysis*, 1994, 6, 317-320.

[118] Van den Berg, CMG; Murphy, K; Riley JP. The determination of aluminium in seawater and freshwater by cathodic stripping voltammetry. *Analytica Chimica Acta*, 1986, 188, 177-185.

[119] Bi, SP; Song, MJ; Xu, D. Chronopotentiometric determination of aluminium by solochrome violet. *Analytical Letters*, 1998, 31, 214-217.

[120] Rurikova, D; Pocuchova, M. Voltammetric determination of antimony in natural waters. *Chemical Papers-Chemicke Zvesti*, 1997, 51, 15-21.

[121] Wagner, W; Sander, S; Henze, G. Trace analysis of antimony(III) and antimony(V) by adsorptive stripping voltammetry. *Fresenius Journal of Analytical Chemistry*, 1996, 354, 11-15.

[122] Li, H; Smart, RB. Catalytic stripping voltammetry of vanadium in the presence of dihydroxynaphthalene and bromate. *Analytica Chimica Acta*, 1996, 333, 131-138.

[123] S. Petrovska-Jovanovic, K; Stojanova, V; Jakimovic, M. Square-Wave Voltammetry Investigation of the Electrochemical Behavior of Mo(VI) in Nitrate Media. *Analytical Letters*, 1997 , 30, 1923 – 1938.

[124] Georgieva, M; Pihlar, B. Determination of palladium by adsorptive stripping voltammetry. *Fresenius' Journal of Analytical Chemistry*, 1997, 357, 874-880.

[125] Ciszkowska, M; Stojek, Z. Determination of traces of EDTA, EGTA and DCTA by adsorptive accumulation of their complexes with hg(II) followed by cathodic stripping. *Talanta*, 1986, 33, 817-823.

[126] Le Gall, AC; Van den Berg, CMG. Determination of folic acid in sea water using adsorptive cathodic stripping voltammetry. *Analytica Chimica Acta*, 1993, 282, 459-470.

[127] Harbin, AM.; Van den Berg, CMG. Determination of ammonia in sea water using catalytic cathodic stripping voltammetry. *Analytical Chemistry*, 1993, 65, 3411-3416.

[128] [128] Le Gall, AC; Van den Berg, CMG. Cathodic stripping voltammetry of glutathione in natural waters. *Analyst*, 1993, 118, 1411 – 1415.

[129] Alekhina, IE; Portalenko, VV; Kuzina, LG; Maistrenko, VN. Stripping-Voltammetric Determination of Sulfur-Containing Pesticides in Water and Soil. *Zhurnal Analaticheskoi Khimii*, 1993, 48, 1189.

[130] Konvalina, J; Švancara, I; Vitřas, K; Kalcher, K. Study of Conditions For Voltammetric Determinations of Iodine at Carbon Paste Electrodes. *Scientific papers of the University of Pardubice, Seria A,* Faculty of Chemical Technology, 1997, 3, 153-162.

[131] Suturović, ZJ; Marjanović, NJ; Dokić, PP. Determination of chloride by stripping chronopotentiometry with silver-film electrode. *Electroanalysis*, 1997, 9, 572-575.

[132] Al-Farawati, R; Van den Berg, CMG. The determination of sulfide in seawater by flow-analysis with voltammetric detection. *Marine Chemistry*, 1997, 57, 277-286.

[133] Redinha, JS; Paliteiro, C; Pereira, JLC. Determination of sulfide by square-wave polarography. *Analytica Chimica Acta*, 1997, 351, 115-125.

[134] Krista, J; Kopanica, M; Novotný, L. Voltammetric determination of nitrates using silver electrodes. *Electroanalysis*, 2000, 12, 199-204.

[135] Riso, RD; Waels, M; Monbet, P; Chanmary, CJ. Measurements of trace concentrations of mercury in sea water by stripping chronopotentiometry with gold disc electrode - influence of copper. *Analytica Chimica Acta*, 2000, Vol. 410, 1-2, 97-105.

[136] [136] Beinrohr, E; Čakrt, M, Dzurov, J; Kottaš, P. Calibrationeless determination of mercury by flow-through stripping coulometry. *Fresenius Journal of Analytical Chemistry*, 1996, 356, 253-258.

[137] Beinrohr, E; Dzurov, J; Annus, J; Broekaert, AC. Flow-through stripping chronopotentiometry for the monitoring of mercury in waste waters. *Fresenius Journal of Analytical Chemistry*, 1998, 362, 201-204.

[138] Reilley, CN; Everet, GW; Johns, RN. Voltammetry at constant current: experimental avaluation. *Analytical Chemistry*, 1955, 27, 283-292.

[139] Švarc-Gajić J; Suturović, Z; Marjanović, N; Kravić, S. Determination of mercury by chronopotentiometric stripping analysis using glassy carbon vessel as a working electrode. *Electroanalysis*, 18, 2006, 513-516.

[140] Brainina, Kh; Neyman, E. Electroanalytical Stripping Methods. USA, John WileyandSons, 1993.

[141] Turner, DR. Lead, mercury, cadmiun and arsenic in the environment. Hutchinson, TCI. Meeme Wiley, J. Chichester; 1987.

In: Drinking Water: Contamination, Toxicity and Treatment ISBN: 978-1-60456-747-2
Editors: J. D. Romero and P. S. Molina

Chapter 10

IDENTIFICATION AND QUANTIFICATION OF ^{224}RA, ^{226}RA AND ^{228}RA IN THE HYDROGEOCHEMICAL ENVIRONMENT OF SOUTHERN NEW JERSEY: A REVIEW

J.A. Armstrong, S.A. Katz and T.L. Paulson
Department of Chemistry, Rutgers University,
Camden, New Jersey 08102-1411, U.S.A.

ABSTRACT

Investigations employing revised procedures that were able to measure short-lived radionuclides detected elevated levels of radioactivity in some potable water supplies from Dover Township, Ocean County, New Jersey. Subsequently, the New Jersey Department of Environmental Protection and the United States Geological Survey initiated collaborative investigations to better understand the source(s) and level(s) of radioactivity in the hydrogeochemical environment of Southern New Jersey. The results of these surveys confirmed some parts of the Kirkwood - Cohansey Aquifer in Atlantic, Burlington, Camden, Cape May, Cumberland, Gloucester, Mannmouth, Ocean and Salem Counties were found to contain ^{228}Ra, ^{226}Ra and ^{224}Ra concentrations in excess of 3 pCi/L.

The radiological origins of ^{228}Ra and ^{224}Ra were attributed to the decay of ^{232}Th while the decay of ^{238}U was the proposed source of ^{226}Ra. The hydrogeological origin of both thorium and uranium was ground water leaching of the quartz sand found in the Kirkwood-Cohansey Aquifer. The hydrogeo-chemistry of radium paralleled that of calcium in the relatively soft (from 30 - 60 ppm $CaCO_3$), slightly acidic (pH from 4 to 5) ground water.

Samples were collected from nearly a hundred wells in the Kirkwood-Cohansey Aquifer for measuring of the concentrations of ^{228}Ra, ^{226}Ra and ^{224}Ra as well as gross alpha activity. Results for ^{228}Ra ranged from < 0.5 to 12.8 pCi/L. Those for ^{226}Ra ranged from < 0.5 to 17.4 pCi/L, and those for ^{224}Ra ranged from < 0.5 to 16.8 pCi/L. Gross α activity measurements made within 48 hours of collection exceeded 15 pCi/L in nearly half of the samples examined.

The short lived ^{224}Ra ($t_{½}$ = 3.66 d) made significant contributions to the observed values for gross alpha activity. These observations are important in improving the understanding of the potential the health effects associated with the consumption of water containing ^{228}Ra, ^{226}Ra and ^{224}Ra.

INTRODUCTION

Naturally occurring radioactivity in ground water has three sources. The ^{238}U, ^{235}U and ^{232}Th in the substrata undergo radioactive decay to form three series of radionuclides. These series of radionuclides begin with ^{238}U, ^{235}U and ^{232}Th and end, respectively, with the stable isotopes of lead, ^{206}Pb, ^{207}Pb and ^{208}Pb. The series beginning with the decay of ^{238}U contains thirteen radionuclides before it terminates with stable ^{206}Pb. Similarly, the decay of ^{235}U to stable ^{207}Pb involves eleven radioactive intermediates, and nine are involved in the decay of ^{232}Th to stable ^{208}Pb. The three decay series are shown, in part, in Figure 1.

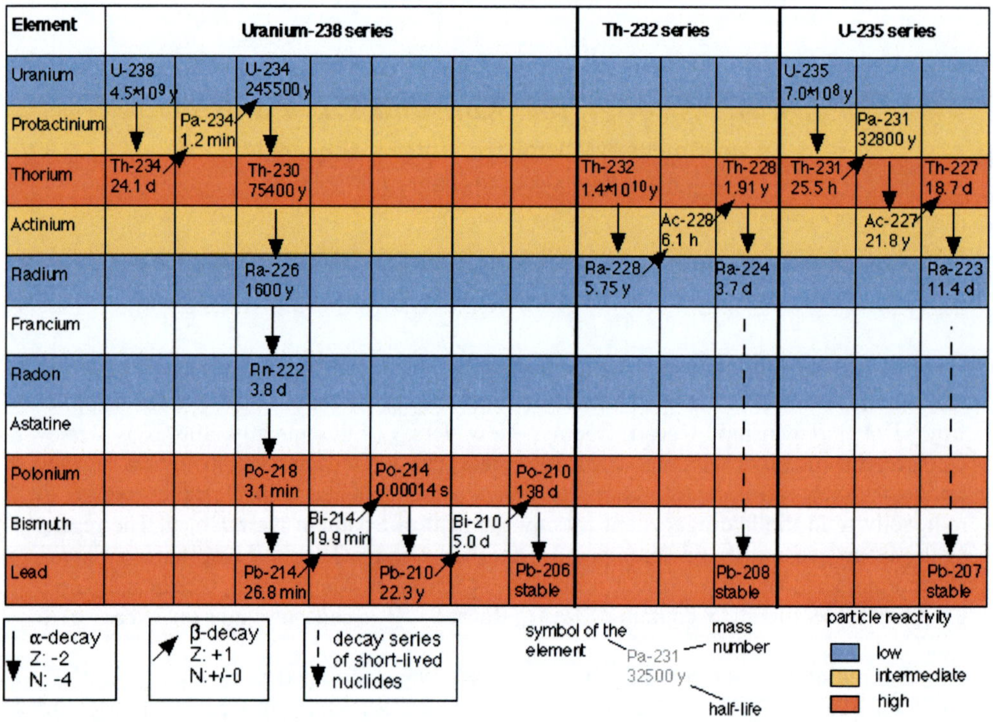

With permission: Alfref Wiegener Institut für Polar und Meersforschung.

Figure 1. ^{238}U, ^{235}U and ^{232}Th Decay Series.

The longest lived of the six radionuclides between ^{224}Ra and ^{208}Pb is the 10½ - hour ^{212}Pb, and the longest lived of the five radionuclides between ^{223}Ra and ^{207}Pb is the 36 - minute ^{211}Pb (Argonne National Laboratory, 2005). If left undisturbed, secular equilibria are established in each series. However, this is rarely the case.

The establishment of a secular equilibrium requires [1] the parent radionuclide must have a radiological half life much greater that that of any of its progeny, and [2] ten or so half lives of decay of the parent must elapse to allow ingrowths of its progeny. While the first requirement is met by ^{238}U, ^{235}U and ^{232}Th, the hydrogeochemistry of the site is the determinate factor for the establishment of the equilibria and the consequent occurrence and distribution of radionuclides in ground water.

The limited solubility of Th(IV) compounds in ground water is responsible for the relative immobility of the thorium component of the geosphere. However, some decay products of ^{232}Th, ^{228}Ra and ^{224}Ra compounds in particular, are mobile. Oxidation of U(IV) to U(VI) in ground water containing carbonate and/or hydrogen carbonate can lead to the formation of the highly mobile tricarbonato complex, $[UO_2(CO_3)_3]^{-4}$. The long lived ^{226}Ra progeny of ^{238}U, like the ^{228}Ra and ^{224}Ra progeny of ^{232}Th, is mobile especially in ground water containing chloride ion and/or organic ligands such as humic acid and fulvic acid (Zapecza and Szabo, 1988). The mobility of radium isotopes in the ground water, and subsequently, into drinking water supplies has been identified as a potential public health hazard from coast to coast in the North America (Albertson, 2003; Ruberu, *et al.*, 2005) as well as South America (Bonotto, 2004) and throughout the Eastern Hemisphere (Rusconi, *et al.*, 2004,; Popit, *et al.*, 2004; Kasztovszky, *et al.*, 1996; Sdrndarski, *et al.*, 1996; Marovi, *et al.*, 1997; Ahmed, 2004).

"The U.S. Geological Survey (USGS) has been assessing and monitoring the natural resources and natural hazards of New Jersey for more than a century" (USGS, 1996). "During routine regulatory monitoring conducted by the NJDEP, naturally occurring radium was detected at concentrations greater than the MCL in water from public supply wells screened in the Kirkwood – Cohansey aquifer system. From 1988 to 1996, the USGS in cooperation with the NJDEP, conducted several sampling programs to determine the distribution of total dissolved radium in the aquifer system and to identify the factors that contribute to the presence of high concentrations of total dissolved radium." (USGS, 1998).

THE HYDROGEOLOGIC SETTING

As shown in Figure 2, the Kirkwood-Cohansey aquifer occupies some 7500 km^2, almost the entire area, of south eastern New Jersey. The aquifer thins to the north and west. Ground water flow is from the northwest to the southeast. The Kirkwood-Cohansey aquifer is less than 15 meters thick near the outcrop of the Kirkwood Formation to more than 150 meters thick near Cape May. This aquifer is composed of two hydraulically connected formations, the Kirkwood Formation and Cohansey Sand. Depending on location, the aquifer is overlain with deposits of Beacon Hill Gravel, the Bridgeton Formation (a discontinuous superficial deposit of feldspathic gravel) and the Cape May Formation. The upper part of the Kirkwood Formation is composed primarily of fine sand to fine gravel and silty clay containing shell and organic material. The Kirkwood Formation is overlain with Cohansey Sand, which is a marginal marine of primarily very fine to coarse grained sand with interbedded layers of clay and silty and clayey sand. The Kirkwood Formation and the Cohansey sand are hydraulically connected. Together, they function as the Kirkwood-Cohansey aquifer. Recharge to the Kirkwood-Cohansey aquifer is predominately from precipitation. Annually the Kirkwood-Cohansey aquifer provides some 2.09×10^{11} liters of water more than half of which, 1.16×10^{11} liters, is drinking water (Vyas, et al., 2004; Szabo, et al., 2004).

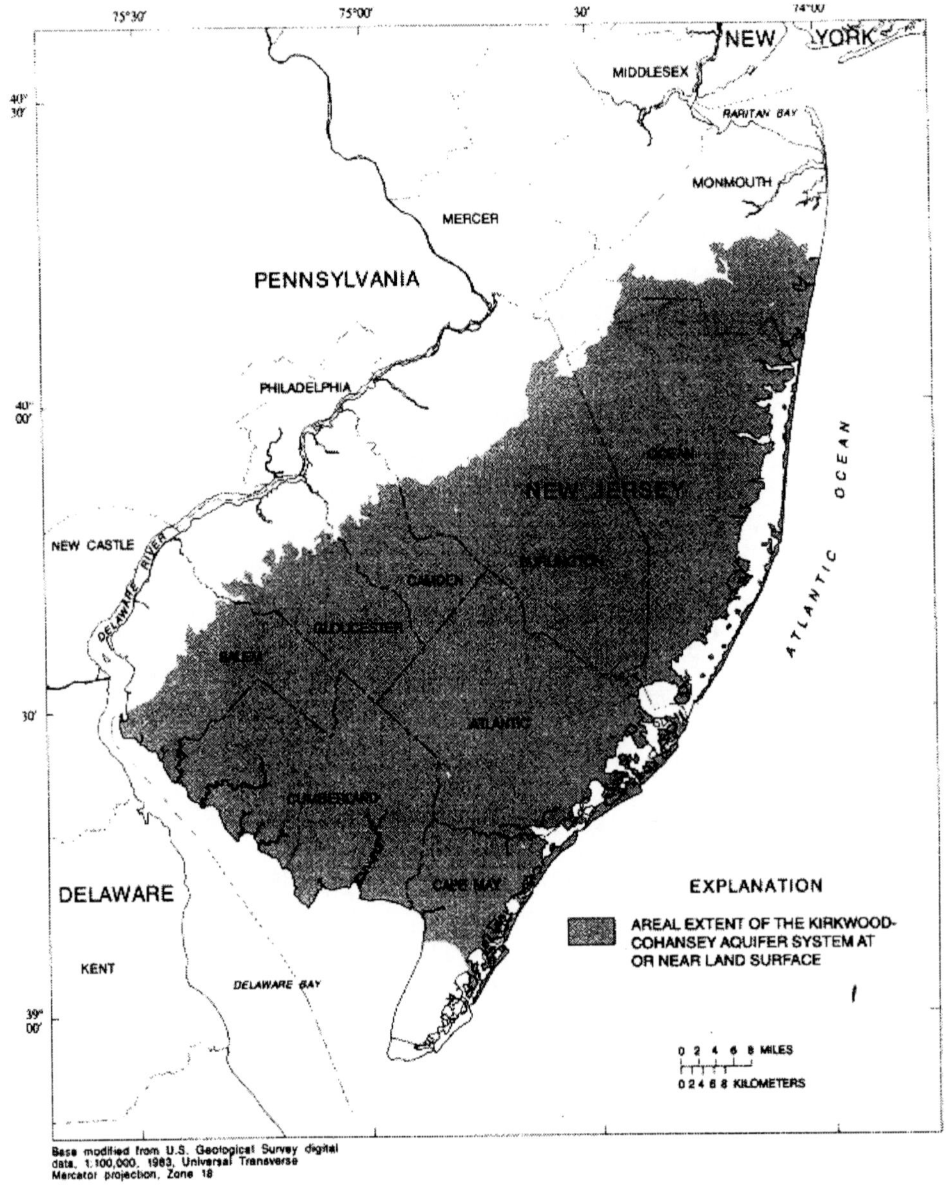

Outcrop of the Kirkwood-Cohansey aquifer system.

With permission: Rutgers University.

Figure 2. Kirkwood-Cohansey Aquifer System.

Radiological Survey

The US EPA established maximum contaminant levels for ^{226}Ra, ^{228}Ra and gross alpha radioactivity in community water supplies with interim guidelines as early as 1976 (US EPA,

1976). These maximum contaminant levels, MCL, have remained essentially unchanged in the final rules (US EPA, 2000):

- Combined ^{226}Ra and ^{228}Ra ≤ 5 pCi/L (0.185 dps/L).
- Gross alpha particle activity including ^{226}Ra but excluding radon and uranium ≤ 15 pCi/L (0.555 dps/L).

Some of the methods approved for determining ^{226}Ra, ^{228}Ra and gross alpha particle activities are listed in Table 1.

Table 1. US EPA Approved Methods for Some Radionuclides in Drinking Water

Parameters	Techniques	References
gross α activity	evaporation	1, 2, 3, 4, 5, 6
gross α activity	co-precipitation	7, 8
^{226}Ra	radon emanation	9, 10, 11, 12, 13, 14, 15, 16, 17
^{226}Ra	radiochemical	18, 19, 20, 21, 22, 23
^{228}Ra	radiochemical	24, 25, 26, 27, 28, 29, 30, 31

References	
1	Prescribed Procedures for Measurement of Radioactivity in Drinking Water, EPA 600/4-80-032, Aug., 1980, Method 900.0
2	Interim Radiochemical Methodology for Drinking Water, revised, EPA 600/4-75-008, March, 1976, p.1
3	Radiochemistry Procedures Manual, EPA 520/5-84-006, Dec., 1987, Method 00-01
4	Radiochemical Analytical Procedures for Analysis of Environmental Samples, March, 1979, p. 1
5	Standard Methods for the Examination of Water and Wastewater, 13th, 17th, 18th, 19th ed. 1971, 1989, 1992, 1995, American Public Health Association, Washington, DC, Method 302, Method 7110B
6	Methods for Determination of Radioactive Substances in Water and Fluvial Sediments, Book 5, Chapter A5, Techniques of Water Resource Investigations of the US Geological Survey, 1977, Method R-1120-76
7	Radiochemistry Procedures Manual, EPA 520/5-84-006, Dec., 1987, Method 00-02
8	Standard Methods for the Examination of Water and Wastewater, 13th, 17th, 18th, 19th ed. 1971, 1989, 1992, 1995, American Public Health Association, Washington, DC, Method 302, Method 7110C
9	Prescribed Procedures for Measurement of Radioactivity in Drinking 903.1 Water, EPA 600/4-80-032, August, 1980, Method
10	Interim Radiochemical Methodology for Drinking Water, revised,
	EPA 600/4-75-008, March, 1976, p.16

Table 1. (Continued)

References	
11	Radiochemistry Procedures Manual, EPA 520/5-84-006, Dec., 1987, Method Ra-04
12	Radiochemical Analytical Procedures for Analysis of Environmental Samples, March, 1979, p. 19
13	Standard Methods for the Examination of Water and Wastewater, 13th, 17th, 18th, 19th ed. 1971, 1989, 1992, 1995, American Public Health Association, Washington, DC, Method 302, Method 7500-Ra C
14	Annual Book of ASTM Standards, vol. 11.02, American Society for Testing and materials, West Conshohocken, PA, 1994, Method D 3454-91
15	Methods for Determination of Radioactive Substances in Water and Fluvial Sediments, Book 5, Chapter A5, Techniques of Water Resource Investigations of the US Geological Survey, 1977, Method R-1141-76
16	EML Procedures Manual, 27th ed. US Department of Energy, 1990, Method RA-05
17	Determination of Ra-226 and Ra-228, New York State Department of Health, Albany, June, 1982, Method Ra-02
18	Prescribed Procedures for Measurement of Radioactivity in Drinking Water, EPA 600/4-80-032, Aug., 1980, Method 903.0
19	Interim Radiochemical Methodology for Drinking Water, revised, EPA 600/4-75-008, March, 1976, p.13
20	Radiochemistry Procedures Manual, EPA 520/5-84-006, Dec., 1987, Method Ra-03
21	Standard Methods for the Examination of Water and Wastewater, 13th, 17th, 18th, 19th ed. 1971, 1989, 1992, 1995, American Public Health Association, Washington, DC, Method 302, Methods 304 and 305
22	Annual Book of ASTM Standards, vol. 11.02, American Society for Testing and materials, West Conshohocken, PA, 1994, Method D 2460-90
23	Methods for Determination of Radioactive Substances in Water and Fluvial Sediments, Book 5, Chapter A5, Techniques of Water Resource Investigations of the US Geological Survey, 1977, Method R-1140-76-76
24	Prescribed Procedures for Measurement of Radioactivity in Drinking Water, EPA 600/4-80-032, Aug., 1980, Method 904.0
25	Interim Radiochemical Methodology for Drinking Water, revised, EPA 600/4-75-008, March, 1976, p.24
26	Radiochemistry Procedures Manual, EPA 520/5-84-006, Dec., 1987, Method Ra-05
27	Radiochemical Analytical Procedures for Analysis of Environmental Samples, March, 1979, p. 19
28	Standard Methods for the Examination of Water and Wastewater, 13th, 17th, 18th, 19th ed. 1971, 1989, 1992, 1995, American Public Health Association, Washington, DC, Method 302, Methods 304 and 7500 Ra-D

Table 1. (Continued)

References	
29	Methods for Determination of Radioactive Substances in Water and Fluvial Sediments, Book 5, Chapter A5, Techniques of Water Resource Investigations of the US Geological Survey, 1977, Method R-1142-76
30	Determination of Ra-226 and Ra-228, New York State Department of Health, Albany, NY, June, 1982, Method Ra-02
31	Determination of Radium 228 in Drinking Water, NJ DEP, Trenton, NJ, Aug., 1980.

US EPA 2002.

In addition, the US EPA established minimum requirements for monitoring frequency and analytical methodology. After the initial analysis of a four calendar quarter composite sample, suppliers of water shall monitor at least once every four years. At the digression of the State, suppliers of water shall conduct annual monitoring of any community water system in which the ^{226}Ra concentration exceeds 3 pCi/L (0.111 dps/L). The State of New Jersey adopted the national standards for combined ^{226}Ra and ^{228}Ra and for gross alpha activity (NJ 1983). The World Health Organization recommended a reference level of 0.1 Bq/L (0.1 dps/L) for gross alpha (WHO, 1996).

The concentration of radium was found to be elevated in water from many parts of the Kirkwood-Cohansey aquifer during a USGS survey of private domestic wells, monitoring wells and community water supply wells (USGS, 1998). This survey was initiated in response to finding excessive gross alpha activity levels in samples taken from community water supply wells previously found to be in compliance with the regulatory standard of 15 pCi/L.

SHORT-LIVED GROSS ALPHA PARTICLE ACTIVITY

Significant increases in the gross alpha particle activity of water samples collected from the Kirkwood-Cohansey aquifer have been observed and reported (Parsa, 1997; Parsa, 1998). Water samples analyzed within 48 hours showed higher gross alpha particle activities than did samples collected from the same wells and analyzed at later times. Such variabilities were attributed to the presence of short-lived alpha-emitting radionuclides. Radiochemical separations coupled with gamma spectrometry provided firm evidence confirming the increased gross alpha particle activity was due to ^{224}Ra. Further support for ^{224}Ra and its four alpha-particle-emitting progeny; ^{222}Rn, ^{216}Po, ^{212}Bi and ^{212}Po as the source of the short-lived alpha particle activity came from theoretical considerations involving solutions of the Bateman equations describing the secular equilibria.

Section 3.2 of Method 900.0 (US EPA, 1980a) for gross alpha and gross beta radioactivity in drinking water recommends delivery of unpreserved samples to the laboratory within 5 days. (No mention is made of holding time for samples preserved by adjustment to pH 2 with nitric acid.) The note to Section 8.4 of the same document requires a 72-hour desiccation prior to alpha counting. Method 903.0 (US EPA, 1080b) for alpha-emitting radium isotopes in drinking water like Method 903.1 for ^{226}Ra in drinking water (US EPA, 1980c) refers to Method 900.0 for sample handling and preservation. Method 903.1

recognizes contributions from ^{224}Ra and recommends a 14-day decay period prior to alpha counting to eliminate them. It appears, consequently, the approved monitoring methods tend to underestimate radium radioactivity in drinking water.

Parsa and his colleagues (Parsa, *et al.*, 1999) observed the alpha activity from water samples having low uranium and radium concentrations decayed more rapidly than could be attributed to the 3.66 – day ^{224}Ra. The samples were divided into two groups on the basis of decay rates obtained from repetitive measurements of the alpha activities. One group appeared to decay with a half life of approximately 30 minutes while the other appeared to be longer lived with a half life of approximately 11 hours. Rapid radiochemical separations by coprecipitation with lead sulfate and immediate measurements of the gamma spectra for consecutive two-hour periods over two-day intervals followed the decay of the 238.6 keV photopeak. On the basis of these measurements, the longer lived activities were attributed to 10.6 – hour ^{212}Pb. The shorter lived activities were attributed to ^{214}Pb ($t_{½}$ = 26.8 minutes) and ^{214}Bi ($t_{½}$ = 20 minutes). Subsequently, Parsa and his collaborators (Parsa, *et al.*, 2000) confirmed the presence of the short lived ^{220}Rn ($t_{½}$ = 1 m) from field measurements made with a portable alpha scintillation spectrometer. Additional date for the concentrations of ^{228}Ra, ^{226}Ra and ^{224}Ra were reported, and a mechanism for the presence of unsupported ^{214}Pb was proposed.

^{224}Ra, ^{226}Ra and ^{228}Ra in Southern New Jersey Drinking Water

Recognizing the need for a comprehensive assessment of the radium concentrations in southern New Jersey drinking water and their impacts on public health, Parsa and his coworkers (Parsa, et al., 2004; Parsa, et al., 2005) developed methods for the determination of ^{224}Ra, ^{226}Ra and ^{228}Ra and unsupported ^{212}Pb concentrations and for the determination of gross alpha activities and the ^{224}Ra, ^{226}Ra and ^{228}Ra concentrations. The former employed a rapid lead sulfate precipitation and gamma spectrometry of their progeny, ^{212}Pb, ^{214}Pb (and/or ^{214}Bi) and ^{228}Ac. In the latter, the alpha emitting radionuclides were isolated by using a barium sulfate/ferric hydroxide coprecipitation scheme. ^{133}Ba tracer was added prior to precipitation to evaluate the recoveries of radium. Gross alpha activities were measured by gas flow proportional counting, and the ^{224}Ra, ^{226}Ra and ^{228}Ra activities were quantifies by gamma spectrometry of their progeny, ^{212}Pb, ^{214}Pb (and/or ^{214}Bi) and ^{228}Ac, respectively. Using 3 – Liter samples and 1000 – minute counting times, the minimum detectable concentrations for ^{224}Ra, ^{226}Ra and ^{228}Ra activities were 0.92 pCi/L, 0.56 pCi/L and 0.97 pCi/L, respectively.

Using these methods and confirming the results with those methods approved by the regulatory agencies, Szabo and his collaborators (Szabo, et al., 2004) conducted a survey of the ^{224}Ra, ^{226}Ra and ^{228}Ra concentrations in water from nearly 100 wells located in the Kirkwood-Cohansey Aquifer. The results of their survey are summarized in Table 2.

The gross alpha activities of these samples ranged from 1.37 to 77.3 pCi/L with a median of 14.9 pCi/L. The gross alpha activities in half of the samples exceed the maximum concentration limit of 15 pCi/L.

Table 2. Percent of Samples in which ^{224}Ra, ^{226}Ra and ^{228}Ra Concentrations Equaled or Exceeded 1 pCi/L, 3 pCi/L and 5 pCi/L

Radium Isotope	≥ 1 pCi/L	≥ 3 pCi/L ≥	5 PCi/L
^{224}Ra	76%	32%	13%
^{226}Ra	69%	23%	9%
^{228}Ra	74%	20%	10%

Szabo, et al., 2004.

PUBLIC HEALTH IMPLICATIONS

At the request of the NJ DEP Commissioner, Bradley Campbell, the New Jersey Water Quality Institute considered the adequacy of the state and federal drinking water standards for ^{226}Ra and ^{228}Ra and the establishment of a drinking water standard for ^{224}Ra (New Jersey Water Quality Institute, 2002). The Institute found, "The current federal maximum contaminant level (MCL) for radium addresses two isotopes, radium – 226 and radium – 228, but does not protect the public from exposure to radium – 224, a short lived isotope recently found in New Jersey water supplies. The New Jersey Water Quality Institute has developed recommendations for modifications of the requirements for the existing radium MCL to reduce risk from exposure to radium – 224. The New Jersey Water Quality Institute has concluded that the USEPA MCL of a gross alpha limit of 15 pCi/L, excluding uranium and radon, is protective to the public for unacceptably high exposures to radium – 224 in drinking water, provided that a requirement for rapid (within 48 hours) gross alpha particle analysis is incorporated. Adoption of such requirements has been found to be feasible both analytically and in regard to treatment technology."

Table 3a. Total Lifetime Cancer Incidence Risk of Three Radium Isotopes

Isotope	Concentration	Total Lifetime Cancer Incidence Risk
^{224}Ra	1 pCi/L	1.96 x 10–5
^{226}Ra	1 pCi/L	5.31 x 10–5
^{228}Ra	1 pCi/L	8.51 x 10–6

Table 3b. Concentrations of ^{224}Ra, ^{226}Ra and ^{228}Ra in pCi/L at Three Risk Levels

^{224}Ra	^{226}Ra	^{228}Ra	Total Lifetime Cancer Incidence Risk
0.012	0.012	0.012	1.0 x 10–6
0.12	0.12	0.12	1.0 x 10–5
1.2	1.2	1.2	1.0 x 10–4

The Institute (New Jersey Water Quality Institute, 2002) considered the morbidity risk associated with radium in drinking water. Risk was calculated assuming two liters of water was consumed each day for 70 years. The total lifetime cancer risks are summarized in Table 3a, and the concentrations of ^{224}Ra, ^{226}Ra and ^{228}Ra for total lifetime cancer incidence risks of one in a million, one in one hundred thousand and one in ten thousand are presented in Table 3b.

The Institute concluded the US EPA considered 3 x 10^{-4} to be an acceptable risk for radionuclides. The risks associated with a gross alpha maximum contaminant level of 15 pCi/L could range from 4.3 x 10^{-4}, if all of the gross alpha were attributed to ^{224}Ra, to 2.1 x 10^{-4}, if the combined ^{226}Ra + ^{228}Ra were just under 5 pCi/L, and the ^{224}Ra concentration was determined assuming the nuclides were present in a ratio of 1.0 : 1.0 : 1.6. The Canadian Federal-Provincial-Territorial Committee on Drinking Water (Canadian Federal-Provincial-Territorial Committee on Drinking Water, 2006), however, did not recommend specific numerical values for gross alpha activity for assessing the risks to human health. This committee voiced the opinion, "There has been a tendency to misinterpret these levels as rigid guidelines rather than as screening tools."

The Commissioner accepted the Institute's findings, and the State revised the New Jersey Safe Drinking Water Requirements (N.J.A.C. 7:18-6.6) to require the "48 – Hour Rapid Gross Alpha Test". In New Jersey, US EPA Method 900.0 is no longer approved for the measurement of gross alpha activity. Current New Jersey gross alpha monitoring requirements must include uranium and radium and exclude radon, and testing "... must be done within 48 hours of sample collection by a laboratory certified to perform the approved method of N.J.A.C. 7:18-6.4."

The New Jersey Department of Health and Senior Services (Clifton Lacy, Commissioner) conducted an epidemiological investigation seeking correlations of the incidences of osteosarcomas with exposures to radium from drinking water (New Jersey Department of Health and Senior Services, 2003). Data from studies conducted earlier in Ontario (Finkelstein, 1994) and in Wisconsin (Guse, *et al.*, 2002) led to opposed conclusions. Radium exposures were classified in two ways: [1] water systems in which the combined radium (^{226}Ra + ^{228}Ra) concentrations exceeded the MCL of 5 pCi/L or the gross alpha activities exceeded the MCL of 15 pCi/L and [2] cancer potency expresses as ^{228}Ra equivalents. The ^{228}Ra equivalents were based on the respective cancer potencies of the individual radium isotopes/Bq; i.e., the ^{228}Ra equivalent was calculated as [(0.16 x ^{224}Ra) + (0.35 x ^{226}Ra) + 1.0 x ^{228}Ra)] where the cancer potencies are ^{224}Ra = 4.5 x 10^{-9}, ^{226}Ra = 1.0 x 10^{-8} and ^{228}Ra = 2.8 x 10^{-8}. The implications of the assessment are summarized in Table 4.

It appears from the data in the Table 4 that the incidence of osteosarcoma among males was more than three times greater in the areas where the drinking water contained combined radium (^{226}Ra + ^{228}Ra) in excess of the 5 pCi/L MCL or where the gross alpha activities exceeded the MCL of 15 pCi/L. Similar observations were reported when the exposures were expresses in terms of ^{228}Ra equivalents. The gender difference was not explained.

The results of this study provide additional evidence that exposure to radium at the concentrations found in drinking water poses a measurable added risk of osteosarcoma. "Risks calculated from these observations are within an order of magnitude of that predicted by the US EPA by linear extrapolation from highly exposed historical cohorts." (Cohn, 2002). The findings are quantitatively consistent with the risk estimates based on the linear extrapolation from studies of occupational and medial exposures, which provided the basis for regulatory actions.

Table 4. Incidence of Osteosarcoma in Relation to Radium Exposures

Demographic Group	Exposure Group	Population per Million	Incidence Ratio <MCL : >MCL
All	< MCL	1287336	2.4
	> MCL*	135638	4.8 : 2.0
Males	<MCL	617223	2.3
	>MCL*	65257	7.7 : 3.3
Females	<MCL	670113	2.5
	>MCL*	70381	2.1 : 0.9

* > MCL signifies combined radium (^{226}Ra + ^{228}Ra) concentrations exceeded the MCL of 5 pCi/L or the gross alpha activities exceeded the MCL of 15 pCi/L.

While they are all based on the available toxicokinetic data for radium, the various risk models were developed using different assumptions about the processes of radium absorption, distribution, metabolism and excretion. The experimental data for radium toxicokinetics are, however, limited.

Measurements of body radium in teenage and adult males exposed only to ambient concentrations of radium in food and water indicated radium retention in the younger cohort was approximately twice that of the older cohort. Based on excretion rates measured for subjects residing in regions where either food or water was the predominate source of radium, it appeared that absorption of from water was greater that absorption from food (ATSDR, 1990). Oral administration of radium to rats showed 1 to 7% of the dose was retained (primarily in the skeleton) in contrast to 77% retention in rats receiving radium by intradermal injection (ATSDR, 1990).

Skeletal deposition of radium has been demonstrated in the exhumed remains of the watch dial painters (the "Radium Girls" from Orange, New Jersey) (ATSDR, 1990). Whole body counting of two of the surviving watch dial painters showed residual body burdens of 72.5 kBq and 1.9 KBq, respectively. The duration of employment of the former was three years while that of the latter was 4½ months (Woodburn and Lengemann, 1964). Between 1957 and 1960, the New Jersey Radium Research Project identified nearly 1000 former radium workers in New Jersey and in the metropolitan areas of New York and Philadelphia. Approximately half, 520, were former watch dial painters. Of these, 340 were alive and 180 were deceased. Approximately 200 of the 340 surviving "Radium Girls" underwent comprehensive medical, dental and clinical laboratory and radiologic studies. Additional data on the body burden for the "Radium Girls" and for others known to have ingested radium compounds have been collected, reviewed and reported by Rowland (Rowland, 1994.).

Following exposure to radium by the oral route, excretion was observed to occur in two phases. In the first phase, approximately 80% of the dose was rapidly excreted in the feces. In the second phase, most of the remaining 20% was ultimately excreted in the feces after absorption into the circulatory system. Biliary excretion was proposed to be a factor in the second phase (ATSDA, 1990). Lucas, *et al.* determined the radium concentrations in blood from four of the surviving "Radium Girls" some 40 years after their exposures. The average rate of excretory plasma or serum clearance and the corresponding coefficient of radium excretion were 80 L/d and 2.5 ± 0.9 %/y, respectively, for these four cases (Lucas, *et al.* 1970).

Some More Recent Activities

DePaul and Szabo (DePaul and Szabo, 2007) have reported on a survey of the ^{224}Ra, ^{226}Ra and ^{228}Ra concentrations and the gross alpha activities in ground water samples collected from 39 wells in the Vincentown and Wennona – Mount Laurel Aquifers, the Englishtown Aquifer System, and the Hornerstown and Red Bank Sands. All are located in regions to the north and to the west of the Kirkwood-Cohansey aquifer. In these samples, the ^{224}Ra concentrations ranged from < 0.5 to 2.7 pCi/L. Their medium was, 0.5 pCi/L. For concentrations of ^{226}Ra, the range and medium were from < 0.5 pCi/L to 3.2 pCi/L and < 0.5 pCi/L, respectively. The corresponding results for ^{228}Ra were < 0.5 pCi/L to 4.3 pCi/L and 0.5 pCi/L. Gross alpha activities in excess of 15 pCi/L were observed only in one sample. This is in contrast to the Kirkwood-Cohansey aquifer where the gross alpha activities from nearly half of the samples were in excess of 15 pCi/L.

The median pH of the water in the Vincentown and Wennona – Mount Laurel Aquifers, the Englishtown Aquifer System, and the Hornerstown and Red Bank Sands was 5.2 while that in the Kirkwood-Cohansey aquifer was 4.7. On the basis correlations of between the ^{224}Ra, ^{226}Ra and ^{228}Ra concentrations and pH, the higher gross alpha activities in the water from the Kirkwood-Cohansey aquifer were attributed to this half pH unit difference. The known pH influence on the geochemistry of radium was cited in support of this proposal.

In addition to the organized ground water surveys, information on radium concentrations and gross alpha activities is collected from the annual water quality reports required from every public distribution system in the state. The "48 – Hour Rapid Gross Alpha Test" was a requirement in the collection of the data for these (and the other) annual water quality reports.

New Jersey American Water provides potable water to some two million people in 175 communities throughout the state. Gross alpha activities ranged from none detected to 10.2 pCi/L and combined radium concentrations ranged from < 1 to 5.6 pCi/L in their Atlantic County system during 2006 (New Jersey American Water, 2006a). For their Burlington, Camden, Gloucester Counties system, the 2006 gross alpha activities ranged from none detected to 1.2 pCi/L and combined radium concentrations ranged from none detected to 4.6 pCi/L (New Jersey American Water, 2006b). New Jersey American Water serves three communities in Cape May County: Ocean City, Strathmere and Cape May Court House. The gross alpha activities in the water from these three systems ranged from none detected to 4.78 pCi/L, from none detected to 8.5 pCi/L and from none detected to 3.99 pCi/L, respectively. The corresponding concentrations of combined radium were from 1.5 to 3.0 pCi/L, from 0.66

to 1.5 pCi/L and from 1.5 to 3.48 pCi/L (New Jersey American Water, 2006c). In their Monmouth County system, the gross alpha activities ranged from none detected to 8.1 pCi/L, and the combined radium concentrations ranged from none detected to 3.30 pCi/L (New Jersey American Water, 2006d). A range of from 1 to 4.7 pCi/L was reported for gross alpha activities in the Mount Holly system (New Jersey American Water, 2006e). The ranges of ^{226}Ra and ^{228}Ra concentrations reported for this system were from 0.24 to 0.74 and from 0.21 to 0.98 pCi/L, respectively.

The Deptford Township Municipal Utilities Authority reported gross alpha activities ranged from 0.519 pCi/L to 5.34 pCi/L and combined radium concentrations ranged from 1.083 pCi/L to 5.544 pCi/L (Deptford Township Municipal Utilities Authority, 2006). Monroe Municipal Utilities Authority reported gross alpha activities ranged from < 3 pCi/L to 10.7 pCi/L and combined radium concentrations ranged from 1.02 pCi/L to 4.60 pCi/L (Monroe Municipal Utilities Authority, 2006). United Water reported gross alpha activities ranged from 0.73 to 19.8 pCi/L and combined radium ranged from 0 to 9.6 pCi/L in the Toms River system (United Water, 2007). Corresponding values from the City of Vineland were from none detected to 16.57 pCi/L for gross alpha activity and from none detected to 6.05 pCi/L for combined radium concentrations (City of Vineland Water Utility, 2005). Washington Township MUA reported the range of gross alpha activities was from 0.9 to 4.4 pCi/L, that of the ^{226}Ra concentrations was from 0.02 to 0.44 pCi/L, and that of the ^{228}Ra concentrations was from 0 to 3.3 pCi/L (Washington Township, 2006).

Where needed for radionuclide removal, private domestic wells can be retrofitted with point of entry remediation technologies, and enhanced treatment technologies are available for public distributions systems. Vigilance continues throughout the state to protect the health and welfare of its residents.

Summary

The detection of short lived alpha activity in potable water supplies in southern New Jersey led to the development of the "48 – Hour Rapid Gross Alpha Test" which has become a requirement in the State of New Jersey. The inclusion of the contribution from ^{224}Ra and its progeny in the gross alpha activity measurements resulted in enhanced assessments of the health effects for the southern New Jersey residents, as well as for the citizens of other states, consuming drinking water containing radium. These efforts are on-going throughout the state.

References

Agronne National Laboratory, 2005, Natural Decay Series: Uranium, Radium, and Thorium, Human Health Fact Sheet.

Ahmed, N. K. 2004. Natural Radioactivity of Ground and Drinking Water in Some Areas of Upper Egypt, *Turkish J. Eng. Env. Sci.*, 28, 345 – 354.

Albertson, P. N. 2003. Naturally Occurring Radionuclides in Georgia Water Supplies: Implications for Community Water Systems, Proc. 2003 Georgia Water Resources Conf. 23 - 24 April, Athens GA.

ATSDR, 1990, Agency for Toxic Substances and Disease Registry, Toxicological Profile for Radium.

Bonotto, D. M. 2004. Doses from ^{222}Rn, ^{226}Ra, and ^{228}Ra in Groundwater from Guarani Aquifer, South America, J. Environ. Radioactivity, 76(3), 319 – 335.

Canadian Federal-Provincial-Territorial Committee on Drinking Water, 2006, Radiological Characteristics of Drinking Water, Water Quality Health Bureau, Ottawa, September 29, 2006.

Cohn, P.D. 2002. Radium in Drinking Water and Osteosarcoma, 130th Annual Meeting of the American Public Health Association, November 11, 2002.

City of Vineland Water Utility, 2005, Annual Water Quality Report, Vineland, NJ.

DePaul, V.T. and Szabo, Z. 2007. Occurrence of Radium – 224, Radium – 226 and Radium – 228 in Water from the Vincentown and Wennona – Mount Laurel Aquifers, the Englishtown Aquifer System, and the Hornerstown and Red Bank Sands, Southwestern and South Central New Jersey, *U.S. Geological Survey Scientific Investigations Report,* 2007 – 5064.

Deptford Township Municipal Utilities Authority, 2006, Consumer Confidence Report, Deptford, NJ.

Finkelstein, M.M. 1994. Radium in Drinking Water and the Risk of Death from Bone Cancer among Ontario Youths, *Can. Med. Assoc. J.*, 151, 565-571.

Guse, C.E., Marbella, A.M., George, V., Layde, P.M. 2002. Radium in Wisconsin Drinking Water: An Analysis of Osteosarcoma Risk, *Arch. Environ. Health*, 57(4), 294-303

Kasztovszky, Z., Kuczi, R., and Szerbin, P., 1996, On the Natural Radioactivity of Waters in Hungary, Central European J. Occupat. Environ. Med., 2(4), 335 – 347.

Lucas, H.F., Marshall, J.H. and Barrer, L.A. 1970. The Level of radium in Human Blood Forty Years After Exposure, Radiation Res., 41(3), 637 – 645.

Marovi, G., Senčar, J., and Franič, Z. 1997. ^{226}Ra in Tap and Mineral Water and Related Health Effects in the Republic of Croatia, Environ. Monitoring and Assessment, 46(3), 233 – 239.

Monroe Municipal Utilities Authority, 2006, Water Quality Test Results, Williamstown, NJ.

New Jersey American Water, 2006a, Annual Water Quality Report, Atlantic County, Cherry Hill, NJ.

New Jersey American Water, 2006b, Annual Water Quality Report, Burlington, Camden and Gloucester Counties, Cherry Hill, NJ.

New Jersey American Water, 2006c, Annual Water Quality Report, Cape May County, Cherry Hill, NJ.

New Jersey American Water, 2006d, Annual Water Quality Report, Mount Holly, Cherry Hill, NJ.

New Jersey American Water, 2006e, Annual Water Quality Report, Monmouth, Middlesex and Ocean Counties,, Cherry Hill, NJ.

New Jersey Department of Health and Senior Services, 2003, Radium in Drinking Water and the Incidence of Osteosarcoma, Trenton, NJ, September 19, 2003.

New Jersey Water Quality Institute, 2002, Maximum Contaminant Level Recommendations for Radium in Drinking Water, Trenton, NJ, May 20, 2002.

NJ, 1983, New Jersey Safe Drinking Water Act, NJ SA 58:12A1, *et seq*.

Parsa, R. 1997. Contribution of Short-Lived Radionuclides to Alpha Particle Radioactivity in Drinking Water and Their Impact on the Safe Drinking Water Act Regulations, 43rd

Annual Conference on Bioassay, Analytical and Environmental Radiochemistry, November 9-13, 1997, Charleston, SC, USA.

Parsa, B. 1998. Contribution of Short-Lived Radionuclides to Alpha-Particle Radioactivity in Drinking Water and Their Impact on the Safe Drinking Water Act Regulations, *Radioact. Radiochem.*, 9(4), 41-50.

Parsa, B., Nemeth, W.K. and Obed, R.N. 1999. The Role of Radon Progenies in Influencing Gross Alpha-Particle Determinations in Drinking Water, 45th Annual Conference on Bioassay, Analytical and Environmental Radiochemistry, October 18-22, 1999, Gaithersburg, MD, USA.

Parsa, B., Nemeth, W.K., Obed, R.N., Szabo, Z. and DePaul, V.T. 2000. Investigation of Factors Contributing to the Presence of Unsupported Lead – 212 in Acidic Drinking Water Samples, Southern New Jersey, 46th Annual Conference on Bioassay, Analytical and Environmental Radiochemistry, November 12-17 2002, Seattle, WA, USA.

Parse, B., Obed, R.N., Nemeth, W.K. and Suozzo, G. 2004. Concurrent Determination of ^{224}Ra, ^{226}Ra and ^{228}Ra and Unsupported ^{212}Pb in a Single Analysis for Drinking Water and Wastewater: Dissolved and Suspended Fractions, *Health Physics,* 86(2), 145-149.

Parse, B., Obed, R.N., Nemeth, W.K. and Suozzo, G. 2005. Determination of Gross Alpha, ^{224}Ra, ^{226}Ra and ^{228}Ra Activities in Drinking Water Using a Single sample Preparation Procedure, *Health Physics*, 89(6), 145-660-666.

Popit, A., Vaupotič, J., and Kukar, N. 2004. Systematic Radium Survey in Spring Waters of Slovenia, *J. Environ. Radioactivity*, 76(3), 337 – 347.

Rowland, R.E. 1994. Radium in Humans: A Review of U.S. Studies, ANL/ER-3 UC-408, Agronne National laboratory, Argonne, IL.

Ruberu, S. R., Liu, Y.G and Perera, S. K. 2005. Occurrence of ^{224}Ra, ^{226}Ra, ^{228}Ra and Gross Alpha, and Uranium in California Groundwater, *Health Physics*, 89(6), 667 – 678.

Rusconi, R., Forte, M., Badalamenti, P., Bellinzona, S., Gallini, R., Maltese, S., Romeo, C and Sgorbati, G, 2004. The Monitoring of Tap Water in Milano: Planning, Methods and Results, *Radiation Protection Dosimetry*, 111(4), 373 - 376.

Sdrndarski, N., Golobočanin, D., and Tutanovi, I. 1996, Natural Radioactivity of Fresh Waters in Serbia, J. Radioanalyt. *Nuc. Chem.*, 213(5), 361 – 368.

Szabo, Z., DePaul, V. T., Kraemer and T. F., Parsa, B. 2004, Occurrence of Radium – 224, Radium – 226, and Radium – 228 in the Water of the Unconfined Kirkwood-Cohansey Aquifer System, Southern New Jersey, *U.S. Geological Survey Scientific Investigations Report*, 2004 – 5224.

United Water, 2006. Consumer Confidence Report 2006, Toms River, NJ, June, 2007.

US EPA, 1976, Environmental Protection Agency, Drinking Water Regulations, Radionuclides, Fed. Reg., vol. 41, no. 133, Friday, July 9, 1976, pp. 28401 - 9.

US EPA, 1980a, Prescribed Procedures for Measurement of Radioactivity in Drinking Water, EPA 600/4-80-032, August, 1980, Method 900.0.

US EPA, 1980b, Prescribed Procedures for Measurement of Radioactivity in Drinking Water, EPA 600/4-80-032, August, 1980, Method 903.0.

US EPA, 1980c, Prescribed Procedures for Measurement of Radioactivity in Drinking Water, EPA 600/4-80-032, August, 1980, Method 903.1.

US EPA, 2000, Environmental Protection Agency, National Primary Drinking Water Regulations, Radionuclides, Fed. Reg., vol. 65, no. 236, Thursday, December 7, 2000.

US EPA, 2002, 40 CFR Ch. 1, §141.25, July 2002, p. 373.

U.S. Geological Survey, 1996, U.S. Geological Survey Programs in New Jersey, New Jersey, Fact Sheet FS-030-96.

U.S. Geological Survey, 1998, Radium – 226 and Radium 228 in Shallow Ground Water, Southern New Jersey, Fact Sheet FS 062-98.

Vyas, V. M., Tong, S. N., Uchrin, C., Georgopoulos P. G., and Carter, G. P., 2004. Geostatistical Estimation of Horizontal Hydraulic Conductivity for the Kirkwood – Cohansey Aquifer, J. Am. Water Resources Assoc., 40(1), 187 – 195.

Washington Township, 2006, Water Quality Report, June 2006, Turnersville, NJ.

WHO, 1996, Water and Sanitation, Guidelines for Drinking Water Quality, Radiological Aspects, pp. 908 – 915.

Woodburn, J.H. and Lengemann, F.W., 1964, *Whole Body Counters*, US AEC, Division of Technical Information, Oak Ridge, TN, p. 16.

Zapecza, O.S. and Szabo, Z., 1988, Source of Natural Radioactivity in Ground Water – A Review, in National Water Summary 1986 – Ground Water Quality: Hydrologic Conditions and Events: U.S. Geological Survey Water – Supply Paper 2325, p. 50.

In: Drinking Water: Contamination, Toxicity and Treatment ISBN: 978-1-60456-747-2
Editors: J. D. Romero and P. S. Molina

Chapter 11

PUMPING TEST DATA ANALYSIS AND CHARACTERIZATION OF DEEP AQUIFER IN THE LOWER DELTA FLOODPLAIN AT MADARIPUR, SOUTHERN BANGLADESH

Anwar Zahid [a, b, *], ***Jeffrey L. Imes*** [c], ***M. Qumrul Hassan*** [b], ***David W. Clark*** [c] ***and Satish C. Das*** [a]

[a] Ground Water Hydrology, Bangladesh Water Development Board, Dhaka 1205, Bangladesh
[b] Department of Geology, Dhaka University, Dhaka 1000, Bangladesh
[c] United States Geological Survey, National Drilling Company-USGS Ground-Water Research Program, P.O. Box 15287, Al Ain, UAE

ABSTRACT

Because of arsenic contamination in shallow (5-50m depth) groundwater in the deltaic and floodplain areas of Bangladesh, characterization of deep aquifers for sustainable drinking water use is becoming an important issue. At Madaripur town, the municipal water supply is facing quality degradation problems because of high arsenic, iron and chloride content in the upper aquifer, and salinity tendency even at greater depth. In this study a 30-m clay layer, identified as a confining unit, is encountered from 130m to 160m below land surface and separates the deep aquifer (generally containing small or no appreciable concentrations of arsenic) from the shallow problematic aquifer. The Hantush-Jacob solution (1955) for a leaky confined aquifer was determined to be the best-fit solution to aquifer-test data collected from a deep borehole. The best fit curve match to two observation wells yielded an average aquifer storage of 0.0003 to 0.006, aquifer transmissivity of 4500 m^2/day, vertical to lateral aquifer hydraulic conductivity ratio of 1.0, and a confining unit vertical hydraulic conductivity of 0.044 to 0.93 m/day. The analysis verifies that the aquifer can yield significant quantities of water to wells with small drawdown and little potential to draw possibly contaminated water from the shallow aquifer.

Keywords: deep aquifer, aquifer test, leaky confined aquifer, transmissivity, storage coefficient

NOTATION

c	D'/K': hydraulic resistance of the semi-pervious layer (day)
L	$\sqrt{Tc}$: leakage factor (meter)
Q	constant discharge rate at the well (meter^3/day)
r	radial distance from the pumping well (meter)
s	drawdown at the well (meter)
S	aquifer storage coefficient
s_m	maximum drawdown (meter) in a piezometer at distance r (meter) from the pumped well
T	transmissivity of the aquifer (meter^2/day)
t	time from the start of pumping (minutes)

1. INTRODUCTION

To protect the population of Bangladesh from water-borne diseases, a concerted effort has been made during the past two decades to replace surface-water drinking supplies with ground-water drinking supplies. Millions of shallow wells (< 80m deep) that have been drilled in the unconfined to semi-confined aquifer system throughout southern Bangladesh now supply water to an estimated 97-98 percent of the Bangladesh population (WARPO, 2000). Since the early 1990s, high concentrations of arsenic were noticed and detected in water pumped from many of these shallow tube wells. About 28 to 35 million Bangladeshis have been exposed to drinking water containing arsenic that exceeded the Bangladesh arsenic standard of 50 micrograms per liter (µg/L), and that from 46 to 57 million people have been exposed to drinking water containing arsenic that exceeded World Health Organization (WHO) arsenic standard of 10 µg/L (DPHE-BGS, 2001). Deeper tube wells (100 to < 300m) have been installed in recent years in an attempt to find groundwater that has little or no arsenic concentration. However, the wells often contain high concentrations of iron and high salinity. As a consequence, characterization of deep aquifers and understanding the natural distribution and movement of this resource for long term sustainability is becoming important issue now.

The term deep aquifer is used in a number of ways depending on the target of its use. Generally, it is defined geologically and hydro-stratigraphically when separated by impermeable or leaky clay/silty clay layer from the upper aquifer. Vertical leakage in alluvial aquifer-aquitard systems may occur through aquitard windows which allow rapid advective flux under stress. The depth of deep aquifers varies depending on the geology and depositional environment of the sediment. Where the deep aquifer (Holocene/ Pleistocene) is separated from the shallow aquifer by substantially thick (>10m) impervious layer, the aquifer can sustainably yield drinking water. Sustainable yield from aquifers, effective use of the water stored in aquifers considering recharge in mind, and preservation of water quality as part of a comprehensive management system must be considered for exploration and development. Aquifer pumping tests can be performed to determine the hydraulic characteristics of aquifers or water-bearing layers, and may provide basic information for the solution of many regional, local, groundwater flow problems. An aquifer test also provides information about the yield and drawdown of the well. These data can be used for determining the specific capacity (discharge-drawdown ratio) of the well, for selecting the type of pump, and for estimating the cost of pumping. In this case, an aquifer test was performed to determine the properties of aquifer sediment to realize the storage and movement of groundwater in the studied aquifer.

A considered and informed approach to the development of deep aquifer resources may prevent damage to the aquifer from over pumping and from inappropriately located and poorly constructed wells. The well drilling and aquifer test was completed for the deep aquifer at Madaripur town to determine the lateral and vertical extent of the aquifer and the response of the deep aquifer to development stresses. In particular, it is essential to understand whether pumping stresses can induce arsenic contaminated water from the shallow aquifer or high salinity water from depth or the shallow aquifer to migrate into the deep aquifer.

2. Study Area

Most of Bangladesh, including the study area, lies within Bengal basin which was initiated during the Late Mesozoic as the continental landmass of Gondwana fragmented (Lindsay et al, 1991) and formed during the Tertiary when the Indian plate collided with the Eurasian plate. At the last phase of sedimentation during the early and mid Miocene large scale regression of the sea commenced and the delta building progressed with continuous westward and southwestward progradation (DPHE-BGS, 1999). Madaripur area is covered by a sequence of fluvio-deltaic deposits of Quaternary age (Figure 1). The Ganges (Padma)-Brahmaputra (Jamuna) river carries the single largest sediment flux and the fourth highest water discharges to the oceans (Coleman, 1969). The alluvium is conventionally subdivided into the older alluvium (Pleistocene) and the newer alluvium (Holocene) (Acharyya, 2005). Saline water from the Bay of Bengal penetrates about 100km or more inland along distributary channels during the dry season (October-April), which defines the lowland floodplain boundary with the lower delta plain (Allison and Kepple, 2001). The study area mostly falls in the lower delta plain separated by a network of tidally influenced distributary channels.

Groundwater is the main source for drinking and irrigation in the study area and is mainly withdrawn from shallow depths in the rural areas except Madaripur town, where water abstracts from deeper depths for urban water supply. Yearly data from two Bangladesh Water Development Board (BWDB) piezometers installed in the upper aquifers at Rajoir and Madaripur (Figure 2 a-b) show that the depth to the groundwater table is high in the dry season (February to May) because of the huge abstraction for irrigation. The water table regains it's static water level in July-September when pumps stop during the monsoon. The seasonal fluctuation of water level ranges between 3.5 and 5.5 m. The trend is almost similar in rural agricultural areas since 1985 with increased lowering during elevated irrigation in dry season recent years.

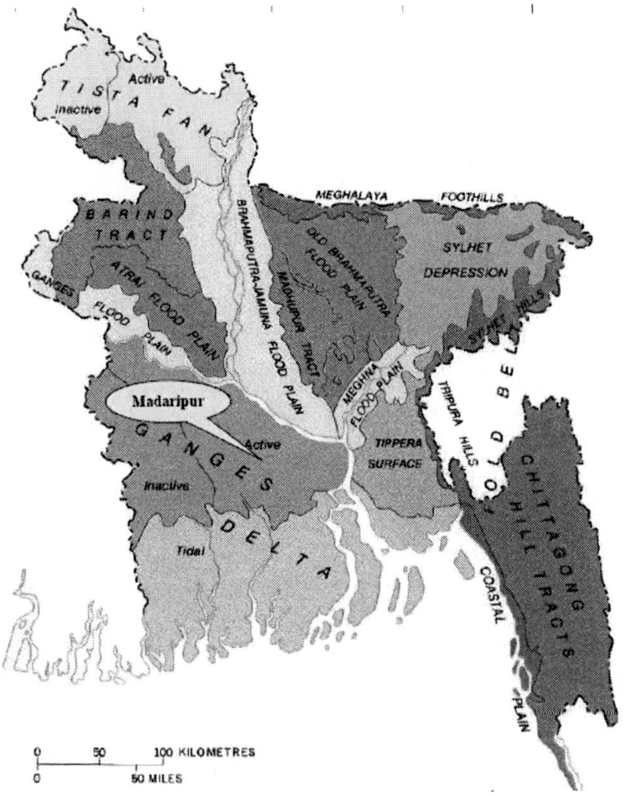

Figure 1. Surface geology map of Bangladesh (Alam et al., 1990) and the study area of Madaripur district.

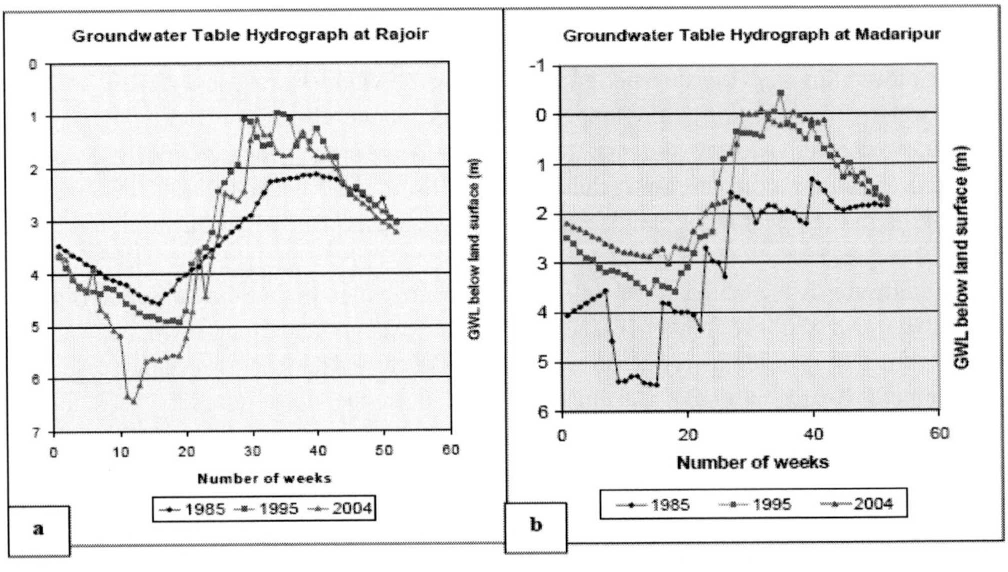

Figure 2. Trend of groundwater table fluctuation (meter) at (a) Rajoir and (b) Madaripur area.

In Madaripur municipal area, depth to water table has been lowered between 1995 and 2004 as compared to 1985. This may be caused by emphasizing the option to use the deeper aquifer, and installing large diameter production wells for urban supply.

Hydrographs show recession during the boro (rice) irrigation season and recovery to a maximum level during August-September in the Madaripur area.. Sudden flattening of the recession or sudden steepening of recovery indicates a sudden change in specific yield. This is often associated with the piezometric level dropping below the base of the upper clay, or rising above it during recovery (WARPO, 2000). A sudden change in the rate of recession during January-February is an indication of the start of the boro season. The sudden rise in groundwater level occurs during rainy periods when abstraction is stopped or reduced, and also at the end of the boro season when irrigation abstraction ends.

3. Methods

3.1. Lithologic Borelogs

Two test holes were drilled by BWDB with the depth of 320m at Rajoir Upazila (sub-district) compound and 241m at BWDB Madaripur office compound to assess sediment lithology, formation/aquifer thickness and aquifer properties in the vicinity. A heavy duty mud pump, FJ casings of 152mm diameter, 76mm diameter NW drilling rod, 140mm diameter carbide and roller bits, split spoon and Denison samplers of 76mm diameter etc were engaged during drilling process. Split spoon sampler was used down to120m depth to collect sediment samples inside PVC liner from selected depth intervals based on the lithology. For the rest depth cuttings were collected from the circulated fluid. Lithologic logs of six piezometric observation wells at Rajoir and two piezometric observation wells installed at Madaripur were also analyzed to cross-check the aquifer formations.

3.2. Installation of Pumping and Observation Wells

The production well, used as the deep aquifer-test pumping well at Madaripur, was drilled to a depth of 233m below land surface and has a well diameter of 0.15m. The screen opening for the pumping well extends from 200m to 230m below land surface. The well annulus was sealed against the clay confining unit. The two observation wells were drilled near the pumping well to monitor changes in water levels as ground water was pumped from the pumping well. One observation well was installed in the deep aquifer. Observation well 1 (PZ-1), drilled 34.0 m west of the pumping well, is screened from 220m to 228m below land surface. The screen opening is near the bottom of the deep aquifer. Observation well 2 (PZ-2), drilled 8.0 m north of the pumping well, is screened above the aquitard from 114m to 123m. The screen opening of PZ-2 is near the bottom of the shallow aquifer, and was used to monitor the response of water levels in the shallow aquifer to stresses applied to the deep aquifer. Both observation wells have a well diameter of 0.15 m. Observation well 3 (PZ-3) was an existing hand tube well located 3.7 m southwest of the pumping well.

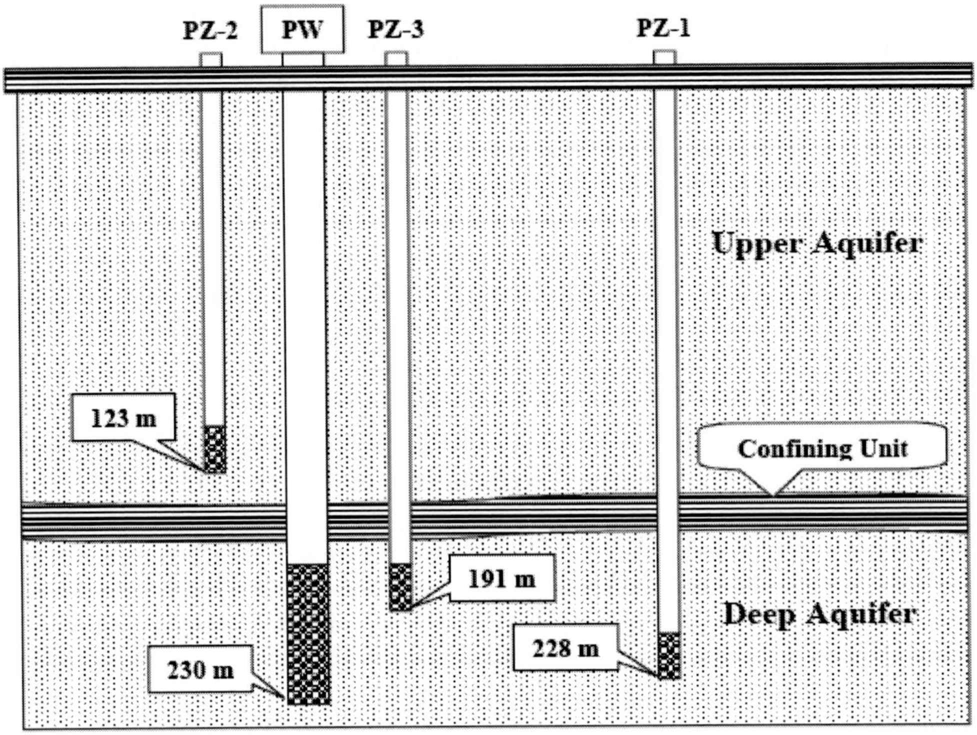

Figure 3. Relative vertical position of pumping and observation wells in the aquifers.

The tube well is open to the middle of the deep aquifer from 188m to 191m below the land surface. The position and design of pumping and observation wells are shown in Figure 3.

3.3. Aquifer Test Data Analytical Method

A pumping test is one of the most useful means of determining hydraulic properties of water-bearing layers and confining beds, and important for a proper understanding of some groundwater system to evaluate groundwater as a water supply. The aquifer was pumped at a constant discharge rate of about 3870 m^3/day. Amongst different analytical methods, it is important to select numerical formula which is more appropriate to actual field conditions. The Hantush-Jacob solution (1955) for a leaky confined aquifer, which was considered to be the best-fit solution to the data, originated with the work of Theis in 1935. Drawdown data for the both deep observation wells can be matched well throughout the 500-minute test period using the Hantush-Jacob solution. Water-level measurements made manually after about 300 min were not used to determine aquifer properties because manual measurements were not sufficiently accurate to track the small water-level changes occurring during the later part of the aquifer test. The following conditions were considered for analyzing Madaripur aquifer-test data valid for confined or leaky-confined aquifer.

- All geologic formations are horizontal and the aquifer has a seemingly infinite areal extent i.e. the cone of depression will never intersect a boundary to the system.
- Aquifer is leaky, horizontal flow stressed aquifer, vertical flow through confining unit.
- An infinite amount of water is stored in the aquifer and groundwater has a constant density and viscosity.
- The aquifer is homogeneous, isotropic and of uniform thickness over the area influenced by the pumping test.
- Prior to pumping, the piezometric surface was nearly horizontal over the area.
- All changes in the position of the potentiometric surface were due to the effect of the pumping well alone.
- Pumping well was 100% efficient
- Drawdown in the water table or unstressed aquifer is negligible.
- Well storage and confining unit storage can be neglected.
- Water instantaneously comes out of storage in the aquifer.

In many aquifer tests, water is contributed from the less permeable confining units, in addition to the aquifer that is pumped (Halford and Kuniansky, 2002). The problem of leakage has been investigated by Hantush and Jacob (1955) and Hantush (1956, 1960, 1964). Hantush and Jacob (1955) presented a solution for drawdown in a pumped aquifer that has an impermeable base and a leaky confining unit above like Madaripur aquifer. For all practical purposes, the time-drawdown behavior before the inflection represents a withdrawal of water from storage from the aquifer, no part of which was contributed from other sources. This part of the curve, along with its straight line extension, may be analyzed with methods discussed for time-drawdown behavior. There are several ways in which this condition may be compromised in field conditions: direct recharge from streams, recharge across bounding low permeable materials etc. The problem of leakage has been extensively investigated by Hantush and Jacob (1955) and Hantush (1956, 1960, 1964). In a leaky aquifer, the drawdown curve will initially follow the nonleaky curve. However after a finite time interval, the lowered hydraulic head in the aquifer will induce leakage from the confining layer. If there is storage in the confining layer, the rate of drawdown will be slower than that in the case where the there is no storage in the confining layer (Hantush, 1960).

The Hantush-Jacob leaky aquifer solution was chosen to analyze the aquifer test data because of the possibility that there was some induced leakage of ground water across the confining unit during the test. The equation is based on the drawdown of a well pumped at a constant discharge rate in a leaky aquifer that requires a curve matching procedure where the drawdown for all time is given as (Hantush and Jacob, 1955; Hantush, 1956).

$$s = \frac{Q}{4\pi T} W(u, {}^{r}\!/_{B})$$

where, *W(u, r/B)* is the tabulated well function for leaky aquifers where long-term drawdowns are described. $u = \dfrac{r^2 S}{4Tt}$ is dimensionless time,

$$\frac{1}{B} = \sqrt{\frac{K_Z / b'}{T}}$$

$K_Z/b'$is the leakance *(1/T)*, where K_Z is vertical hydraulic conductivity of the confining unit *(L/T)* and b' is thickness of the confining unit *(L)*.

Early drawdown data from the field curve match the non-leaky part of the curve, but they soon deviate and follow one of the leaky *r/B* curves. The match point method yields values of *w(u, r/B), 1/u, t* and *s*. In addition, the *r/B* curve followed by the field data is noted. *T* and *S* are readily determined from

$$T = \frac{Q}{4\pi s} W(u, \frac{r}{B}) \text{ and } S = \frac{4uTt}{r^2}$$

Unaware of the work done many years earlier by De Glee (1930, 1951), Hantush and Jacob (1955) also derived the same equation, which expresses the steady-state distribution of drawdown in the vicinity of a pumped well in a semi-confined aquifer in which leakage takes place in proportion to the drawdown. Hantush (1956, 1964) noted that if *r/L* is small *($r/L \leq 0.05$)*, the equation may, for practical purposes, be approximated by

$$s_m \approx \frac{2.30Q}{2\pi T} (\log 1.12 \frac{L}{r})$$

Thus a plot of s_m against *r* on semi-logarithmic paper, with *r* on the logarithmic scale, will show a straight-line relationship in the range where *r/l* is small. The slope of the straight portion of the curve, i.e. the drawdown difference Δs_m per log cycle of *r*, is expressed by

$$\Delta s_m = \frac{2.30Q}{2\pi T}, \text{ consequently } T = \frac{2.30Q}{2\pi \Delta s_m}$$

4. Results and Discussion

4.1. Aquifer System at Madaripur Area

The study area consists primarily of fluvio-deltaic deposits of the Ganges river. There is a complex aerial variability of the hydrogeologic characteristics as a consequence of the floodplain depositional history of the formations that comprise the aquifer.

Simplified lithological sections are drawn based on borelogs conducted at Rajoir and Madaripur town to assess formation thickness as well as aquifer system of the area.

The upper clay and silty clay layer, thickness varies from 30m at Rajoir area to 5m to 7m at Madaripur test site, is generally grey to light grey in color and characterized by high porosity and low permeability. This near surface aquiclude sediments are identified as clayey silt, silty fine sand, sandy silt, silty clay and sand. The lithological logs of the test hole (Figure

4) and piezometric wells at Rajoir reveal that two confining clay layers are encountered in between 165m to 175m and 207m to 212m depths. Below this clay is another silty clay layer intercalated with silty sand layers between 253m to 293m that separates the deeper aquifer from upper aquifer system. Below this 40m thick silty clay bed the deeper aquifer has been identified till drilled depth of 320m, composed mainly of greenish grey to grey fine to medium sand.

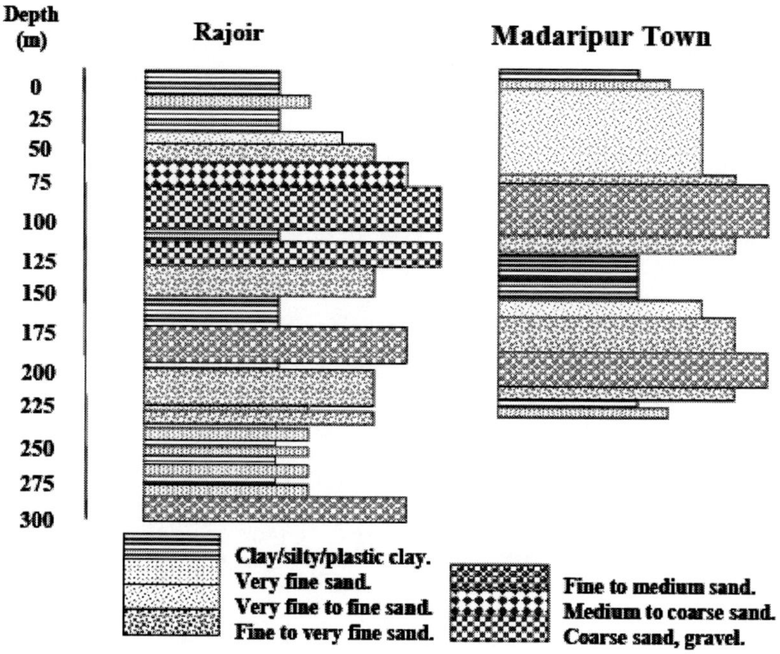

Figure 4. Lithologic sections at Rajoir and Madaripur test holes.

At Madaripur town, a 28m clay layer, identified as a probable deep-aquifer confining unit, is encountered from 130m to 160m below land surface (Figure 4). Sediments above the confining unit form a shallow aquifer predominately composed of very-fine to fine sand with traces of silt and mica. The confining unit consists of a 2.5m cap of peat covering 25m of bluish grey and brown clay. Sediments below the confining unit form the deep aquifer, may also defined as the main aquifer (BWDB-UNDP, 1982), which predominately contains fine to medium sand. The top of the deep aquifer is encountered at 160m below land surface. The bottom of the deep aquifer is apparently at 233m, where a 5m thick layer of grey silty clay and bluish plastic clay is intersected till the drilled depth. The aquifer is about 75m thick at the drill site.

Water collected from a test well drilled in the Rajoir Upazilla compound contained unacceptable salinity and high iron concentrations. The test well drilled in the BWDB compound at Madaripur, about 20 km south of Rajoir presented an aquifer that contained water of acceptable quality and was encountered at a depth of about 230m. The deep aquifer is hydrologically separated from the shallow aquifer by 30m of clay and peat, and generally contains small or no appreciable concentrations of arsenic.

4.2. Monitoring of Groundwater Level

Daily groundwater levels were measured in the three observation wells at Madaripur test site (Figure 5) from April 11, 2005 until May 2, 2005. Water levels in the three observation wells slowly decline from about 12 days before the aquifer test in response to climatic conditions. Linear least squares fits of water levels during the 12-day interval show that water levels in PZ-1 decline at a rate of 0.013 m/day; water levels in PZ-2 decline at a rate of 0.012 m/d; and water levels in PZ-3 decline at a rate of 0.019 m/d. During the entire monitoring interval, water levels in PZ-2 were higher than water levels in PZ-1 and PZ-3, indicating that a downward hydraulic gradient existed from the shallow aquifer to the deep aquifer.

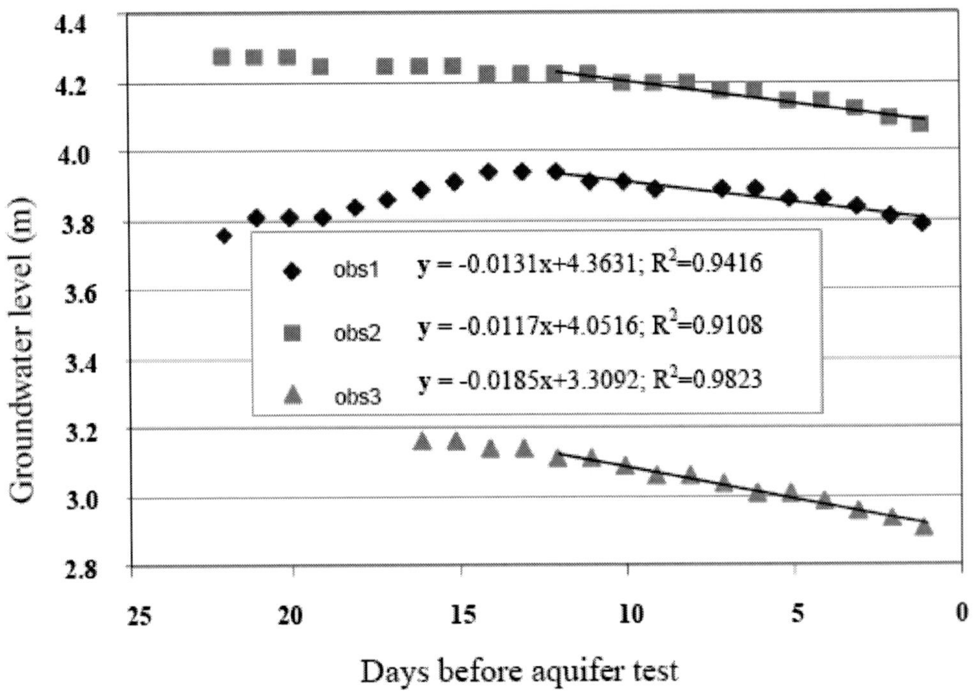

Figure 5. Elevation of groundwater level before aquifer test was started.

Water levels in PZ-1 remained about 0.8 m higher than water levels in PZ-3 during the monitoring interval, indicating that there is some vertical stratification within the zone identified as the deep aquifer. PZ-3 is shallower than PZ-1. The bottom of the screened interval in PZ-3 is 30m higher than the top of the screened interval in PZ-1. The lower water levels in PZ-3 indicates that water preferentially flows in the middle part of the aquifer. This may be indicative of the natural flow system, or may be caused by extraction of groundwater by municipal wells in and around Madaripur.

The water from the shallow arsenic contaminated aquifer generally has a higher head than water in the deeper zones, indicating that water from the shallow aquifer may move downward into the deeper fresh water zones through poorly constructed wells.

4.3. Recharge to Aquifers

Actual recharge that contributes to the groundwater reservoir depends on the degree of depletion (seasonal fluctuation of water table) of this reservoir during the dry irrigation season. In Rajoir area, thick surface silty clay retards the vertical percolation. However, the thin subsurface aquiclude towards Madaripur is favorable for vertical recharge in the upper aquifer, whereas the 28m thick plastic clay layer above the deeper aquifer is a strong barrier against vertical recharge to deeper aquifer. However, this aquiclude is protecting the groundwater of deeper aquifer against vertical infiltration of brackish water from upper aquifer. Generally, groundwater withdrawal from upper aquifer during dry period is balanced with the vertical percolation and inflow from surrounding aquifers in monsoon. Assessment done by WARPO (2000) indicates that when the upper clay is thick and laterally continuous, like Rajoir aquiclude, the piezometric level in the aquifer will remain above the base of the clay and withdrawal from the aquifer is balanced by downward vertical leakage through the clay into the underlying aquifer. When the clay is relatively thin, but exhibits low vertical permeability, the piezometric level may drop below the base of the clay. The resulting piezometric level in the aquifer remains below the base of the clay throughout the year. In this case the total gross quantity of groundwater abstracted from the aquifer is balanced by the leakage flow through the upper clay.

4.4. Aquifer-Test Measurements and Data Analysis

The aquifer test began on 3 May 2005 at 8:00 pm. Before the test began, water-level transducers were placed in the pumping well and observation wells. In addition, a transducer was placed above ground to monitor changes in barometric pressure during the test. The pumping rate was estimated using the height of the water column in a manometer attached to the side of the pumping-well discharge tube. The flow rate of 3870 m^3/day was determined using standard tables for the discharge pipe diameter.

Water levels were measured manually and by using transducers in the observation wells and the pumping well during the aquifer test. Water-level changes measured by transducer and manually in PZ-1 were analyzed, and water-level changes measured manually in PZ-3 were analyzed to determine aquifer properties. The data collected by transducer was corrected to remove barometric effects before analysis. Barometric pressure varied by about 2.0 centimeter (cm) during the 500-minute (min) test, and the correction was small. The data was not corrected for long-term trends, because the magnitude of the estimated long-term water-level decline during the 500 min test (estimated at about 0.006 m for PZ-3) was insignificant. Water levels in PZ-2 varied within a range of about 1.5 cm during the aquifer test and did not exhibit drawdown in response to pumping stresses applied to the deep aquifer. This small water-level change indicates that the 28m thick confining unit hydraulically separates the shallow and deep aquifers, and that there is little likelihood of contaminated water in the shallow aquifer being drawn into the deep aquifer.

The best fit curve match (Figure 6) to PZ-1 (manual measurements) yielded an aquifer storage of 0.0002, aquifer transmissivity of 4550 m^2/day, vertical to lateral aquifer hydraulic conductivity ratio of 1.0, and a confining unit vertical hydraulic conductivity of 0.026 m/day ($r/B = 0.01546$; $K' = (T*b')/B^2$). The best fit curve match (Figure 6) to PZ-1 (transducer measurements) yielded an aquifer storage of 0.0003, aquifer transmissivity of 4500 m^2/day, vertical to lateral aquifer hydraulic conductivity ratio of 1.0, and a confining unit vertical hydraulic conductivity of 0.044 m/day (r/B = 0.02017).

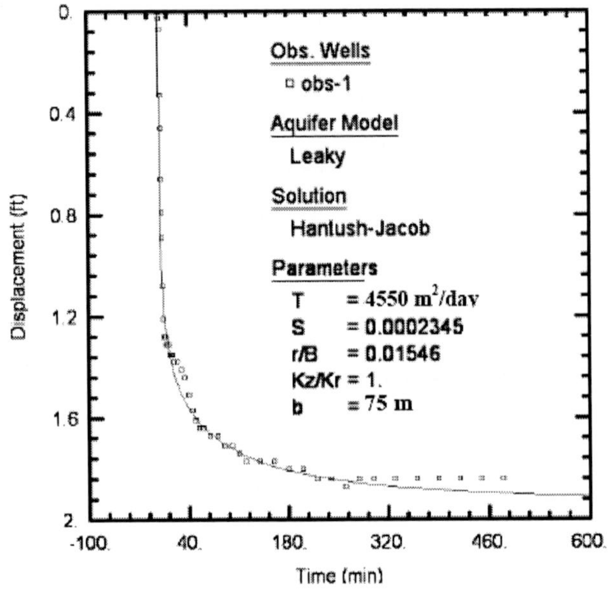

a)

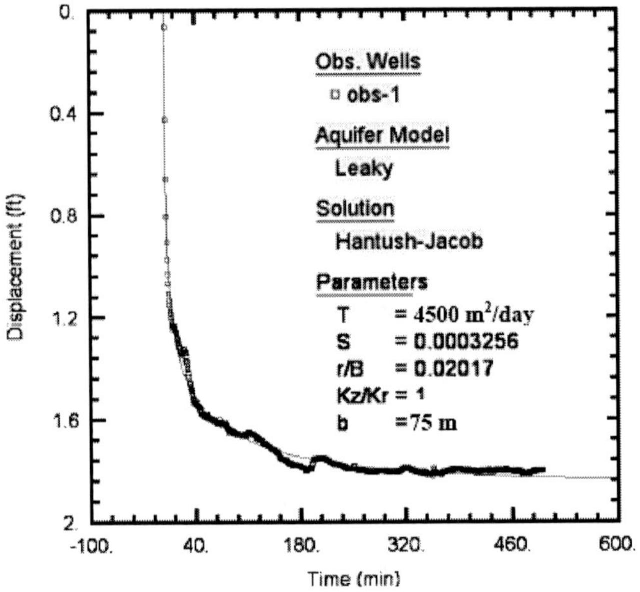

b)

Figure 6. Best-fit Hantush-Jacob solution to drawdown data for observation well 1. Graphs are drawn using (a) manually measured data and (b) transducer recorded data.

The best fit curve match (Figure 7) to PZ-3 yielded an aquifer storage of 0.006, aquifer transmissivity of 4500 m^2/day, vertical to lateral aquifer hydraulic conductivity ratio of 1.0, and a confining unit vertical hydraulic conductivity of 0.93 m/day (r/B = 0.01002). The analysis verifies that the aquifer can yield significant quantities of water to wells with small drawdown and little potential to draw possibly contaminated water from the shallow aquifer. Estimated aquifer transmissivities are consistent among the three analyses, and storage coefficients are consistent with the generally accepted 0.000001 per foot of confined aquifer, or about 0.0002 for the 75m aquifer tested.

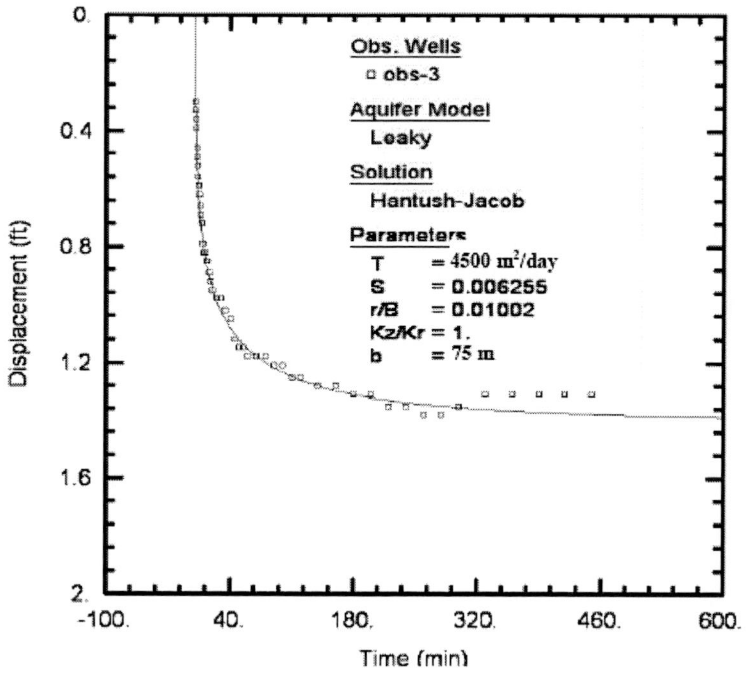

Figure 7. Best-fit Hantush-Jacob solution to drawdown data for observation well 3.

In reality, hydrogeologic settings rarely have aquifers that can be considered of infinite areal extent, rather they change laterally in grain size, shape or lithology that affect the shape of a time-drawdown curve. Although such aquifers do not exist in lower delta, many aquifers are of such wide extent that for all practical purposes they can be considered infinite. Homogenous aquifers seldom occur in the lower delta of Bengal basin and most aquifers are stratified to some degree. As a result of this stratification the drawdown observed at a certain distance from the pumped well may be different at various depths within the aquifer, because of differences in hydraulic conductivity in vertical and horizontal direction. The pumped well penetrates almost the entire aquifer and thus receives water from the entire thickness of the aquifer by horizontal flow.

The maximum measured drawdown in PZ-3 (about 0.45m) is less than the maximum measured drawdown in PZ-1 (about 0.58m) despite the fact that PZ-3 is about 30m closer in lateral distance to the pumping well. It is likely that the data from PZ-1 offers the more accurate analysis, and that the aquifer transmissivity may be 5 percent to 10 percent higher than the original analysis (4550 m^2/day) predicts.

5. Conclusions

Though different hydrogeologic factors control the aquifer parameters, solution results can also vary because the graphical methods of solution depend upon the accuracy of the graph construction and subjective judgments in matching field data to type curves. The Hantush-Jacob solution for a leaky confined aquifer accurately fits the measured drawdown data for observation wells PZ-1 and PZ-3, installed in deep aquifer, and yields reasonable aquifer transmissivity and storage values. The measured 0.8-m water-level difference between the middle part and the lower part of the aquifer, indicates that vertical gradients are present in the deep aquifer and suggests that the analysis of the aquifer-test data for PZ-1 may be more appropriate because the pumping well and PZ-1 are open to the same stratgraphic interval. Water-level data from PZ-1 will also fit a solution with a vertical to lateral aquifer hydraulic conductivity ratio of 0.5. However, the solution does not fit the data as well if the vertical to lateral aquifer hydraulic conductivity ratio is lowered farther. The analysis verifies that the aquifer can yield significant quantities of water to wells with small drawdown and little potential to draw possibly contaminated water from the shallow aquifer.

Acknowledgments

The authors are grateful to the respective authorities of Ground Water Hydrology (GWH) of BWDB and Bangladesh Arsenic Mitigation Water Supply Project (BAMWSP) for undertaking deep aquifer investigation and aquifer pumping test component at Madaripur. Gratitude goes to Mr. Md. Ejar Uddin, Director, Mr. Alamgir Hossain, Former Director, and Mr. MA Karim, Deputy Director, GWH, BWDB for supporting authors to participate during the aquifer test. Thanks are due to the World Bank for providing fund for the study under BAMWSP.

References

Acharyya, S.K., (2005) Arsenic levels in groundwater from Quaternary alluvium in the Ganga plain and the Bengal Basin, Indian Subcontinent: in sights into influence of stratigraphy. International Association for Gondwana Research, Japan. Gondwana Res 8 (1): 55-66.

Allison, M.A., Kepple EB (2001) Modern sediment supply to the lower delta plain of the Ganges-Brahmaputra river in Bangladesh. Geo-marine Letters 21:66-74.

BWDB, 2006. Report on Deep Aquifer Characterization and Mapping Project, Phase-I (Madaripur), Ground Water Hydrology Division-I, Bangladesh Water Development Board.

BWDB-UNDP, 1982. Groundwater Survey: the Hydrogeological Conditions of Bangladesh. UNDP Technical Report DP/UN/BGD-74-009/1, P. 113.

Coleman, J.M., 1969. Brahmaputra river: channel processes and sedimentation. Sedimentary Geology 3, 129– 239.

De Glee, G.J., 1930. Over grondwaterstromingen bij wateronttrekking door middle van putten. Thesis, J. Waltman, Delft (The Netherlands) P. 175.

De Glee, G.J., 1951. Berekeningsmethoden voor de winning van groundwater. In: Drinkwatervoorziening, 3e Vacantie cursus: 38-80. Moorman's periodieke pers. The Hague (The Netherlands).

DPHE-BGS, 2001. Arsenic contamination of groundwater in Bangladesh. British Geological Survey and Department of Public Health Engineering, Govt. of Bangladesh; rapid investigation phase, Final Report.

DPHE-BGS, 1999. Groundwater studies for arsenic contamination in Bangladesh, phase-I: Rapid investigation phase.

Halford, K.J., Kuniansky, E.L., 2002. Documentation of spreadsheets for the analysis of aquifer test and slug test data. U.S. Geological Survey, open-file report 02-197, p. 51.

Hantush, M.S., 1956. Analysis of data from pumping tests in leaky aquifers. Transactions of the American Geophysical Union 37, 702-714.

Hantush, M.S., 1964. Drawdown around wells of variable discharge. *Journal of Geophysical Resources* 69, 4221-4235.

Hantush, M.S., 1960. Modification of the theory of leaky aquifer. Journal of Geophysical Resources 65, 3713-25.

Hantush, M.S., Jacob, C.E., 1955. Non-steady flow in an infinite leaky aquifer. Transactions of the American Geophysical Union 36, 95-100.

Lindsay, J. F., Holliday, D.W., and Hulbert, A. G., 1991. Sequence stratigraphy and the evolution of the Ganges-Brahmaputra Delta complex: American Association of Petroleum Geologists Bulletin, 75, 1233–1254.

Theis, C.V., 1935. The relation between the lowering of the piezometric surface and the rate and duration of discharge of a well using ground water storage. Transaction of American Geophysical Union 16, 519-524.

WARPO, 2000. National Water Management Plan Project, Draft Development Strategy, Annex C Appendix 6, Estimation of Groundwater Resources, Water Resources Planning Organization.

In: Drinking Water: Contamination, Toxicity and Treatment ISBN: 978-1-60456-747-2
Editors: J. D. Romero and P. S. Molina

Chapter 12

THE INFLUENCE OF PH ON SORPTION/DESORPTION EQUILIBRIUM OF Cd^{2+} AND Pb^{2+} IN A BOTTOM SEDIMENT-WATER SYSTEM

***Vladislav Chrastný*[1,*], *Michael Komárek*[2] *and Aleš Vaněk*[3]**
[1]Department of Applied Chemistry and Chemistry Teaching, University of South Bohemia, Studentská 13, 370 05, České Budějovice, Czech Republic
[2]Department of Agrochemistry and Plant Nutrition, Czech University of Life Sciences Prague, Kamýcká 129, 165 21, Prague 6, Czech Republic
[3]Department of Soil Science and Soil Procetion, Czech University of Life Sciences Prague, Kamýcká 129, 165 21, Prague 6, Czech Republic

ABSTRACT

The behavior of metal/metalloid pollutants can play an important role in water quality, especially in the case of drinking water reservoirs. Bottom sediments are a good sorption medium for many organic and inorganic pollutants. Metals and metalloids of natural or anthropogenic origin present in aquatic environments can be bound to sediments and preconcentrated there. Changes of geochemical conditions of the water environment can trigger metal/metalloid release from the sediments. Among many, pH is a very important factor influencing the sorption/desorption equilibrium. The aim of this work was to quantify the influence of pH on the sorption/desorption equilibrium of Cd and Pb. For this purpose, laboratory batch experiments were carried out. The scheme consisted of pH- and concentration gradients of the studied metals added in soluble forms to the water-sediment suspension. The sorption/desorption behavior of the studied metals under a pH gradient of 2-12 was shown in appropriate concentration gradients of 0.001 – 0.1 and 0.01 – 1.0 ppm for Cd and Pb, respectively.

* e-mail: vladislavchrastny@seznam.cz.

INTRODUCTION

The fact that chemical, biological and physical characteristics of bottom sediments reflect the behavior of contaminants in water column upon them presents an important base for further research concerning surface water contaminations with metals. It has been shown that the bottom sediments can act as carriers and/or sources of possible metal contaminations. The basic question is what is the conditions of water/sediment system influenced the metal behavior (e.g. sorption/desorption processes)?

The main sources of metal contaminations can be divided into two main groups: (i) metals from weathering bedrocks (Merian, 1991) and (ii) metals originating from anthropogenic sources, e.g., mining and smelting industry, manufacturing, textile and boot industry, use of fossil fuels, fertilizers, sewage and sludge application (Adriano, 2001; Smith, 1996). Some metals are toxic under relatively higher concentrations. For example, Cu, Zn, Co, Mo or Ni can play an important role in many biochemical processes in living organisms (Adriano, 2001; Merian, 1991). It has been commonly accepted that Cd, Hg and Pb belong among the most toxic metals with respect of anthropogenic character of contamination and human health (Ahumada and Vargas, 2005; Ikem et al., 2003; Fytianos and Lourantou, 2004; Lima et al., 2001; Ho and Egashira, 2000). Their natural occurrence and geochemical cycles are given in Adriano (2001), Merian (1991) summarizes possible toxic effects.

The metals present in water can react with several sediment compounds: (i) clay minerals, (ii) Fe, Mn oxides and hydroxides and (iii) organic matter (Barrow, 1999). Clay minerals having colloidal character are products of alumosilicate weathering. Due to existence of surface hydroxide (–OH) groups, the clay minerals can serve as natural ion exchangers (specific surface about tens $m^2 g^{-1}$ with cation exchange capacity from tenths to units mmol H^+ per 1 g). In the pH range from 6 to 8 (normal pH range of natural waters), clay minerals exchange cations because of their negatively charged surface. The sorption capacity depends directly on a space structure of clay minerals. For example zeolits, minerals with complicated space structure are very efficient ion exchangers often used in industry. The structure and functions of clay minerals in relation to the above mentioned fact is available in other works, e.g., Barrow (1999), Shumann (1985), Hayes and Leckie (1987), Okazaki et al. (1986). Metal oxides (e.g., Fe, Mn, Al-oxides) incorporate the disturbance places in clay minerals structure. The oxides have unsaturated bonds where the hydratation of water molecules can stabilize their structure. Hydrated metal oxides surface can bind or remove H^+ with respect of water pH and therefore can be negatively or positively charged. It is distinguished between six types of Fe oxides, five Al oxides and thirteen Mn oxides with different space structures.

The organic matter has a very complicated composition with number of functional groups, reaction centers (e.g. –OH groups or –COOH groups). Carboxylic acids can be dissociated depending on their pK value. While lower pH causes a positively, higher pH leads to a negatively charged surface of organic matter (Alloway, 1994). The affinity of metals to organic matter is different. McBridge (1989) stated the decerasing order of metal affinities to organic matter as follows: Cu > Ni > Co > Ca > Zn > Mn > Mg. McBridge (1989) supposes that the reaction centers with higher affinity to metals have sulphydrylic groups. These findings are in a good agreement with the fact that the affinity of metals to organic matter decreases with increasing metal concentration because the reaction centres with higher affinity can be quickly saturated.

The metal in a solution can be present in free ion form or bound to inorganic or organic complexes (Elzenga et al., 1999). Boekhold et al. (1993) suppose that sorption of metal bound to complexes is relatively insignificant compared to free metal form. The metal cations can react with charged surface by specific or non-specific adsorption. If some sorption of metal cations in water/sediment system occurs, equilibrium of charges has to be kept. Either change of surface charge or releasing of H^+ has to take place. Specific sorption is characterised with changing the surface potential of sorbent.

The distribution of reactants between solid and liquid phases is very important (Barrow, 1999). A construction of adsorption isotherms is often a result of such a study. Nevertheless one can not be sure, that all accomplishing reactions are adsorptions or whether another reaction type takes place. Moreover, the character of a reaction in a soil chemistry is not isothermal. Linear behaviour of sorption curves (quantity/intensity) can be obtained only for very small concentrations of metals. In this case of higher concentrations, the Langmuir or Freundlich mathematical model by can be applied (Barrow, 1999).

The remobilisation of metals from contaminated sediments is controlled by the following factors (Merian, 1991): (i) pH, (ii) competitive effect on the sorption site, (iii) the occurrence of natural or synthetic complexing agents and (iv) the changing of the redox potential.

The aim of this study was to find the influence of pH on sorption/desorption equilibrium of Pb and Cd in a bottom sediment-water system.

MATERIALS AND METHODS

Study Area, Sampling and Sample Preparation

The composite sample of bottom sediment was collected from a drinking water reservoir in South Bohemian region near the city of České Budějovice, Czech Republic. The reservoir is 12 km long, covers an area of 210 ha and holds about 34.5 mil m^3 of water. The 45 m high dam is situated to the north of the reservoir. The water reservoir lies on the river Malše in a rural area without any major industry. A contaminated stream flowing into the reservoir is situated near the dam. The stream is contaminated by waste waters originated from a factory producing components for aircrafts. The electrotype department of this factory emits into the stream high concentration of Zn and Cd. The concentration of Cd in stream sediments exceeds 200 mg Cd kg-^{1}(Chrastný, unpublished data).

Samples were collected using a gravity corer from five independent sampling profiles near the dam. Sediment samples were homogenised and stored in Teflon containers that were pre-treated with 10 % HNO_3 for 24 h. The samples were air dried at room temperature to constant weight, sieved through a 2 mm stainless sieve, homogenised again and stored in Teflon containers. Samples were not ground because the sediments were already fine-grained and sample grinding could results in a more reactive material that does not reflect natural behaviour.

Table 1 summarises basic physico-chemical properties of the sediment sample. Particle size distribution was determined by the hydrometer method.

Table 1. Basic sediment characteristics

Grain size <0.1 mm	0.1 –0.5 mm	0.5 –2 mm	CEC	TOC	pH	$CaCO_3$	Total Fe	Total Mn	Total Cd	Total Pb
(%)	(%)	(%)	cmol kg^{-1}	(%)		mg g^{-1}	mg g^{-1}	mg g^{-1}	mg kg^{-1}	mg kg^{-1}
46.1	52.7	5.21	15.2	13.1	7.2	126	18.1	2.23	0.05	120

Cation exchange capacity (CEC) was determined as sum of basic cation and Al extracted with 0.1 M $BaCl_2$ (Kalra and Maynard, 1994). Total organic carbon content (TOC) was determined by catalytic oxidation under high temperature (1250 °C) using ELTRA Metalyt CS1000S elemental analyser (Neuss, Germany). Total metal concentrations were determined in digests prepared from 0.5 g sediment sample digested with concentrating acid mixtures ($HF/HNO_3/HClO_4$). Digestion process was carried out in platinum pots and samples were evaporated till the almost solid phase and consequently dissolved in 3 ml of HNO_3. Carbonate content was determined by an alkalinity titration to pH 4.5 with 0.1 M HCl.

Sorption Experiments

In order to study sorption/desorption behaviour of Cd and Pb batch experiments were carried out. A set of 12 sediment samples (approx. 2 g) were placed into polypropylene pots and an amount of 50 ml of 0.01 KNO_3 (basic electrolyte) was added. Each of the pots represented a point within a pH range (from 2 to 12, basic unit was 1). The pH range was adjusted using an automatic titrator (Mettler Toledo, DL 25, Switzerland). The titrating agent for acidic range was 0.1 M and 0.01 M HNO_3 and for alkaline range 0.1 and 0.01 M NaOH. The samples were shaken at an end-over-end shaker overnight (16 h) and the pH was measured after the shaking process. The water/sediment suspension was centrifuged, the supernatant was removed into polypropylene tubes and the samples were treated with 1 ml of concentrated HNO_3. The same scheme, as mentioned above, was carried out with the standard addition of the studied metals to basic electrolyte in following concentration: 0.001, 0.01, 0.1 mg Cd l^{-1} and 0.01, 0.1, 1.0 mg Pb l^{-1} in order to distinguish between the sorption and the precipitation processes at given pH. All the solutions presented here were prepared using chemicals of analytical grade and deionised water (Milli-Q Element, Millipore, France).

Metal Determinations

An inductively coupled plasma mass spectrometer under standard conditions with a Meinhard concentric nebulizer (ICP-MS, PQ ExCell, ThermoElemental, UK) was used for Pb and Cd determination. For all measurement presented, quantification was performed using an aqueous multielement standard solution Merck VI (CertiPUR, Merck, Germany). The calibration procedure consisted of ten measurement replicates of the instrumental blank and five measurement replicates of the standard solutions. In order to eliminate non-spectral interferences internal standardization was used.

Results and Discussion

Sorption Curves of Cadmium

Figures 1a, 1b and 1c are constructed as sorption curves which correspond to the standard addition of Cd concentrations as 0.001, 0.01 and 0.1 μg Cd l^{-1}, respectively. The presented sorption curves reflect the behaviour of Cd in three separate schemes: (A) sediment with the standard Cd addition; (B) basic electrolyte with the standard Cd addition and (C) sediment without the standard Cd addition (only Figure 1 a).

All the presented sorption curves have a similar waveform for the all concentration levels of the standard Cd addition mainly in the acidic area (pH between 2 and 4) (Figs. 1a, b and c). This part is characterised by steep decrease of free Cd^{2+} cations present in solution in scheme A (sediment with the standard Cd addition). The same situation is clearly demonstrated in Table 2 using recovery values computed as a ratio of A/B × 100 (schemes sediment with the standard addition/basic electrolyte with standard addition) and consequently the difference between the next and previous recovery values which can characterise the trend within the pH gradient. The relatively high difference means that the rapid decrease of free Cd^{2+} in solution takes place due to the efficient sorption by sediment. While a negative value of difference informs about the sorption process, a positive value corresponds to Cd release from the sediment (compared to previous state) due to the desorption process (Table 2). The differences between the next and previous recovery values show the mentioned steep decrease of Cd concentration (Table 2). With the increasing pH values the differences between recovery values decrease. It means that with the increasing pH values the sorption reaction increases and therefore decreases the difference between the two consequent pH values (Table 2). The sorption reaction increases to certain pH values (approx. 6.6, 6.4 and 7.3 for standard addition of 0.001, 0.01 and 0.1 mg Cd l^{-1}, respectively) and over these values, the sorption reaction remains almost constant (*plateau* phase) (Figure 1, Tab. 2). The concentration of Cd slightly increases at pH > 10 for relatively higher standard Cd addition (Fig 1a compared to 1b and 1c). The sorption maximum (the minimum of Cd present in basic electrolyte) lies between the pH values of 8 – 9 for all presented schemes (Figure 1a, 1b, 1c). The area with the constant sorption/desorption equilibrium becomes shorter with increasing Cd concentration due to a constant sorption capacity of sediment (well demonstrated in Fig 1a, 1b and 1c).

If the sediment is not present (scheme B), a steeper decrease of Cd concentration in solution can be seen at higher pH values (minimum of Cd concentration about approx. pH > 6) as well. The curve waveform corresponding to a standard addition concentration of 0.001 mg l^{-1} is quite flat compared to the scheme with higher concentrations added (Fig 1a compared to 1b and 1c). In other words, it is possible to assume that when the value of pH is > 2.5, a rapid precipitation process of Cd takes place (Figure 1) which is available as the differences calculated from Cd concentrations recoveries (Table 2). If there is not any sediment in the reaction scheme, the area of the steeper Cd decrease in the basic electrolyte lies between the pH values of 6 – 10. The addition of sediment to the reaction scheme causes expansion of this area (mainly for higher standard Cd addition, Figure 1 b and c).

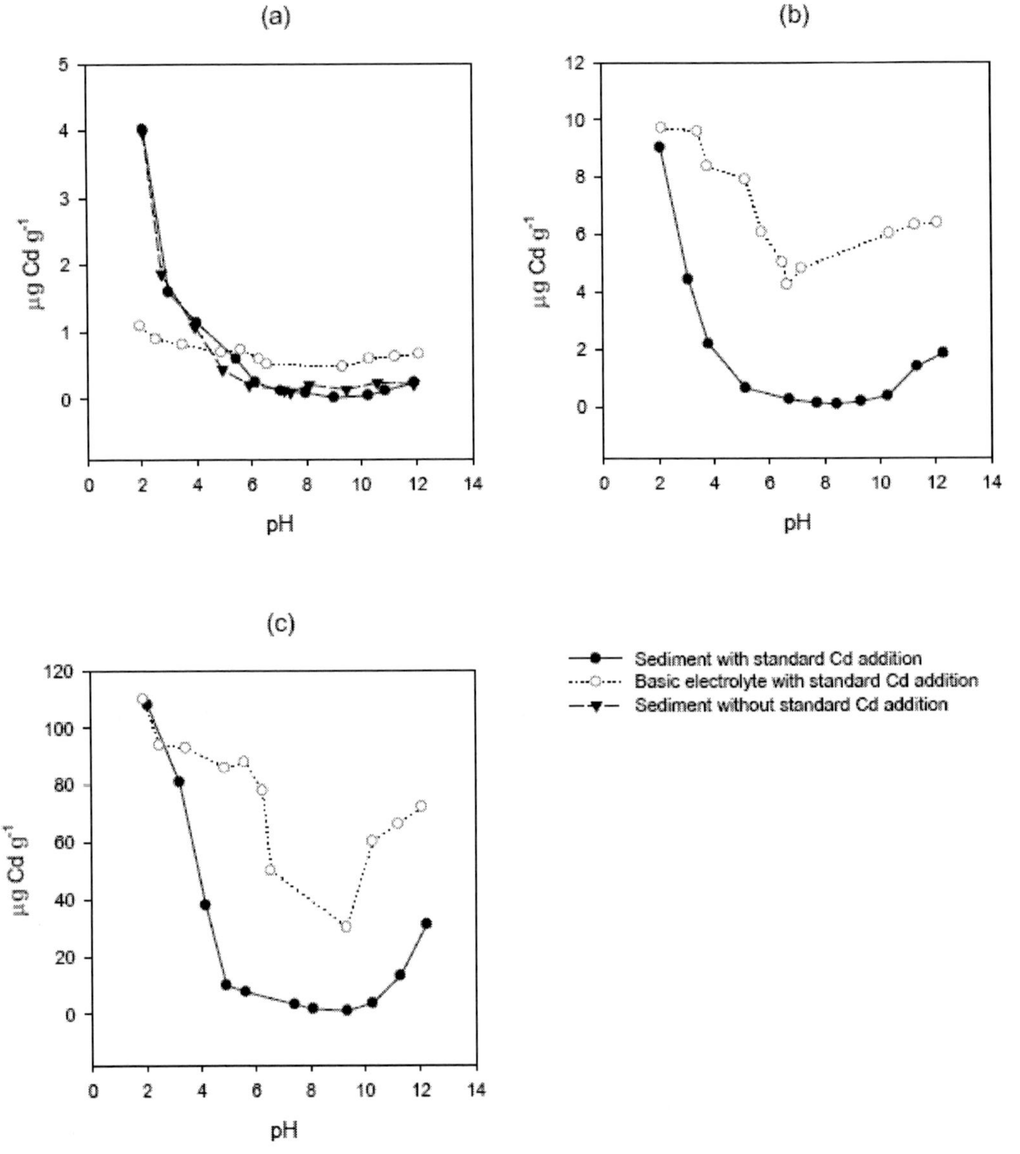

Figure 1. Sorption curves of Cd in pH gradient. Figures a, b and c correspond to Cd standard addition of 0.001, 0.01 and 0.1 mg l-1, respectively.

The presented results show that the highest sorption capacity of the sediment is obtained at pH values ranging from 4 to 6 where the differences the Cd concentration with and without sediment sample reach a maximum. At pH ~ 6-10, the sorption passes the maximum and over the pH of 10 the Cd concentration slightly increases which is more evident at higher Cd concentrations added (Figure 1c). Moreover, from the comparison between Figure 1a, 1b and 1c, it can be concluded that with increasing Cd concentration in solution, the pH value of maximum sorption is shifted to more alkaline areas (Table 2).

Table 2. The recovery for Cd sorption curves

Std. addtion of 0.001 mg Cd l^{-1}			Std. addtion of 0.01 mg Cd l^{-1}			Std. addtion of 0.1 mg Cd l^{-1}		
pH	recovery[a]	difference[b]	pH	recovery[a]	difference[b]	pH	recovery[a]	difference[b]
1.9	365	-188	2.1	93.0	-46.7	2.1	98.2	-12.0
2.5	178	-38.8	3.3	46.3	-20.2	3.3	86.2	-45.3
3.5	139	-53.3	3.8	26.1	-18.1	4.2	40.9	-29.2
4.9	85.7	-51.9	5.1	8.08	-3.97	4.9	11.7	-2.87
5.7	33.8	-13.8	5.8	4.11	-1.73	5.7	8.82	-4.50
6.3	20.0	-2.69	6.4	2.39	-0.50	7.3	4.32	-0.71
6.6	17.3	-13.1	6.8	1.88	1.85	8.1	3.61	0.13
9.3	4.17	4.17	7.1	3.73	2.24	9.3	3.74	2.42
10	8.33	10.7	10	5.97	16.0	10	6.16	13.9
11	19.0	16.8	11	22.0	6.76	11	20.0	23.1
12	35.8		12	28.8		12	43.1	

[a]recovery computed as a ratio of version A/version B × 100 [b]difference between next and previous recovery values

Sorption Curves of Lead

The sorption curves constructed for sediment samples with standard Pb addition (concentrations of 0.01, 0.1 and 1.0 mg Pb g^{-1}) are given in Figure 1a,1b, and 1c, respectively. The highest Pb concentrations in solution are in the acidic area under the lowest pH for both schemes (sediment suspension with the standard Pb addition in basic electrolyte (A), the standard Pb addition in basic electrolyte (B), sediment suspension in basic electrolyte (C)). The difference between the two reaction schemes A and B is shown clearly when the pH value increases. The minimum Pb concentrations are achieved at the pH values of approx. 4, 6 and 8 for the standard addition of 0.01, 0.1 and 1.0 mg Pb l^{-1}, respectively. If the sediment is present (schemes A and C, Figure 2), the pH value corresponding the lowest Pb concentration is lower about approx. 4, and seems to be constant for all the studied standard additions (Figure 2). The presence of sediment causes the rapid sorption of free Pb^{2+} cations mainly at lower pH values (from 2 to 4). At the certain pH values, the Pb concentration achieves a minimum presented as a *plateau* phase on the curve waveform (Figure 2), where Pb desorption takes place very slightly. When no sediment is present in the reaction system, the mentioned *plateau* phase is shorter; moreover, with higher concentration of free Pb^{2+} cations in solution, this *plateau* phase is reduced (Fig 2 b and 2 c compared to Figure 2 a). A very sharp minimum of the sorption curve is obtained after the highest Pb concentration added (Figure 2 c). The presence of the sediment in the reaction system causes the expansion of this area with constant Pb desorption (Figure 2 a, b and 2 c).

Some discrepancies can be seen from the Table 3 within the mentioned *plateau* phase. For the lowest Pb addition at pH between 4 and 6 (Table 3) the recovery values rapidly increased (about 447%). A similar situation is shown for the standard addition concentration of 0.1 mg Pb l^{-1} at pH of about 8 (Table 3).

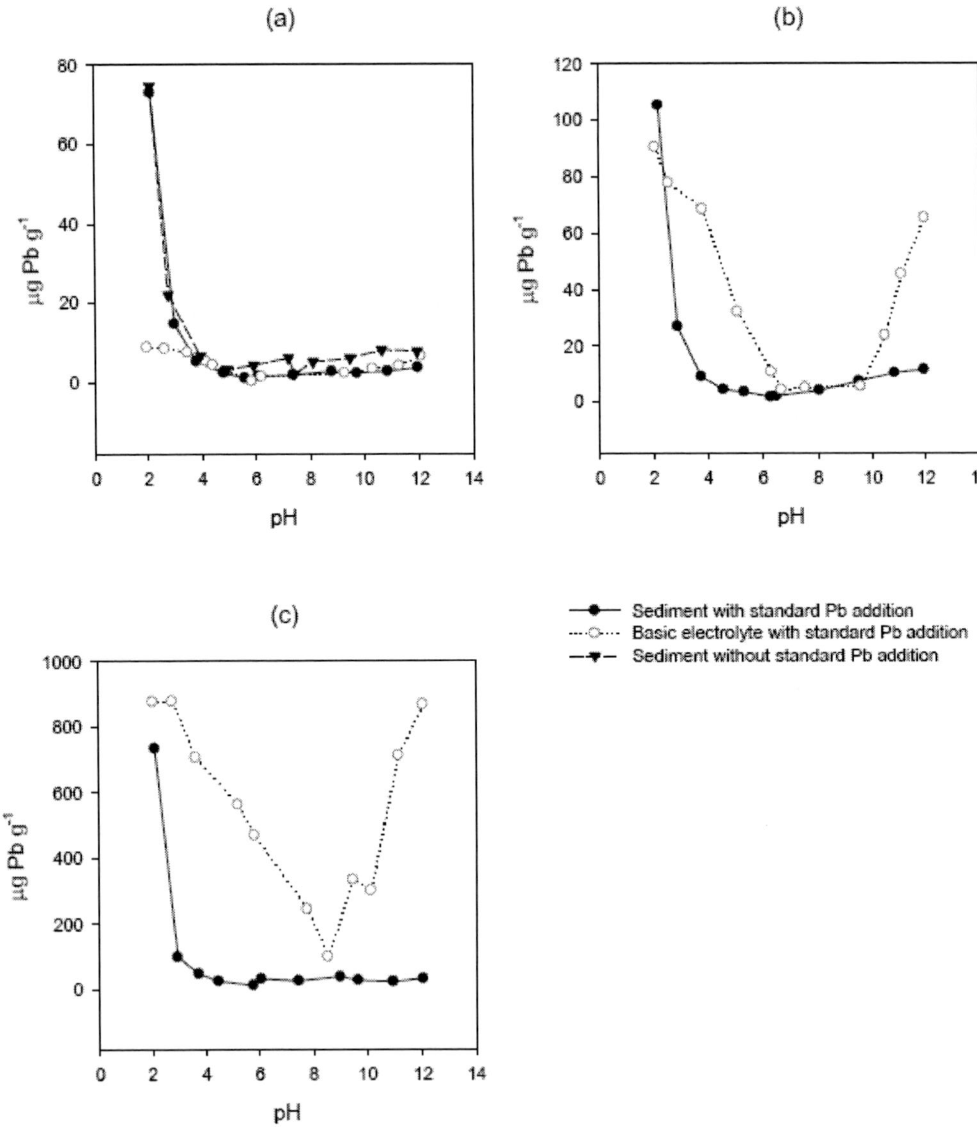

Figure 2. Sorption curves of Pb in pH gradient. Figures a, b and c correspond to Pb standard addition of 0.01, 0.1 and 1.0 mg l-1, respectively.

A strong desorption process is evident here from the recovery value of 74 (Table 3), which means, that Pb concentration released from the sediment reached 74% of concentration obtained at the same pH in basic electrolyte (scheme B). Data for the highest Pb addition confirm previous observations at pH of about 7.5 (Table 3). This can be explained by the fact that under a given pH, the desorption of Pb from sediments is prevailing over the precipitation (scheme B).

Bottom sediments are able to sufficiently adsorb Pb^{2+} cations in a very large pH range with the only exception of highly acidic conditions (pH $\leq$ 2, Figure 2).

Table 3. The recovery for Pb sorption curves

Std. addtion of 0.01 mg Pb l^{-1}			Std. addtion of 0.1 mg Pb l^{-1}			Std. addtion of 1.0 mg Pb l^{-1}		
pH	recovery[a]	difference[b]	pH	recovery[a]	difference[b]	pH	recovery[a]	difference[b]
1.9	822	-649	2.2	116	-81.8	2,1	83.9	-72.6
2.6	173	-102	2.9	34.5	-21.4	3,0	11.3	-4.52
3.5	70.1	-25.3	3.7	13.2	0.51	3,7	6.78	-2.25
4.1	44.9	-20.2	4.6	13.7	19.3	4,5	4.53	-1.71
4.4	24.7	423	5.3	32.9	9.19	5,8	2.82	10.5
5.9	447	-321	6.3	42.1	-0.85	6,1	13.4	13.3
6.2	126	-11.3	6.5	41.3	32.7	7,5	26.7	-15.0
9.3	115	-45.4	8.1	74.0	-42.7	9,0	11.7	-2.28
10	69.6	-2.90	9.5	31.3	-8.60	9,7	9.38	-6.00
11	66.7	-11.0	11	22.7	-5.06	11	3.39	0.50
12	55.7		12	17.7		12	3.89	

[a]recovery computed as a ratio of version A/version B × 100.
[b]difference between next and previous recovery values.

It must be pointed out that under neutral pH, the sorption process is minimal (pH about 7–8) and desorption is prevailing. In the area of more alkaline conditions, the sorption of the sediment is almost at the same level as in the neutral area (Figure 2).

Conclusion

Bottom sediments can serve as sufficient sorption media for both Cd and Pb. The maximum sorption efficiency lies for both Cd and Pb approximately at pH>2. The lowest concentration of both studied metals is present at neutral pH. In the alkaline area the behaviour of Cd and Pb is different. The relatively higher desorption rate of Cd is observed while the concentration of Pb in the alkaline area remains almost constant. A relative decrease of Pb sorption takes place at pH between 7–8. It can be concluded that at pH values corresponding to natural water conditions, sediments can play an important role as sinks for Cd and Pb, but both metals can be easily mobilisable due to changes of physico-chemical characteristics of the sediments (mainly the excavation of sediment, its aeration and consequent oxidization of sulphides to sulphates). Moreover, Cd reflects a higher desorption rate in the alkaline area as well.

Acknowledgments

The presented study was supported by the research projects MSM 6007665806, MSM 6046070901 Ministry of Education of Czech Republic.

REFERENCES

Adriano, D. C. (Ed.) Trace elements in the terrestrial environments: biogeochemistry, bioavailability, and risks of metals. Springer, New York-Berlin-Heidelberg-Tokyo, 2001, 867 p.

Ahumada, R.; Vargas, J. Trace metals: inputs, sedimentation and accumulation in San Vicente Bay, Chile. Environmental Monitoring and Assessment, 2005, 100, 11-22.

Alloway, B. J. (Ed.) Heavy metals in soils. Springer; 2nd ed. Edition, 1994, 384.

Barrow, N. J. The four laws of soil chemistry: the Leeper lecture 1998. *Australian Journal of Soil Research*, 1999, 37, 787-829.

Boeckhold, A. E.; Temminghoff, E. J. M.; Vanderzee, S. E. A. T. M. Influence of electrolyte composition and pH on cadmium sorption by an acid sandy soil. *Journal of Soil Science*, 1993, 44, 85-96.

Elzenga, J. E.; van Grinsven, J. J. M.; Swartjes, F. A. General purpose Freundlich isotherms for cadmium, copper and zinc in soil. *European Journal of Soil Science*, 1999, 50, 139-149.

Fyttianos, K.; Lourantou, A. Speciation of elements in sediment samples collected at lakes Volvi and Koronia, N. Greece. *Environmental International*, 2004, 30, 11-17.

Hayes, K. F.; Leckie, J. O. Modelling ionic strength effcet on cation adsorption at hydrous oxide/solution interfaces. *Journal of Colloid and Interface Science,* 1987, 115, 564-572.

Ho, T.L.T.; Egashira, K. Heavy metal characterization of river sediment in Hanoi, Vietnam. *Communications in Soil Science and Plant Analysis*, 2000, 31, 2901-2916.

Ikem, A.; Egiebor, N.O.; Nyavor, K. Trace elements in watcr, fish and sediment from Tuskegee Lake, Southeastern USA. *Water Air and Soil Pollution*, 2003, 149, 51-75.

Kalra, Y.P.; Maynard, D.G. A comparison of extractants for the determination of cation exchange capacity and extractable cations by a mechanical vacuum extractor. *Communications in Soil Science and Plant Analysis,* 1994, 25, 1505-1515.

Lima, M.C.; Giacomelli, M.B.O., Stupp, V., Roberge, F.D., Barrera, P.B. Speciation analysis of copper and lead in Tubarao River sediment using the Tessier sequential extraction procedure. Quim. *Nova* 24, 2001, 734-742.

McBridge, M. B. Reactions controlling heavy metal solubility in soils. *Advances in Soil Science*, 1989, 10, 1-56.

Merian, E. (Ed.) Metals and their compounds in the environment. Analysis and biological relevance. VCH Weinheim, 313-398, 1991, 1438 p.

Okazaki, M.; Takomidoh, K.; Yamate, I. Adsorption of heavy metal cations on hydrated oxides of iron and aluminium with different crystalinites. *Soil Scinece*, 1986, 142, 523-533.

Shuman, L. M. Fractionation methods for soil micronutrients. Soil Science, 1985, 140, 11-22.

Smith, S. R. Agricultural recycling of sewadge sludge and the environment. CAB International, 1996, Wallingford, 382p.

INDEX

A

Aβ, 251
AA, 109, 115, 148
AAS, xi, 243, 245, 260
abdominal, 127
abdominal cramps, 127
aberrant, 112
abiotic, 237
abnormalities, 110, 271
absorption, xi, 80, 187, 201, 204, 243, 245, 258, 268, 307, 308
Abundance, 80
AC, 115, 117, 178, 294, 295
access, 122, 226, 229
accidental, 123
accidents, 200
accommodation, 52, 54, 63, 65
accounting, 129
accuracy, xi, 135, 243, 247, 254, 272, 275, 326
ACE, 112
acetate, 104, 108, 109, 115, 116, 176, 201, 202, 203, 204, 205, 209
acetone, 256
acetonitrile, 87
Ach, 104
acid, 78, 87, 96, 109, 111, 119, 125, 127, 134, 173, 182, 216, 225, 232, 233, 237, 239, 240, 259, 261, 262, 267, 270, 271, 272, 285, 299, 332, 338
Acid mine drainage, 233, 238
acidic, xii, 2, 80, 225, 228, 233, 245, 258, 261, 297, 332, 333, 335, 336
acidification, 225, 257
acidity, 226, 261
Actinobacteria, 127
activated carbon, 94, 206
activation, 256
active site, 279
acute, 78, 79, 125, 128, 147, 272
AD, 109, 110, 248
Adams, 77, 95, 133, 142, 147
adaptation, 81
additives, 139
adducts, 111
adhesion, 171, 174, 176, 182, 264
adjustment, 155, 303
administration, 105, 109, 114, 115, 307
adsorption, x, 69, 126, 162, 168, 171, 189, 194, 195, 196, 199, 201, 202, 204, 205, 208, 228, 234, 241, 246, 249, 254, 256, 257, 264, 269, 282, 331, 338
adsorption isotherms, 331
adult, 78, 106, 107, 114, 115, 118, 162, 268, 307
adulthood, 105
adults, 110, 112, 125, 258, 271
aerobic, 127, 214, 233, 283
aerosol, 122, 132, 194
aerosols, 196
AFM, 189, 190, 209
Africa, 125, 128, 145
Ag, 155, 247, 254, 257, 259, 268, 277, 278, 281
agar, 133, 134, 144
age, 2, 7, 69, 93, 105, 126, 127, 229, 230, 231, 315
agent, 110, 124, 128, 145, 176, 246, 249, 259, 269, 270, 285, 332
agents, ix, 60, 63, 77, 122, 123, 124, 125, 129, 146, 245, 249, 255, 270, 271, 285, 331
aggregates, 225
aggregation, 190, 194
aggression, 258
aggressive behavior, 106
agricultural, 54, 55, 59, 60, 70, 71, 72, 77, 83, 102, 103, 138, 316
agriculture, 83, 102
aid, 210
AIDS, 132, 164
AIP, 211
air, xi, 16, 173, 178, 196, 239, 243, 258, 269, 270, 273, 283, 331
AJ, 116, 294

AL, 289
alcohol, 127
algae, 42, 79, 81, 95, 99, 131, 146
algal, vii, 1, 89, 94, 98, 100
alkali, 5
alkaline, 168, 190, 228, 231, 264, 266, 332, 334, 337
alkalinity, 332
allele, 135
allosteric, 111
alluvial, 7, 314
alpha, xii, 263, 297, 298, 300, 301, 303, 304, 305, 306, 307, 308, 309
alpha activity, xii, 297, 298, 303, 304, 306, 309
Alps, 80
alternative, 28, 147, 174, 235
alternatives, 134, 241
alters, 107, 112
aluminium, 103, 270, 294, 338
aluminum, 109, 116, 239
Alzheimer, 109, 116, 117, 270
AM, 288, 295
amalgam, 248
Amazon, 143
AMD, 225
amendments, 149
amino, 78, 81, 99, 261, 268
amino acid, 78, 81, 99, 261, 268
amino acids, 78, 81, 261, 268
amino groups, 268
ammonia, 41, 286, 295
ammonium, 271
ammonium hydroxide, 271
amplitude, 252
Amsterdam, 144, 236
amylase, 106
amyloid, 109, 116
amyloid angiopathy, 109
amyloid beta, 109
AN, 293
anaemia, 105, 258, 269, 271
anaerobic, 263, 265, 270
anaerobic bacteria, 263, 265
analysts, xi, 243, 278
analytical techniques, 246
anemia, 267
angiosarcoma, 110
animal models, 110, 112
animal studies, 108
animal waste, 42
animals, viii, x, 78, 103, 107, 109, 111, 117, 122, 123, 125, 126, 128, 129, 139, 148, 213, 214, 224, 228, 234, 269
anion, 103, 201
anions, 201, 205, 208, 263, 271
ANOVA, 186
anoxic, 33
anterior pituitary, 114
anthracene, 219
anthropogenic, xi, xiii, 39, 82, 83, 102, 128, 240, 243, 244, 245, 265, 329, 330
antibiotics, 178, 255
antigen, 111
antimicrobial, viii, 122, 128, 136
antimony, 259, 261, 294
antioxidant, 105, 106, 114, 115, 116
antioxidative, 112
AP, 293
apatite, 42
Apicomplexa, 147
apoptosis, 107, 111, 116
apoptotic, 109, 111, 116
application, ix, xi, xii, 103, 114, 150, 161, 162, 164, 176, 201, 209, 210, 229, 243, 244, 245, 248, 253, 254, 255, 257, 258, 260, 261, 262, 264, 266, 267, 269, 270, 271, 272, 273, 280, 287, 290, 330
aquaculture, 235
aquatic, xiii, 94, 102, 128, 131, 135, 142, 214, 225, 226, 244, 258, 266, 272, 329
aquatic systems, 102
aqueous solution, x, 106, 199, 200, 201, 203, 208
aqueous solutions, 106, 200, 201, 208
Aquifer, vi, xii, 297, 300, 304, 308, 310, 311, 312, 313, 314, 315, 318, 319, 320, 323, 326
aquifers, xiii, 7, 8, 10, 16, 47, 51, 72, 112, 124, 224, 232, 234, 235, 236, 313, 314, 315, 318, 319, 323, 325, 327
AR, 115, 178, 291
Argentina, 101, 200
arginine, 115
aromatic, 244
aromatic hydrocarbons, 244
arsenic, viii, ix, x, xii, xiii, 101, 104, 107, 110, 111, 112, 113, 115, 117, 118, 119, 199, 200, 201, 236, 237, 238, 240, 241, 244, 245, 259, 265, 266, 272, 283, 284, 285, 286, 290, 291, 292, 295, 313, 314, 315, 321, 322, 327
arsenic metabolites, 119
arsenic poisoning, 117
arsenic trioxide, 119
arsenite, ix, x, 112, 113, 199, 202, 203, 204, 205, 208, 228, 240
ART, 59, 60
arteriosclerosis, 118
artificial, 40, 118, 236
AS, 115, 117, 292
ascorbic, 109, 116, 266, 285

ascorbic acid, 109, 116, 266, 285
aseptic, 124
ash, 7, 226
Asia, 125, 128
Asian, 241
assessment, 93, 99, 102, 138, 235, 236, 238, 239, 289, 304, 306
associations, 110, 239
assumptions, 13, 174, 307
ASTM, 302
Athens, 309
atherosclerosis, 112, 118
Atlantic, xii, 154, 197, 282, 283, 297, 308, 310
Atlantic Ocean, 197, 282, 283
Atlas, 132, 142
atmosphere, 181, 246, 265
atmospheric deposition, 102
atomic absorption spectrometry, xi, 243, 244, 288
atomic force, 209
atomic force microscope, 209
atoms, 278
attachment, 173, 174, 180, 181, 183, 195
attention, viii, xi, 78, 101, 107, 131, 139, 214, 228, 243, 256, 272, 286
atypical, 132
atypical pneumonia, 132
Australia, 78, 96, 148
Austria, 147
autoimmune, viii, 101, 107, 115
autoimmune disease, viii, 101, 107, 115
automation, 137
AV, 288
availability, 45, 68, 80, 81, 113
avoidance, 253, 260
awareness, 95, 138

B

β-D-glucuronidase, 133
bacilli, 149
Bacillus, 164
bacteria, viii, ix, 79, 81, 102, 122, 123, 127, 128, 131, 132, 133, 142, 146, 149, 161, 164, 172, 191, 192, 195, 196, 197, 215, 263, 264, 265, 269, 286
bacterial, viii, ix, 16, 122, 124, 126, 127, 128, 131, 132, 133, 134, 136, 146, 147, 150, 191, 195
bacterial contamination, viii, 122
bacterial infection, 132
bactericides, 129
bacteriophage, 190, 197
bacteriophages, 176, 189, 190, 191, 192, 193, 196
bacterium, 127, 132, 134, 220
Bangladesh, vi, xiii, 111, 119, 200, 265, 272, 313, 314, 315, 316, 326, 327
barium, 304
barrier, ix, 105, 161, 162, 191, 323
barriers, 138
basal ganglia, 267
basidiomycetes, 222
baths, 179
bathymetric, 13
batteries, 103, 263
battery, 105
Bax, 109
B-cell, 111
Bcl-2, 107, 109, 115
beef, 126
behavior, xiii, 65, 79, 106, 165, 168, 194, 196, 258, 319, 329, 330
benzo(a)pyrene, 219
beta, 109, 116, 145, 194, 303
bias, 136
bicarbonate, 80, 81
bile, 133
binding, 111, 216
bioaccumulation, 116
Bioanalytical, 290
bioassay, 289
bioavailability, 107, 240, 260, 263, 289, 338
biochemical, 74, 114, 257, 272, 330
biochemical action, 257
bioconversion, 272
biodegradable, xi, 143, 243, 244, 286
biodegradation, 215, 216
biodiversity, 244, 287
biofilm formation, 128, 175
biofilms, viii, 122, 123, 126, 138, 142, 146, 148
biogeochemical, 231, 240
biological, xi, 93, 98, 103, 110, 116, 123, 137, 139, 144, 214, 220, 221, 233, 243, 261, 269, 291, 330, 338
biological activity, 233
biological stability, 123
biologically, 214
biology, 128, 135, 144, 147
biomass, 42, 69, 76, 94
Biometals, 115
bioremediation, 216, 219
Biosensor, 137
biosensors, 137, 146
biosynthesis, 96, 214, 258
biota, 225
biotechnology, 236
biotic, 237
birth, 119, 264

birth weight, 119
bismuth, 259, 266, 267, 269, 275, 277, 291, 292, 293
bisphenol, 222
bivalve, 145
black, 6, 57, 85, 87, 267, 268
bladder, 110, 118, 265
bleeding, 78
blindness, 264
blocks, 7, 129
blood, 104, 105, 107, 108, 112, 114, 118, 128, 258, 266, 289, 308
blood pressure, 108, 112, 118, 258
blood vessels, 104
body weight, 107, 270
bogs, 59
Boltzman constant, 176, 181
bonding, 170, 256, 260, 274
bonds, 247, 249, 257, 262, 278, 330
bone, 107, 258, 270
bone marrow, 107
Boron, 104
Boston, 241
BP, 119
BPA, 216, 217, 219
brain, 104, 105, 108, 109, 114, 115, 116, 124, 270
brain injury, 109
branching, 106
Brazil, 95, 98, 100, 154, 159
breakdown, 36
breast, 113, 119
breast milk, 113, 119
breeding, 42, 51, 52, 59, 70, 82, 83
brevis, 137
British, 149, 327
broilers, 149
Bromide, 272
bromine, 272
bronchiectasis, 111, 117
Brownian motion, 176
bubbles, 253, 274, 283, 285
buffer, 155, 178, 194
buildings, 54, 235
bulbs, 263
business, 59
by-products, 102

C

cadmium, viii, 101, 103, 104, 105, 106, 107, 113, 114, 115, 245, 256, 258, 259, 260, 288, 338
calcium, xii, 106, 244, 297
caldera, 5, 6, 7, 22, 51, 93
calibration, 15, 155, 156, 247, 253, 262, 282, 283, 285, 294, 332
California, 108, 146, 311
Campylobacter jejuni, 146, 149
Canada, 110, 121, 214
cancer, 100, 107, 108, 110, 111, 112, 115, 117, 265, 306
Cancer, 115, 117, 118, 145, 305, 310
cancers, 118
capacity, vii, 43, 129, 193, 202, 244, 249, 251, 315, 330, 332, 333, 334, 338
capillary, 45
capital, 200
carbide, 317
carbohydrates, 81, 256
carbon, viii, 42, 80, 81, 87, 100, 121, 143, 145, 249, 254, 255, 256, 259, 260, 263, 264, 266, 267, 268, 269, 271, 273, 274, 275, 276, 277, 278, 279, 280, 281, 282, 289, 290, 291, 293, 294, 295, 332
carbon dioxide, 80, 81
carbonates, 225, 232, 234, 256, 268
carboxyl, 208
Carboxylic acid, 330
carcinogen, 106, 110
carcinogenesis, 79, 94, 112, 115, 117, 118
carcinogenic, viii, 101, 107, 108, 112, 200, 231, 234, 269
carcinogenicity, 79, 107, 108, 115, 118
cardiovascular, 110, 265, 269, 291
cardiovascular disease, 291
cardiovascular system, 110, 269
carotenoids, 80
carrier, 282
case study, 114
catalase, 108
catalytic, 209, 288, 289, 291, 295, 332
catchments, 227
catechol, 270, 294
cathodic process, 268
cation, 249, 267, 330, 332, 338
cations, 200, 208, 249, 330, 331, 333, 335, 336, 338
cats, 129
cattle, 129
cavities, 230
CB, 148
CD, 106, 115
CEC, 332
cell, 43, 78, 80, 81, 85, 106, 107, 109, 110, 111, 112, 118, 127, 131, 132, 135, 143, 147, 196, 246, 260, 262, 263, 264, 266, 268, 269, 273, 274, 275, 276, 277, 278, 281, 282
cell culture, 135
cell cycle, 111

cell death, 109
cell division, 112
cellulose, x, 176, 213, 255
cement, 129
Centers for Disease Control (CDC), 122, 123, 124, 126, 132, 134, 139, 142, 143
Central Europe, 310
central nervous system, 109, 110
ceramic, 168, 169, 176
cerebellum, 108, 109, 116
cerebral cortex, 109
cerebrovascular, 112
cerebrovascular disease, 112
cesium, 178
CH4, x, 223, 225
channels, vii, 1, 315
chaotic, 7
charged particle, 181
chelates, 246
chemical, ix, x, xi, 16, 17, 19, 27, 42, 65, 69, 83, 98, 101, 103, 110, 113, 125, 127, 129, 161, 162, 164, 165, 170, 183, 186, 187, 188, 193, 200, 208, 209, 223, 224, 225, 226, 232, 243, 244, 247, 249, 252, 253, 256, 259, 261, 270, 272, 292, 330
chemical agents, 249
chemical composition, 165, 224
chemical properties, xi, 243
chemicals, x, 102, 103, 113, 214, 216, 220, 221, 223, 224, 244, 332
chemiluminescence, 133
chemistry, xii, 5, 237, 266, 288, 297, 331, 338
chemokine, 111
chemokines, 111
Chemotherapy, 147
chicks, 110
childhood, 117, 143, 162
children, viii, 109, 121, 126, 127, 258
Chile, 119, 200, 338
China, vii, 78, 100, 108, 262
chitin, 129
Chl, 84
chloride, xiii, 105, 106, 115, 178, 201, 225, 247, 258, 266, 268, 271, 295, 299, 313
chlorination, viii, 122, 123, 127, 138, 146
chlorine, viii, 122, 129, 138, 162, 247
chlorophenols, 215
chlorophyll, 80
chloroplast, 80
cholera, 128, 131, 145, 164
cholesterol, 106, 109, 110, 116, 271
chromatographic technique, 262
chromatography, 154
chromium, 104, 107, 108, 115, 240, 256, 288, 289
chromosome, 264
chronic, 79, 103, 105, 110, 112, 114, 115, 117, 128, 130, 154, 267
chronic cough, 110
cigarette smoking, 111, 117
Cincinnati, 178, 180
circadian, 105
circulation, 2, 8, 11, 41, 230, 233
cirrhosis, 109, 263
citizens, 309
CK, 46
CL, 113, 117, 118
classes, 54
classical, 260, 273, 280
classification, 27
classified, 54, 56, 106, 127, 168, 169, 173, 229, 306
clay, xiii, 178, 183, 187, 200, 202, 209, 224, 226, 299, 313, 314, 317, 320, 321, 323, 330
clays, 3, 7
cleaning, 255, 256, 277
climatic factors, 29, 82
clinical, 124, 139, 307
closure, x, 223, 226, 230, 231, 235
clouds, 170
clusters, 117
CNS, 116
Co, xi, 99, 121, 178, 180, 224, 229, 230, 231, 238, 254, 257, 259, 288, 330
CO2, 235
coagulation, 138, 162, 169, 188, 208, 268
coal, x, 168, 223, 224, 225, 226, 238, 241, 265
coal mine, 238, 241
cobalt, 269, 293
codes, 56
coenzyme, 111
cohesion, 274
cohort, 307
colic, 258
coliforms, viii, 122, 123, 133, 138, 144, 146, 147, 150
colleges, 59
collisions, 181, 183
colloidal particles, 246, 247, 256, 260, 261
colloids, 228
colonisation, 139
colonization, 132, 139, 149
Colorado, 239, 240
colorectal, 100
colorectal cancer, 100
combined effect, 162
combustion, 201
commercial, viii, 93, 121, 247
commodities, 226

communication, 134
communities, ix, 153, 308
community, 102, 123, 132, 142, 148, 149, 150, 300, 303
compaction, 192
competence, 128
complex behaviors, viii, 101
complexity, 255, 286
compliance, 69, 303
complications, 127
components, viii, 46, 74, 101, 102, 113, 173, 204, 206, 249, 262, 331
composite, 154, 155, 158, 272, 303, 331
composites, 155, 158
composition, 80, 87, 107, 153, 173, 195, 221, 227, 233, 240, 255, 261, 330, 338
compounds, x, xi, 82, 83, 87, 107, 117, 118, 125, 204, 207, 214, 216, 219, 221, 223, 225, 238, 240, 243, 246, 249, 256, 260, 262, 263, 267, 268, 283, 286, 290, 299, 307, 330, 338
computation, 45, 46, 47, 52, 74
computer, 137, 209
computing, 229
concentration, xi, xiii, 37, 42, 63, 74, 75, 76, 77, 93, 94, 102, 104, 106, 107, 108, 109, 110, 115, 125, 135, 137, 154, 159, 162, 163, 173, 174, 178, 180, 181, 182, 183, 184, 185, 186, 187, 188, 195, 206, 207, 208, 214, 216, 226, 228, 229, 231, 233, 243, 245, 246, 247, 248, 249, 250, 251, 253, 254, 257, 258, 259, 261, 262, 263, 264, 265, 266, 268, 271, 272, 273, 275, 276, 277, 278, 279, 280, 282, 284, 285, 292, 303, 304, 306, 314, 329, 331, 332, 333, 334, 335, 336, 337
conceptual model, 45
conditioning, 60, 246
conductance, 133, 144
conductivity, xi, xiii, 16, 17, 19, 23, 29, 36, 84, 232, 234, 236, 243, 257, 313, 320, 323, 324, 325, 326
congress, 238
Congress, iv, 194, 236, 238
coniferous, 59
connective tissue, 268
consent, 19
conservation, 40, 173, 226, 238
construction, ix, 59, 122, 138, 247, 326, 331
consumers, 102, 123, 138
consumption, viii, xii, 94, 109, 118, 121, 124, 125, 128, 130, 143, 148, 196, 226, 270, 298
contact time, 125
contaminant, ix, 113, 115, 139, 153, 154, 159, 226, 228, 300, 305, 306
contaminants, ix, xi, 102, 104, 114, 117, 126, 146, 153, 154, 196, 214, 226, 232, 235, 243, 244, 256, 272, 286, 330
contaminated food, 125, 127
contamination, vii, viii, ix, x, xiii, 1, 2, 16, 65, 83, 91, 98, 101, 102, 103, 122, 123, 124, 125, 133, 138, 139, 145, 154, 161, 162, 178, 187, 188, 200, 213, 223, 224, 228, 233, 235, 237, 239, 240, 241, 245, 246, 253, 260, 265, 270, 272, 273, 283, 286, 313, 327, 330
continuity, 40
control, ix, xii, 99, 106, 107, 108, 110, 112, 122, 131, 134, 135, 137, 138, 142, 148, 210, 226, 227, 244, 252, 256, 263, 270, 272, 326
controlled, 15, 170, 178, 291, 331
convective, 247, 248, 249, 257, 273, 279, 282, 283
conversion, 142
cooking, 268
cooling, 235, 241, 272
coordination, 200, 201, 202, 205
copper, 103, 109, 116, 200, 230, 239, 247, 259, 260, 261, 263, 266, 268, 273, 275, 276, 282, 284, 285, 287, 288, 289, 290, 291, 292, 295, 338
corn, 41
cornea, 194
correction factors, 69
correlation, 38, 41, 110, 187, 228, 279, 284
correlation analysis, 187
correlation coefficient, 187, 284
correlations, 69, 306, 308
corrosion, 103, 108, 258, 271
cortex, 116
Corynebacterium, 164
Corynebacterium diphtheriae, 164
cost-effective, 133, 150
costs, 102, 130, 137, 139, 162, 200
coupling, 219, 259, 260
coverage, 28
covering, 321
cows, 128
coxsackievirus, 125
CR, 134
Crassostrea gigas, 150
CRC, 97, 142, 144, 197, 221
cretinism, 271
Croatia, 310
crops, 41, 54, 56, 59, 60, 67, 68, 126, 149
CRS, 288
crust, 226, 265
crustaceans, 235
cryptosporidium, 147
crystal, 201
crystals, 230

cultural, 102
culture, 80, 133, 134, 135, 146, 147
culture media, 133, 147
current prices, 235
customers, 188
cyanide, 225
cyanobacteria, 78, 79, 80, 84, 96, 97, 99, 100, 125, 148
Cyanobacteria, 77, 95, 100
cyanobacterium, 95, 97, 125, 148
cycles, 23, 74, 80, 127, 148, 330
cycling, 231
cyclohexane, 269
cyclooxygenase, 115
cyst, 132, 149
cysteine, 285
cysts, 129
Czech Republic, 329, 331, 337

D

DA, 294
dairy, 226
dairy products, 226
danger, 138
data analysis, 137
data base, 41, 42, 52, 54, 56, 139
data collection, 139
database, 228
DD, 117
death, viii, 78, 95, 121, 122, 125, 162, 265, 271
deaths, 107, 126, 162
Debye, 181, 184
decay, xii, 297, 298, 299, 304
decisions, ix, 102, 122, 138
decomposition, 42, 245, 247
defects, 192, 264
deficiency, 80, 81, 271
deficit, 51
definition, 83, 168, 282
defreezing, 85
degradation, x, xiii, 125, 213, 214, 215, 262, 313
degrading, x, 213, 215, 219, 220, 222
degree, ix, 109, 133, 136, 161, 164, 171, 176, 189, 191, 232, 267, 323, 325
dehydration, 130
dehydrogenation, 201
delivery, viii, 122, 123, 137, 303
delta, vi, 313, 315, 325, 326, 327
demand, 51, 68, 69, 70, 231
dendrites, 185
denitrification, 69
Denmark, 128
density, 56, 85, 89, 94, 136, 137, 145, 171, 174, 180, 181, 260, 279, 319
Department of Energy, 302
Department of Health and Human Services, 142
dependant, 262
deposition, 6, 102, 109, 174, 181, 195, 246, 247, 249, 252, 253, 254, 259, 263, 264, 273, 274, 275, 276, 278, 282, 283, 284, 285, 292, 307
deposits, x, 2, 3, 6, 7, 8, 9, 103, 108, 150, 223, 225, 228, 232, 238, 240, 262, 268, 272, 277, 299, 315, 320
depression, 7, 9, 13, 264, 271, 319
deprivation, 117
derivatives, 265
dermal, 118, 122
desalination, 162
desert, 196
desiccation, 303
desorption, x, xiii, 199, 200, 201, 249, 329, 330, 331, 332, 333, 335, 336, 337
destruction, 246
detachment, 78, 174, 175
detection, 125, 133, 134, 135, 136, 137, 138, 142, 143, 144, 145, 146, 147, 149, 150, 185, 216, 221, 229, 247, 253, 264, 266, 267, 281, 282, 284, 286, 295, 309
detergents, 42, 83
deterministic, 183
detoxification, x, 108, 213, 219
Detoxification, v, 213
detoxifying, x, 213, 215
developed countries, viii, ix, 121, 122, 125, 129, 130, 138, 139, 161
developing brain, 105
developing countries, viii, 110, 121, 125, 127, 128, 129, 200
deviation, 35, 183
DG, 115
diabetes, 110
diagnostic, 146
dialysis, 78, 95, 110
diamonds, 203
diarrhea, 126, 148, 258, 263
diarrhoea, 127, 129, 130, 134, 258
dielectric, 181
dielectric permittivity, 181
diet, 103, 110
dietary, 103, 258, 268
diets, 110
differentiation, 112
diffraction, 201
diffusion, 171, 172, 174, 176, 177, 193, 248, 250
digestion, 131, 246, 293

dimeric, 201
dioxins, 215, 244
discharges, 9, 15, 102, 233, 237, 315
discriminatory, 136
diseases, 123, 126, 128, 138, 146, 162
disinfection, viii, 103, 104, 122, 123, 129, 131, 138, 144, 146, 147, 162, 164, 176, 194, 268, 272
dispersion, 68, 174, 180, 185, 232, 233
displacement, 136
dissociation, 111
dissolved oxygen, 16, 17, 29, 32, 33, 34, 84, 253, 259, 285
distillation, ix, 153, 235
distilled water, 16, 109, 110, 155, 285
distress, 109
distribution, viii, 41, 43, 56, 57, 60, 72, 73, 103, 122, 123, 130, 137, 138, 142, 143, 146, 147, 150, 154, 168, 192, 195, 196, 240, 248, 260, 265, 299, 307, 308, 314, 320, 331
diversity, 99, 123, 272
dizziness, 258
DNA, 107, 108, 111, 112, 118, 136, 137, 142, 147, 255, 258
DNA damage, 108
DNA repair, 111
dogs, 78, 129
DOI, 238
dominance, 94
dopamine, 105, 114
dosage, 154, 186
DP, 114, 326
drainage, xi, 10, 52, 69, 73, 93, 224, 229, 232, 233, 234, 237, 239, 240, 241
drinking, vii, viii, ix, x, xiii, 1, 78, 93, 94, 97, 98, 100, 101, 102, 103, 104, 105, 106, 107, 108, 109, 110, 111, 112, 113, 114, 115, 116, 117, 118, 119, 122, 123, 124, 125, 126, 127, 128, 129, 130, 132, 133, 138, 142, 143, 144, 145, 146, 147, 148, 150, 153, 154, 161, 162, 176, 196, 199, 200, 223, 225, 226, 227, 228, 231, 234, 235, 236, 246, 256, 258, 262, 263, 264, 265, 267, 269, 271, 272, 286, 287, 291, 299, 303, 304, 305, 306, 309, 313, 314, 315, 329, 331
drinking water, vii, viii, ix, x, xiii, 1, 78, 93, 94, 97, 98, 100, 101, 102, 103, 104, 105, 106, 107, 108, 109, 110, 111, 112, 113, 114, 115, 116, 117, 118, 119, 122, 123, 124, 125, 126, 127, 128, 129, 130, 132, 133, 138, 142, 143, 144, 145, 146, 147, 148, 150, 153, 154, 161, 162, 176, 196, 199, 200, 223, 225, 226, 227, 228, 231, 234, 235, 236, 256, 258, 262, 263, 264, 265, 267, 269, 271, 272, 286, 287, 291, 299, 303, 304, 305, 306, 309, 313, 314, 329, 331
drug resistance, 127
drug-resistant, 128
dry, 25, 74, 80, 102, 231, 233, 263, 271, 315, 323
drying, 16, 190
DTA, 201
duodenum, 108
durability, 273, 282
duration, 103, 273, 307, 327
dust, 194, 224, 225, 226
dyes, 135, 215, 220, 221

E

E. coli, ix, 123, 126, 133, 134, 137, 161, 189, 190, 191
EA, 288, 290, 292
early warning, 148
earth, 168, 200
earthworms, 236
eating, 126
EB, 326
ecological, 82, 113, 240
ecology, 97, 98, 123, 138, 142
economic, 56, 82, 102, 123, 129, 130, 139, 144, 200, 240
economic losses, 130
ecosystem, xi, 243, 264
ecosystems, 100, 226, 287
education, 97, 138
efficacy, 123, 128, 248, 259
effluent, 126, 164, 178, 185, 188, 196, 214, 220, 221
effluents, 26, 41, 54, 68, 83, 102, 146, 214, 215, 220, 221, 224, 228, 232, 264
egg, 129
eggs, 129
Egypt, 309
elderly, 110, 126, 132
elderly population, 110
electric potential, 184
electrical, 170, 231, 234, 251, 255
electrical conductivity, 231, 234, 255
electricity, 235
Electroanalysis, 287, 288, 289, 290, 291, 292, 293, 294, 295
electrochemical, xi, xii, 137, 243, 244, 245, 246, 247, 254, 255, 256, 260, 261, 262, 266, 271, 272, 273, 274, 275, 276, 277, 278, 279, 280, 281, 282, 287, 288, 293
electrochemical deposition, 262
electrochemistry, 246
electrodes, xi, 243, 248, 249, 251, 252, 253, 254, 255, 258, 263, 264, 266, 269, 270, 271, 272, 273,

278, 279, 280, 282, 283, 284, 290, 291, 293, 294, 295
electrolysis, 246, 248, 250, 252, 253, 254, 257, 259, 264, 266, 268, 274, 279, 280, 283, 284
electrolyte, 181, 249, 251, 253, 264, 267, 283, 285, 332, 333, 335, 336, 338
electrolytes, 255
electron, 80, 165, 195, 201, 202, 209, 216
electron microscopy, 165
electronic, iv, 200, 201, 262, 266
electronic structure, 200
electrophoresis, 134, 137
electrophysiological, 105
electrostatic, iv, 168, 170, 171, 185, 187, 191
ELISA, 93
embryogenesis, 107
emission, 2, 6, 201, 203, 204, 206, 207
emphysema, viii, 101
employment, 307
encephalitis, 124
encephalopathy, 110
endocrine, x, 213, 214, 216, 220, 221
endocrine system, x, 213, 214
endocrine-disrupting chemicals, 216, 220
endogenous, 22, 23, 24, 125
endonuclease, 134
endothelial cell, 117
energy, xi, 77, 80, 81, 85, 173, 181, 201, 204, 205, 206, 207, 224, 227, 236, 263, 271, 274
energy recovery, 236
engineering, 17, 164
England, 227, 238
enlargement, 192, 193, 274, 283
enteric, ix, 124, 125, 126, 135, 139, 145, 161, 176, 197
enterovirus, 124, 143
enteroviruses, 123, 124, 194
ENV, 237
environment, x, xii, xiii, 41, 56, 65, 81, 82, 97, 103, 107, 124, 125, 128, 135, 139, 171, 190, 196, 200, 213, 214, 224, 228, 229, 231, 232, 233, 235, 236, 237, 238, 265, 267, 269, 273, 287, 295, 297, 314, 329, 338
environmental, vii, ix, x, xi, 1, 13, 41, 43, 78, 81, 82, 94, 102, 103, 115, 118, 124, 125, 131, 132, 135, 137, 139, 142, 143, 144, 145, 150, 151, 199, 200, 206, 210, 213, 215, 219, 220, 226, 229, 231, 235, 236, 237, 238, 239, 240, 243, 272, 287, 292
environmental conditions, 78, 94
environmental contamination, 139, 237
environmental control, 238
environmental degradation, 125
environmental effects, 240
environmental impact, 41, 226, 235
Environmental Information System, 42
environmental protection, 235, 236
environmental regulations, 235
environmental standards, 239
environmental temperatures, 144
enzymatic, 106, 216, 217, 218, 219, 258
enzyme, 81, 105, 109, 111, 112, 133, 216, 219, 258, 262, 267
enzymes, x, 80, 111, 114, 115, 116, 133, 213, 215, 219, 220, 221, 222
Environmental Protection Agency (EPA), 86, 109, 113, 123, 130, 153, 155, 159, 162, 178, 191, 194, 196, 228, 238, 241, 262, 264, 266, 267, 269, 271, 300, 301, 302, 303, 306, 311
epidemic, 100, 125
epidemics, 126
epidemiologic studies, 107
epidemiological, 112, 130, 134, 135, 139, 196, 291, 306
epidemiology, 118, 130, 134, 142, 150
epigenetic, 115, 118, 232
epigenetic mechanism, 115, 118
epithelial cell, 78, 132
epithelial cells, 78
epithelium, 112, 118
epoxy, 266, 273
equilibrium, xiii, 202, 205, 298, 329, 331, 333
equilibrium sorption, 205
equipment, 16, 87, 137, 226, 263, 266
ER, 311
erosion, 82, 83, 103, 108, 230, 232
erythrocytes, 107
ESA, xi, 243, 245, 246, 247, 257, 261, 279
Escherichia coli, 123, 126, 133, 142, 144, 145, 146, 150, 164, 195
ester, 118
estimating, 41, 43, 54, 98, 133, 158, 315
estradiol, x, 213, 214, 220, 221
estrogen, 214, 215, 219
estrogen receptor modulator, 214
estrogens, x, 213, 214, 215, 216, 219, 220, 221, 222
estuaries, 128
estuarine, 237, 258, 262, 286, 287
ethanol, 87, 201, 202, 203, 256, 277
ethylenediamine, 259
etiology, 108, 109, 117, 130, 131, 147
EU, 287
eukaryotes, 130
Europe, 80, 124, 129, 146, 237
European, 65, 80, 98, 99, 124, 128, 130, 133, 135, 226, 237, 338
European Commission, 237

European Community, 65
European Union, 226, 237
eutrophic, 11, 82, 97
eutrophication, vii, 2, 42, 82, 95
evaporation, 10, 45, 51, 87, 93, 230, 301
evapotranspiration, 43, 44, 45, 46, 47, 49, 230
evidence, 22, 104, 109, 115, 130, 131, 134, 144, 148, 192, 193, 214, 228, 291, 303, 306
evolution, 96, 118, 239, 255, 274, 283, 327
EXAFS, 201, 202, 211
exclusion, 165, 191
excretion, 113, 119, 258, 307, 308
exogenous, 129
experimental condition, 182, 248, 259, 261, 277, 279
expert, iv
exploitation, 225, 226, 235, 245
explosive, 2, 5, 6
explosives, 225
exponential, 132, 135, 136
exposure, viii, 78, 97, 99, 101, 102, 103, 104, 105, 106, 107, 108, 109, 110, 111, 112, 113, 114, 115, 116, 117, 118, 119, 122, 125, 229, 232, 239, 241, 258, 262, 263, 265, 266, 267, 270, 272, 291, 305, 306, 307, 308, 310
external environment, 129
extracellular, 85, 86, 220
extraction, 86, 87, 102, 220, 224, 227, 235, 246, 267, 292, 322, 338
extrapolation, 306
eye, 271
eyes, 268

F

fabric, 59
facies, 3, 7, 8
factor analysis, 130
faecal, 144
failure, 138
false, 262, 275, 277
family, 123, 124, 127, 128, 164
FAO, 100, 114
farm, viii, x, 26, 59, 122, 139, 213, 214
farmers, 77, 95
farming, viii, 2, 41, 51, 60, 82, 102, 123
farmland, 38, 41, 42, 51, 52, 55, 56, 68, 69, 83
farms, 25, 59, 70
fat, 264
fatigue, 262
fatty acid, 111
faults, 2, 230
fauna, 78, 94
fax, 101
February, 16, 17, 19, 22, 24, 26, 28, 29, 30, 32, 34, 43, 46, 50, 76, 77, 85, 88, 94, 315, 317
fecal, 123, 125, 131, 133, 137, 138, 142, 196
fecal indicating bacteria, 137
feces, 126, 128, 308
feeding, 7, 54, 63, 73, 114, 128
females, 108
ferric oxide, 269
ferrous ion, 269
fertility, 262
fertilization, 77
fertilizer, 60, 68, 69, 103
fertilizers, 42, 60, 64, 67, 68, 69, 70, 77, 83, 93, 102, 103, 330
fetal, 112, 113
fetal growth, 113
fetus, 113
fetuses, 107, 115
fever, 124
fiber, 289
fibrosis, 105, 110
filament, 165
film, 248, 251, 254, 258, 259, 260, 261, 264, 266, 267, 270, 271, 273, 278, 279, 283, 284, 285, 286, 287, 290, 291, 292, 294, 295
film formation, 279
film thickness, 251
films, 278
filters, ix, 85, 86, 94, 150, 161, 175, 187, 188, 195, 196, 246
filtration, ix, 76, 126, 133, 138, 161, 162, 163, 164, 165, 168, 169, 170, 171, 173, 174, 175, 176, 177, 178, 179, 180, 182, 183, 184, 187, 188, 191, 193, 194, 195, 196, 197, 246, 261, 268
Finland, 126, 150, 238
fires, 59
firms, 60, 63
fish, vii, 1, 2, 51, 78, 92, 93, 94, 98, 200, 214, 220, 221, 235, 257, 264, 292, 338
fishing, 60, 94, 227, 229, 234
FL, 142, 144, 221
flame, 288
flexibility, 176, 260
flocculation, 138, 162, 169, 174, 180, 188
flood, 194
flooding, 235
flora, 287
flora and fauna, 287
flotation, x, 223, 225
flow, viii, xi, xii, 7, 13, 53, 63, 87, 93, 102, 106, 114, 121, 123, 137, 146, 168, 169, 171, 174, 176, 178, 180, 184, 186, 196, 206, 224, 229, 232, 234, 244, 245, 248, 259, 260, 263, 264, 266, 268, 269, 273,

281, 282, 287, 288, 289, 290, 291, 292, 293, 294, 295, 299, 304, 315, 319, 322, 323, 325, 327
flow rate, 53, 63, 87, 106, 178, 266, 323
fluctuations, 24, 97, 286
fluid, 128, 173, 240, 317
fluid mechanics, 173
fluorescence, 133, 136, 201, 204, 205, 206, 207
fluorescent light, 263
fluoride, ix, 153, 154, 155, 156, 157, 158, 159, 176, 268
Fluoride, v, 103, 104, 153, 158, 159
fluorine, 219
fluorogenic, 133, 144, 147
fluorophores, 137
focusing, x, 213
folic acid, 294
food, xi, 42, 63, 67, 102, 114, 125, 126, 127, 128, 133, 150, 162, 231, 243, 244, 245, 256, 257, 258, 268, 270, 272, 307
food additives, 114
food products, 128
foodborne illness, 126
foodstuffs, 133
Ford, 117
forest fire, 258
forest fires, 258
forests, 59
forgetfulness, 271
fossil, 265, 330
fossil fuel, 265, 330
fossil fuels, 265, 330
fouling, viii, 122, 123, 193, 194
Fourier, 201
fractionation, 220
fracture, 234
fractures, 7, 232, 234
France, 117, 220, 332
fresh water, 77, 84, 124, 129, 130, 257, 258, 272, 287, 291, 322
freshwater, vii, 1, 148, 294
Freundlich isotherm, 338
Friday, 311
frontal cortex, 108
FS, 85, 116, 312
FT-IR, 201
fuel, x, 223, 225, 226
funding, 139
fungi, x, 131, 164, 172, 213, 215, 219, 220, 221
fungicides, 264
fungus, 219, 221
fusion, 132

G

gallium, 259
gangrene, 291
gas, x, 35, 81, 100, 102, 125, 175, 181, 201, 223, 225, 253, 259, 304
gasoline, 200
gastroenteritis, 78, 95, 124, 126, 128, 130, 142, 147
gastrointestinal, 108, 109, 128, 132, 162, 196, 258, 265, 266, 270, 272
gastrointestinal tract, 108, 128, 132, 162, 258, 270, 272
Gaussian, 204, 206, 250
GE, 289, 292
gel, 134, 135
gender, 115, 306
gene, 105, 106, 112, 114, 117, 118, 135, 136
gene arrays, 136
gene expression, 105, 106, 114, 118
generation, 40, 112, 227, 235, 236, 247
genes, 96, 105, 106, 112, 132, 135, 136
genetic, 112, 118, 135, 139, 149, 230
genetics, 117
Geneva, 114, 117, 150
genistein, 219, 221
genome, 135, 150, 164
genomes, 144
genomic, 111, 136, 137
genotoxic, 78, 107, 112
genotypes, 130
geochemical, xiii, 103, 227, 228, 234, 236, 238, 329, 330
geochemistry, 226, 237, 238, 239, 240, 308
geology, 226, 230, 240, 314, 316
geophysical, 98
Georgia, 309
geothermal, 7, 96, 241
Germany, 79, 100, 195, 241, 332
gestation, 108, 112
GIS, 56, 99
glaciers, 77
gland, 106, 115
glass, 16, 150, 168, 262
global warming, 200
glucose, 113, 119, 143
glutathione, 105, 262, 271, 295
goals, 139, 188
gold, 209, 234, 237, 239, 254, 258, 261, 263, 264, 266, 267, 269, 273, 283, 284, 285, 286, 289, 290, 291, 292, 295
Gore, 200
government, 113, 139
GPS, 84

grades, 232
grain, 171, 173, 181, 325
grains, 180, 185
Gram-negative, 123
gram-negative bacteria, 149
Gram-positive, 127
grants, 210
granules, 80
graph, 14, 22, 23, 36, 326
graphite, 249, 254, 263, 264, 266, 272, 273, 290
gravitational constant, 181
gravity, 87, 168, 171, 331
grazing, 59
Great Britain, 100
Greece, 148, 338
groundwater, vii, x, xii, xiii, 1, 2, 8, 9, 13, 22, 23, 24, 25, 26, 27, 41, 51, 65, 72, 73, 74, 75, 93, 94, 102, 110, 130, 136, 142, 150, 154, 200, 223, 224, 226, 228, 231, 233, 234, 235, 236, 237, 240, 241, 244, 268, 270, 287, 297, 298, 299, 308, 313, 314, 315, 316, 317, 318, 319, 322, 323, 326, 327
groups, 27, 106, 111, 162, 170, 189, 255, 256, 264, 304, 330
growth, viii, 80, 81, 113, 122, 123, 130, 132, 146, 147, 191, 270
guidelines, vii, 97, 113, 195, 300, 306
Guinea, 239
gums, 267

H

habitat, 128
haemoglobin, 258, 269
hair loss, 266
half-life, 103, 270
Hamaker constant, 182
handling, 137, 245, 303
hanging, 287, 289, 291
harbour, 122, 132
hardness, 258, 263
harm, 235
harmful, vii, x, 146, 162, 199, 226
harvesting, 145
hazards, 138, 145
haze, 194
head, 178, 190, 319, 322
health, vii, viii, ix, xi, xii, 1, 66, 94, 98, 101, 102, 103, 104, 109, 110, 113, 116, 121, 122, 123, 130, 131, 132, 139, 144, 148, 150, 161, 196, 199, 200, 224, 228, 229, 231, 237, 239, 240, 243, 244, 267, 269, 271, 272, 298, 306, 309, 330
health care, 66, 102, 123, 130, 132
health care costs, 102
health effects, xii, 102, 104, 110, 131, 148, 196, 228, 239, 267, 298, 309
health problems, vii
heart, 124, 269, 271
heart rate, 271
heat, 224, 235, 241
heating, ix, 122, 127, 235, 241
heavy metal, viii, xi, 101, 103, 106, 113, 116, 240, 243, 244, 256, 258, 263, 267, 286, 338
heavy metals, viii, xi, 101, 103, 240, 243, 244, 256, 258, 267, 286
height, 43, 178, 227, 274, 279, 323
helium, 259
hematite, 209
hematopoietic, 115
hematopoietic cells, 115
heme, 111, 215, 271
hemicellulose, x, 213
hemodialysis, 109
hemoglobin, 111, 272
hepatic necrosis, viii, 101
hepatitis, 125
Hepatitis A, 124, 125, 143, 147, 164
hepatocellular, 110
hepatocellular carcinoma, 110
hepatocytes, 78, 96
hepatomegaly, 110
hepatotoxins, 99
herbicides, 102
heterogeneity, 165
heterogeneous, 226, 245
heterotrophic, 81, 123, 127, 143, 149
high temperature, 332
high-speed, 178
hippocampus, 108, 109, 116
histogram, 19
histological, 78, 105
HIV, 127
HK, 119
holistic, xi, 224
Holocene, 3, 314, 315
homeostasis, 111
homes, 54, 55, 59, 63, 72
homogeneity, 37, 38
homogeneous, 43, 45, 174, 248, 319
homogenisation, 248
homogenized, 82
homogenous, 278
homolog, 147
hormone, viii, 101, 106, 114, 214, 216, 220, 221, 258
hormones, x, 105, 213, 219
horse, 70
hospital, 63, 143, 196

hospitalized, 132
hospitals, 60
host, x, 93, 128, 131, 132, 136, 139, 143, 147, 149, 223, 225, 230, 235
hostile environment, 123
hot water, 127
hotels, 59, 63, 65
House, 308
household, 103, 108, 269
households, 130
housing, 54, 56
HPLC, 216, 220
Hsp70, 111
HTLV, 164
human, vii, viii, ix, xi, 2, 41, 83, 94, 98, 101, 102, 103, 106, 107, 108, 109, 110, 113, 116, 117, 118, 119, 121, 122, 124, 125, 126, 128, 129, 130, 131, 132, 135, 138, 142, 143, 145, 148, 149, 162, 164, 194, 199, 200, 224, 226, 228, 229, 235, 240, 243, 244, 257, 258, 269, 272, 306, 330
human activity, 200, 228
human brain, 257
human development, 119
human exposure, 107
human genome, 135
humans, vii, x, 78, 102, 103, 109, 110, 112, 122, 124, 125, 128, 129, 132, 138, 213, 214, 226, 228, 268, 269
humic acid, 196, 256, 261, 299
Humic acid, 187
humic substances, 187, 260
humidity, 227, 230
Hungary, 310
HUS, 126
hybrid, 201, 216, 217
hybridization, 136, 137
hydration, 165
hydrazine, 266, 285
hydro, x, 9, 168, 177, 189, 223, 225, 261, 262, 314
hydrocarbons, x, 223, 225, 261, 262
hydrochemical, 16, 233
hydrochloric acid, 267, 271
hydrodynamic, 171, 172, 173, 174, 176, 177, 183, 185, 190, 191, 247, 248
hydrogen, 170, 247, 255, 270, 274, 283, 291, 294, 299
hydrogeochemical, xii, 297
hydrogeology, 238
hydrological, 13, 14, 19, 41
hydrolysis, 111, 261
Hydrometallurgy, 238
hydrophilic, 168, 177, 189
hydrophobic, 168, 171, 177, 188, 189, 193, 197, 260
hydrophobic interactions, 171
hydrophobicity, 165, 168, 173, 176, 189, 191, 192
hydroquinone, 249
hydroxide, ix, x, 199, 200, 201, 208, 304, 330
hydroxides, 228, 234, 249, 268, 330
hydroxyl, 189, 253
hyperactivity, 106, 258
hypermethylation, 112, 118
hyperplasia, 112
hypertension, 103, 104, 108, 112, 272
hypertensive, 114
hyperthyroidism, 271
hypomethylation, 112
hypoplasia, 107
hypothalamic, 105
hypothalamus, 114
hypothesis, 107, 174, 175, 187, 192, 193
hypothyroidism, 271

I

IARC, 110, 117
icosahedral, 125, 190
id, 106
identification, 54, 79, 85, 130, 134, 135, 137, 138, 145, 146, 149, 154
IFA, 178
images, 135, 190, 209
imaging, 209
immobilization, 68
immune function, 107, 115
immune response, 111
immune system, 262
immunoassays, 137
immunocompetent cells, 111
immunocompromised, 123
immunodeficiency, 143
immunological, 93, 135
impairments, 271
implementation, 113, 134, 139, 169, 184, 244
impulsive, 258
impurities, 246, 251, 252, 286
in situ, 16, 81, 245, 266, 267, 273, 291
in utero, 112, 117
in vitro, 96, 110, 214, 220
in vivo, 96, 116, 214, 220, 221
inactivation, ix, 125, 138, 161, 170, 171, 173, 175, 196
inactive, 235
incidence, viii, 122, 125, 127, 146, 147, 306
inclusion, 136, 309
independent variable, 186
India, 104, 117, 237

Indian, 104, 117, 144, 315, 326
Indiana, 80, 98
indication, 168, 317
indicators, 107, 123, 133, 138, 142, 194
indigenous, 131
indirect measure, 80
induction, 106, 220
industrial, viii, x, 17, 41, 42, 51, 52, 54, 63, 71, 77, 83, 101, 102, 103, 110, 121, 175, 213, 216, 226, 228, 235, 241, 244, 272, 287
industrial application, 175
industrial wastes, 272
industrialization, 103
industrialized countries, 235
industry, 52, 123, 200, 226, 231, 235, 236, 244, 258, 282, 286, 330, 331
inert, 246, 248, 254, 259, 261, 273
inertia, 171, 172, 173, 176
inertness, 255, 273
infancy, 175, 271
infant mortality, 119
infants, 271
infection, 123, 125, 126, 128, 130, 132, 134, 135, 138, 143, 147, 148, 162, 164
infections, 122, 123, 124, 130, 131, 134, 144
infectious, 122, 124, 125, 132, 144, 148
Infiltration, 73
infinite, 319, 325, 327
inflammation, 125
inflammatory, 111
inflammatory response, 111
Information System, 99
infrared, 201
ingestion, 78, 122, 123, 125, 265
inhalation, 107, 122, 132
inhibitory, 109, 114
inhibitory effect, 109
initiation, 185
injection, 263, 266, 289, 292, 307
injury, iv, 103, 258
inorganic, xiii, 42, 80, 81, 83, 102, 110, 112, 117, 118, 119, 204, 207, 244, 246, 256, 257, 260, 261, 263, 265, 269, 282, 283, 286, 289, 290, 291, 329, 331
inorganic salts, 246, 257
insecticide, 87
Inspection, 158
instability, 233
instruments, 113
insulin, 258
integration, 138
integrity, 104, 192
intensity, 29, 31, 81, 94, 106, 201, 204, 206, 249, 255, 331
interaction, 81, 82, 107, 127, 170, 171, 181, 185, 216, 220, 261
Interaction, 143, 150, 170, 261
interactions, viii, ix, 106, 122, 128, 131, 149, 168, 170, 171, 173, 185, 186, 194, 238
interface, 33, 87, 148, 173, 182, 195, 196
interference, x, 199, 249, 257
international, viii, ix, 94, 101, 113, 161, 240, 241
International Agency for Research on Cancer, 117
international standards, viii, 101
interpretation, 11, 56
interrelationships, 103
interstitial, 174
interval, 184, 185, 247, 319, 322, 326
intervention, 139, 142
intervention strategies, 139
interviews, 77
intestinal malabsorption, 129
intoxication, 95, 103, 114
inventories, 93
invertebrates, 130
Investigations, xii, 297, 301, 302, 303, 310, 311
investment, 164
Iodine, 271, 295
ion adsorption, 189
ion exchangers, 330
ionic, 16, 155, 170, 173, 181, 184, 195, 246, 257, 290, 338
ions, 155, 170, 184, 189, 202, 244, 253, 260, 261, 268, 271, 292, 293
IRA, 262
Ireland, 121
iron, xiii, 104, 228, 232, 233, 234, 268, 269, 270, 271, 286, 288, 293, 294, 313, 314, 321, 338
irradiation, 80, 81, 246, 262
irrigation, vii, 41, 51, 97, 129, 130, 149, 234, 315, 317, 323
irritability, 262, 264
irritation, 78, 266
IS, 85
ischemic, 112, 118, 291
ischemic heart disease, 112, 118, 291
ISO, 133, 145
isoelectric point, 168, 173, 194
isoenzymes, 221
isolation, 133, 135
isothermal, 331
isotherms, 202, 203
isotope, 305
isotopes, 270, 298, 299, 303, 305, 306
isotropic, 174, 319

Israel, 161
Italy, v, vii, 1, 2, 64, 79, 96, 98, 99, 214, 237, 240

J

JAMA, 117
January, 17, 29, 30, 34, 35, 36, 37, 38, 40, 46, 50, 88, 89, 317
Japan, 199, 200, 201, 213, 220, 326
Japanese, 220, 221
Jordan, 116
judge, x, 199
Jun, 116
Jung, 131, 143

K

kaolinite, 232
Kenya, 99
keratin, 129
KH, 293
kidney, 104, 109, 110, 112, 114, 258, 262, 265, 266, 267, 270
kidneys, 78, 105, 114, 126, 258, 267, 270
kinetics, 173, 174, 246
King, 131, 132, 146
KL, 115
Kobe, 213

L

L1, 28, 76
L2, 28, 29, 36, 76, 166
LA, 85, 117
labile zinc, 293
labour, 135, 136
laccase, 215, 216, 217, 218, 219
lack of confidence, 123
lactating, 105
lactation, 105, 108, 109, 114
lakes, vii, viii, 1, 2, 10, 41, 43, 72, 78, 80, 81, 94, 122, 130, 148, 224, 228, 235, 256, 265, 338
lamina, 174
laminar, 174
land, xiii, 40, 41, 43, 45, 59, 60, 82, 83, 93, 102, 233, 266, 313, 317, 318, 321
land disposal, 102
land use, 41, 43, 60, 82, 93
landfill, 102
Langmuir, 203, 331
Langmuir equations, 203
laser, 137
Latin America, 125
laundry, 268, 269
law, 35, 173, 264
laws, 338
LC, 85, 87, 220
leach, 103
leachate, 231
leachates, x, 223, 226, 227, 229, 231, 232, 233, 236
leaches, 287, 291
leaching, x, xii, 51, 66, 67, 68, 69, 70, 77, 83, 103, 223, 224, 230, 231, 235, 297
lead, viii, ix, x, 101, 103, 104, 108, 109, 114, 115, 116, 126, 162, 164, 176, 178, 199, 200, 206, 207, 208, 245, 256, 258, 259, 260, 261, 271, 287, 288, 298, 299, 304, 338
leakage, 314, 319, 320, 323
leaks, 224
learning, viii, 101, 106, 109, 258
Legionella, ix, 122, 123, 131, 132, 135, 142, 143, 144, 146, 147, 149
Legionella pneumophila, ix, 122, 132, 143, 144, 146, 147, 149
Legionella spp, 131
legislation, 54, 60, 228
leisure, viii, 121, 122
Leprosy, 164
lesions, 78
lettuce, 97, 126, 130, 149
leukocyte, 98
LH, 115, 236
liberation, 125
life cycle, 129
life-threatening, 127, 130
lifetime, 176, 204, 207, 262, 266, 306
ligand, 216
ligands, 256, 260, 299
lignin, x, 213, 215, 216, 219, 220, 221, 222
lignins, 269
likelihood, 323
limestones, 3, 8, 228, 230, 232, 234
limitation, 81, 134, 186
limitations, viii, 81, 122, 137, 261
linear, 166, 191, 250, 252, 253, 254, 258, 259, 264, 266, 271, 275, 280, 284, 306
linear dependence, 253, 254, 275
linear polymers, 191
lipase, 111
lipid, 105, 106, 107, 108, 111, 115, 116, 129
lipid metabolism, 111
lipid peroxidation, 105, 108, 115, 116
lipid profile, 106
lipids, 106, 115, 129
lipoprotein, 111

liquid chromatography, 85, 87, 221
liquid phase, 181, 331
Listeria monocytogenes, 131
literature, 13, 68, 83, 228, 273, 284
lithologic, 232
liver, 78, 96, 98, 100, 104, 107, 109, 110, 112, 114, 117, 118, 125, 258, 262, 263, 266, 267, 270
liver cancer, 98, 100, 125
liver disease, 267
livestock, 26, 41, 51, 52, 59, 68, 70, 78, 102, 122, 126, 127, 128, 130, 138, 139, 143, 226
LM, 290
lobby, 226
location, 4, 11, 14, 23, 41, 52, 299
locomotor activity, 105
LOD, 267, 281
London, 96, 100, 195, 236, 237, 238, 239
long period, 51, 268
long-term, vii, viii, 80, 97, 101, 107, 111, 118, 148, 226, 238, 258, 270, 291, 319, 323
loss of appetite, 258
losses, 10, 45, 64, 69, 93, 130, 246
low molecular weight, 78
low temperatures, viii, 121, 128
low-level, 114, 136
low-permeability, 7
lung, 108, 110, 111, 112, 117, 118
lung cancer, 108, 117
lung function, 111
lungs, 78, 265
lying, 7
lymphocytes, 111, 118
lysis, 85, 131
lysosome, 132

M

M.O., 240
macrophage, 149
macrophages, 107, 115, 132
magma, 96
magmatic, 7, 8
magnesium, 244, 271, 272
magnetic, iv, 137, 202, 248
magnetic resonance, 202
magnetite, 168
mainstream, 178
maintenance, 200, 273
males, 108, 306, 307
malignancy, 118
malignant, 110
Mammalian, 119
mammals, 128, 129
management, 15, 82, 83, 102, 138, 145, 286, 315
manganese, 104, 215, 216, 221, 222, 256, 267, 268, 277, 286, 293
manipulation, 96, 246
manufacturing, 330
manure, 126, 127
mapping, 41
marketing, 60
marrow, 107
mass spectrometry, 259
mass transfer, 247, 248, 249, 251, 257, 273, 279, 282, 283
mastitis, 128
maternal, 105, 107, 115, 118
mathematical, 81, 173, 331
matrix, 136, 193, 227, 230, 254, 260, 275, 280, 282, 287, 291
maximum sorption, 334, 337
MCL, 113, 228, 299, 301, 305, 306, 307
measurement, xii, 35, 45, 133, 136, 144, 154, 190, 191, 201, 207, 209, 220, 244, 247, 254, 275, 279, 306, 332
measures, 17, 19, 28, 72, 130, 134, 236
meat, 128, 226
mechanical, iv, 17, 232, 248, 255, 256, 338
media, ix, 131, 133, 144, 161, 168, 170, 171, 173, 174, 178, 180, 181, 182, 184, 185, 188, 193, 290, 337
median, 105, 304, 308
mediators, 219
medium composition, 262
melanterite, 232
melatonin, 105, 106, 114
membranes, 168, 176, 188, 189, 191, 192, 195, 196
memorizing, 109
memory, 106
memory processes, 106
men, xi, 243, 269, 287
mental retardation, 271
Merck, 332
mercury, xi, xii, 102, 104, 200, 224, 231, 237, 238, 239, 240, 241, 244, 245, 246, 248, 249, 250, 251, 252, 253, 254, 256, 257, 258, 259, 260, 261, 262, 263, 264, 265, 266, 267, 269, 270, 271, 272, 273, 274, 275, 276, 277, 279, 280, 281, 282, 283, 284, 285, 287, 289, 290, 291, 292, 294, 295
mesoporous materials, 201, 202, 210
Mesozoic, 315
metabolic, 42, 81, 101, 103, 107, 111, 116, 133, 228
metabolic pathways, 81, 107
metabolism, 83, 100, 106, 108, 112, 117, 200, 219, 269, 307
metabolite, 113

metabolites, 110, 112
metal content, x, 223, 225, 231, 244, 245, 256, 260
metal ions, 154, 196, 256, 260, 261
metal oxide, 330
metal oxides, 330
metal salts, 244
metallic taste, 269
metalloids, xiii, 224, 226, 231, 256, 271, 329
metallurgy, 226, 228, 229
metals, viii, x, xiii, 101, 102, 103, 104, 107, 114, 223, 224, 226, 228, 229, 230, 233, 237, 239, 240, 244, 245, 246, 248, 256, 259, 260, 261, 271, 287, 288, 290, 329, 330, 331, 332, 337, 338
meteoric water, 224
methane, 221, 264
methanol, 87
methemoglobinemia, 272
methionine, 110, 262
methyl group, 87
methylation, 112, 118, 237
methylene, 87
methylene chloride, 87
metropolitan area, 122, 307
mica, 190, 321
mice, 105, 106, 107, 108, 111, 112, 114, 115, 117, 118, 132, 178
microarray, 136, 137, 145, 146, 147, 150
microarray technology, 147
Microarrays, 136, 137
microbes, viii, 122, 129, 131, 134, 145, 149, 162
microbial, v, viii, ix, 121, 122, 123, 131, 135, 136, 137, 138, 139, 142, 144, 146, 147, 149, 150, 161, 162, 168, 170, 171, 175, 194, 196, 200, 227, 237
microbial communities, 131, 136, 137, 144
microbial community, 123, 136, 150
microbiota, 126
microelectrode, 263
microelectrodes, 259, 287
microfluidic devices, 137, 143
micrograms, 271, 314
micrometer, 136
micronucleus, 107
micronutrients, 338
microorganism, 125, 191
microorganisms, ix, x, 76, 77, 122, 123, 131, 133, 151, 161, 162, 163, 164, 166, 168, 175, 180, 191, 193, 213, 215, 219, 257, 260, 272
microscope, 80, 85, 98, 189, 191, 201, 202, 209
microscopy, 195, 209
microspheres, 137
microsystem, 137
microwave, 266, 285, 291
microwave heating, 266, 285
migration, 98, 229, 237
milk, 107, 115, 126, 128, 200, 260
mine tailings, 236
mineral resources, 226
mineral water, 144, 258, 271
mineralization, x, 68, 69, 223, 227, 230, 234, 240, 246
mineralized, x, 23, 223, 224, 230, 231, 236
mineralogy, 226
minerals, x, 83, 172, 223, 224, 226, 227, 230, 232, 233, 236, 237, 263, 330
mines, 200, 224, 225, 226, 231, 233, 235, 236, 237, 238
mining, x, 102, 110, 223, 224, 226, 227, 228, 229, 230, 231, 232, 233, 234, 235, 236, 237, 238, 239, 241, 287, 330
Ministry of Education, 210, 337
Miocene, 7, 315
misinterpretation, 178, 187
mixing, 29, 34, 94, 155, 180, 200, 201
ML, 116
mobility, 228, 232, 240, 245, 299
models, 52, 94, 132, 173, 174, 180, 307
modulation, 143
moisture, 226
moisture content, 226
molar ratio, 200
molecular biology, 134
molecular oxygen, 216
molecular weight, 191
molecules, 170, 171, 181, 191, 246, 254, 330
molybdenum, 239, 271
monochromator, 201
monoclonal, 135
monoclonal antibody, 135
monocyte, 132
monolayer, 278
monomeric, 110
Monroe, 309, 310
monsoon, 315, 323
Montenegro, 282
montmorillonite, x, 199, 200, 202, 203, 204, 205, 208, 209, 267
morbidity, viii, 121, 128, 130, 306
morphological, 79
morphology, 12, 41, 43, 77, 106, 107, 115, 165
morphometric, 13, 14
mortality, viii, 108, 115, 117, 118, 121, 127, 128, 291
mortality rate, 108, 127, 128
Mössbauer, 202
motor activity, 105, 114
mouse, 110, 111, 115, 116, 117

movement, 33, 173, 230, 314, 315
mRNA, 116
MS, 85, 87, 115, 193, 220, 332
Mt. Amiata, 237
MUA, 309
mucoid, 127
mucous membrane, 132
mucous membranes, 132
multicellular organisms, 95
multidisciplinary, vii, 1, 286
multiplexing, 135
multiplication, 131, 270
municipal area, 317
municipal sewage, 102, 146, 219, 222
murine model, 115
muscarinic receptor, 104
muscle, 93, 124, 258
muscular tissue, 93
mutants, 118
mutations, 111
mycobacteria, 123, 127, 131, 132, 148
Mycobacterium, 127, 131, 143, 146, 147, 148, 149, 150, 164
myelin, 106
myelin basic protein, 106
myocarditis, 124

N

NA, 290, 294
NaCl, 271
nafion, 263, 269
Namibia, 240
nanoparticles, 200, 201, 202, 205
nanotechnology, 139, 201, 209
nation, 214
national, xi, 224, 303
National Academy of Sciences, 143, 148, 195
National Research Council, 121
natural, vii, x, xi, xiii, 1, 2, 39, 40, 41, 43, 59, 80, 81, 82, 83, 96, 97, 99, 103, 108, 126, 127, 128, 131, 132, 154, 159, 162, 213, 214, 219, 221, 224, 230, 233, 237, 238, 240, 243, 246, 256, 257, 260, 262, 263, 264, 266, 268, 269, 270, 271, 272, 273, 277, 282, 283, 286, 288, 289, 290, 291, 292, 294, 295, 299, 314, 322, 329, 330, 331, 337
natural environment, 131, 132, 162, 233, 238
natural hazards, 299
natural resources, 299
nausea, 109, 258, 269
Nd, 283
nebulizer, 17, 332
necrosis, 78
nematodes, 129
neonatal, 115, 130, 150
nerve, 266
nervous system, 258, 265
nervousness, 271
Netherlands, 326, 327
network, 16, 52, 72, 196, 227, 234, 315
neurobehavioral, viii, 101, 105
neurodegeneration, 109, 116
neurodegenerative, 109, 270
neurodegenerative disease, 109, 270
neurodegenerative diseases, 109, 270
neurofibrillary tangles, 110
neuropathological, 109
neurotoxic, 105, 109, 116
neurotoxic effect, 116
neurotrophic, 116
neutralization, 184
neutrophils, 111
New England, 147, 149
New Jersey, vi, ix, xii, 153, 154, 159, 297, 299, 303, 304, 305, 306, 307, 308, 309, 310, 311, 312
New York, iii, iv, 95, 159, 195, 197, 211, 236, 237, 239, 240, 287, 302, 303, 307, 338
New Zealand, 99
Ni, xi, 224, 229, 230, 231, 254, 257, 259, 330
nickel, 103, 104, 107, 115, 293
Nielsen, 117
Nile, 287
nitrate, 27, 38, 64, 74, 99, 100, 200, 268, 271, 288
nitrates, 16, 19, 26, 29, 38, 41, 42, 69, 268, 271, 272, 295
nitric acid, 303
nitric oxide, 104, 109, 115
nitric oxide synthase, 104, 109
Nitrite, 104
nitrogen, 19, 26, 29, 41, 42, 52, 64, 66, 67, 68, 69, 71, 72, 73, 74, 75, 76, 77, 80, 81, 82, 87, 88, 91, 93, 94, 99, 225, 259, 285
nitrogen compounds, 65, 225
NMR, 202
NO, 104
noble metals, 248, 253
non-crystalline, 201
non-destructive, xi, 244
nonlinear, 246
non-linear, 191
non-uniform, 254
normal, 128, 129, 330
North America, 128, 241, 299
Northern Ireland, 121, 150
Norway, 80
NTU, 163

nuclear, 103, 111, 216, 220, 270
nuclear power, 103
nucleic acid, 125, 135, 136, 143, 268
nuclides, 306
nutrient, viii, 2, 60, 77, 80, 81, 113, 121, 123, 131, 262, 268
nutrients, xi, 27, 42, 60, 81, 82, 94, 128, 131, 243

O

obligate, 130
observations, xii, 51, 191, 214, 298, 306, 336
obsolete, 87
occupational, 107, 307
oceans, 315
ODS, 216
OECD, 83, 99
Ohio, 178
oil, x, 102, 223, 225
oligomeric, 219
oligonucleotides, 136
olive, 59
omnibus, 186
on-line, xi, 244, 273, 287
optical, 17, 137, 201
optimization, 188
OR, 59, 60
oral, 105, 108, 109, 125, 128, 132, 268, 308
ores, 225, 228, 232, 262, 263, 266
organ, 78, 105, 107, 110
organic, viii, x, xi, xiii, 42, 63, 68, 69, 77, 81, 82, 83, 102, 121, 143, 145, 173, 187, 193, 196, 197, 200, 213, 216, 243, 244, 246, 249, 255, 256, 257, 260, 261, 262, 269, 282, 286, 289, 299, 329, 330, 331, 332
organic compounds, 77, 261, 262
organic matter, 42, 173, 196, 197, 200, 246, 257, 262, 269, 282, 330
organic polymers, x, 213, 255
organic solvent, 246
organism, 134, 135, 137, 138
organization, 81
organotin compounds, 244
orthophosphates, 88
oscillations, 227
osmosis, 163, 175
osteoporosis, viii, 101
osteosarcoma, 306
Ottawa, 121, 310
outliers, 158
ovarian, 118
ovaries, 112
ovary, 118
oviduct, 112
oxalate, 215
oxidants, 253
oxidation, x, 111, 162, 205, 214, 215, 219, 220, 223, 225, 227, 229, 230, 232, 236, 245, 247, 249, 253, 254, 255, 259, 260, 269, 271, 272, 279, 290, 332
oxidation products, 219, 229
oxidative, 105, 107, 108, 113, 114, 201, 204, 205, 222, 262
oxidative damage, 105, 262
oxidative stress, 105, 107, 108, 113, 114
oxide, 233, 255, 269, 338
oxides, 228, 232, 234, 249, 268, 330, 338
oxygen, 31, 32, 33, 34, 35, 69, 111, 246, 249, 253, 254, 258, 259, 263, 269, 271
oxygen saturation, 35
oxygenation, 82
oysters, 150
ozone, 129, 138, 162

P

p53, 107, 112, 115, 118
PA, 117, 238, 302
Pacific, 124, 150
palladium, 271, 294
palpitations, 271
PAN, 176
paper, xi, 102, 168, 223, 226, 236, 320
Paper, 312
parabolic, 248
paralysis, 124, 257, 264
parameter, viii, 82, 122, 260
parasite, viii, 122, 138, 144
Parasite, 147
parasites, 128, 129, 144
Paris, 99, 240
Parkinson, 267, 270
Parkinsonism, 267
particles, 83, 87, 123, 165, 172, 173, 174, 178, 183, 185, 190, 192, 196, 225, 226, 229, 232, 246, 260, 261
particulate matter, 224
pathogenesis, 164
pathogenic, ix, 122, 128, 134, 142, 146, 149, 161, 162, 163, 164, 168, 175, 194
pathogens, vii, viii, ix, 102, 121, 122, 123, 124, 126, 127, 128, 129, 130, 131, 134, 135, 136, 137, 138, 139, 143, 144, 146, 147, 148, 149, 150, 161, 162, 164, 165, 168, 170, 171, 173, 174, 175, 176, 181, 188, 286
pathology, 110, 116, 142
pathways, 112, 115, 231

patients, 109, 110, 112, 123, 132, 147, 267, 270
Pb, ix, x, xiii, 105, 108, 109, 115, 199, 200, 201, 206, 207, 208, 223, 225, 231, 236, 249, 257, 258, 259, 260, 287, 288, 329, 330, 331, 332, 335, 336, 337
PCBs, 102
PCR, 134, 135, 136, 137, 143, 145, 146, 149
PDC, 290
PE, 72
peat, 59, 321
Pennsylvania, 193
PEP, 80, 81
peptides, 125
percolation, 10, 43, 64, 323
performance, 145, 235, 248
perinatal, 106, 114
peripheral blood, 107
peripheral nervous system, 262
peritoneal, 107, 115
peri-urban, 113
permeability, 7, 8, 46, 93, 227, 234, 320, 323
permit, 70, 72, 93
permittivity, 181
peroxidation, 108, 116
peroxide, 247
personal, 79
personal communication, 79
personality, 271
person-to-person contact, 127
perturbations, 108, 193
pesticide, 110
pesticides, 102, 244, 264, 265, 271, 272
petroleum, 102
Petroleum, 327
pets, 122
PF, 290, 292
pH, vi, x, xi, xii, xiii, 16, 17, 19, 20, 21, 24, 26, 28, 29, 30, 31, 32, 80, 84, 125, 168, 170, 171, 173, 175, 176, 177, 178, 188, 189, 190, 191, 194, 208, 223, 225, 227, 228, 231, 233, 243, 244, 246, 249, 253, 255, 258, 261, 262, 263, 267, 297, 303, 308, 329, 330, 331, 332, 333, 334, 335, 336, 337, 338
pH values, x, 29, 188, 223, 225, 333, 334, 335, 337
phage, 171, 189, 190
phagocytosis, 132
pharmaceutical, xi, 243, 245, 272
pharmaceuticals, 219
phenol, 219, 255
phenolic, 216
phenolic compounds, 216
Philadelphia, 307
phorbol, 118
phosphatases, 78, 125
phosphate, 26, 36, 80, 81, 99, 178
phosphates, 16, 19, 24, 27, 29, 36, 70
phospholipids, 106
phosphorus, 19, 25, 26, 29, 41, 42, 70, 72, 81, 82, 83, 88, 92, 93, 94
photochemical, 262
photoemission, 209
photosynthesis, 79, 80, 81
photosynthetic, 80, 81
phycocyanin, 80
phycoerythrin, 80
physical interaction, 170
physical treatments, 162
Physicians, 117
physico-chemical characteristics, 235, 337
physico-chemical properties, 245, 246, 248, 331
physiological, 103
physiology, 104
phytoplankton, 78, 79, 85, 94, 95, 97, 268
phytoplanktonic, 83
piezoelectric, 137
pigments, 80, 81, 97, 98
pigs, 78
pituitary, 105
PL, 294
placenta, 113, 258
placental, 113
planar, 137, 254, 255
plankton, 84
planning, 235
plants, x, 45, 46, 102, 129, 138, 213, 214, 219, 221, 222, 224, 232, 244, 260, 269
plasma, xi, 111, 201, 243, 308, 332
Plasmodium malariae, 164
plastic, 321, 323
platinum, 209, 247, 254, 264, 266, 278, 289, 290, 332
play, xiii, 106, 127, 129, 132, 149, 171, 329, 330, 337
Pleistocene, 3, 93, 314, 315
Pleurotus ostreatus, 219
Pliocene, 2, 3, 7
PLUS, 17
pneumonia, 132
PO, 161
poison, 265
poisoning, 78, 228, 258, 264, 272
poisonous, 200
polar media, 168
polarity, 252
policy makers, 139
polio, 124, 143
poliovirus, 125, 142, 193, 194, 195

political, 286
pollutant, 236, 263, 265
pollutants, xiii, 11, 51, 63, 102, 215, 219, 228, 244, 245, 286, 329
pollution, 2, 28, 41, 42, 65, 102, 103, 114, 133, 210, 229, 231, 233, 235, 236, 238, 239
polycyclic aromatic hydrocarbon, 215
polyethylene, 155, 285
polygons, 56
polymer, 176, 193, 255
polymer materials, 193
polymerase, 135, 146, 151
polymerase chain reaction, 135, 146, 151
polymerization, 219
polymers, 176
polymorphism, 134, 148
polymorphonuclear, 98
polynomial, 275
polypropylene, 176, 180, 246, 332
polysaccharide, 129
pond, 196
pools, 59, 60, 129, 130, 272
poor, viii, 25, 81, 93, 121, 131, 183, 188, 224, 226, 236, 283, 286
population, vii, viii, 1, 2, 41, 52, 54, 56, 63, 80, 83, 89, 94, 96, 97, 98, 99, 101, 108, 115, 118, 121, 122, 127, 131, 158, 162, 232, 270, 272, 286, 314
pore, 86, 165, 176, 191, 192, 193, 196, 201
pores, 166, 174, 191, 192, 260, 261
porosity, 7, 174, 178, 180, 230, 320
porous, 174, 180, 193, 194, 195, 197, 201, 208, 254, 260, 266
porous media, 174, 194, 195
portability, 137
portal hypertension, 110
ports, 178, 184
potassium, 266, 270, 285, 294
potato, 149
poultry, 126, 139, 143
powder, 260, 266, 272
power, 136, 259
PPM, 109, 110
precipitation, 14, 40, 43, 44, 45, 46, 47, 48, 50, 51, 69, 82, 83, 162, 230, 240, 245, 299, 301, 304, 332, 333, 336
predators, 131
prediction, 94, 100, 184, 193
preference, 94
pregnancy, 105, 108, 109, 113, 119, 271
pregnant, 105, 107, 269, 272
pregnant women, 269, 272
premature infant, 110
preneoplastic lesions, 112
preparation, iv, xi, 87, 137, 234, 243, 245, 246, 272, 281, 286, 291
preservatives, 155, 265
pressure, 24, 81, 100, 168, 169, 178, 191, 192, 194, 323
prevention, 130, 135
preventive, 134, 236
prices, 231
priorities, 228
prisons, 59
pristine, viii, 122, 123
private, 51, 60, 130, 303, 309
proactive, 139
probability, 183
probe, 17, 84, 201, 202, 209
procedures, viii, xii, 100, 122, 245, 246, 256, 259, 261, 262, 286, 297
process control, 144
production, x, 41, 82, 83, 98, 102, 108, 136, 176, 214, 223, 224, 226, 227, 228, 231, 233, 236, 255, 269, 286, 287, 317
productivity, 130
profit, 235
progeny, 298, 299, 303, 304, 309
program, 82
progressive, 235
prokaryotic, 77, 79, 80
prolactin, 105, 114
proliferation, 107, 112, 118, 143
promote, xii, 103, 168, 173, 244, 282
promoter, 112, 118
promoter region, 112
propane, 201
property, iv, 128, 168
prophylaxis, 132
proportionality, 46
prostate, 106, 115
protection, x, 104, 113, 125, 138, 144, 162, 186, 187, 199, 235, 236
protective role, 106
protein, 78, 80, 103, 104, 106, 107, 108, 109, 111, 118, 125, 129, 194, 214, 269
proteins, 80, 81, 129, 221, 255, 258
protocol, 145, 154, 178
protocols, 16, 138
protozoa, viii, ix, 122, 123, 126, 131, 132, 138, 142, 146, 149, 150, 161, 164, 172, 192
protozoan parasites, 129, 146
proximal, 105
PSA, 247, 249, 252, 253, 254, 257, 259, 260, 263, 267, 282
pseudo, 44, 204, 206
Pseudomonas, 131

Pseudomonas aeruginosa, 131
psychological, 108
PT, 21, 22, 293
public, vii, 1, 41, 54, 55, 56, 63, 72, 78, 103, 122, 130, 135, 138, 139, 144, 147, 148, 150, 154, 187, 228, 235, 258, 299, 304, 305, 308, 309
public health, 78, 103, 122, 135, 144, 148, 150, 187, 299, 304
pulp, 102, 219
pulse, 251, 262, 266, 267, 269, 287, 289, 290, 291, 292
pumice, 7
pumping, 54, 63, 197, 314, 315, 317, 318, 319, 323, 325, 326, 327
pumps, 179, 236, 315
purification, 17, 41, 52, 54, 63, 72, 124, 137, 195, 210
PVC, 16, 255, 317
pyrene, 219
pyrite, x, 223, 225, 227, 230, 232, 236

Q

quartz, xii, 182, 227, 230, 232, 297
questioning, 185

R

RA, 115, 117, 302
race, 60
radar, 99
radial distance, 314
radiation, 80
Radiation, 310, 311
radical, 219
radiological, xii, 113, 270, 297, 298
radionuclides, xii, 297, 298, 299, 303, 304, 306
radium, xii, 297, 299, 303, 304, 305, 306, 307, 308, 309, 310
radius, 191, 250
radon, 301, 305, 306
rain, 116, 245, 259, 267, 270, 282, 285, 287
rainfall, 9, 25, 38, 42, 43, 45, 46, 47, 72, 73, 74, 82, 93, 99, 227, 230, 232
rainwater, 83, 230, 269
Raman, 202
random, 16
random errors, 16
range, 23, 27, 30, 35, 36, 45, 47, 51, 98, 110, 123, 125, 130, 136, 154, 170, 175, 189, 190, 191, 227, 228, 229, 233, 234, 249, 253, 254, 255, 262, 265, 267, 271, 275, 276, 280, 282, 306, 308, 309, 320, 323, 330, 332, 336
rash, 124
rat, 96, 106, 109, 114, 115, 116
rationalisation, 286, 287
rats, 104, 105, 106, 107, 108, 109, 110, 112, 114, 115, 116, 118, 307
RB, 292, 294
RB1, 68
RC, 289, 292
REA, 134
reactants, x, 223, 224, 331
reaction center, 330
reactive arthritis, 128, 148
reactivity, xii, 244
reading, 12
ready to eat, 126, 127
reagent, 137, 246
reagents, x, 137, 223, 225, 246
reality, 325
real-time, 134, 135, 137, 144, 146, 151
recalling, 13
receptors, 106
recession, 317
reclamation, 226
recognition, 111, 125, 134
recollection, 259
recovery, 88, 145, 285, 317, 333, 335, 336, 337
recreational, viii, 60, 78, 97, 99, 122, 123, 138, 142, 143, 145, 235
recreational areas, 60
recycling, 338
red blood cell, 126, 269
red blood cells, 126, 269
redox, 233, 248, 253, 331
Redox, 293
reduction, viii, 2, 47, 101, 108, 109, 128, 164, 186, 188, 195, 246, 249, 253, 259, 260, 262, 266, 268, 283, 285, 291, 292
regional, 9, 93, 116, 224, 230, 235, 237, 315
Registry, 310
regression, 40, 294, 315
regular, 256, 282
regulation, ix, 98, 100, 147, 199, 200, 206
regulations, viii, 2, 95, 116, 122, 154, 244
rehabilitation, 226, 235
rehydration, 128
rejection, ix, 161, 166, 189, 191, 195, 196
relationship, 110, 112, 116, 118, 148, 320
relationships, 118, 123, 131, 139
relaxation, 104
relevance, 287, 338
reliability, ix, 51, 153

Reliability, v, 153
remediation, viii, 2, 94, 102, 224, 231, 309
remodeling, 111, 117
renal, 95, 103, 104, 109, 114, 258
renal failure, 109, 258
replicability, 88
replication, 132, 149, 258
reporters, 136
reproductive organs, viii, 101
research, x, xii, 96, 113, 150, 153, 193, 213, 228, 244, 245, 266, 272, 273, 286, 330, 337
researchers, 11, 82
reservoir, vii, xi, 1, 126, 188, 224, 227, 228, 229, 236, 323, 331
reservoirs, xiii, 99, 128, 131, 142, 329
residential, viii, 2, 41, 42, 51, 52, 53, 54, 55, 59, 60, 63, 65, 71, 72, 77, 83, 102, 216
residuals, 123
resilience, 83
resin, 262, 264, 266, 267, 272, 273, 292
resins, 168, 249, 255, 262
resistance, viii, 94, 122, 123, 132, 177, 190, 314
resolution, 47, 201, 204, 206, 207, 209, 251
resources, x, 213, 263, 315
respiratory, 107, 124, 131, 132, 162
restaurant, 54
restaurants, 59, 63
restriction fragment length polymorphis, 134
retardation, 130, 173
retention, ix, 73, 161, 162, 163, 164, 165, 167, 168, 170, 171, 175, 176, 185, 189, 191, 192, 193, 197, 234, 258, 307
Reynolds, 99, 115
RF, 117
RFLP, 134
rice, 116, 147, 200, 317
Rio de Janeiro, 98
riparian, 16, 17, 32
risk, vii, viii, xi, 1, 28, 79, 83, 94, 98, 102, 107, 108, 110, 111, 115, 116, 117, 119, 121, 123, 124, 129, 130, 138, 144, 148, 149, 162, 178, 187, 188, 196, 200, 223, 229, 231, 232, 234, 241, 244, 245, 253, 262, 265, 271, 305, 306, 307
risk assessment, xi, 223, 229, 231, 234, 241
risk factors, vii, 1, 107, 111, 149
risks, 102, 113, 117, 142, 306, 338
Rita, 2, 5, 6, 96, 97
rivers, vii, 1, 122, 128, 148, 214, 220, 225, 228, 238, 240, 256, 258, 265
RL, 116
RNA, 136, 195
rodents, 105
Rome, 16, 17, 93
room temperature, 175, 180, 266, 331
Royal Society, 97
RP, 237
rubber, 262
runoff, 41, 43, 44, 45, 46, 47, 49, 50, 51, 64, 72, 73, 74, 82, 83, 93, 102, 227, 229, 231
rural, xi, 41, 147, 149, 224, 315, 331
rural areas, xi, 41, 224, 315
Russian, 195

S

SA, 115, 119, 238, 310
safe drinking water, 122
Safe Drinking Water Act (SDWA), 153, 159, 310, 311
safeguard, 82, 138
safety, viii, xi, 97, 122, 139, 162, 232, 243, 270
saline, 178, 196, 235
salinity, x, xiii, 223, 225, 313, 314, 315, 321
saliva, 106
salivary glands, 106
salmon, 42
Salmonella, 123, 126, 137, 146, 147
salt, 36, 130, 225, 261, 271
salts, 16, 17, 29, 37, 102, 170, 195, 224, 246, 258, 263, 267, 271
sample, ix, x, xi, xii, 76, 77, 87, 88, 134, 137, 139, 153, 155, 184, 185, 186, 199, 201, 204, 209, 214, 243, 244, 245, 246, 252, 254, 257, 261, 262, 266, 272, 273, 279, 281, 282, 285, 303, 306, 308, 311, 331, 332, 334
sampling, ix, 16, 17, 19, 28, 29, 69, 83, 84, 86, 153, 154, 155, 157, 158, 178, 186, 188, 246, 252, 299, 331
sampling error, ix, 153, 158
sand, ix, xii, 142, 161, 163, 168, 169, 178, 197, 224, 268, 297, 299, 320, 321
sandstones, 7, 230, 232
sanitation, 102, 104, 113, 139, 150
saturation, 12, 256
scanning tunneling microscopy, 209
scattering, 202
school, 59
science, 113, 116, 194, 195, 210
scientific, 138, 272, 286
scientists, 139, 286
SD, 88, 115
SE, 2, 117
sea level, 7, 14, 16, 19, 20, 21, 24, 93, 229
searching, 41
seasonal variations, 31
seasonality, 138

Seattle, 97, 311
seawater, 129, 197, 258, 265, 268, 272, 288, 289, 291, 292, 293, 294, 295
SEC, 111
secretion, 105, 106, 114, 132
secular, 298, 303
sediment, xiii, 9, 93, 237, 253, 314, 315, 317, 326, 329, 330, 331, 332, 333, 334, 335, 336, 337, 338
sedimentation, 7, 85, 138, 162, 169, 171, 172, 176, 180, 197, 224, 315, 326, 338
sediments, viii, xiii, 2, 5, 10, 93, 102, 122, 227, 230, 231, 232, 233, 234, 237, 264, 267, 287, 320, 329, 330, 331, 336, 337
seeds, 82
selecting, 204, 205, 207, 315
selectivity, xi, 154, 210, 243, 246, 254, 260, 263, 264, 272, 274, 282, 287, 290
selenium, 103, 107, 259, 262, 266, 288, 289
senile, 110
senile plaques, 110
sensitivity, xi, xii, 104, 136, 137, 154, 209, 243, 244, 245, 246, 247, 248, 249, 253, 254, 259, 260, 262, 264, 266, 267, 272, 273, 274, 276, 277, 279, 282, 283, 284, 287
sensors, 255, 270
separation, 134, 137, 145, 245, 261
septic tank, 103
sequencing, 135, 149
Serbia, 243, 311
series, 7, 96, 108, 144, 180, 209, 210, 298
serotonin, 105, 114
serum, 107, 220, 308
services, iv
settlements, 63
sewage, viii, x, 2, 26, 41, 42, 54, 63, 102, 103, 126, 196, 213, 214, 215, 219, 220, 221, 244, 245, 287, 330
sex, 106, 114
sexual development, 258
SFT, 287
SH, 114, 289, 294
shape, 165, 167, 168, 191, 325
sharing, 219
shear, 123, 146, 173, 174
sheep, 78, 129
shellfish, 130
shelter, 123
Shigella, 123, 126, 127, 134, 137, 144
shock, 258
short period, 137, 258, 262
shortage, 235
shortness of breath, 272
short-term, 139, 258, 266
shoulder, 208
Siberia, 237
signal transduction, 112
signaling, 104, 111, 112
signaling pathway, 104
signals, 201, 278, 279
signs, 105, 106, 124
silica, 87
silver, 259, 268, 271, 272, 277, 293, 295
sites, ix, x, xi, 59, 63, 65, 102, 153, 155, 157, 158, 199, 201, 202, 209, 223, 224, 226, 227, 228, 231, 232, 233, 234, 235, 236, 237, 239, 241
skeleton, 307
skin, 78, 110, 112, 118, 132, 265, 268, 271, 272
skin cancer, 112
skin diseases, 110
slag, 225, 226
Slovenia, 237, 239, 240, 311
sludge, 196, 197, 214, 219, 220, 222, 263, 330, 338
smelting, xi, 224, 227, 232, 239, 330
SO2, 285
social, 102, 106, 272
sodium, 107, 108, 112, 145, 200, 201, 249, 267, 272
sodium arsenate, 112
sodium hydroxide, 200
software, 17, 229
soil, xi, 2, 26, 41, 42, 43, 45, 46, 47, 60, 63, 68, 69, 70, 73, 77, 83, 92, 93, 103, 110, 113, 114, 126, 148, 149, 193, 194, 195, 196, 197, 200, 228, 229, 236, 238, 240, 243, 244, 268, 269, 271, 289, 331, 338
soil erosion, 103
soils, 52, 65, 67, 83, 100, 127, 194, 227, 228, 229, 230, 231, 236, 237, 239, 266, 287, 338
solar, 235
solid phase, 182, 255, 267, 332
solid surfaces, 137, 173
solid waste, 126
solubility, 35, 227, 233, 237, 239, 240, 244, 248, 266, 267, 268, 299, 338
solutions, 106, 155, 204, 205, 207, 208, 253, 273, 275, 277, 279, 284, 285, 291, 303, 332
solvents, 87, 246
sorbents, ix, x, 199, 200, 201, 210
sorbitol, 134
sorption, xiii, 103, 193, 196, 200, 201, 206, 209, 214, 234, 329, 330, 331, 332, 333, 334, 335, 337, 338
sorption curves, 331, 333, 335, 337
sorption process, 103, 333, 337
South America, 299, 310
SP, 294
Spain, xi, 214, 223, 225, 227, 229, 230, 231, 234, 235, 236, 237, 238, 239, 240

spatial, 47, 54, 70, 96, 174, 232
speciation, xii, 110, 240, 244, 245, 246, 260, 261, 262, 289, 290, 291
species, vii, x, 1, 77, 78, 79, 81, 83, 93, 94, 107, 123, 125, 126, 127, 128, 130, 131, 134, 135, 136, 138, 139, 149, 154, 199, 201, 205, 207, 208, 214, 233, 245, 246, 249, 255, 257, 260, 262, 265, 272, 289, 290
specific adsorption, 249, 264, 331
specific gravity, 129
specific surface, 182, 201, 202, 330
specificity, 66, 133, 137
spectra, 201, 204, 205, 207, 208, 304
Spectrophotometer, 17
spectroscopy, x, 99, 100, 199, 200, 201, 203, 209
spectrum, 80, 81, 201, 203, 204, 206, 207
speed, 29, 137, 139, 154
spills, 224
spinach, 126
spindle, 165
spleen, 110, 272
sporadic, 125, 148, 149, 229
sports, 130
Sprague-Dawley rats, 106, 109
spreadsheets, 327
springs, 225, 232, 234
square wave, 252, 292
SR, 117, 119
stability, 36, 111, 177, 187, 194, 227, 237, 239, 240, 273
stabilization, 187, 196
stabilize, 16, 330
stages, 5, 6, 7, 129, 186, 230
stainless steel, 281
standard deviation, 88, 155, 217, 286
standard error, 158
standardization, 332
standards, 107, 116, 153, 155, 196, 225, 303, 305
Standards, 302
Staphylococcus, 144
Staphylococcus aureus, 144
State Department, 302, 303
statistical analysis, ix, 139, 153, 186
statistics, 123
steady state, 188
steel, 102
steroid, 111, 214
STM, 209
stomach, 108
stomatitis, 267
storage, vii, xiii, 2, 41, 137, 235, 313, 314, 315, 319, 323, 324, 326, 327
strain, 96, 127
strains, 99, 123, 126, 128, 133, 135, 136, 148
strategies, 135, 138
stratification, 29, 30, 33, 34, 37, 38, 89, 322, 325
streams, viii, 40, 42, 80, 122, 224, 225, 230, 232, 233, 234, 238, 319
strength, 155, 170, 173, 176, 184, 191, 195, 246, 263, 278, 338
stress, 82, 105, 111, 257, 314
structural changes, 105
structural characteristics, 219
subacute, 115
subjective, 102, 326
subjective judgments, 326
substances, vii, x, 1, 11, 39, 41, 42, 51, 70, 71, 93, 94, 95, 102, 223, 224, 237, 246, 247, 249, 256, 271
substrates, 133
sugars, 256, 258, 261
sulfate, 304
sulphate, x, 109, 223, 225, 240, 270, 272, 285
sulphur, 226, 266, 268, 271
summer, 29, 35, 56, 63, 82, 93, 124, 227, 233
Sun, 291
sunlight, 125, 173, 197
Superfund, 228
superiority, 184, 206
supernatant, 332
superoxide, 108
superoxide dismutase, 108
supplemental, 128
suppliers, ix, 154, 161, 228, 303
supply, vii, viii, x, xiii, 25, 68, 69, 70, 80, 94, 101, 102, 122, 123, 147, 154, 178, 188, 200, 223, 226, 231, 234, 235, 236, 299, 303, 313, 314, 315, 317, 318, 326
suppression, 224
surface area, 14, 56, 60, 87, 169, 280
surface layer, 268
surface properties, 170
surface water, viii, 2, 29, 41, 42, 65, 72, 102, 122, 123, 126, 130, 138, 139, 145, 168, 175, 178, 224, 229, 231, 233, 240, 244, 256, 259, 262, 268, 269, 270, 286, 330
surfactants, 145, 193, 247, 257, 262, 293
surplus, 235
surrogates, 142
surveillance, 124, 138, 154
survival, viii, 121, 123, 132, 145, 173, 193, 194, 196, 197
surviving, 307, 308
susceptibility, 107
suspensions, 195
sustainability, 83, 314

sustainable development, 244
swamps, 59
Sweden, 95, 146, 147
Switzerland, 114, 332
symbiotic, 139
symmetry, 201
symptoms, 109, 124, 125, 127, 258, 264, 267
syndrome, 126
synthesis, 81, 99, 106, 202
synthetic, x, 102, 168, 169, 213, 214, 215, 219, 266, 331
systematic, 15, 16, 208, 245, 291
systematic review, 291
systems, viii, xii, 2, 41, 54, 63, 72, 81, 94, 102, 106, 107, 109, 112, 121, 122, 123, 127, 129, 136, 137, 138, 142, 145, 150, 195, 196, 241, 244, 245, 248, 258, 260, 262, 267, 269, 273, 279, 280, 281, 282, 287, 290, 306, 308, 309, 314
systolic pressure, 114

T

T cell, 112
T cells, 112
Taiwan, 114, 117, 291
tallium, 266
tamoxifen, 117, 118
tandem mass spectrometry, 85, 221
tanks, 54
tannins, 269
targets, 28, 107, 134, 136, 137
taste, 104, 268, 269, 271, 286
taxonomic, 79
taxonomy, 150
T_c, 314
TDI, 78
TE, 115
technological, 255
technology, 136, 138, 247, 305
teeth, 103
Teflon, 281, 331
TEM, 189, 190, 201, 202, 209
temperature, 16, 17, 19, 22, 28, 29, 30, 33, 35, 43, 47, 84, 86, 94, 124, 155, 171, 173, 175, 176, 181, 182, 194, 224, 227, 230, 235, 249, 253, 255, 258, 263
temporal, 45, 70, 144, 209, 221
temporal distribution, 70
test data, xiii, 313, 318, 319, 326, 327
testes, 107
testosterone, 220
Tetanus, 164
textile, 262, 330
textile industry, 262
Thailand, 143, 241
thallium, 256, 266, 292
thawing, 45
theoretical, 180, 181, 182, 184, 303
theory, 165, 173, 175, 181, 183, 195, 204, 327
thermal, 22, 30, 33, 34, 60, 77, 89, 125, 137, 201, 227, 235, 262
thermal analysis, 201
thermal energy, 235
thermodynamic, xi, 244
thermodynamic properties, xi, 244
Thomson, 237
thorium, xii, 297, 299
threat, 135, 145, 162, 229, 231, 236
three-dimensional, 255
thyroid, 106, 114, 271
thyroid cancer, 271
thyroid gland, 271
Ti, 259
time, vii, ix, xii, 1, 16, 28, 45, 47, 79, 87, 107, 109, 112, 131, 133, 134, 135, 137, 139, 153, 173, 174, 175, 180, 184, 185, 186, 187, 188, 191, 193, 226, 244, 245, 246, 247, 248, 249, 250, 251, 252, 253, 254, 258, 259, 261, 262, 264, 266, 267, 268, 272, 273, 274, 275, 277, 278, 279, 280, 284, 285, 303, 314, 319, 325
time consuming, 253
tin, 209, 241, 259, 271
tissue, 78, 107, 257, 262, 264
titanium, 270, 294
titration, 332
TM, 288, 289, 292
TMP, 191, 192
tobacco, 103, 117
tobacco smoking, 103
TOC, 332
Tokyo, 220, 338
tolerance, 116, 176
topographic, 42, 227
total cholesterol, 106
tourism, 82, 83
tourist, 25, 82
toxic, vii, viii, x, xi, 1, 78, 79, 83, 94, 95, 101, 102, 103, 105, 107, 109, 110, 113, 125, 148, 199, 203, 208, 209, 210, 224, 225, 228, 229, 231, 236, 240, 243, 244, 245, 256, 257, 258, 260, 261, 263, 265, 267, 268, 269, 287, 330
toxic effect, viii, 78, 101, 105, 109, 110, 113, 224, 225, 228, 257, 267, 330
toxic metals, xi, 231, 243, 256, 287, 330
toxic substances, 240, 256

toxicity, vii, x, xii, 42, 78, 79, 103, 109, 110, 115, 117, 213, 231, 240, 244, 245, 258, 260, 263, 268, 269, 270, 286
toxicological, 231, 264, 272, 287
toxicology, 97
toxin, 85, 96, 126, 128
toxins, vii, 1, 2, 78, 87, 94, 96, 97, 98
toys, 200
trace elements, 103, 107, 117
tracers, 185
tracking, 186
tradition, 231
training, 147
trajectory, 173
transcript, 112
transcription, 135
transcripts, 111
transducer, 323, 324
transfer, 27, 51, 247, 248, 249, 257, 270, 273, 279
transformation, 128, 201, 265
Transgenic, 117
transition, 187
transitions, 81
translocation, 128
transmembrane, 191
transmission, 115, 124, 125, 127, 129, 130, 136, 138, 139, 145, 146, 148, 201, 205, 207
transmission electron microscopy, 115
transparent, 178
transport, ix, xii, 43, 102, 105, 111, 113, 119, 154, 161, 170, 171, 172, 173, 175, 176, 181, 183, 193, 194, 196, 197, 219, 240, 241, 244, 245, 260
transportation, 139
travel, 102, 129
treatment methods, 104
tremor, 271
trend, 26, 32, 38, 74, 94, 185, 315, 333
trial, 196
triggers, 105, 114
triglyceride, 111, 117
triglycerides, 106
Trojan horse, 131
trout, 214, 220, 221
tuberculosis, 127, 148, 150, 164
tubular, 105, 266, 273
tuff, 6, 193
tumor, 78, 79, 111, 112
tumors, 108, 112, 118
turbulent, 123, 146
turgor, 100
typhoid, 126
typhoid fever, 126

U

U.S. Geological Survey, 299, 310, 311, 312, 327
UAE, 313
ubiquitous, 102, 110, 131
UK, 129, 332
ultra-fine, 7
ultrasound, 17
ultrastructure, 105
ultraviolet, 173
Umbria, 95
UN, 326
uncertainty, 96, 127
UNDP, 321, 326
uniform, 173, 178, 248, 255, 319
uniformity, 178, 248
United Kingdom, 121, 220
United Nations, 113
United States, ix, xii, 124, 126, 132, 140, 143, 148, 159, 161, 162, 224, 238, 241, 291, 297, 313
universality, 287
unpredictability, 232
updating, 19
uranium, xii, 270, 288, 294, 297, 301, 304, 305, 306
urban, vii, 41, 55, 59, 82, 83, 102, 149, 232, 315, 317
urban areas, 55, 59, 83, 232
urban settlement, 82
urbanisation, 122
urbanization, 103
urbanized, 56
urinary, 110, 112, 117, 118
urinary bladder, 110, 112, 117, 118
urine, 105, 114, 128
USEPA, 123, 305
uterus, 112, 114
UV, 17, 77, 123, 130, 138, 145, 146, 147, 150, 162, 173, 175, 246, 261, 262, 288
UV irradiation, 138, 246, 261, 262, 288
UV light, 162
UV radiation, 77

V

vacation, 56
vaccines, 124
vacuum, 168, 181, 338
vagina, 112
Valdez, 149
valence, x, 181, 199, 201, 204, 267
validity, 155, 191
values, xii, 2, 8, 15, 17, 22, 26, 29, 30, 31, 32, 35, 37, 38, 40, 43, 46, 47, 51, 63, 68, 69, 70, 73, 74, 76,

77, 83, 88, 93, 94, 104, 154, 155, 158, 167, 168, 181, 182, 184, 185, 186, 187, 189, 190, 191, 204, 228, 231, 233, 248, 277, 278, 298, 306, 309, 320, 326, 333, 335, 337
van der Waals, 170
vanadium, 209, 271, 294
variability, 29, 36, 45, 136, 186, 320
variable, 7, 8, 29, 47, 78, 79, 94, 102, 232, 236, 269, 327
variables, ix, 52, 145, 161, 168
variance, 185
variation, 180, 228, 241, 281
vascular, 104, 112, 265, 291
vascular disease, 112, 291
vascular diseases, 112
VBNC, 135
vegetables, 41, 126, 128
vegetation, 43, 45, 47, 59, 225, 269
vehicles, 125, 258
vein, 227
velocity, 168, 171, 174, 176, 178, 180, 184, 186, 187, 188, 206, 208
vertebrates, 130
vesicle, 81, 97
vessels, 175, 246
veterinarians, 78
Vibrio cholerae, 123, 128, 142, 143, 145, 147, 164
Vietnam, 200, 338
village, 147, 231
Vineyards, 59
violent, 2
viral, 124, 125, 144, 192, 193
viral meningitis, 124
Virginia, 101
virulence, 143
virus, 123, 125, 135, 143, 145, 147, 164, 165, 171, 172, 190, 191, 192, 193, 194, 195, 196
virus infection, 143, 147
viruses, ix, 102, 123, 124, 125, 133, 135, 142, 161, 164, 165, 166, 168, 171, 176, 191, 192, 193, 194, 195, 197, 286
viscera, 93
viscosity, 173, 176, 181, 319
visible, 6, 79, 173, 190
vision, 269
visual, 22
vitamin B1, 264
vitamin B12, 264
vitamin E, 103
voids, 230, 234, 235
volatility, 272
volatilization, 69
volcanic activity, 9
voltammetric, xii, 244, 247, 249, 250, 251, 253, 254, 257, 258, 260, 263, 267, 268, 270, 272, 273, 278, 288, 290, 291, 292, 293, 294, 295
vomiting, 109, 258, 263, 269

W

Washington, 80, 97, 117, 196, 241, 301, 302, 309, 312
waste, viii, 95, 101, 103, 196, 224, 226, 229, 231, 232, 233, 235, 236, 237, 239, 259, 264, 266, 282, 287, 295, 331
waste disposal, viii, 101
waste products, 226
waste water, viii, 95, 101, 226, 264, 266, 282, 295, 331
wastes, xi, 224, 226, 227, 228, 230, 231, 232, 235, 236, 237, 238
wastewater, 99, 102, 123, 136, 144, 146, 197, 214, 220, 221
wastewater treatment, 221
water quality, vii, viii, x, xii, xiii, 82, 97, 104, 116, 122, 123, 125, 133, 135, 143, 150, 154, 195, 223, 224, 231, 235, 236, 244, 245, 256, 308, 314, 329
water quality standards, 225
water resources, viii, 101, 113, 244, 286
water supplies, ix, xii, 125, 126, 130, 132, 136, 153, 154, 159, 270, 297, 300, 305, 309
water table, 224, 237, 315, 317, 319, 323
watershed, vii, 2, 7, 9, 12, 14, 15, 40, 42, 43, 47, 51, 82, 113, 138, 162, 225
watersheds, 40, 44, 47, 50, 52, 54, 63, 72, 73
water-soluble, 80, 191, 262
weathering, xi, 103, 224, 225, 227, 232, 234, 258, 270, 330
weight gain, 271
weight loss, 130
welfare, 309
wells, vii, xii, xiii, 1, 2, 7, 9, 15, 16, 17, 18, 19, 22, 23, 24, 25, 26, 27, 51, 76, 84, 85, 86, 88, 90, 91, 92, 94, 137, 200, 234, 269, 297, 299, 303, 304, 308, 309, 313, 314, 315, 317, 318, 321, 322, 323, 325, 326, 327
Western Hemisphere, 124
wet, 102, 231, 234
wetlands, 59
wildlife, x, 102, 122, 126, 128, 138, 148, 213, 214, 235
wind, 29, 31
windows, 314
wine, 260
winning, 327
winter, 2, 29, 32, 34, 74, 89, 93, 94, 227, 233

Wisconsin, 130, 143, 237, 306, 310
Wistar rats, 104, 106, 112
withdrawal, 15, 319, 323
WM, 292
women, 113, 119
wood, x, 213, 265
workers, 307
working hours, 139
World Bank, 326
World Health Organization (WHO), 78, 79, 88, 96, 100, 103, 104, 105, 110, 113, 114, 117, 125, 150, 238, 270, 303, 312, 314

X

XANES, 201, 202, 204, 205, 207, 211
XLPE, 178
XPS, 209
x-ray, x, 195, 199, 201, 202, 209, 210, 211
X-ray absorption, x, 199, 201, 202
x-ray diffraction, 195
X-ray diffraction, 202
XRD, 201, 202, 210

Y

yeast, 216, 217
yield, xiii, 43, 102, 313, 314, 317, 325, 326
young adults, 117

Z

Zen, 292
zeta potential, 168, 170, 189, 190, 191
zinc, 103, 106, 110, 115, 116, 256, 258, 259, 260, 261, 288, 338
Zn, 106, 208, 225, 236, 257, 258, 259, 260, 287, 288, 330, 331
zoonosis, 128, 130
zoonotic, 128, 129, 138, 139, 144, 145
zooplankton, 78